S0-AAH-766

SECOND EDITION

BIODEFENSE RESEARCH METHODOLOGY AND ANIMAL MODELS

SECOND EDITION

BIODEFENSE RESEARCH METHODOLOGY AND ANIMAL MODELS

EDITED BY
JAMES R. SWEARENGEN

CRC Press is an imprint of the
Taylor & Francis Group, an **informa** business

Cover credits: Top two photos courtesy of Dr. Chad J. Roy and Dr. Xavier Alvarez. Bottom two photos courtesy of Dr. Tom Geisbert.

CRC Press
Taylor & Francis Group
6000 Broken Sound Parkway NW, Suite 300
Boca Raton, FL 33487-2742

Printed in the United States of America on acid-free paper
Version Date: 20111115

International Standard Book Number: 978-1-4398-3632-3 (Hardback)

Visit the Taylor & Francis Web site at
http://www.taylorandfrancis.com

and the CRC Press Web site at
http://www.crcpress.com

In the world of biodefense research, there exists a cadre of men and women who have dedicated their lives to protecting the world from those who would use infectious biological organisms and toxins for nefarious purposes. The scientific community has banded together across many organizational lines to bring new technology, information, and countermeasures into the biodefense portfolio to better prepare against these threats. In addition to the devoted scientists, I want to acknowledge the people whose critical contributions made these advances possible. These are the professionals who maintain the facilities, make sure the research is done safely, oversee the use of animals and ensure they are used humanely in accordance with regulatory requirements; and the laboratory and veterinary technicians who are the heart and soul of this research. The vigilance and remarkable talents of these teams of professionals are our best defense.

Contents

Preface

The evolution of biodefense research has made significant advances in animal model development since the publication of the first edition of this book in 2006. The Food and Drug Administration's (FDA) Animal Efficacy Rule (read more about this in Chapter 3) has begun to mature in both understanding by the scientific community and the expectations of the FDA. Like the first edition, this edition continues to span the spectrum from basic research to advanced development of medical countermeasures. The return reader will most likely notice an increase in discussions about the FDA animal efficacy rule as it applies to animal model development and research directions for the various biological agents and toxins. As we all know, redundant efforts often waste more than just time and fiscal resources—they also result in the unnecessary use of animals. Animals have been and will continue to be an invaluable and absolutely necessary part of infectious disease research, but we all have the ethical and moral obligation to ensure that each animal is used in the most humane manner possible and to obtain the maximum benefit in advancing science and human health. It should be understood that much work precedes moving to the use of animal models, and the models presented in this book were developed in conjunction with many *in vitro* techniques including computer modeling, cell culture systems, hollow fiber systems, and other *in vitro* laboratory procedures. All of these techniques have replaced or reduced the use of animals for certain purposes, but as questions arise that require an intact, more complex biological system to answer, animal use becomes essential. The primary aims of this edition remain true to the first edition in an effort to share science, to advance science, and to minimize the number of animals required for use by reducing unnecessary duplication of effort in animal model development and use. The participation of all the chapter authors and coauthors is a testament to their belief in these values and dedication to advancing science, and protecting the health of our world's population.

Editor

Dr. James R. Swearengen, following retirement from the U.S. Army after 21 years of service, served for 4 years as the senior director at the Association for Assessment and Accreditation of Laboratory Animal Care International before joining the National Biodefense Analysis and Countermeasures Center as their comparative medicine veterinarian in 2009. Dr. Swearengen obtained his DVM degree from the University of Missouri-Columbia in 1982 and joined the Army after 2 years of private practice. After tours in Texas and Germany, Dr. Swearengen completed a residency in laboratory animal medicine at the Walter Reed Army Institute of Research from 1990 to 1994, during which period he attained board certification in the specialties of both Laboratory Animal Medicine and Veterinary Preventive Medicine and is a past-president of the American College of Laboratory Animal Medicine.

He began working at the U.S. Army Medical Research Institute of Infectious Diseases (USAMRIID) in 1994 as the assistant director, and then director, of the Veterinary Medicine Division. He gained extensive experience in providing veterinary and husbandry support to infectious disease animal research at all levels of biocontainment and spent many hours working under biosafety level 3 and 4 conditions. Dr. Swearengen became intimately involved with the existing animal models used in biodefense research, provided veterinary expertise in the development of new models, and coauthored publications utilizing animal models for Ebola virus and monkeypox virus infections. In 1996, he was selected to serve on the United Nations Special Commission (Biological Group) and spent 3 months in Iraq performing monitoring and verification functions of Iraq's former biological weapons program. Since 2007, Dr. Swearengen has served on the National Academies of Science National Research Council Standing Committee on Biodefense for the U.S. Department of Defense and the National Academies of Science Institute for Laboratory Animal Research Committee on Animal Models for Assessing Countermeasures to Bioterrorism Agents.

In 1997, Dr. Swearengen provided part-time support for a Defense Threat Reduction Agency program by evaluating and modernizing animal care and use programs in infectious disease research institutes in the former Soviet Union. His expertise was recognized in 2003 as he was selected as the Laboratory Animal Medicine Consultant to the Surgeon General of the U.S. Army. Dr. Swearengen's military career culminated in 2003 as he was chosen to serve as the Deputy Commander of USAMRIID, a position he held until his retirement from the U.S. Army in 2005.

Contributors

Jeffrey J. Adamovicz
Midwest Research Institute
Frederick, Maryland

Arthur O. Anderson
U.S. Army Medical Research Institute of Infectious Diseases
Fort Detrick, Maryland

Jaime B. Anderson
U.S. Army Medical Research and Material Command
Fort Detrick, Maryland

Donald L. Fine
Dynport Vaccine Company LLC
Frederick, Maryland

David L. Fritz
Bacteriology Division
U.S. Army Medical Research Institute of Infectious Diseases
Fort Detrick, Maryland

Thomas W. Geisbert
Galveston National Laboratory
and
Department of Microbiology and Immunology
University of Texas Medical Branch
Galveston, Texas

Stephen B. Greenbaum
BAI, Inc.
Alexandria, Virginia

Mary Kate Hart
DynPort Vaccine Company LLC
Frederick, Maryland

Peter B. Jahrling
Integrated Research Facility
National Institute of Allergy and Infectious Diseases
Fort Detrick, Maryland

Teresa Krakauer
U.S. Army Medical Research Institute of Infectious Diseases
Fort Detrick, Maryland

Frank J. Lebeda
U.S. Army Medical Research Institute of Infectious Diseases
Fort Detrick, Maryland

Elizabeth K. Leffel
Director of Non-clinical Sciences
PharmAthene, Inc.
Annapolis, Maryland

James W. Martin
DiLorenzo Healthcare Clinic
Walter Reed Army Medical Center
Washington, District of Columbia

Shannon S. Martin
Dynport Vaccine Company LLC
Frederick, Maryland

M. Louise M. Pitt
U.S. Army Medical Research Institute of Infectious Diseases
Fort Detrick, Maryland

William D. Pratt
U.S. Army Medical Research Institute of Infectious Diseases
Fort Detrick, Maryland

Bret K. Purcell
U.S. Army Medical Research Institute of Infectious Diseases
Fort Detrick, Maryland

Nelson W. Rebert
Division of Laboratory Sciences
U.S. Army Public Health Command Region-Pacific
Camp Zama, Japan

Douglas S. Reed
Center for Vaccine Research
University of Pittsburg
Pittsburg, Pennsylvania

Robert Rivard
U.S. Army Medical Research Institute of Infectious Diseases
Fort Detrick, Maryland

Chad J. Roy
Tulane National Primate Research Center
Tulane School of Medicine
Covington, Louisiana

Bradley G. Stiles
U.S. Army Medical Research Institute of Infectious Diseases
Fort Detrick, Maryland

James R. Swearengen
National Biodefense Analysis and Countermeasures Center
Fort Detrick, Maryland

Kenneth Tucker
Tauri Group
Alexandria, Virginia

David M. Waag
Bacteriology Division
U.S. Army Medical Research Institute of Infectious Diseases
Fort Detrick, Maryland

Victoria Wahl-Jensen
Integrated Research Facility
National Institute of Allergy and Infectious Diseases
Fort Detrick, Maryland

Kelly L. Warfield
Vaccine Development
Integrated Biotherapeutics, Inc.
Gaithersburg, Maryland

Patricia L. Worsham
U.S. Army Medical Research Institute of Infectious Diseases
Fort Detrick, Maryland

1 History of Biological Agents as Weapons

*James W. Martin**

CONTENTS

The earliest use of biological weapons in warfare resulted from the use of corpses first to contaminate water sources and subsequently as a terror tactic, hurling bodies over the wall of fortified cities. From these crude beginnings were to develop national programs for biological weapons development, stockpiling, and deployment that would rival all other weapons systems in scope and magnitude as well as potential to cause human harm. Recent unveiling of these programs as well as recognition of the failure of the Biological Weapons Convention to prevent some countries from engaging in biological weapons development has made the public aware, if not frightened, of the possibilities. Ergo, use of biological agents as weapons of warfare, methods of terrorism, or means for engaging in criminal activity has come to the forefront of public attention in recent years. Widespread understanding of the biological threat in terms of biological agents' historic use is vital for those who endeavor to find ways to protect society from those who intend to use these agents. It is important to have some common agreement of definitions of terminology used in this discussion. *Biological agent* refers to any living organism or substance produced by an organism that can be used as a weapon to cause harm to humans. Broadly speaking, this includes any living organism or biologically derived substance, but in practical terms (for the classical biological warfare agents), this list is limited to viruses, bacteria, and toxins. *Biowarfare* in its broadest sense refers to any use of these agents to harm others. However, *biowarfare* in more common usage ascribes a narrower definition—use in the context of war, that is, it refers to the use of a biological agent by a nation-state as an act of war. *Bioterrorism* refers to the use of biological agents by a political group, religious group, or cult (group not otherwise recognized as an extension of the government of a state) to achieve some intended political or ideological objective. However, even this definition is fraught with confusion because it does not preclude use by an organization with state sponsorship which can be covert. The term *biocrime* refers to the use of biological agents in the perpetration of criminal activity in which the perpetrator's

* The views expressed in this chapter are those of the author and do not reflect the official policy of the Department of the Army, Department of Defense, or the U.S. Government.

motivation appears to be personal in nature, as opposed to some broader ideological, political, or religious objective. Although specific circumstances and events can blur the distinction, it is helpful to keep these three definitions in mind as we review the world's experience with biological agent use.

De Mussis provides a dramatic record of the use of plague victims in an attempt to engage in biological warfare. After war broke out between the Genoese and the Mongols in 1344 over control of access to the lucrative caravan trade route from the eastern shores of the Black Sea to the Orient, the Mongols laid siege to the Genoese port city of Caffa. The plague, which was later to become known as the Black Death, was spreading from the Far East and reached the Crimea in 1346. The Mongols besieging the city were severely affected and had come close to lifting their siege when they changed their tactics and hurled bodies of plague victims over the city wall, probably with the use of a trebuchet. Eventually, plague did spread to the city, though more likely from rats fleeing the Mongol encampment than as a consequence of the spread of the disease by contamination of the city with plague-infected corpses. After plague struck, the residents of Caffa, who had been successfully withstanding the siege, abandoned their defense and fled to ports in Italy, carrying the plague on board the ships with them. As a consequence, the Black Death began its scourge across Europe [1].

Along with contamination of water sources, another ancient tactic was to allow the enemy to take sanctuary in an area endemic for an infectious agent in anticipation that the enemy force would become infected and weakened by the resulting disease. Most prominent examples were the allowance of unimpeded access to malarious areas, where disease transmission was highly likely to occur [2].

The Carthaginian leader Hannibal is credited with the first use of biological toxins in warfare, in the naval battle of Eurymedon in 184 BC. He ordered earthen pots filled with serpents hurled onto the decks of the Pergamene ships, creating panic and chaos. The Carthaginians exploited the situation, with Hannibal defeating King Eumenes of Peragamum in the battle that ensued [2].

Smallpox was particularly devastating to the Native Americans. Cortez's introduction of smallpox to the Aztecs, whether intentional or not, played a major role in allowing for their defeat and subjugation by the Spanish conquistadors. Sir Jeffery Amherst, British commander of forces in the American colonies during the French and Indian War, provided Indians loyal to the French with blankets and other articles contaminated by smallpox. Native American Indians defending Fort Carillon (subsequently named Fort Ticonderoga) experienced an epidemic of smallpox that contributed to their defeat and the loss of the fort to the British. Subsequently, a smallpox epidemic broke out among the Indians in the Ohio River valley [3].

During the American Revolutionary War, successive smallpox epidemics affected major Continental Army campaigns early in the conflict and resulted in the aborted attempt to capture Quebec City early in the war. The British forces, which were immune to the disease because of their exposure to the natural infections endemic in much of Europe, were relatively protected from smallpox, whereas the colonists, living in more rural and isolated settings, were nonimmune. Because of his recognition of the consequences of this disparity of immunity between the two forces, General George Washington ordered the variolation (inoculation with smallpox) of all nonimmune recruits in 1778. This was a controversial procedure that predated vaccination and carried a potential mortality of 1–3%; it was the first time in world military history that such a measure had been ordered by a commander and it set the precedence for military immunization programs of today [2].

The Germans undertook a covert biological campaign in the United States in the first part of World War I, before the United States had entered the war. The Allies had been purchasing draft animals from the United States for use by their military forces. German operatives infected animals awaiting shipment overseas with glanders and anthrax organisms [4]. The Germans also conducted similar operations in Romania, Russia, Norway, Mesopotamia, and Argentina, with varying levels of success. Attempts were also made to infect the grain production in Spain with wheat fungus, but without success [5].

An international protocol, known as the 1925 Geneva Protocol [for the Prohibition of the Use in War of Asphyxiating, Poisonous, or Other Gases, and Bacteriological (*Biological*) Methods of Warfare], was created in response to the use of chemical agents during World War I. The 1925 Geneva Protocol created by the League of Nations' Conference for the Supervision of the International Trade in Arms and Ammunition concerned use only between nation-states. It has no verification mechanism and relies on voluntary compliance. Many of the original signatory states held reservations to the protocol for the right to retaliatory use, making it effectively a no-first-use protocol [2]. After the Japanese defeat of Russia in the 1905 Russo-Japanese War, Japan had become the dominant foreign power in Manchuria. The Kwantung Army was created to maintain Japanese economic interests in the region. During the 15 months from September 1931 to the end of 1932, the Japanese military seized full control of all of Manchuria, setting the stage for its complete exploitation. It was in 1932, just as Japan obtained military control, that Major Ishii Shiro, a Japanese Army physician with a confirmed interest in biological agents, came to Harbin to exploit Manchurian human resources in the name of research. He established his initial laboratory in the industrial sector of Harbin known as the Nan Gang District, but he soon came to realize that his more controversial involuntary human research could not be conducted without scrutiny there and moved the human research to a secret facility at Beiyinhe, which was 100 km south of Harbin. Unobserved by the outside world, Major Ishii began human experimentation on a more dramatic scale. Each victim, once selected for study, continued to be a study subject until his or her death as part of the study—or through live vivisection. There were no survivors among the research study subjects. These studies continued until the occurrence of a prisoner riot and escape, which resulted in closure of the facility in 1937. Not to be deterred, the closure of the Beiyinhe facility was followed by the creation of even larger, more extensive facilities [6].

In August 1936, Lt. Col. Ishii was made Chief of the Kwantung Army Boeki Kyusui Bu (Water Purification Bureau). That autumn, the Japanese appropriated 6 km^2 of farmland, which encompassed 10 villages located 24 km south of Harbin, displacing 600 families from their ancestral homes. It was here that Ishii built the massive Ping Fan research facility, where 200 prisoners were always on hand to become the expendable subjects of further experimentation. A minimum of 3000 Chinese prisoners were killed and cremated consequent to these experiments, but most of the evidence was destroyed at the end of the war—in all likelihood the actual number of victims of this ghastly research was much greater [6].

The Unit 100 facility at Changchun was run by an equally ruthless veterinary officer, Major Wakamatsu Yujiro. In 1936, the Japanese appropriated 20 km^2 of land near Mokotan, a small village just 6 km south of Changchun, the capital of Japanese-occupied Manchuria. Unit 100 was a predominantly veterinary and agricultural biowarfare research unit—a completely independent operation from Unit 731 at Ping Fan. The principal focus of Unit 100 was to develop biological weapons useful in sabotage operations. Although animals and crops were the focus of most of the research, a tremendous number of human studies were also conducted that were very similar in nature to those conducted at Ping Fan by Unit 731 [6].

In April 1939, a third major research facility, known as Unit Ei 1644, was established in an existing Chinese hospital in Nanking under the command of one of Ishii's lieutenants, Lt. Col. Masuda. On the fourth floor of the hospital were housed prisoners, many of them women and children, who became the subjects of grisly experimentation. The human experimental subjects were cremated after the studies in the camp incinerator, usually late at night. A gas chamber with an observation window was used to conduct chemical warfare experiments. Unit Ei 1644 supported the research efforts of Unit 731, with support responsibilities that included production of bacterial agents as well as cultivation of fleas [6]. At the end of the war, in a move that has now become controversial, Ishii, then a lieutenant general, and his fellow scientists were given amnesty in exchange for providing information derived from their years of biological warfare research [2].

In contradistinction to Japanese efforts during World War II, German interest seemed to be more focused on developing an adequate defense against biological agents. Although German researchers experimentally infected prisoners with infectious agents, there were no legal actions taken after the war, and no German offensive biological warfare program was ever documented. The Germans, however, accused the British of attempting to introduce yellow fever to the southern Asian subcontinent as well as of an Allied introduction of Colorado beetles to destroy the German potato crops. These claims were never substantiated [5].

During the Korean conflict, numerous allegations of use of biowarfare by the United States were made by North Korean and Chinese officials. Many of the allegations appear to be based on experiences that the Chinese had in Manchuria with the "field testing" done by Unit 731. Polish medical personnel were sent to China to support the Communist war effort, accompanied by Eastern European correspondents. Numerous allegations based on anecdotal accounts of patients came from these correspondents and other sources. These accounts were not supported with scientific information. Some of the accounts, such as the use of insects for vectors of cholera and the spread of anthrax with infected spiders, had dubious scientific validity [7].

After World War I, Major Leon Fox, Medical Corps, U.S. Army, wrote an extensive report in which he concluded that modern improvements in health and sanitation made use of biological agents unfeasible and ineffective. Some mention was made of the ongoing Japanese offensive biological program in his report, but it was, ironically, his erroneous concerns about German biological weapons' development that led to serious U.S. interest in the subject. In the autumn of 1941, before U.S. entrance into World War II, opinions differed as to the validity of biological warfare potential: "Sufficient doubt existed so that reasonable prudence required that a serious evaluation be made to the dangers of a possible attack" [8, p. 1]. As a consequence, the Secretary of War asked the National Academy of Sciences to appoint a committee to study the question. The committee concluded in February 1942 that biowarfare was feasible and that measures were needed to reduce U.S. vulnerability [2].

President Roosevelt established the War Reserve Service, with George W. Merck as director, with the initial task of developing defensive measures to protect against a biological attack. By November 1942 the War Reserve Service asked the Chemical Warfare Service of the Army to assume the responsibility for a secret large-scale research and development program, which included the construction and operation of laboratories and pilot plants. The Army selected the small National Guard airfield at Camp Detrick, Frederick, Maryland, as a site for new facilities in April 1943. By the summer of 1944, the Army had a testing site at Horn Island, Mississippi, which was subsequently moved to Dugway Proving Grounds, Utah, and a production facility in Terre Haute, Indiana, which was soon closed. The War Reserve Service was disbanded and the Research and Development Board established under the War Secretary to supervise the biological research programs. An assessment of the biological warfare situation was provided to the Secretary of War by George Merck in January 1946. The report concluded that the United States clearly needed to have a credible capability to retaliate in kind if ever attacked with biological weapons [7].

Only after the end of World War II did the United States learn of the extent of Japanese biological weapons research. Gradually, in the late 1940s, the scope of the Japanese program became known, along with an awareness of Soviet interest in the program. War broke out on the Korean peninsula in June 1950, adding to concerns about Soviet biological weapons development, and the possibility that the North Koreans, Chinese, or Soviets might resort to biological weapons use in Korea. The Terre Haute, Indiana, production facility, which was closed in 1946, was replaced with a large-scale production facility in Pine Bluff, Arkansas. During the 26 years of biological weapons development, the United States weaponized eight antipersonnel agents and five anticrop agents [9].

Field testing was done in the United States in which the general public and the test subjects themselves were uninformed, and these studies have unfortunately tainted the history of the offensive biological warfare program. The first large-scale aerosol vulnerability testing was the San Francisco Bay study conducted in September 1950. *Bacillus globigii* and *Serratia marcescens* were used as stimulants for biological agents. Unfortunately, a number of *Serratia* infections occurred subsequently in one of the hospitals in the study area, and although none of the infections was ever documented to be the 8UK strain, many people held on to their perceptions that the U.S. Army study had caused the infections [10].

Serratia marcescens, then known as *Chromobacter*, was thought to be a nonpathogen at the time. Several controversial studies included environmental tests to see whether African Americans were more susceptible to fungal infections caused by *Aspergillus fumigatus*, as had been observed with *Coccidioides immitis*, including the 1951 exposure of uninformed workers at Norfolk Supply Center, in Norfolk, Virginia, to crates contaminated with *Aspergillus* spores. In 1966, in New York City subways, the U.S. Army conducted a repeat of studies that had been done by the Germans on the Paris Metro and some of the forts in Maginot Line to highlight the vulnerability of ventilation systems and confined spaces. Light bulbs filled with *Bacillus subtilis* var. *nigeri* were dropped into the ventilator shafts to see how long it would take the organisms to spread through the subway system [11]. The Special Operations Division at Camp Detrick conducted most of the studies on possible methods of covert attack.

After 1954, the newly formed Medical Research Unit conducted medical research separately from the studies done by the Chemical Corps. This research began using human volunteers in 1956 as part of a congressionally approved program known as "Operation Whitecoat." This use of human volunteers set the standard for ethics and human use in research. The program used army active-duty soldiers with conscientious objector status as volunteers to conduct biological agent-related research. All participation was voluntary and was performed with the written informed consent of each volunteer. The program concluded in 1973 with the end of the draft, which had been the source of conscientious objectors [9]. In July 1969, Great Britain issued a statement to the Conference of the Committee on Disarmament calling for the prohibition of development, production, and stockpiling of bacteriological and toxin weapons [12].

In September 1969, the Soviet Union unexpectedly recommended a disarmament convention to the United Nations General Assembly. In November 1969, the World Health Organization of the United Nations issued a follow-on to an earlier report by the 18-nation Committee on Disarmament, on biological weapons, describing the unpredictable nature, lack of control once released, and other attendant risks of biological weapons use. Then, President Nixon, in his November 25, 1969, visit to Fort Detrick, announced new U.S. policy on biological warfare, renouncing unilaterally the development, production, and stockpiling of biological weapons, limiting research strictly to the development of vaccines, drugs, and diagnostics as defensive measures. The 1972 Biologic Weapons Convention, which was a follow-on to the 1925 Geneva Protocol, is more properly known as the "1972 Convention on the Prohibition of the Development, Production, and Stockpiling of Bacteriological (Biological) and Toxin Weapons and their Destruction." Agreement was reached among 103 cosignatory nations and went into effect in March 1975. "The convention prohibits the development, production, stockpiling or acquisition by other means or retention of microbial or other biological agents toxins whatever their origin or method of production of types and in quantities that have no justification of prophylactic, protective or other peaceful purposes, as well as weapons, equipment or means of delivery designed to use such agents or toxins for hostile purposes or in armed conflict"[13].

The U.S. Army, in response to the 1969 presidential directive, did not await the creation of the 1972 Biological Warfare Convention or its ratification. By May 1972, all personnel-targeted agents had been destroyed and the production facility at Pine Bluff, Arkansas, converted to a research facility. By February 1973, all agriculture-targeted biological agents had been destroyed. Fort Detrick and other installations involved in the offensive weapons program were redirected, and the U.S. Army Medical Research Institute of Infectious Diseases was created in place of the U.S. Army Medical Unit, with biosafety level 3 and 4 laboratories dedicated strictly to the development of medical defensive countermeasures [2].

Although a signatory to the 1925 Geneva Convention, the Soviet Union began its weapons development program at the Leningrad Military Academy in Moscow under the control of the state security apparatus, the GPU. Work was initially with typhus, with what was apparently human experimentation on political prisoners during the prewar era conducted at Slovetsky Island in the Baltic Sea and nearby concentration camps. This work was subsequently expanded to include work with Q fever, glanders, and melioidosis, as well as possibly tularemia and plague. Outbreaks of Q fever among German troops resting in Crimea and outbreaks of tularemia among the German siege forces of Stalingrad are two suspected but unconfirmed Soviet uses of biological warfare during World War II [14].

During World War II, Stalin was forced to move his biological warfare operations out of the path of advancing German forces. Study facilities were moved to Kirov in eastern European Russia, and testing facilities were eventually established on Vozrozhdeniya Island on the Aral Sea between the Soviet Republics of Kazakhstan and Uzbekistan. At the conclusion of the war, Soviet troops invading Manchuria captured the Japanese at the infamous Unit 731 at Ping Fan. Through captured documents and prisoner interrogations, the troops learned of the extensive human experimentation and field trials conducted by the Japanese. Stalin put KGB chief Lavrenty Beria in charge of a new biowarfare program, emboldened by the Japanese findings. The production facility at Sverdlosk was constructed using Japanese plans. When Stalin died in 1953, a struggle for control of the Soviet Union ensued. Beria was executed during the struggle to seize power, and Khruschev emerged as the Kremlin leader and transferred the biological warfare program to the Fifteenth Directorate of the Red Army. Colonel General Yefim Smirnov, who had been the chief of army medical services during the war, became the director [14].

Smirnov, who had been Stalin's minister of health, was a strong advocate of biological weapons. By 1956, Defense Minister Marshall Georgi Zhukov announced to the world that Moscow would be capable of deploying biological in addition to chemical weapons in the next war. By 1960, there existed numerous research facilities addressing every aspect of biological warfare scattered across the Soviet Union [14].

The Soviet Union was an active participant in the World Health Organization's smallpox eradication program, which ran from 1964 to 1979. Soviet physicians participating in the program sent specimens back to Soviet research facilities. For the Soviets, participation in the program presented an opportunity not only to rid the world of smallpox but also obtain, as source material for biological weapons development, virulent strains of smallpox virus that could be used subsequently for the more sinister purpose of releasing it as a weapon of war. In 1980 the World Health Organization announced the eradication of smallpox, and the world rejoiced at the elimination of a disease that had caused more human deaths than any other infection. However, the Soviets had another reason to celebrate: Elimination of natural disease meant that, over time, vaccination programs would terminate, and neither natural nor vaccine-acquired immunity would exist for the majority of the world's population [2].

In 1969, President Richard Nixon announced unilateral disengagement in biological warfare research [12]. As mentioned previously, research came to an abrupt halt; production facilities and weapon stockpiles were destroyed. The 1972 Biological

Weapons Convention was signed by the Soviet Union. To the Soviets, this may have seemed like an excellent opportunity to obtain a significant advantage over its adversaries in the West. The Soviets even appear to have increased their efforts [2].

In October 1979, a Russian immigrant newspaper published in Frankfurt, Germany, published a sketchy report of a mysterious anthrax epidemic in the Russian city of Sverdlosk (now known as Yekaterinburg). The military were reported to have moved into the hospitals in Sverdlosk and taken control of the care of reportedly thousands of patients with a highly fatal form of anthrax. Suspicions emerged that there had been an accidental release of anthrax agent into an urban area near the Soviet military installation, Compound 17. The CIA asked the opinion of Harvard biologist, Dr. Matthew Meselson, in what turned out to be a poor choice of experts. He attempted to refute the Soviet weapon release theory—after all, he had been a strong proponent of the Nixon ban on the U.S. biological warfare program. More objective observers reviewing the same evidence have reached different conclusions. Furthermore, satellite imagery of Sverdlosk from the late spring of 1979 showed a flurry of activity at and around the Sverdlosk installation, which was consistent with a massive decontamination effort. The event did, however, raise enough concerns within the Reagan administration and the Department of Defense to seek better military biopreparedness [15].

Debate raged on for the next 12 years, with Meselson testifying before the Senate that the burden of evidence was that the anthrax outbreak was a result of the failure of the Soviets to keep anthrax-infected animals out of the civilian meat supply and not the consequence of an accident at a military weapons facility, as maintained by many U.S. officials. Meselson went on to say that in his opinion the 1972 Biological Weapons Convention had been a total success and that no nation possessed a stockpile of biological weapons. In June 1992, during a brief but open period of detente, Meselson was allowed to take a team of scientists to review autopsy material and other evidence from the Sverdlosk incident. Autopsy specimens for mediastinal tissue represented clear evidence to the team pathologist Dr. David Walker that the disease had been contracted from inhalation of anthrax spores, not from ingestion of tainted meat, as the Soviets had continued to allege. Meselson continued to insist that the evidence was not conclusive that this event was not a natural disease occurrence [15].

Previously, in private conversations with President George H. W. Bush, Russian leader Boris Yeltsin admitted that the KGB and military had lied about the anthrax deaths and that he would uncover the explanation. In the meantime, several Soviet defectors, including Ken Alibek, confirmed not only the Sverdlosk incident as an accidental release of weaponized anthrax but also the extensive nature of the Soviet biological weapons program [11]. Subsequently, in a press release, Yeltsin admitted to the offensive program and the true nature of the Sverdlosk biological weapons accident [15].

The Soviet biological weapons program had been extensive, comprising a range of institutions under different ministries, as well as the commercial facilities collectively known as Biopreparat. The Soviet Politburo had created Biopreparat to carry out offensive research, development, and production under the concealment of legitimate civil biotechnology research. Biopreparat conducted clandestine activities at 52 sites, employing over 50,000 people. Annualized production capacity for weaponized smallpox, for instance, was 90–100 tons [14].

Seth Carus from the National Defense University studied all biological agent use in the twentieth century and found 270 alleged cases involving illicit biological agents; of 180 cases of confirmed agent use, 27 were bioterrorism and 56 were biocrimes. In 97 situations, the purpose or intent of the perpetrator was unknown. Ten fatalities were caused by the criminal use of biological agent [5].

An example of state-sponsored bioterrorism occurred in 1978, when a Bulgarian exile named Georgi Markov was attacked in London with a device concealed in the mechanism of an umbrella. This weapon discharged a tiny pellet into the subcutaneous tissue of his leg. He died mysteriously several days later. At autopsy, the pellet was found; it had been drilled for filling with a toxic material. That material turned out to be ricin [9].

In 1995, Dr. Debra Green pleaded no contest to charges of murder and attempted murder. The murder charges stemmed from the deaths of two of her children in a fire for which she was thought to have been the arsonist. The attempted murder charges stemmed from the poisoning of her estranged spouse with ricin. Green was sentenced to life imprisonment [5].

Another example of criminal activity occurred in 1996, when Diane Thompson deliberately infected 12 coworkers with *Shigella dysenteriae.* She sent an e-mail to her coworkers, inviting them to partake of pastries she had left in the laboratory break room. Eight of the 12 hospital personnel who became ill tested positive for *Shigella dysenteriae type* 2, and one of the muffins also grew the same pathogen. During their investigation, police were to learn that a year before this incident, her boyfriend had suffered similar symptoms and had been hospitalized at the same hospital facility and that Thompson had falsified his laboratory test results. Thompson was sentenced to 20 years in prison [5].

The first episode of bioterrorism in the United States occurred in 1984. The Rajneeshee cult was founded by an Indian guru named Bhagwan Shree Rajneesh in the 1960s. Rajneesh was a master at manipulating people and was highly successful in attracting followers from the upper-middle classes and accumulating vast amounts of money from donations and proceeds from the sale of books and tapes. Because of the cult's radical beliefs the ashram became unwelcome in Poona (now Pune), India. Rajneesh acquired the Big Muddy Ranch near The Dalles, Oregon. Here he built a community for his followers, named Rajneeshpuram, which became an incorporated community. Within a few years, the Rajneeshees came into conflict with the local population pertaining to land use and development. To take control of the situation, the Rajneeshees realized that they needed to control the Wasco County government. To accomplish this, they brought in thousands of homeless people from cities around the country through their share-a-home program, counting on their votes in the upcoming elections. The Rajneeshees also plotted to make the local population sick so that they would not participate in the election [5].

The first documented incident of the Rajneeshee use of biological agents involved provision of water contaminated with *Salmonella typhimurium.* Two of the Wasco County commissioners visiting Rajneeshpuram on August 29, 1984, consumed the contaminated water. Both became sick, and one required hospitalization. In trial runs in the months leading up to the November 1984 elections, several attempts at environmental, public water supply, and supermarket food contamination were unsuccessful. In September 1984, the Rajneeshees began contaminating food products at local restaurants. A total of 10 restaurants suffered attacks involving pouring slurries of *S. typhinurium* into food products at salad bars, into salad dressing, and into coffee creamer. As a consequence of this attack, much of The Dalles community became sick—there were 751 documented cases of *S. typhimurium* infection, resulting in several hundred hospitalizations. Despite the success of the restaurant contamination, the Rajneeshee cult abandoned its efforts to take over Wasco County. No further attacks were conducted. Interestingly, the Centers for Disease Control and Prevention investigated the outbreak and concluded its cause was poor sanitation and hand-washing practices. Only a year later when several cult members defected and revealed the internal operations of the cult was the sinister nature and cause of the epidemic finally established [5].

In 1995, the Aum Shinrikyo Cult released sarin gas in the Tokyo subway system, resulting in 12 deaths and thousands of persons presenting for emergency medical care. The Aum Shinrikyo Cult, founded by Shoko Asahara, had grown into a massive organization with a membership of approximately 10,000 and financial assets of $300,000,000. Aum Shinrikyo mimicked the organization of the Japanese government, with "ministries and departments." The department of "Health and Welfare" was headed by Seichi Endo, who had worked in genetic engineering at Kyoto University's Viral Research Center. "Science and Technology" was headed by Hideo Murai, who had an advanced degree in astrophysics and had worked in research and development for Kobe Steel Corporation. Endo attempted to derive botulinum toxin from environmental isolates of *Clostridium botulinum* at the cult's Mount Fuji property. There, a production facility was built and horses were stabled for the development of a horse sera antitoxin. It is uncertain whether Endo was able produce potent botulinum toxin successfully [5,15].

In 1993 Aum Shinrikyo built a new research facility on the eighth floor of an office building owned by the cult in eastern Tokyo. At this location, the cult grew *Bacillus anthracis* and installed a large industrial sprayer to disseminate the anthrax. The cult was also believed to have worked with *Coxiella burnetti* and poisonous mushrooms, and they sent a team to Zaire (now Democratic Republic of the Congo) in the midst of an Ebola epidemic to acquire the Ebola virus, which they claimed to have cultivated. According to press accounts from 1990 to 1995, the cult attempted to use aerosolized biological agents against nine targets—three with anthrax and six with botulinum toxin. In April 1990, the cult equipped three vehicles with sprayers targeting (with botulinum toxin) the Japan's parliamentary Diet Building in central Tokyo, the city of Yokahama and the Yosuka U.S. Navy Base, and Nairta International Airport. In June 1993, the cult targeted the wedding of Japan's Crown Prince by spraying botulinum toxin from a vehicle in downtown Tokyo. Later that month, the cult spread anthrax using the roof-mounted sprayer on the same eight-story office building used as their research and production facility. In July 1993, the cult targeted the Diet Building in central Tokyo again, this time with a truck spraying anthrax, and later the same month they targeted the Imperial Palace in Tokyo. On March 15, 1995, the cult planted three briefcases designed to release botulinum toxin in the Tokyo subway. None of these numerous attacks were successful; none are known to have produced any casualties from biological weapons. Ultimately, Aum Shinrikyo gave up on its biological weapons program and released sarin in the Tokyo subway on March 20, 1995, with results that shocked and horrified the world [5].

Reasons given for the cult's failure to produce effective biological attacks include use of a nontoxin-producing (or low-yield) strain of *C. botulinum*; use of a vaccine strain (low pathogenicity) of *Bacillus anthracis*; use of inappropriate spraying equipment, on which nozzles clogged; and perhaps subversion by some cult members reluctant to follow through with the planned operation [5].

On October 4, 2001, just two weeks after the United States had been made dramatically aware of its vulnerability to international terrorism with the September 11 attacks on the World Trade Center and the Pentagon, health officials in Florida reported a case of pulmonary anthrax. During the first week of September, American Media, Inc. received a letter addressed to Jennifer Lopez, containing a fan letter and a "powdery substance." The letter was passed among employees of American Media, Inc., including Robert Stevens. Retrospectively, investigators would consider that it was not this letter, but possibly a subsequent letter, that was the source of his infection [16]. Stevens was admitted to a Palm Beach, Florida, hospital with high fever and disorientation on October 2, 2001. By October 5, 2001, Robert Stevens was dead from inhalational anthrax—the first such case in the United States in over 20 years. An autopsy performed the following day revealed hemorrhagic pleural effusions and mediastinal necrosis. Soon other anthrax mailings and resultant infections became known, first at civilian news media operations in New York City, and then in the Congressional office buildings in Washington, DC, with concurrent contamination of U.S. postal facilities in the national capital area and Trenton, New Jersey [16].

At least five, and theoretically as many as seven, letters (four of which were recovered) containing anthrax spores had been mailed, perhaps in two mailings, on September 18 and the October 9, 2001. A total of 22 people were infected with anthrax, with 11 pulmonary cases resulting in five deaths. Issues of contamination and screening for anthrax exposures resulted in significant disruption of operations at the Congressional office building and U.S. postal facilities, not to mention millions of dollars spent in the cost of decontamination. Probably the most important issue and lesson learned, however, was related to the importance of effective and accurate communication about the nature of the threat and the response efforts to the public [16].

In 2003, at least four ricin-related incidents took place. On January 5, 2003, six Algerians, thought to be part of the Chechen network linked to Al Qaeda and Iraq, Ansar al-Islam, were arrested in North London by British security agencies. They were in the possession of ricin as well as castor seeds and equipments to make ricin. In March 2003, traces of ricin were found by the police in a locker at a railway station in Paris. On October 2003 a container with ricin was discovered at a postal facility in South Carolina, United States. A November 2003 disclosure confirms that traces of ricin were also found in mail bound for the White House [17].

In South Asia, Tamil rebel groups had threatened to use biological materials against the native Sinhalese in the early 1980s and resurfaced again in March 2008. The rebels threatened to spread bilhariasis and yellow fever in the country and allegedly laid out plans to attack rubber plantations and tea gardens using antiplant agents [17]. Feasibility of these threats are at best uncertain, but demonstrate that disparate terrorist organizations contemplate biological agents, if only as threats, as potential weapons in their campaign of terror.

Al Qaeda's Abdur Rauf, a Pakistani microbiologist, reportedly has searched Europe to obtain anthrax spores and equipment for Al Qaeda's biological weapons laboratory in Afghanistan. Menad Benchellali, an Al Qaeda-trained terrorist, engaged in covert activities including weaponization of ricin in his biological and chemical laboratory in Lyon, France, before his arrest in 2004 [17]. This case represents a credible example of handling of biological agents in a small laboratory at the disposal of terrorists confirming the feasibility of Al Qaeda-sponsored bioterrorism.

The use of biological agents has increased dramatically in the last two decades, and the threat of bioterrorism took on dramatically new proportions after the anthrax mail attacks of September–October 2001. Groups with political objectives, religious groups, and apocalyptic cults have become important players in the world of terrorism [18]. Increasingly, these terrorist organizations have taken an interest in biologic agents. One of the more alarming recent trends has been the increased motivation of terrorist groups to inflict mass casualties [19]. The possibility of a major biological agent release as an act of terrorism resulting in massive casualties looms ever more likely, which is all the more reason that medical personnel, public health officials, and government agencies that deal with emergency response must be prepared for such an event.

REFERENCES

1. Derbes, V. J. 1966. De Mussis and the great plague of 1348. *JAMA*, 196(1): 59–62.
2. Martin, J. W., G. W. Christopher, and E. M. Eitzen. 2007. History of biological weapons: From poisoned darts to intentional epidemics. In *Medical Aspects of Chemical and Biological Warfare*, Z. F. Dembek, ed., pp. 1–20. Washington, DC: Borden Institute.
3. Christopher, G. W., T. J. Cieslak, J. A. Pavlin, and E. M. Eitzen. 1997. Biological warfare: A historical perspective. *JAMA*, 278: 412–417.
4. Jacobs, M. K. 2004. The history of biologic warfare and bioterrorism. *Derm. Clin.*, 22(3): 231–246.
5. Carus, W. S. 1998 (February 2001 Revision). Bioterrorism and biocrimes (Working paper), Center for Counterproliferation Research. Washington, DC: National Defense University.
6. Harris, S. H. 1994. *Factories of Death*. London: Routledge.
7. Rolicka, M. 1995. New studies disputing the allegations of bacteriological warfare during the Korean War. *Military Med.*, 160(3): 97–100.
8. Department of the Army. 1977. Special Report to Congress: U.S. Army Activity in U.S. Biological Warfare Programs, 1942–1977. Vol 1(1). Washington, DC: DA. p.1.
9. Eitzen E. M. and E. T. Takafuji. 1997. Historical overview of biological warfare. In *Medical Aspects of Chemical and Biological Warfare*, F. R. Sidell, E. T. Takafuji, and D. R. Franz, eds Washington, DC: Borden Institute. pp. 415–423.
10. Malloy, C. D. 2000. A history of biological and chemical warfare and terrorism. *J. Pub. Health Manag. Pract.*, 6(4): 30–37.
11. Bacon, D. 2003. Biological warfare: A historical perspective. *Semin. Anesthesia, Perioperative Med. Pain*, 22: 224–229.
12. Smart, J. K. 1997. History of chemical and biological warfare: An American perspective. In *Medical Aspects of Chemical and Biological Warfare, Textbook of Military*

Medicine, F. R. Sidell, E. T. Takafuji, and D. R. Franz, eds, pp. 9–86. Washington, DC: Borden Institute.

13. Bodell, N. 2010. Armaments, disarmaments and international security. In *Stockholm International Peace Research Institute Yearbook 2010*, p. 486. Oxford, U.K.: Oxford University Press.
14. Alibek, K. 1999. *Biohazard*. New York: Random House.
15. Miller, J. S., S. Engelberg, and W. J. Broad. 2001. *Germs, Biological Weapons and America's Secret War*. New York: Simon & Schuster.
16. Center for Counterproliferation Research. 2002. Anthrax in America: A chronology and analysis of the fall 2001 attacks (Working paper). Washington, DC: National Defense University.
17. Roul, A. 2009. Is bioterrorism threat credible? *CBW Magazine*, 1(3) (April); http://www.idsa.in/cbwmagazine/IsBioterrorismThreatCredible_aroul_0408.
18. Tucker, J. B. 1999. Historical trends related to bioterrorism: An empirical analysis. *Emerg. Infect. Dis.*, 5: 498–504.
19. Noah, D. L., K. D. Huebner, R. G. Darling, and J. F. Waeckerle. 2002. The history and threat of biological warfare and terrorism. *Emerg. Med. Clin. N. Am.*, 20(2): 255–271.

2 Bioterrorism and Biowarfare

Similarities and Differences

Nelson W. Rebert

CONTENTS

2.1 INTRODUCTION

Biological threat agents have often been called the poor man's nuclear weapon [1]. This chapter explores the concepts involved in biowarfare (BW) and bioterrorism (BT) and compares the strategies used. Biological threat agents, such as bullets, bombs, and chemicals, can result in a large number of people who are dead or injured. In this chapter, attacks against personnel are primarily discussed. Similar considerations would be involved in a discussion of antianimal, antiplant, and antimaterial agents. It is useful to start with a discussion of some of the common attributes of both BW and BT.

For example, if a tactical nuclear weapon is designed to attack a city block and a strategic weapon an entire city, then an operational nuclear device would affect a neighborhood. Both state and nonstate actors can potentially use biological threat agents on similar scales and can even more finely hone the attack to affect a single house or a single person in the house. Even the threat of using biological threat agents can be devastating. State actors and nonstate actors may have different moral constraints, resources, and motivations that influence the choice of agents, potential delivery methods, maximum quantity of agent available for use, and threshold for use.

A single incident of intentional release or threatened release of a biological threat agent will have a number of dimensions. One obvious example is scale (large or small numbers of casualties). The victims may be civilian, military, or paramilitary; the perpetrators may also be civilian, military, or paramilitary. The motivation for the attack is yet another dimension. This is a harder dimension to define, but generally, the motivations can be placed in one of three categories. The first is the traditional motivations used by states for waging war. Examples are conquest or defense of territory, defense of national interests, regime survival, and so on. Another category is those motives generally associated with criminal activity, such as greed, revenge, and so forth. The final category is the motivations associated with nonstate terrorist groups: these include religious or ideological motivations of groups such as the Al Qaeda [2], self-defense motives as in the case of the Aum Shinrikyo [3], or manipulative motives such as exhibited by the Rajneeshee [4]. However, the latter two could be combined, for what is defense but an attempt to manipulate your attacker's behavior.

One key dimension to study is the perpetrator. The resources of the perpetrating group or individual determine the maximum level of agent sophistication, delivery method, and quantity of agent. These parameters determine the potential magnitude of events, and hence the required response. For example, an event involving 100 kg of weaponized anthrax might cause roughly a couple of hundred thousand deaths and tens of thousands more intensive care unit patients. However, an event during which salad bars are contaminated with a foodborne agent will most likely result in a few deaths, and maybe a couple of thousand sick, most of whom should not be intensive care unit patients [4]. Therefore, any predictions that can be made regarding the probable sophistication of an attack can help determine the response that one needs to be able to execute.

State-sized perpetrators, such as military organizations, potentially have access to sufficient resources to use the most sophisticated agents and delivery systems and to have the largest quantities. This does not mean that, for a given mission, the most sophisticated agent or the largest quantities would be used. Just as every bombing does not require the use of a nuclear weapon, every attack with a biological agent does not require the most sophisticated agent or largest quantities (consider all of the state-sponsored assassinations) [5]. However, a state program would most likely be required to execute a massive anthrax attack. At the other extreme, individuals committing biocrimes will generally have limited access to bioagents and delivery methods and, of course, would have only small quantities. Even the salad bar example may be out of the reach of biocriminals. Nonstate actors, such as terrorist groups, would usually have more resources than a biocriminal, but not as many resources as those of most states. As a result, a BT incident is unlikely to reach the scale of a multikilogram anthrax attack, but the salad bar example would be well within a terrorist's reach.

Considering these factors, it is appropriate to break the topic of the use of biological threat agents into sections on the basis of the two factors that most influence the choice of agent: the scale and the state versus nonstate actor. The former would include military organizations and state-sponsored terrorists. The latter would include criminals and nonstate-sponsored terrorist groups. The scale and state sponsorship play a significant role in agent choice. The agent used plays a significant role in ease of detection, diagnosis, and so on, and hence it plays a role in how easily one can defend against or mitigate a BW attack.

The scale of the desired effect plays a large role in agent choice. Diseases like anthrax, plague, or smallpox are well suited to city attacks but are generally poor choices if only a city block is the desired target, especially the latter two. However, brucellosis or Venezuelan equine encephalitis are much better choices if the desired effect is to incapacitate everyone in a stadium or city block. Many agents can be used at multiple scales. All the criteria involved in agent choice are beyond the scope of this chapter. However, a key criterion that may override all others is what agent is available to the perpetrators. Because state actors probably have access to more sophisticated weaponized agents than nonstate actors, and are also concerned about their own troops or citizens, their attacks could be more devastating than those of the terrorists, whereas the latter are generally more concerned with the body count or how the event plays on the 6 o'clock news. The exception might be if the state sponsor could place all of the blame on the sponsored terrorist group. The major difference between BW and BT is the perpetrator's intended end effect.

A significant difference between a biological weapon of mass destruction and other weapons of mass destruction is the potential for disguising the attack as a natural outbreak. The detonation of a nuclear weapon cannot be a natural event, although attempts may be made to disguise the identity of the perpetrators. The same can be said for an incident involving chemical agents such as mustard gas or a nerve agent. However, it is possible that a perpetrator would wish to escape detection by trying to fool people into believing that a BW or BT attack was a natural outbreak. This would have the advantage of preventing a search for the perpetrator, much as disguising a murder as a heart attack may allow the perpetrator to "get away" with murder. These are some of the dimensions and concerns that need to be addressed in building definitions of BW and BT.

2.2 BIOWARFARE

Biowarfare (BW) denotes the hostile use of biological agents against an enemy in the context of a formally declared war [6]; it is the intentional use of biological threat agents to kill or incapacitate adversaries on the battlefield or in a theater of operations, usually requiring agents that are fast-acting; in other words, agents that produce pathogenesis rapidly (i.e., botulinum toxin) or within a few days (i.e., *Bacillus anthracis* [anthrax], *Yersinia pestis* [plague], etc.). The requirement for rapidly acting BW agents is truer today than in past wars, in which battles were fought over extended periods of time rather than in a matter of hours or days.

Therefore, organisms such as *Mycobacterium tuberculosis*, which has a long incubation period, would have little effect on the outcome of a modern battle of a few days' duration at most. Hence, some of the questions arise about the Iraqi weaponization of alfatoxin [7], an agent whose major mode of action is carcinogenic. Other factors for use on the battlefield are the abilities or attributes of a biological agent that render it suitable to be weaponized for delivery as an aerosol or in other forms to be used against an adversary. However, other biological threat agents that produce long-term sequelae (i.e., brucellosis, viral infections, etc.) can put heavy demands on logistical and medical support personnel. By affecting the supply lines, front-line military performance is dramatically affected by interrupted supply lines and possible exposure from contagious supply personnel who are in the prodromal period. Finally, the use of certain threat agents requires, in many cases, that the adversaries have a means to protect their own troops with prophylactic measures such as vaccines and personal respiratory protection.

BW differs from BT in that the military enters into the theater of operations prepared for the possible use of BW or chemical agents. Personnel are trained to respond to BW agents, they are provided with mission-oriented protective posture (MOPP) gear to protect themselves from threats at the lowest (MOPP1) up to the highest (MOPP4) levels. The use of the gear does not come without a price with regard to performance, in that the use of MOPP4 gear places tremendous physical burdens on the soldier, such as causing heat exhaustion and being cumbersome. Like BT, the mere notion that a BW agent may have been released upwind of military personnel may force them to don MOPP gear, resulting in impaired function and giving the adversary a decided advantage. This shows one of the similarities between BT and BW—instilling fear into military personnel that a biological attack is imminent will, in effect, achieve a military advantage.

In general, the use of biological threat agents in a theater of operations during conflicts with an adversary is considered BW, whether or not the adversary is a state actor. Such use of threat agents will tend to be more overt in their use, and the military are trained and equipped to cope. However, military forces can undertake covert operations with the intent of performing surprise attacks on unsuspecting military or civilian populations, designed to create fear or intimidate governments or societies, and so on.

Biological terrorism is used to generate terror or fear in a society. Through coverage by the media, medical cases, and other factors, the fear of biological threat agents elicits significant responses, whether as a hoax or as an actual release. It has been proposed that this tactic is preferable to the actual use of a threat agent, in that the adversary gains an advantage without later being subject to international disdain and retaliation, not to mention the cost. If the perpetrators can accomplish their goals using a hoax, there is a tremendous cost-saving over agent development or acquisition. The other aspect of this tactic is that if a military opponent is made to think that a biological threat agent has been released, yet see that all detectors and assays are negative, the military will find it difficult to know when to come out of protective posture and when to stop prophylaxis. They may also think that it is an unknown agent or a heavily modified agent. Once again, it is the fear of the unknown that gives credence to the use of biological threat agents in both BW and BT. In a battlefield scenario, there actually may be less fear of BW by well-trained and equipped military, because they are ready, they are expecting a release, and they are often vaccinated and have other forms of prophylaxis and therapeutics. Where fear comes more into play is when there are large gatherings of nonequipped, nontrained civilian populations with limited prophylactic and therapeutic measures available.

It has been argued by proponents of BW use that BW agents are simply another method of killing that is little different from other methods [5]. There are, however, attributes that make biological agents different. These include the delay in killing—in other words, the incubation period plus the time required for the disease to kill. Most other methods used on the battlefield generally kill faster than biological agents. Another difference is that some biological threat agents are capable of self-propagation, and hence they are likely to attack unintended targets. The effect on the local civilian population includes the psychological blow of not knowing whether and to what extent the affected area is a hazard. There is also the psychological effect on the families of BW victims, who may have difficulty in having the body of their loved one returned to them. Methods of combat that are too indiscriminate and kill inhumanely (by the standards of the time) have long been considered inappropriate weapons by civilized societies, including at one time the crossbow (forbidden for use against Christians by Pope Innocent II in 1139 CE) [8].

Uses of antiplant, antianimal, or antimaterial agents are more likely to be used in BW than in BT. A state sponsor may very well wish to cripple an opponent's economy or ability to wage war without bringing down either the condemnation or possible nuclear retaliation that an act of BW against personnel may cause. Terrorists are unlikely to disguise an attack in this manner. First, they are unlikely to have the sophistication to accomplish such a subtle attack, though their choice of unconventional agents may lend itself to disguising the attack. Their success in disguising the act is just as likely to be a result of chance or incompetence on the part of the country attacked or the terrorists as a result of efforts on the part of the terrorists (e.g., the infamous salad bar incident perpetrated by the Rajneeshee [4], which was not recognized as a terrorist act until a member of the cult confessed a year later). Second, much of the desired terror is lost if the attack is seen as a natural occurrence. Although automobile accidents and influenza kill thousands of people every year, they do not significantly affect the daily lives of the population. A new disease, even if it killed thousands a year, would cause some initial fright, but it would eventually be accepted and would not cause any of the changes desired by the terrorists, such as in the case of HIV. It may indeed create an economic burden and could affect a country's ability to fight the terrorists, but without the terrorists taking credit, it is unlikely to influence the outcome of the struggle.

2.3 STRATEGIC BW

2.3.1 Strategic Level of War

The Department of Defense defines the strategic level of war as that "at which a nation, often as a member of a group of nations, determines national or multinational (alliance or coalition) security objectives and guidance, and develops and uses national resources to accomplish these objectives. Activities at this level establish national and multinational military objectives; sequence initiatives; define limits and assess risks for the use of military and other instruments of national power; develop global plans or theater war plans to achieve these objectives; and provide military forces and other capabilities in accordance with strategic plans" [9].

2.3.2 Strategic Objectives and Requirements

Strategic BW implies large scale in terms of both geography and time. Strategic nuclear weapons are designed to destroy cites and surrounding areas. The strategic bombing of Germany in World War II was designed to break the industrial base and the people's will to support the war. In fact, use of anthrax was seriously considered for use against Germany by the allies [5]. This was not something that was accomplished in days or weeks, but over months and years. Nor was it conducted on the scale normally associated with a battlefield, but on a country-sized scale. Strategic BW would be conducted on a similar scale, by attacking whole populations with intercontinental ballistic missiles, cruise missiles, and so on, as the Soviet Union planned [10]. There would most likely not be just a single attack but multiple attacks, spanning weeks, if not months or years. Antiplant or antianimal agents would most likely be used to attack agriculture. Antimaterial agents could be used to attack industry. These requirements imply certain desirable characteristics of any agent. For the purposes of this book, we will focus on antipersonnel agents, but similar lists could be made for antiplant, antianimal, or antimaterial agents.

2.3.3 Agent Characteristics

Primarily, the agent needs to be able to be distributed on an appropriately large scale or be able to self-distribute over the same scale. Few agents fit these criteria. Most agents are insufficiently stable to withstand distribution over the required scale. On this scale, it is impractical to distribute a bomblet at each intersection of a 1- or 2-km grid covering hundreds of square kilometers. Therefore, the agent must be able to cover hundreds, if not thousands, of square kilometers from a single distribution point. Fortunately, from the defensive standpoint, this agent list is short. At present, there are only three diseases on this list: anthrax, smallpox, and plague. That does not mean that agents such as Ebola cannot be engineered to meet these criteria. Anthrax, as was shown at Sverdlosk [11], is capable of being blown in spore form for up to 50 km downwind without losing its effectiveness. Smallpox and plague are self-distributing (i.e., contagious from person to person). According to Alibek [10], the Soviet Union fielded plague-, anthrax-, and smallpox-containing intercontinental ballistic missiles with the intent of causing epidemics in the surviving immune-compromised populations in the event of nuclear war. The intent was to destroy the ability of the U.S. population to wage war, if not to destroy the population.

2.4 OPERATIONAL BW

The Department of Defense defines the operational level of war as that "at which campaigns and major operations are planned, conducted, and sustained to accomplish strategic objectives within theaters or other operational areas. Activities at this level link tactics and strategy by establishing operational objectives needed to accomplish the strategic objectives, sequencing events to achieve the operational objectives, initiating actions, and applying resources to bring about and sustain these events. These activities imply a broader dimension of time or space than do tactics; they ensure the logistic and administrative support of tactical forces, and provide the means by which tactical successes are exploited to achieve strategic objectives" [9].

2.4.1 Operational Objectives and Requirements

The operational level is the most similar of all BW levels to the terrorist level. It is also the level for which the best argument can be made for using BW on the battlefield—particularly incapacitating agents. At the operational level, potentially doomsday plagues such as smallpox or plague would not be appropriate—they would kill too many people. Unlike the tactical level, which requires very fast-acting agents, at the operational level agents working in weeks or a month would work very well. The idea at this level would be to disrupt the supply of either personnel or material. Either infecting the combat reserves or the personnel staffing the supply train could accomplish this. Without food, fuel, ammunition, and so on, the warfighters cannot do their job. The agent need not be fatal, however, as simply by rendering a majority of the rear echelon troops unfit for duty and clogging the medical chain could cause the required disruption.

2.4.2 Agent Characteristics

The agent would need to be able to cover roughly neighborhood-sized spaces—not necessarily cities, but areas larger than single blocks. However, the agent should not spread beyond the desired area, which would eliminate such agents as smallpox and plague from the list. Anthrax would still be on the list. Agents such as tularemia, brucellosis, Venezuelan equine encephalitis virus, and toxins would be added to the list. In fact, just about the entire traditional BW list would be included.

2.5 TACTICAL BW

The Department of Defense defines the tactical level of war as that "at which battles and engagements are planned and executed to accomplish military objectives assigned to tactical units or task forces. Activities at this level focus on the ordered arrangement and maneuver of combat elements in relation to each other and to the enemy to achieve combat objectives" [9].

2.5.1 Tactical Objectives and Requirements

Biological threat agents do not readily lend themselves for use in tactical warfare. In fact, some experts contend that there are no tactical biological threat agents. Tactical warfare as defined above is the action taken on the scale of battalions and lower to accomplish missions such as conquering a hill. Biological threat agents, unlike most conventional, chemical, or nuclear weapons, have a latent period before clinical effects are visible. This period can be as short as a few hours, particularly with some toxins—which in the Russian doctrine are considered chemical agents. However, the wait is generally several days or more. A battlefield commander is not going to want to wait about a week for his weapon to take effect before attacking.

2.5.2 Agent Characteristics

Continuing the above analogy, a tactical BW weapon needs to destroy or affect a city block and needs to work in hours, or days at most. This, as in the strategic incident, leaves a short list of potential agents. These would include toxins and a few other fast-acting agents.

2.6 BIOASSASSINATION OBJECTIVES AND REQUIREMENTS

Bioassassination is the use of biological agents to commit an assassination [12]. It is yet a further reduction in the size of the area or number of personnel needed, below the tactical level. A famous example is the assassination of Georgi Markov with ricin by the Bulgarians [13]. The South African program also operated at this level [7]. Many of the constraints operating at other levels do not apply here. Agent stability is little if any problem, as the circumstance of administration is tailored to fit agent stability. It can be administered indoors, by injection, by contamination of food or drink, and so on. The incubation time need not be a problem either. Available agents or the desired mode of death—quick and relatively painless, or drawn out and excruciating—would very likely drive the agent choice. This level of attack has many commonalities with both BT and biocrime. The level of murdering a single individual (or a handful of people) differs from BT or biocrime only by the political importance of the individuals murdered. It is included here because it may be state-sponsored. Hence, the assassin would have access to more sophisticated agents and tools for administering the agent.

There are few if any constraints on agent choice. An agent causing a high mortality rate would most likely be desirable, as the death of the victims would generally be the desired outcome. However, it is conceivable that the desired outcome would be to cause the person or persons to be too sick to perform their job for a given period. One possibility would be making a politician sufficiently sick to miss an important legislative vote or to miss an important meeting with a foreign dignitary. However, this would not be bioassassination. Assassination implies killing the person or persons. It is a method of using biological agents to attempt to influence the course of events while maintaining an extremely low profile. This is the house or single-occupant level. The list of potential agent balloons as the operation would be tailored to fit the agent characteristics. The good news is this is not a mass casualty event. Extensive medical facilities could and likely would be mobilized to treat the affected individuals.

One possible example of an attempt to use a biological agent or the threat of an agent to influence a vote is the Amerithrax incident that occurred shortly after 9/11. It has been hypothesized that one motive for the anthrax letters was to influence the vote on the patriot act [14].

2.7 BIOTERRORISM

BT is the threat or use of biological agents by individuals or groups motivated by political, religious, ecological, or other ideological objectives [12,15] as well as in some cases attempts to manipulate an election [4] or to defend against police actions [3]. BT is considered to differ from BW primarily by generally not being state-sponsored. However, that does not preclude a state from sponsoring the perpetrators or engaging in terrorism directly. Another, perhaps better-distinguishing, characteristic might be an attack on civilians or noncombatants as primary targets. This definition, however, does run into the problem of defining civilians or noncombatants. In Western society, civilians or noncombatants are considered to be everyone but members of the military. They are easily recognizable by not wearing distinctive uniforms. Other entities such as the Al Qaeda consider everyone who pays U.S. taxes to be a legitimate combatant, thereby eliminating the entire noncombatant category [2]. Because nonstate entities have generally fewer resources than states, most of these differences will be resource driven. These individuals or groups are motivated by more ideological objectives. The purpose of a terrorist attack historically has been to draw attention to a specific cause and to cause terror and fear. In recent years, with many terrorist organizations, such as the Al Qaeda, cloaking their objectives in religious terms, there has been a growing desire for a large body count. Biological threat agents can provide this and can tap into the inherent terror caused by various historical plagues.

2.7.1 BT Objectives and Requirements

Bioterror events would probably be on the scale of operational or tactical BW, in part because of the difficulty of nonstate actors in obtaining the sophisticated agents and delivery systems of state actors. Virtually any disease could be on the list of potential actions, depending on the terrorists' desired effect, from using *Salmonella* to contaminate salad bars, to influence an election [4], to unleashing an Armageddon plague of smallpox. The latter is unlikely because of the difficulty of obtaining the agent and probable lack of desire to destroy the world (an unlikely outcome even of a worldwide smallpox outbreak, but an Armageddon-style world-ending battle was part of the Aum Shinrikyo doctrine) [16]. Unfortunately, the threshold for use of such agents, if obtained by a terrorist, is much lower than for state actors. There does not seem to be a lack of ability to find suicide bombers. The ultimate suicide bomber could be an infectious mobile smallpox sufferer. They may truly believe that their god will protect them; however, there is no historical record of prayer or any type of sacrifice affecting the course of a plague. The terrorists' belief in supernatural protection or a desirable after life may lead them to operate in ways that a state-sponsored program would not—they may be willing to take much greater risks in agent choice, preparation, and dissemination than a state.

2.7.2 Agent Characteristics

The primary characteristic of an agent for use by a terrorist would be availability of the agent in the desired volume. Few groups would necessarily attempt to produce the classical BW agents. The Rajneeshee cult, for instance, used *Salmonella* in a salad bar partly because it was what they could easily obtain. It had the added benefit from their point of view of being generally nonlethal [4]. The Aum Shinrikyo, in contrast, went for botulinum toxin and anthrax [16]. This cult seems to be the outlier that deliberately chose to attempt the more difficult task of weaponizing both chemical and BW agents. It is unlikely that other groups could or would expend that much effort.

2.8 BIOCRIME

Biocrime is the threat or use of biological agents by individuals or groups to commit a crime, such as robbery or murder, or to further their criminal intent [12]. This level of the use of biological agents or the threat of use has degenerated to the level of ordinary crime. A couple of examples are the revenge on coworkers by contaminated donuts [17] or, in a more general example, robbing a bank by threatening to spray anthrax in the lobby. This level bears many similarities to bioassassination without the benefit of state sponsorship. In some ways, it is simpler to obtain an Erlenmeyer flask of finely sifted flour that would probably work for the bank robbery than to obtain the actual agent and is safer for the perpetrator.

This level of attack is similar to state-sponsored bioassassination. The agents used at this level are more likely to be determined by what the perpetrator can obtain than any other criteria. In many instances, any substance that the perpetrator can convince people is a dangerous agent will work just as well as the actual agent, with far less risk to the perpetrator. This level of attack, like bioassassination, is unlikely to result in mass casualties. It should be possible to bring to bear much of modern medicine's capabilities. In many cases, the use of fake material is much more likely than the use of the actual agent. This is because of both the difficulty in obtaining the agent and the ability to accomplish the perpetrator's goal with a fake agent. The historically more prevalent biohoaxes would fall into this category [18].

2.9 CONCLUSIONS

Disease can be used to deliberately cause casualties in numbers ranging from one to millions. The former is relatively easy to accomplish and has been practiced by states, terrorist organizations, and criminals. At small scales, the choice of killing by disease instead of by knife or bullet causes little if any change in how the victims are treated. The problems arise when the killing is attempted on the tactical to strategic scales. The differences in the effects of BT or BW events are insignificant to medical practitioners who must deal with them. A natural outbreak of the plague, such as that which has occurred in the past, would cause similar problems. Unfortunately, the first sign of a biological threat agent attack (BT or BW) or an outbreak of a natural disease will be unusual numbers of patients turning up at the emergency rooms. Therefore, an efficient disease surveillance network is vital. It will have the added benefit of catching the next "acquired immunodeficiency syndrome" (AIDS) epidemic as well as the terrorist or state attack. The real differences among a natural outbreak, BT, or BW attack will be in the actions taken after the event has been dealt with. In the case of either a BT or a BW attack, the state that is responsible or that is harboring the perpetrators will be attacked. The terrorists or state leaders, when caught, will be tried for their crimes and punished.

One of the difficulties in this topic is that there is considerable overlap between both BT and BW. Both involve the use or threat of use of biological threat agents to harm humans. These difficulties stem from the lack of universally accepted definitions of both terrorism and noncombatants. The Federal Bureau of Investigation (FBI) defines terrorism as "the unlawful use of force and violence against persons or property to intimidate or coerce a government, the civilian population, or any segment thereof, in furtherance of political or social objectives" [19, p. i]. In contrast, the State Department defines terrorism as "premeditated, politically motivated violence perpetrated against noncombatant targets by subnational groups or clandestine agents, usually intended to influence an audience" [12, p. xvi]. The FBI definition includes guerrillas fighting an occupation—perhaps legitimately. The State Department definition would not include the attacks by the insurgents in Iraq against U.S. military but would include those actions targeting civilians or noncombatants. It would also not include attacks on noncombatants by military forces. However, that brings us back to the question of defining noncombatants or civilians. Most people would not think this a problem. In fact, the conventions on war did not even attempt to define this concept until 1977 [20]. Most people consider the 9/11 attacks on the United States a terrorist event, but the Islamic terrorists justified their attacks on the civilians in the two towers by claiming that because they paid taxes and voted, among other justifications [20], they had become combatants. This is their interpretation of certain passages in the Quran. Therefore, by that line of reasoning, 9/11 was an act of war not terrorism, and thus the difference between BW and BT comes down to defining the difference between warfare and terrorism. Until there is universal agreement on the definitions of warfare, terrorism, civilians, and noncombatants, there will be disagreement on whether a particular incident is a BW or BT incident. Although there may be disagreement over naming a given incident, consideration of the resources available to a given perpetrator can lead to insights into the possible agents that would be used and into the scale of the attack. This can lead to reasonable preparations to respond to the attack.

REFERENCES

1. Roberts, B., Controlling chemical weapons. *Transnational Law and Contemporary Problems*, 2(2), 435–452, 1992.
2. Wiktorowicz, Q. and J. Kaltner, Killing in the name of God: Al-Qaeda's justification for September 11. *Middle East Policy*, 10, 76, 2003.
3. Murakami, H., *Underground: The Tokyo Gas Attack and the Japanese Psyche*, Vintage International, New York, 2002.
4. Török, T. J. et al., A large community outbreak of Salmonellosis caused by intentional contamination of restaurant salad bars. *JAMA*, 278, 389, 1997.
5. Harris, R. and J. Paxman, *A Higher Form of Killing*, Random House, New York, 2002.
6. Parker, H. S., *Bioterrorism, Biowarfare, and National Security*, National Defence University, U.S. G.P.O., Washington DC, 2002, Chapter 1, http://www.ndu.edu/inss/McNair/mcnair65/05_cho1.htm (accessed on December 22, 2004).

7. Mangould, T. and J. Goldberg, *Plague Wars: A True Story of Biological Warfare*, St. Martin's Press, New York, 1999.
8. Chronology of the crossbow, http://www.thebeckoning.com/medieval/crossbow/chronology.html (accessed on August 5, 2010).
9. Glossary of military terms, http://www.militaryterms.info/about/glossary-s.shtml (accessed on August 8, 2005).
10. Alibek, K., *Biohazard*, Random House, New York, 1999.
11. Meselson, M. et al., The Sverdlovsk anthrax outbreak of 1979. *Science*, 266, 1202, 1994.
12. United States, Department of State. Office of the Coordinator for Counterterrorism, Patterns of Global Terrorism 2001, U.S. G.P.O., Washington, DC, 2002, p. xvi.
13. U.S. Army Medical Research Institute of Infectious Diseases, *Medical Management of Biological Casualties Handbook*, 2nd edn. USAMRIID, Fort Detrick, MD, 1996.
14. Graysmith, R., *Amerithrax: The Hunt for the Anthrax Killer*, Jove Books, New York, 2004.
15. Carus, W. S., *Bioterrorism and Biocrimes*, Center for Counterproliferation Research, National Defense University, Washington, DC, 1998.
16. Smith, R. J., Japanese cult had network of front companies. *The Washington Post*, November 1, 1995, p. A8.
17. Kolavic, S. A. et al., An outbreak of *Shigella dysenteriae* type 2 among laboratory workers due to intentional food contamination. *JAMA*, 278, 396, 1997.
18. Leitenberg, M., An assessment of the biological weapons threat to the United States, a white paper prepared for the conference on emerging threats assessment: Biological terrorism, at the Institute for Security Technology Studies, Dartmouth College, July 7–9, 2000.
19. United States, Federal Bureau of Investigation, Counterterrorism Threat Assessment and Warning Unit, *Terrorism in the United States 1999*, U.S. G.P.O., Washington, DC, 1999, p. i.
20. The 1977 protocols additional to the Geneva conventions of 1949, December 12, 1977, 16 I.L.M. 1391.

3 Scientific and Ethical Importance of Animal Models in Biodefense Research

James R. Swearengen and Arthur O. Anderson

CONTENTS

Contagion and catastrophic illnesses have affected the outcome of wars throughout history. Military officers have duties to protect their soldiers from becoming disease casualties, conserve their fighting strength, and ensure the success of the mission. Discharging those duties requires more than site sanitation and encouraging personal cleanliness. Armies need to have, and should have, at their disposal the best available vaccines and medicines directed against specific disease hazards. The ability to provide procedures, remedies, antidotes, and medical countermeasures has been almost as important as good military training and advanced weaponry in the success of military operations.

Historically, decisions to institute a new medical practice, use a vaccine, or adapt the use of a drug to protect soldiers from disease often were arbitrary and fraught with risks. Applying new forms of protection from disease was often made compulsory by commanders because military doctrine recognizes the interdependence of soldiers on each other for safety and support and requires that all participate, or the mutual support chain might break. This broadly utilitarian ethic of involuntary vaccination or treatment has been critical to protecting soldiers facing battlefield biological hazards during the Revolutionary War and throughout successive conflicts both inside and outside our hemisphere until the recent past [1]. In order to adequately address the topic of the importance of animal models of human disease in biodefense research, we have chosen to trace the development of military medical countermeasures from the time of George Washington to the present, with an eye on the ethical, moral, and legal tensions that led to the recent implementation of an animal efficacy rule by the U.S. Food and Drug Administration (FDA).

Smallpox was epidemic during the French and Indian War, and outbreaks continued to plague George Washington's army during the Revolutionary War. When George Washington ordered his entire army variolated, there had been no study carried out in animals to determine whether it was safe and effective for him to do so. The use of animals in medical research was not a practice until late in the nineteenth century. However, this anecdote about Washington's lucky decision is significant in showing that commanders of armies who have limited available information need to be able to have the discretionary authority to make health and safety decisions on behalf of soldiers in wartime.

Variolation had become accepted among European aristocrats, who believed that a "mild case" of smallpox would grant immunity. The method of variolation was not vaccination as we know it today. Variolation involved inoculating a person with smallpox scabs obtained from someone who had survived the disease. Many recipients suffered only mild illness, but there was a risk that variolation might cause serious illness, or even death in some. Washington ordered all his soldiers to undergo variolation without knowing with certainty that his men would be protected. His men became simultaneously the subjects of "research" and recipients of benefit—if they survived their deliberately induced disease outbreak. Variolation did cause some deaths among his soldiers and among members of the communities where his soldiers were encamped. Although both the idea of variolation and Washington's decision to make it involuntary among his troops became very controversial, his choice protected his troops from smallpox, which was critical to America securing independence from England. Indeed, the importance of smallpox and its mitigation of the outcome of the Revolutionary War figures prominently in Hugh Thursfield's "Smallpox in the American War of Independence" [2] and Elizabeth Fenn's *Pox Americana* [3]. The negative outcomes of variolation and the need to provide continuous and sustainable progress in providing the means to protect the health and safety of soldiers in any future battlefield in all likelihood led Congress to create the Army Medical Department in the spring of 1818.

The story about the discoverers of mosquito transmission of yellow fever is important to describe because it is a milestone leading up to the need for using animals to prove efficacy of medical countermeasures against serious biological hazards [4–6]. The story involves numerous connections with William Welch and William Osler, two of the first four physician professors of the new Johns Hopkins Hospital, who would become important advocates for medical ethics and the use of animals in research.

Yellow fever epidemics frequently broke out in the Caribbean and the southern United States, and it especially plagued American soldiers during the Spanish American War. Outbreaks were so prevalent that President Roosevelt asked the Army Surgeon, General George M. Sternberg to create a commission to study yellow fever in Cuba. Stenberg selected Walter Reed and James Carroll as the first and second officers in command of the Yellow Fever Commission. Reed and Carroll were highly regarded by Drs. Osler and Welch at Johns Hopkins, who also recommended that Jesse Lazear, the Hopkins clinical laboratory officer, be added to the commission. Sternberg, Reed, and Carroll received research training in Welch's laboratory at the Johns Hopkins Hospital. Sternberg was the first bacteriologist trained in Welch's laboratory in the late 1880s, before being appointed Army Surgeon General.

Major Reed's research into the cause and transmission of yellow fever in Cuba did not involve experiments in animals primarily because there was confusion about what kind of agent actually was the cause of the disease. Some felt the disease spread through the air in fomites from the bedding of previously ill patients. Others—Sternberg, Reed, and Carroll included—thought the disease was caused by a new form of bacterium that needed to be discovered. Several of the members of the commission, especially Aristides Agramonte and Jesse W. Lazear, had other reasons to include still-unknown causes for the disease. Agramonte proved that patients suffering from yellow fever were not infected with the bacterium widely believed to be the cause. Lazear, who had been a student of malaria and knew about mosquito transmission, allowed himself to be bitten by a mosquito that had been feeding on a patient suffering from yellow fever. He subsequently died from the illness resulting from this mosquito bite, thus fixing mosquito transmission of yellow fever as the leading hypothesis [6]. Human volunteers were recruited from among Major Reed's military detachment, and Cuban civilians also came forward and volunteered. Reed's use of volunteer contracts that spelled out the full extent of risk and possible benefits of participation in this research is regarded as an ethics milestone, introducing to medicine the concept of voluntary consent. The research risks the patients accepted enabled the discovery that yellow fever was transmitted by mosquitoes, which contributed immeasurably to public health because now the spread of the disease could be prevented by mosquito control [4–6].

Experimental use of animals was becoming popular in the laboratories of Osler, Welch, and other physician scientists of the Johns Hopkins Hospital. Their approach would revolutionize medicine by demonstrating that experimental evidence could be obtained to support the scientific practice of medicine. Rather than being immediately recognized by the public as a good development, Reed's experiments in humans and experimental use of animals for medical research attracted criticism by antivivisectionists, who had become very influential in England and the United States during the early 1900s [4,5].

In 1907, Osler was invited to address the Congress of American Physicians and Surgeons about the evolution of the idea of experiment in medicine. Osler was a strong proponent of academic medicine and a well-respected medical philosopher. He had been busy testifying during the last several years in legislative forums in the United States and abroad about the value of research, because it was under attack by antivivisectionists [4]. This is what he said about the need for animal experimentation and also about the voluntary nature of the participation of soldiers in Reed's yellow fever experiments:

> The limits of justifiable experimentation upon our fellow creatures are well and clearly defined. The final test of every new procedure, medical or surgical must be made on man, but never before it has been tried on animals.... For man absolute safety and full consent are the conditions which make such tests allowable. We have no right to use patients entrusted to our care for the purpose of experimentation unless direct benefit to the individual is likely to follow. Once this limit is transgressed the sacred cord which binds physician and patient snaps instantly.... Risk to the individual may be taken with his consent and full knowledge of the circumstances, as has been done in scores of cases, and we cannot honor too highly the bravery of such men as the soldiers who voluntarily submitted to the experiments on yellow fever in Cuba under the direction of Reed and Carroll [4, pp. 7–8].

When Osler testified before the U.S. Congress and British Parliament to protect medical research from being blocked by legislation triggered by the activities of antivivisectionists, his presentations often paired the medical fruits of research conducted with human volunteers with the benefits of testing drugs and vaccines in animals [5], which may have ensured that these paired concepts would endure. Thus, at the dawn of the twentieth century, medicine had arrived at two truths that would help define what was required for research with humans to be regarded as ethical. The first was the need for experiments in animals to assess the risk or validate the disease causality before involving human subjects in tests. The second was that participation of human subjects in tests of efficacy must be voluntary and can take place only after human subjects are told the risks and benefits of participation in the research.

During the first third of the twentieth century, research with animals was becoming an important vehicle for scientific biomedical discovery across the globe. Animal experimentation would become even more important as World War II approached, and the United States was not prepared to deal with a biological warfare threat. Facing the emergency of war in 1941, Secretary of War Henry L. Stimson asked the president of the National Academy of Sciences, Frank B. Jewett, to appoint a committee that would recommend a course of action "because of the dangers that might confront this country from potential enemies employing what may be broadly described as biological warfare" [7]. This committee, chaired by Edwin B. Fred, reported to Secretary Stimson that, "There is but one logical course to pursue, namely, to study the possibilities of such warfare from every angle, make every preparation for reducing its effectiveness, and thereby reduce the likelihood of its use" [7].

To accelerate the development of programs to respond to the biological warfare threat, the War Research Service was established, under George W. Merck Jr., inside the civilian Federal Security Agency to begin development of the U.S. Biological Warfare program, with both offensive and defensive objectives. The first major objective of War Research Service was to develop defensive measures against possible biological weapons attack [8]. Under the guidance of Ira Baldwin, the Army Chemical Warfare Service commenced operation of a large-scale research and development program, and the facility at Camp Detrick was the first of the laboratories and pilot plants to be constructed, starting in April 1943 [8,9].

Most of the serious infectious diseases regarded as biological warfare threats were natural diseases of agricultural animals that could also cause devastating illness in humans. Indeed, one may make the argument that all serious infectious diseases come about by interaction of humans with animals, even those diseases with limited host-range specificity [10]. The risk that humans might die because of infection or intoxication by biological threat agents was so great that a major commitment to testing in animals was incorporated into program objectives and the design of the Camp Detrick laboratories. Animal models of disease figured prominently in validating what could be learned about human disease diagnostics and medical countermeasures. These serious risks also prompted a major commitment to developing safe working environments, occupational health practices, and on-site medical care in a station hospital to reduce the risk of injury or death to workers [9].

Animals stood in for humans in most of the offensive and defensive biological warfare research conducted at Camp Detrick during the war. In addition, research in animals was directed by prominent civilian medical researchers, at universities, companies, and research institutes, who received Federal Security Agency grants after review by the War Research Service Committee on Medical Research. A list of persons directing specific contract protocols included eminent scientists, future Nobel Prize winners, and corporate leaders who shaped modern biology, pharmaceuticals, and medicine [8–14].

Any involvement of humans at Camp Detrick was limited to epidemiological studies of workers, with occupational exposures seen in the dispensary or treated at the station hospital. One study involving human subjects was carried out so that data from animal models of aerosol exposure could accurately be extrapolated to humans. *Serratia marcescens* was used as a putatively nonpathogenic simulant in humans instead of the more hazardous pathogen that would be used for animal exposures [14]. Nonpathogenic simulants were used in model human aerosol exposures so that risks of harming human volunteers would be held to a minimum [11]. In contrast, the Nazi doctors who used Holocaust victims and prisoners of war in research at concentration camps made no attempt to minimize risk because genocide was a major objective. The details of the immoral Nazi experiments became known to the world through the media and were further revealed at the War Crimes Tribunal held in Nuremberg at the end of World War II.

In December 1946, shortly before he left for Germany to participate in the tribunal, Dr. Andrew Ivy released to the American Medical Association a draft of his list of conditions required for research in healthy subjects to be regarded as ethical. Ivy and Dr. Leo Alexander, the court's medical consultant, testified as to the ethical standards of medical practice and compiled for the tribunal 10 conditions that must be met for research involving human subjects to be permissible [15]. This list of conditions, now referred to as the Nuremberg Code [16], included a requirement for prior animal experimentation validating the possible risks and benefits of the research to be completed before humans would be involved.

In 1952, the Armed Forces Medical Policy Council noted that tests at Fort Detrick with biological warfare simulants showed U.S. vulnerability to biological attack. Similar experiments with virulent disease agents in animal models attested to incapacitating and lethal effects of these agents when delivered as weapons [8,15]. However, a long time had passed without any human testing, and there was doubt among Armed Forces Medical Policy Council members that extrapolation of animal data to humans was valid.

Human vulnerability to actual biological agents delivered under realistic scenarios was not known, and human studies were strongly encouraged to prove that continuation of the biological warfare program was justified; however, military medical scientists assigned to Fort Detrick were reluctant to pursue human testing without thorough discussion of the ethical, moral, and legal basis for such studies [17–19]. A memorandum dealing with human experimentation was issued to the military branches by Secretary of Defense Charles Wilson on February 26, 1953. Referred to as the Wilson Memorandum, this memorandum adopted the 10 principles of the Nuremberg Code, including the need for prior animal experimentation, as official guidance promulgating ethical research involving human subjects [20].

The consequences of the Nazi war crimes and availability of the code principles motivated military medical researchers to find the moral high ground while developing medical countermeasures against nuclear, biological, and chemical agents during the Cold War [9,15]. Military physicians assigned to develop medical countermeasures against biological weapons were reluctant to put humans at risk in experiments without first obtaining sufficient information from other sources that could be used to mitigate the danger. Responding to the need to conduct human studies, *ad hoc* meetings of scientists, Armed Forces Epidemiology Board advisors, and military leaders took place at Fort Detrick during the spring of 1953 [18,19]. The depth and breadth of these discussions resulted in the design of several prototype research protocols and the creation of an institute heavily invested in animal experimentation aimed at modeling human infectious diseases so that pathogenesis and response to vaccines and therapeutics could be studied. The Army Chief of Staff issued on June 30, 1953 a directive (cs-385) derived from the Wilson Memorandum. The directive (cs-385) contained additional safeguards not mentioned in the Wilson Memorandum that were proposed by the scientists who had attended the *ad hoc* meetings [21].

Under cs-385, the only studies of human infections and of the efficacy of vaccines in protection, or the efficacy of drugs in treatment of a biological warfare agent, that scientists felt were ethical were the diseases Q fever (*Coxiella burnettii*) and tularemia (*Francisella tularensis*). These disease agents were able to be made less likely to result in mortality by limiting infectious dose; substantial information on disease pathogenesis and vaccine efficacy in animals was already known, and there were drugs available that could be used to quickly end the infections for the safety of the volunteers. This left all of the other agents on the biological warfare threat list ineligible for testing in humans on the grounds that to do so would be immoral.

Vaccines or drugs against most of the agents on the threat list were tested for efficacy in animal challenge models, whose responses could be compared to the responses of humans tested in the safety trials that included an assessment of markers of immunity or drug metabolism and kinetics. Except for tularemia and Q fever, which were regarded as ethically acceptable, no threat agent challenges to prove efficacy of medical countermeasures were performed in humans.

The idea that medical countermeasures against hazardous viruses, bacteria, and toxins would be tested for prophylactic or therapeutic efficacy in valid animal models was intrinsic to all military research programs for developing products that would be used in humans. The investigational products that showed efficacy in animals were tested in humans for safety and, if determined to be safe, were used to protect or treat workers after approval by the appropriate members of the chain of command, up to the most senior level, as defined in regulations. This could be the Army Surgeon General or go as high as the secretary of the military service sponsoring the study, depending on the level of risk or the military organization structure at the time the study was conducted.

Human research volunteers were recruited from among Seventh Day Adventist conscientious objectors who were being trained as medics at Fort Sam Houston, Texas. These men, who were willing to serve at Fort Detrick as noncombatants, participated as volunteers in reviewed and approved studies testing human vulnerability to biological warfare agents in realistic scenarios. Multiple new products for defense against biological warfare and hazardous infectious diseases were developed and tested for human safety and for surrogate markers of efficacy with their participation.

Using animals as surrogates for humans in efficacy trials came under regulatory pressures in the late 1950s. The FDA strengthened their drug regulations because of new drugs that were being introduced that either were not effective or had serious but undiscovered side effects. Thalidomide, a new sedative drug that was already introduced in Europe, was blocked by an FDA reviewer because there was evidence that its use was associated with birth abnormalities in the limbs. U.S. Senate hearings followed, and in 1962, the so-called Kefauver–Harris Amendments to the Food, Drug, and Cosmetic Act were passed into law to ensure drug efficacy and greater drug safety. For the first time, drug manufacturers were required to prove to the FDA the human clinical efficacy of their products before marketing approval would be granted [22,23].

The Army replaced cs-385, which had guided the ethical use of humans in drug and vaccine research, with a more widely distributed Army Regulation 70-25 (AR 70-25) on March 26, 1962 [24]. The new FDA requirements to prove human clinical efficacy caused the Army to introduce the following exemptions in paragraph 3 of the new AR 70-25.

3. The following categories of activities and investigative programs are exempt from the provisions of these regulations:
   a) Research and non-research programs, tasks, and tests which may involve inherent occupational hazards to health or exposure of personnel to potentially hazardous situations encountered as part of training or other normal duties, e.g., flight training, jump training, marksmanship training, ranger training, fire drills, gas drills, and handling of explosives.
   b) That portion of human factors research which involves normal training or other military duties as part of an experiment, wherein disclosure of experimental conditions to participating personnel would reveal the artificial nature of such conditions and defeat the purpose of the investigation.
   c) Ethical medical and clinical investigations involving the basic disease process or new treatment procedures conducted by the Army Medical Service for the benefit of patients. [This exemption permitted use of FDA-unapproved products in clinical studies, force health protection, experimental infections and vaccine challenge studies.]

Having recognized that the 1962 Food, Drug, and Cosmetic amendments now required that there be substantial evidence of human clinical efficacy, a requirement that would perilously put humans in harm's way, the Department of Defense (DoD) negotiated a memorandum of understanding (MOU) with the FDA in 1964 so that it could continue to provide its troops with the best available products for the protection from or treatment of biowarfare hazards, irrespective of their FDA approval status [25]. This MOU was important because the DoD had no intention of conducting hazardous challenge studies in humans to prove human clinical efficacy. The MOU allowed the DoD to continue to approve its own use of these products without having to comply with FDA requirements for providing investigational products to soldiers under a clinical trial format when this would confuse the intent to benefit in emergency operations with an unintended objective—that of conducting an experiment for marketing approval [24].

This MOU permitted the DoD to use investigational products in classified clinical investigations and nonclassified research programs. The term "nonclassified research programs" included "ethical medical and clinical investigations involving the basic disease process or new treatment procedures conducted by the Army Medical Service for the benefit of patients" [24]. Clinical research with drugs and biologics required submission of an Investigational New Drug (IND) application to the FDA or Public Health Service. Because of "intent to benefit," the Special Immunizations Program, which provided laboratory workers with investigational vaccines intended to provide additional protections above environmental safety considerations, were also permitted by this MOU [25].

During the war in Vietnam, an investigational plague vaccine that had been tested for safety in human volunteers and for efficacy in experimental animals was given to troops without investigational labels or data collection requirements [8,19]. Plague was a serious battlefield hazard, and epidemiological data subsequently showed that the plague vaccine provided a benefit and reduced the incidence of plague in vaccine recipients. Under the MOU, these data were submitted to the Public Health Service, which subsequently approved the vaccine.

In 1972, Congress added Title 10, U.S. Code 980 to the defense appropriation bill [26]. This public law mandates that informed consent must be obtained from subjects or their guardians, irrespective of levels of risk, for all research including that intended to benefit the patient. Also in 1972, the authority for regulating biologics including serums, vaccines, and blood products was transferred from the Public Health Service/National Institutes of Health to the FDA [23].

The U.S. Public Health Service syphilis study also created a major public controversy in 1972. This was a study of indigent black men from Tuskegee, Alabama, who were prevented from receiving treatment so that the natural course of syphilis could be studied. The program ran from 1932 until it was exposed in 1972. Irrespective of the physicians' ability to cure syphilis with penicillin, the subjects were never told about it, nor were they treated after penicillin became widely available. These revelations led to passage of the National Research Act of 1974 [27], which added additional restrictions and oversight to research involving human subjects.

Among the new requirements of the National Research Act were the requirement for informed consent and the need for a review committee, knowledgeable in the basic ethical principles of beneficence, respect for persons, and justice, to assess the risk–benefit criteria and the appropriateness of research involving human subjects. This committee, referred to as an Institutional Review Board (IRB), was expected to be independent of the chain of command, so that no conflict of interest would exist between the need to develop a product and the need to protect the rights and welfare of the human volunteer subjects. The act provided no guidance on the structure or operation of the IRB, however. AR 70-25 already specified these conditions, so no specific changes were needed at this time, other than the consideration of moving the IRB function out of the commander's office and into a more independent forum.

The Army revised AR 70-25 in 1974 to account for a reorganization within the DoD that resulted in transfer of final approval authority from the Chief of Research, Development; Testing, and Evaluation to the Surgeon General of the Army Medical

Department for all research using volunteers [28]. It distinguished between research conducted in Army Medical Services and that conducted by the Army Medical Research and Development Command, and it identified the requirements for use of active duty military personnel as volunteers.

Although it does not appear that these changes in the regulation were associated with either of the previous regulatory developments, it did necessitate negotiating a new MOU with the FDA because of the transfer of authority [29]. The FDA had also undergone changes during the period between the 1964 MOU and the MOU signed with DoD on October 24, 1974, so it included additional FDA review requirements [23]. Again, the additional restrictions provided improved protections for human volunteer subjects who participated in research, but the restrictions would also prevent use by the military of well-studied potentially beneficial products that had not completed all the tests needed for FDA approval unless agreed to in the MOU.

The 1974 MOU [29] restricted DoD authority to use investigational products for health protection of armed forces. Classified clinical investigations could be exempted from the Food, Drug, and Cosmetic Act. However, both DOD and FDA would need to review and approve use of products in military personnel that were not approved by the FDA but were "tested under IND regulations sufficiently to establish with reasonable certainty their safety and efficacy" [29]. All other clinical testing of investigational drugs sponsored or conducted by the DoD required submission of an IND application to the FDA.

The National Commission for the Protection of Human Subjects of Biomedical and Behavioral Research, created by the National Research Act of 1974, published its report, entitled Ethical Principles and Guidelines for the Protection of Human Subjects of Research (popularly referred to as the Belmont Report), on April 18, 1979 [30]. In 1981, the Department of Health and Human Services and the FDA published convergent regulations based on the Belmont principles, adding additional restrictions to what may be regarded as research versus treatment and to the use of unapproved drugs and biologics [31,32].

Furthermore, the FDA underwent a reorganization during the late 1980s that vastly increased its position in a department whose director, the Secretary of Health and Human Services, held a cabinet office [22]. This change necessitated that the DoD negotiate a new MOU with the FDA if it wished to continue to provide unapproved drugs or vaccines for health protection of armed forces or for medical use under an intent to benefit. The new MOU between the DoD and FDA that was signed on May 21, 1987, removed the discretionary privileges that enabled the DoD to use FDA-unapproved products in wartime or to protect at-risk personnel who worked in hazardous environments [33]. DoD no longer had FDA-permitted authority to self-approve use of drugs, vaccines, and devices that remained in IND/IDE status.

The 1987 MOU is still in effect. This MOU requires FDA review for the use of any IND by the DoD (i.e., no exemptions from the Food, Drug, and Cosmetic Act). In the case of classified research, the DoD must submit a "classified IND application or investigational device exemption (IDE) application" [33] for review and approval by the FDA. The FDA is responsible for having reviewers with the security clearance needed to assess these activities. The new MOU also ended exempt status for the use of IND vaccines in the Special Immunization Program for vaccinating workers at risk of occupational exposure to hazardous disease agents. The new MOU with the FDA required complete compliance with FDA regulations on products labeled investigational. The DoD could no longer exclude from FDA requirements products they wished to use in contingency situations or for health protection of armed forces.

These new changes occurred as the situation in the Persian Gulf heated up, and it became clear that U.S. forces would be deployed against an enemy who had a large program for developing chemical and biological weapons and who had used such weapons on opposing factions within his own country. The U.S. preparations to enter Iraq during Desert Shield/Desert Storm produced a moral dichotomy because some of the medical countermeasures that might be used to protect or treat soldiers for chemical or biological hazards were still not approved by the FDA because of the lack of substantial evidence of human clinical efficacy. It was once possible for the DoD to use intent as a means of determining how a product would be used and under what kind of restrictions. Was the intended use "research," or was it "intent to benefit"? The new MOU made the ability to use products labeled IND to benefit war fighters and laboratory personnel less clear.

It was expected that exploratory research with IND products to discover new treatment uses and to generate data requirements to apply for new drug marketing approval would continue to rigorously follow DoD and FDA requirements. But what if investigational status prevented lifesaving use of the only available product to protect against anticipated mass casualties produced by biological weapons? If drugs or vaccines still under IND status were needed for protecting or treating persons during a national emergency, they would have to be given according to research protocols. Creating the pretense of experiment to get around the moratorium on use of unlicensed products seems disingenuous when the basis of the intent is really based on knowledge of human safety and animal efficacy of the product. Furthermore, the need to quickly provide prophylaxis or treatment of whole populations who could be suffering from nuclear, chemical, and biological injuries might require drastic emergency actions not anticipated by clinical trial protocols. The alternative choice of not providing soldiers prophylaxis or treatment because a product is not FDA approved was also an unsatisfactory solution. The DoD decided to apply for a waiver from the FDA.

The FDA grants waivers of certain requirements of its regulations usually because it is unfeasible or impracticable to comply. A waiver of the requirement for "informed consent" was requested, and IND products that were the only drugs or vaccines developed to a degree that might enable the DoD to conclude that it would be protective or beneficial could be used. However, this waiver was removing an essential ethical principle. The principle of "respect for persons" respects a person's self-determination and autonomy and is a component of his or her dignity. It is understood that an unconscious person in need of lifesaving treatment may be unable to give consent, in which case providing an IND drug or antidote to prevent death would be acceptable. The FDA regulations already had such an allowance, but it limited its use to a small number of subjects or a single incident. In a military emergency, hundreds or thousands of subjects would need to receive the IND product—and that is not allowed by regulation. Through ethical analysis, one could conclude that it should be allowed. But the FDA regulations are law, which is immutable. The decision of whether or not to use a waiver of informed consent is a difficult choice [34–38].

There are other waivers that could have resolved this conflict without having to abandon an ethical principle for utility. At a symposium convened on September 30, 1988, at Fort Detrick by the Post Chaplain, the rhetorical question was asked of whether it would ever be legal, moral, or ethical to do a real test of the safety and effectiveness of an antidote developed to protect humans from a lethal nerve gas exposure [34,35]. Carol Levine, speaking as Executive Director of the Citizens Commission on AIDS for New York City and Managing Editor of IRB: A Review of Human Subjects Research, gave a "yes" or "no" answer to the question of whether there are ethical exceptions for military medical research. Levels of risk and the voluntariness of participation affected which answer she would give. Further, Levine allowed that it would be easier to say "yes" if the difficult affirmative choice were made by "regulators" [34]. Richard Cooper, who had been the chief counsel for the FDA between 1977 and 1979, also struggled with these choices. However, he recommended choosing to request a waiver of the requirement to provide substantial evidence of human clinical efficacy over a waiver of informed consent for reasons similar to what has been expressed earlier in this chapter [35]. In 1990, the DoD sent the U.S. military into the Persian Gulf having chosen a waiver of informed consent, and over the first half of the 1990s, we codified this in regulation [39–42].

The DoD was one of the 17 federal departments and agencies that agreed to adopt the basic human subject protections of 45 CFR 46, referred to as the "Common Rule." Thus, all federally sponsored research involving human subjects was now covered by a common set of policies, assurances, and protections (the DoD uses 32 CFR 219). The current version of AR 70-25, published on January 25, 1990, already complied with the Common Rule [43].

Additional FDA enforcement laws enacted in 1993–2002 made compliance with FDA IND and Good Clinical Practices Act requirements during diagnosis, treatment, or prophylaxis for emergencies related to domestic or biological warfare virtually impossible unless relief from these rules was obtained. Clearly, it was not possible to efficiently and effectively obtain FDA approval for important military medical countermeasures without relief from the additional requirements. It would be immoral to conduct valid challenge trials to prove human clinical efficacy needed for FDA approval, and it was also not feasible to fully comply with all the Good Clinical Practices requirements [34–36]. There was no other choice but to resolve to comply with all of them as best as possible and to use the IND products needed to protect soldiers in the battlefield [39–42].

Most of the IND products the DoD may want to use in contingency situations were supported by a great deal of animal efficacy and human safety data but could not be licensed until they also had substantial evidence of human clinical efficacy. These biological agents were hazardous, and performing clinical challenge studies was certain to cause deaths. Thus, the Food, Drug, and Cosmetic law required immoral efficacy studies to achieve licensure without relief from this requirement [37,38,44–46]. Fortunately, the argument was made that FDA approval on the basis of human safety and substantial evidence of efficacy in animal models might suffice in circumstances where it would be unfeasible or immoral to attempt to obtain substantial evidence of human clinical efficacy. This concept was utilized to a great extent in August 2000, before the animal efficacy rule was made final, when the FDA approved the first antimicrobial drug for treatment of infection due to a biological agent used intentionally as a weapon [47]. Ciprofloxacin hydrochloride (Bayer) was approved by the FDA under the accelerated approval process for use in the medical management of inhalational anthrax. This approval was based on existing knowledge of the drug's safety profile and efficacy data for other bacterial agents, having a surrogate marker of serum concentrations of ciprofloxacin in humans and extensive animal efficacy data. The animal model for inhalational anthrax and subsequent efficacy and pharmacological data obtained from the model played a key role in the approval by the FDA.

Mary Pendergast was the Deputy Commissioner of the FDA at the time the reconsideration of the DoD waiver of informed consent was coming up for review. At the time, she was also considering a new rule for allowing emergency medical device research for development of new lifesaving devices that would have to be tested in civilian emergency rooms. Memoranda submitted to the docket on reconsideration of the DoD waiver of informed consent included several that proposed an ethical construction that threaded its way between the need to prove safety and efficacy and the need to provide lifesaving products for extremely hazardous conditions under which it would be unfeasible or immoral to conduct human clinical efficacy trials [36–38,44–46]. A draft animal efficacy rule was prepared by the FDA Commissioner's office and had been published for public comment two years before the terrorist attacks of fall 2001. The FDA recognized the acute need for an "animal efficacy rule" that would help make certain essential new pharmaceutical products—those products that because of the very nature of what they are designed to treat cannot be safely or ethically tested for effectiveness in humans—available much sooner [48].

The FDA amended its new drug and biological product regulations so that certain human drugs and biologics that are intended to reduce or prevent serious or life-threatening conditions may be approved for marketing on the basis of evidence of effectiveness from appropriate animal studies when human efficacy studies are not ethical or feasible. The agency took this action because it recognized the need for adequate medical responses to protect or treat individuals exposed to lethal or permanently disabling toxic substances or organisms. This new rule, part of the FDA's effort to help improve the nation's ability to respond to emergencies, including terrorist events, will apply when adequate and well-controlled clinical studies in humans cannot be ethically conducted because the studies would involve administering a potentially lethal or permanently disabling toxic substance or organism to healthy human volunteers. It is important to remember that the animal efficacy rule cannot be used when clinical studies in humans can still be performed, such as when the accelerated approval process can be used when surrogate markers or clinical endpoints other than survival or morbidity are available [49].

Under this new rule, certain new drug and biological products used to reduce or prevent the toxicity of chemical, biological, radiological, or nuclear substances may be approved for use in humans based on evidence of effectiveness derived only from appropriate animal studies and any additional supporting data. Products evaluated for effectiveness under this rule will be evaluated for safety under preexisting requirements for establishing the safety of new drug and biological products. The FDA proposed this new regulation on October 5, 1999, the final rule was published in the *Federal Register* on Friday, May 31, 2002 [43], and the rule took effect on June 30, 2002. The advent of the animal efficacy rule brings to bear the importance of animals in finding safe and effective countermeasures to the myriad of toxic biological, chemical, radiological, or nuclear threats. With this new opportunity to advance human and animal health and protect our nation, we also have to recognize that a great responsibility comes with it. That responsibility includes being rigorous in searching for the most optimal model that accurately mimics human disease and thoroughly researching potential refinements to animal use and incorporating applicable findings into the research. Refinements, such as developing early endpoints and administration of analgesics, must be discussed ahead of time with the FDA to ensure the animal model will meet the necessary criteria to clearly show a product's effectiveness.

The FDA will consider approval of a new drug product under the auspices of the animal efficacy rule only if four requirements are met. The first requirement necessitates a well-understood pathophysiological mechanism of how the threat of concern causes damage to the body and how damage is prevented or substantially reduced. This requirement goes far beyond a proof-of-concept study that may be designed to strictly look at whether a product shows an obvious benefit to make a determination on future development of that product. The effect of the agent of concern and the response of the animal to treatment should be thoroughly understood for an animal model to be used to submit data for drug approval under the animal efficacy rule. Although a full understanding of the pathophysiological processes of a disease and treatment is not required when human studies are used to support approval of a new product, the requirement for animal studies represents the need for additional assurance that information obtained from animal studies can be applied with confidence to humans.

The second requirement is one that has led to a frequent misperception by scientists and lay persons alike and is why the animal efficacy rule is many times referred to as the "two-animal rule." The animal efficacy rule states that the effect of a product should be demonstrated in more than one animal species whose responses have been shown to be predictive for those of humans. However, the rule goes on to state that a single-animal model may be used if it is sufficiently well characterized for predicting the response in humans. Because using animal efficacy data to approve drugs that have no evidence of efficacy in humans is a significant deviation from previously standard practices, there will likely be extremely close scrutiny of the animal models by the FDA and an expectation of testing to be performed in two species unless a very strong case can be made for use of a single-animal model. As an example, many infectious diseases have been studied in great detail for decades, with very well-characterized animal models. In cases in which a well-defined model is used in conjunction with a product that already has significant human data, using a single animal model may be appropriate. Also, in situations in which there is only one animal model that represents a response predictive of humans, a single animal model may be considered sufficient.

Animal study endpoints come into play as the third requirement of the animal efficacy rule. It is important in developing animal models to use under this rule that endpoints reflect the desired benefit in humans. Survival is one consideration, but the prevention of morbidity may be equally important. Some infectious diseases may have very low mortality—but very high morbidity—in humans, and establishing an animal model for this type of disease in a highly susceptible species with death as an endpoint may be completely inappropriate. Also, since the model should reflect the human clinical condition as much as possible, consideration for supportive care to reflect what would reasonably be expected for people should be incorporated into the animal model and associated control groups [49].

The final condition for approval of a product using the animal efficacy rule requires that the animal model being used allows for the collection of data on the kinetics and pharmacodynamics of the product that will allow for an effective dose in humans to be determined. Using an animal model that does not allow the necessary pharmacodynamics and kinetics studies to be performed with the product being tested should be excluded during early phases of model development.

Once a product receives approval under the animal efficacy rule by the FDA, there remain additional requirements that include postmarketing studies to gather data on the safety and efficacy of the product when used for its approved purpose; labeling requirements that describe how efficacy was determined through the use of animals alone, as well as other relevant product information; and the potential for approval with defined restrictions on the product's use.

Comments on the animal efficacy rule when it was in the proposal stage, and more detailed discussions and responses by the FDA, can be found in the *Federal Register* (2002, Vol. 67, No. 105, 21 CFR Parts 314 and 601). These discussions can provide useful insight into the applicability and limitations of the animal efficacy rule. In an attempt to provide additional clarification, the FDA published a notice in January 2009 announcing the availability of draft guidance designed to identify the critical characteristics of animal models intended for use under the animal efficacy rule [50]. Although this guidance is not binding to sponsors using the animal efficacy rule for product approval, it does give some insight to FDA's current interpretation of developing animal models for use under the rule [49].

In conclusion, the animal efficacy rule should be used for product approval only when it is clear that conducting research trials to find substantial evidence of human clinical efficacy would be unethical or immoral. This is the only appropriate justification because the justification for this rule is not lessening the standards of product approval. Many would argue the standards are much greater, and justifiably so in order to provide additional confidence that products will be efficacious in humans. Rather, the justification is to allow approval of medical countermeasures where it would be extremely dangerous to test for efficacy in humans. Future successes will require the scientific and regulatory communities to work closely together to develop and refine effective animal models that lead to safe and efficacious countermeasures to biological, chemical, and radiological threats.

REFERENCES

1. Christopher, G.W., Cieslak, T.J., Pavlin, J.A. and Eitzen, E.M., Biological warfare: A historical perspective. *JAMA* 278, 412, 1997.
2. Thursfield, H., Smallpox in the American War of Independence. *Ann. Med. History* 2, 312, 1940.
3. Fenn, E.A., *Pox Americana. The Great Smallpox Epidemic of 1775–82*. Hill and Wang, New York, 2001.
4. Osler, W., The historical development and relative value of laboratory and clinical methods in diagnosis. The evolution of the idea of experiment in medicine, in *Transactions of the Congress of American Physicians and Surgeons*, Vol. 7: Seventh Triennial Session, Washington, DC, May 7–9th. New Haven, CT: Congress, 1907, pp. 1–8.
5. Osler, W., The evolution of modern medicine, the rise of preventive medicine. A series of lectures delivered at Yale University on the Silliman Foundation in April 1913. Project Gutenberg, http://biotech.law.lsu.edu/Books/osler/chapvi.htm.
6. Harvey, A.M., Johns Hopkins and yellow fever: A story of tragedy and triumph. *Johns Hopkins Med. J.* 149, 25, 1981.
7. Clendenin, R.M., *Science and Technology at Fort Detrick, 1943–1968*. Technical Information Division, Fort Detrick, Frederick, MD, 1968, pp. 1–68.
8. U.S. Department of the Army, *US Army Activity in the US Biological Warfare Programs*, Unclassified, Vols. 1 and 2, 1977.
9. Covert, N.M., *Cutting Edge: A History of Fort Detrick, Maryland 1943–1993*. Headquarters, United States Army Garrison, Public Affairs Office, Fort Detrick, MD, 1993, pp. 17–19.
10. Torrey, E.F. and Yolken, R.H., *Beasts of the Earth: Animals, Humans and Disease*, Rutgers University Press, New Brunswick, NJ, 2005.
11. Wedum, A.G., Barkley, W.E. and Hellman, A., Handling of infectious agents. *J. Am. Vet. Med. Assoc.* 161, 1557, 1972.
12. Smart, J.K., History of chemical and biological warfare: An American perspective, in *Textbook of Military Medicine: Medical Aspects of Chemical and Biological Warfare*, Office of the Surgeon General, Department of the Army, Washington, DC, 1989, p. 9.
13. Franz, D.R, Parrott, C.D. and Takafuji, E. The U.S. biological warfare and biological defense programs, in *Textbook of Military Medicine: Medical Aspects of Chemical and Biological Warfare*, Office of the Surgeon General, Department of the Army, Washington, DC, 1989, p. 425.
14. Paine, T.F., Illness in man following inhalation of *Serratia marcescens*. *J. Infect. Dis.* 79, 227, 1946.
15. Moreno, J.D., *Undue Risk: Secret State Experiments on Humans*, W.H. Freeman and Company, New York, 2000, p. 68.
16. Permissible medical experiments, in *Trials of War Criminals before the Nuremberg Military Tribunals under Control Council Law No. 10*, Vol. 2, Government Printing Office, Washington, DC, 1949, p. 181.
17. Beyer, D.H., Cathey, W.T., Stover, J.H., Williams W.J. and Green, T.W., Human experimentation in the biological warfare program, Memorandum, Fort Detrick, MD, October 9, 1953.
18. Woodward, T.E. (ed.). *The Armed Forces Epidemiological Board. Its First Fifty Years*. Washington, DC: Borden Institute, Office of the Surgeon General, Department of the Army, 1994.
19. Woodward, T.E. (ed.). *The Armed Forces Epidemiological Board. The Histories of the Commissions*. Section 3 (D. Crozier) Commission on Epidemiological Survey. Center of Excellence in Military Medical Research and Education, Office of the Surgeon General, Department of the Army, Washington, DC, 1990.

20. Wilson, C.E., Memorandum for the Secretaries of the Army, Navy and Air Force, Subject: Use of humans in experimental research. Office of the Secretary of Defense, Washington, DC, February 26, 1953.
21. Oakes, J.C., Cs-385, Memorandum Thru: Assistant Chief of Staff, G-4 For: the Surgeon General. Subject: Use of volunteers in research, Office of the Chief of Staff, Department of the Army, Washington, DC, June 30, 1953.
22. Swann, J.P., History of the FDA, http://www.fda.gov/oc/history/historyoffda/default.Htm; adapted from George Kurian (ed.), *A Historical Guide to the U.S. Government*, Oxford University Press, New York, 1998.
23. Milestones in U.S. food and drug law history, FDA Backgrounder, May 3, 1999, http://www.fda.gov/opacom/backgrounders/miles.html.
24. U.S. Army Regulation 70-25, Research and development: Use of volunteers as subjects of research, March 26, 1962.
25. Memorandum of understanding between the Department of Health, Education and Welfare, and the Department of Defense concerning investigational use of drugs by the Department of Defense, May 12, 1964.
26. Title 10, U.S. Code, Section 980. Limitation on use of humans as experimental subjects, 1972.
27. National Research Act (Pub. L. 93-348) of July 12, 1974.
28. U.S. Army Regulation 70-25, Research and development: Use of volunteers as subjects of research, July 31, 1974.
29. Memorandum of understanding between the Food and Drug Administration and the Department of Defense concerning investigational use of drugs by the Department of Defense, October 24, 1974.
30. The National Commission for the Protection of Human Subjects, The Belmont report, ethical principles and guidelines for the protection of human subjects in research, April 18, 1979.
31. Title 45, Code of Federal Regulations, Part 46, Protection of human subjects.
32. Title 21, Code of Federal Regulations, Parts 50, 56, 71, 171, 180, 310, 312, 314, 320, 330, 430, 601, 630, 812, 813, 1003, 1010, Protection of human subjects: Informed consent, January 27, 1981.
33. Memorandum of understanding between the Food and Drug Administration and the Department of Defense concerning investigational use of drugs, antibiotics, biologics and medical devices by the Department of Defense, May 21, 1987.
34. Levine, C., Military medical research: 1. Are there ethical exceptions? *IRB* 11, 5–7, 1989.
35. Cooper, R.M., Military medical research: 2. Proving the safety and effectiveness of a nerve gas antidote—A legal view. *IRB* 11, 7, 1989.
36. Martin, E.D., Acting Assistant Secretary of Defense for Health Affairs. Letter to Friedman, M.A., Attn.: Documents Management Branch (HFA-305) Re.: Docket No.; 90N-0302, July 22, 1997.
37. Anderson, A.O., Memorandum to U.S. Army Medical Research and Material Command, Subject: Reply to memorandum regarding review and comments about protocol entitled “Administration of pentavalent botulinum toxoid to individuals preparing for contingency combat operations” (Log No. A-6622), December 6, 1994.
38. Anderson, A.O., Letter to Mary K. Pendergast, J.D., Deputy Commissioner and Senior Advisor to the Commissioner, FDA, October 26, 1995.
39. Title 10, Code of Federal Regulations, Section 1107 (10 USC 1107). Notice of use of an investigational new drug or a drug unapproved for its applied use, November 18, 1997.
40. Executive Order 13139 (EO 13139), Improving health protection of military personnel participating in particular military operations. *Federal Register*. 64, 54175, 1999.
41. Title 21, Code of Federal Regulations, Section 50.23(d) (21 CFR 50:23(d)). Determination that informed consent is not feasible or is contrary to the best interests of recipients—New Interim Final Rule. *Federal Register*. 64, 54180, 1999.

42. DoD Directive 6200.2 (DODD 6200.2), Use of investigational new drugs for force health protection, August 1, 2000.
43. Embrey, E., Protecting the nation's military may include the use of investigational new drugs. National defense and human research protections. *Account Res.* 10(2), 85–90, 2003.
44. Howe, E.G. and Martin, E.D., Treating the troops. *Hastings Cent. Rep.* 21(2), 21–24, 1991.
45. Fitzpatrick, W.J. and Zwanziger, L.L., Defending against biochemical warfare: Ethical issues involving the coercive use of investigational drugs and biologics in the military, *J. Phil. Sci. Law.* 3, 1, 2003.
46. Gross, M.L., Bioethics and armed conflict: Mapping the moral dimensions of medicine and war. *Hastings Cent. Rep.* 34, 22, 2004.
47. Meyerhoff, A., Albrecht, R., Meyer, J.M., Dionne, P., Higgins, K. and Murphy, D. US Food and Drug Administration approval of ciprofloxacin hydrochloride for management of postexposure inhalational anthrax, *Clin. Infect. Diseases.* 39, 303, 2004.
48. Title 21, Code of Federal Regulations, Parts 314 and 601. New drug and biological drug products; evidence needed to demonstrate effectiveness of new drugs when human efficacy studies are not ethical or feasible, July 1, 2002.
49. Snoy, P.J., Establishing efficacy of human products using animals: The US Food and Drug Administration's "animal rule," *Vet. Path.* 47, 774, 2010.
50. U.S. Department of Health and Human Services, Food and Drug Administration, Draft Guidance for Industry—Essential elements to address efficacy under the animal rule. *Fed. Regist.* 74, 3610–3611, 2009.

4 Development and Validation of Animal Models

Jaime B. Anderson and Kenneth Tucker

CONTENTS

4.1 INTRODUCTION

Modern states have developed offensive biological weapons research programs. Terrorists have been known to deliberately release infectious agents or toxins to strike at their targets. Many of these agents are highly lethal to humans and also lack antidotes or therapeutic countermeasures. Others strike economically important animals or food crops. Since the deliberate release of anthrax in the letter incidents of 2001, the policy makers have become acutely aware of the perils of these activities, and the U.S. government has significantly increased funding for biological threat agent research aimed to support homeland defense.

Much of what is currently known about human consequences of deliberate release of biological agents comes from animal models research. Development of safe and efficacious vaccines and therapeutics relies on proper understanding of advantages and limitations of models available to researchers. Beyond that, well-designed animal models research provides a sound basis for risk assessments and policy decisions on how to deploy limited resources in time of need. Policy makers need a clear understanding of what can be learned from animal, tissue, and cell culture models. For both scientists and policy makers to draw useful conclusions from past and previous research efforts, they need to appreciate the characteristics and purposes served by relevant animal models of human pathogenesis. To that end, the information presented in this chapter provides a systematic approach to animal model development.

4.2 WHAT IS A MODEL?

A model is meant to resemble something else such as a model airplane resembles an actual airplane. There are many types of biomedical models that range from biological models, such as whole animal models and *ex vivo* models, to models of nonbiological origin, such as computer or mathematical models [1]. Biomedical models simulate a normal or abnormal process in either an animal or a human. This chapter reviews animal models for diseases caused by biological threat agents. In this context, animal models are meant to emulate the biological phenomenon of interest for a disease occurring in humans.

Before entering a discussion of how to develop an animal model, it is necessary to first have an understanding of terminology commonly used when discussing animal models. Important parameters of animal models include the concepts of homology, analogy, and fidelity. Homology refers to morphological identity of corresponding parts with structural similarity descending from common form. Homologous models therefore have genetic similarity. The degree of genetic similarity required for a model to be considered homologous is variable. Analogy refers to the quality of resemblance or similarity in function or appearance, but not of origin or development. Therefore, analogous models have functional similarity. In general, animal models exhibit both of these attributes to various degrees, and so may be considered a hybrid of these. Model fidelity refers to how closely the model resembles the human for the condition being investigated [2]. Another layer of fidelity also may be a measurement of how reproducible the data are within the model itself.

Other important concepts include one-to-one modeling and many-to-many modeling. These terms refer to the general approach to the modeling process itself and not the individual animal model. In one-to-one modeling, the process that is being simulated in a particular animal has analogous features with the human condition. In many-to-many modeling, each component of a process is examined in many species at various hierarchical levels, such as system, organ, tissue, cell, and subcellular [3]. Many-to-many modeling is often used during the development of animal models, whereas one-to-one modeling is more suited for research when the animal model is already well characterized and validated for the specific biological phenomenon being investigated [4].

Conceptually, animal models may be described in a number of ways [5,6]: experimental (induced), spontaneous (natural), genetically modified, negative, orphan, and surrogate. However, these descriptive categories cannot be used as classifications because the descriptions are not exclusive and models may have properties of more than one of the descriptions. Furthermore, as the knowledge of the model and the disease process progresses, the descriptive category of the model may change. Each of these descriptive categories will be discussed in the following paragraphs.

Experimental animal models are models wherein a disease or condition is induced in animals by the scientist. The experimental manipulation can take many forms, including exposure to biological agents such as an infectious virus or bacteria, exposure to chemical agents such as a carcinogen, or even surgical manipulations to cause a condition. In many cases, this approach would allow the selection of almost any species to model the effect. For example, many biological toxins may be assayed for activity in invertebrates as well as vertebrates [7,8]. The model selected would depend on the needs of the researcher. However, many biological agents are selective and cause species-specific responses. This is particularly true of infectious agents including bacteria and viruses. Many infectious agents are limited in the species that they can infect and in which they can cause disease. Some are restricted to a single known host, such as the human immunodeficiency virus causing disease only in humans. Thus, these models are restricted to animals that are susceptible to the induced disease or condition.

The spontaneous model is typically used in research on naturally occurring heritable diseases. There are hundreds of examples of this type of model, including models for cancer, inflammation, and diabetes. As the term "spontaneous" implies, these models require the disease to appear in the population spontaneously. These types of models are not limited to inherited disease but may also apply to inherited susceptibility to disease. For instance, susceptibility to type 1 diabetes is a heritable trait. The nonobese diabetic (NOD) strain of mouse also exhibits a heritable susceptibility to diabetes relative to most strains of mice and has been used as a spontaneous model for type 1 diabetes [9]. Although the appearance of diabetes in the NOD mouse is spontaneous, the occurrence of the disease is associated with environmental factors. Thus, the NOD mouse model is described as a spontaneous model because diabetes arises without experimental intervention, even though the disease is triggered by environmental factors. Although spontaneous models are typically associated with genetically inherited diseases, some of these models may represent diseases for which the inducing agent, such as virus, bacterium, or chemical, has not been identified. Once the inducing agent has been identified and actually applied by the researcher, the model would be described as an induced model. An example would be type 1 diabetes, for which it has been demonstrated that viral infections can either destroy beta cells directly or induce an autoimmune response that destroys the cells [10]. If the researcher uses the virus to induce diabetes in the NOD mouse, then this becomes an experimental model.

The genetically modified animal model is one in which the animal has been selectively modified at the genetic level. Because these models are produced from manipulation by researchers, models using genetically modified animals are actually a special example of the experimental model. In the broadest sense, genetically modified models may result from breeding or chemically induced mutations. These may also include animals that have been modified through the use of recombinant DNA—a subgroup of genetically modified animal models referred to as transgenic animal models. Such transgenic models can involve gene deletions, replacements, or additions. The development of genetically modified animal models has rapidly expanded as technologies for genetic engineering have advanced. For example, a transgenic model for staphylococcal enterotoxin B (SEB) has been developed in mice that were genetically modified to express human leukocyte antigens DR3 and CD4 [11]. These mice have an increased sensitivity to SEB and develop an immune response more similar to that of humans relative to the parent strain of mouse.

In a negative model, the agent that causes disease in humans does not cause disease in the animal. In the early stages of development of an animal model for disease, the lack of disease would often cause the animal to be rejected as a model. However, exploring why an agent does not cause disease can also provide insights into the disease process. This may be applied across species. For example, the resistance of bovines to shiga toxins relative to the sensitivity of rabbits and mice is caused by the relative levels of expression of receptors for the toxin [12–14]. Negative models are particularly powerful when differences are identified between strains of a species, thereby allowing a comparison within the same species. As a recent example, Lyons et al. [15] observed that the sensitivity of mice to anthrax could vary more than 10-fold, depending on the strain of mouse tested. Comparing the response to infection between these strains of mice should provide significant insights into the disease process. The use of transgenic models provides additional power to the negative model; animals may be genetically engineered to create an isogenetic change. This was applied as described earlier for SEB to create a more sensitive animal by inserting the gene for human leukocyte antigens.

Orphan models are those with no known correlation to human disease. However, as we increase our understanding of these animal diseases and human diseases, correlations may become apparent in the future. Some orphan models may have direct comparison to human disease, such as the realization that the enteritis and death caused by administering antibiotics to hamsters was related to antibiotic-associated pseudomembraneous colitis in humans by facilitating the overgrowth of toxigenic *Clostridium difficile* [16]. In other cases, the connection may be indirect but still provide an appropriate model of human disease, such as using the feline leukemia virus in felines as a model for the human immunodeficiency virus infections in humans [17]. Once an orphan model is linked to a human disease, it is no longer considered an orphan model.

A new descriptive category is the surrogate model. In a surrogate model, a substitute infectious agent is used to model a human disease. In some cases, the substitution may be obvious, as when the feline leukemia virus in felines is used to model the human immunodeficiency virus in humans, or *Salmonella typhimurium* in mice is used to model *Salmonella typhi* infection in humans [18]. However, more subtle differences also apply such as a human pathogen adapted to infect the species used for the animal model. For instance, the Ebola Zaire virus can infect and cause disease in mice and guinea pigs after it is serially passaged in these species [19]. The fact that the virus has to be adapted to the new host implies that the virus undergoes a change; the Ebola virus adapted to the mouse and guinea pig cannot be considered identical to the human virus and must be considered a surrogate agent.

More difficult to define are the unintentional changes that occur to a pathogen with the mere passaging of organisms in the laboratory, such as propagation of human viruses in nonhuman primate cell lines or the cultivation of bacteria in artificial media. The potential for genetic drift in the strains of organisms underscores the need to minimize the passage of strains to maintain identity with the original clinical isolate. Incumbent on the interpretation of results in the surrogate model is the understanding that not only does the animal differ from humans but the infectious agent in the animal model differs from the agent that infects humans. This adds an additional layer to the extrapolation of the results from the animal to the human disease.

An animal model can be described in more than one way. For example, the mouse used to analyze SEB can be described as a genetically modified, induced model. Alternatively, a model may be described in a different way depending on the experimental design. For instance, the spontaneous mouse model for diabetes may be described as an induced model if the experimenter uses a virus to cause destruction of the beta cells.

4.3 WHY ARE MODELS NECESSARY?

Animal models provide a critical role in research, as eloquently stated by Massoud et al. [20], "Models are an indispensable manipulation of the scientific method: as deductively manipulatible constructs they are essential to the evolution of theory from observation" (p. 275). Models play an especially important part in biological threat agent research because, in many cases, the agents are potentially lethal or permanently disabling and therefore do not readily lend themselves to research using human subjects. Whenever possible, modeling should use alternatives to animals for ethical considerations. If alternative models are not applicable, then the lowest appropriate phylogenetic animal model should be used, such as the use of *Caenorhabditis elegans* to model infection of human bacterial and fungal pathogens [21]. Although models are essential for scientific advancement, if they are not well characterized and understood, erroneous conclusions may be drawn, hindering scientific advancement.

Animal models have traditionally played critical roles in the safety and efficacy testing of prophylactic and therapeutic products. To reduce risk to humans, the U.S. Food and Drug Administration (FDA) requires that prophylactic and therapeutic products must be shown to be reasonably safe and efficacious in animals before advancing to human safety and efficacy trials. The FDA has recently modified the requirement for testing efficacy in humans when it is not feasibly or ethically appropriate to test efficacy in humans. In these circumstances, efficacy testing is only required in animals. Investigational prophylactics and therapeutics for many biological threat agents fall into this category. These modifications are described in the *Code of Federal Regulations*, Chapter 21, Parts 314, Subpart I, and 601, Subpart H. Specifically, the FDA may grant marketing approval for new drugs and biological products for which safety has been established and the requirements of efficacy can rely on evidence from animal studies for cases in which the following circumstances exist:

1. There is a reasonably well-understood pathophysiological mechanism of the toxicity of the biological substance and its prevention or substantial reduction by the product.
2. The effect is demonstrated in more than one animal species expected to react with a response predictive for humans, unless the effect is demonstrated

in a single animal species that represents a sufficiently well-characterized animal model for predicting response in humans.
3. The animal study endpoint is clearly related to the desired benefit in humans, generally the enhancement of survival or prevention of major morbidity.
4. The data or information on the kinetics and pharmacodynamics of the product or other relevant data or information, in animals and humans, allows selection of an effective dose in humans.

New drugs and biological products that can be assessed on the basis of a surrogate endpoint or on a clinical endpoint other than survival or irreversible morbidity must be approved under the FDA's accelerated approval of new drugs and biological products for serious or life-threatening illnesses, as addressed in the *Code of Federal Regulations*, Chapter 21, Parts 314, Subpart H, and 601, Subpart E. The therapeutic product must be shown to have "an effect on a surrogate endpoint that is reasonably likely, based on epidemiologic, therapeutic, pathophysiologic, or other evidence, to predict clinical benefit or on the basis of an effect on a clinical endpoint other than survival or irreversible morbidity" [22, p. 96].

The modifications to the FDA's approval process for vaccines and therapeutics increase the significance of animal models in the approval process. The approval of treatments based on these models could affect the clinical outcome of potentially millions of people exposed to agents that are covered by the modifications. Further, the perception of effective treatments based on these animal models will affect funding for research and governmental policy. This places a tremendous burden of responsibility on the reliability of these models.

4.4 IDENTIFICATION AND DEVELOPMENT OF AN ANIMAL MODEL

The focus will be on the identification of induced animal models representative of diseases caused by biological threat agents and will include considerations for identifying potential animal models for biological threat agents where none exist. Existing animal models suitable for the researchers' needs also will be evaluated. Although this process begins with a one-to-one comparison of the pathological progression of the disease, conceptually the collective analysis provides a many-to-many perspective. As a model is selected and validated, analysis may focus on a one-to-one approach to modeling. The basic steps to identify and develop an animal model are as follows:

1. Define the research objective.
2. Define the intrinsic factors associated with the biological phenomenon under investigation, such as the pathological progression of the disease process.
3. Define the extrinsic factors associated with the biological phenomenon under investigation such as the method used to prepare the pathogenic bacteria.
4. Create a search strategy and review the literature of previous animal models.
5. Create a biological information matrix.

6. Define unique research resources.
7. Identify preliminary animal models of choice.
8. Conduct research to fill critical gaps of knowledge in the biological information matrix for the preliminary animal models of choice.
9. Evaluate the validity of the animal models of choice.
10. Identify animal models of choice.

Finding a model of disease depends first on identifying animals or tissues that are responsive to the agent. Then the intrinsic factors in humans, such as pathological progression of the disease, must be related to the factors of the disease in the model to support its validity. If a disease-causing agent is novel, and no animal models are described, the researcher must identify and develop animal models. By identifying the relationship of a novel agent to known pathogens with established animal models (e.g., identification and rRNA sequencing), animals for modeling may be initially selected based on known models for the related organisms. In lieu of known models, animals for modeling will have to be identified empirically, and this selection should start with the evaluation of animals that are well supported by reagents for research (e.g., mouse) and progress to less-supported animals only as needed to meet the requirement of mirroring the disease in humans.

4.4.1 Step 1: Define the Research Objective

A single model likely will not be applicable to every situation. The model of choice is the model that best addresses the study's research aim within the research constraints. Therefore, a first step in animal model development is to define explicitly the specific question the research needs to address. The next step is to determine what specific information must be provided by the animal model to accomplish the research objective. This information is critical and will give direction to the remaining animal model development process.

4.4.2 Step 2: Define the Intrinsic Factors

Once the experimental need has been defined, the next step in establishing an animal model is to develop the intrinsic points of reference to the human illness. Intrinsic factors are inherent factors in the interaction between the host and the biological agent or pathogen. Confidence in the model will grow with increasing common points of reference between the animal model and the human. It is critical that all intrinsic factors relevant to the biological process associated with the research question be identified so that they may form the basis for comparing the animal model to the human condition being studied.

Although the early steps in the model development process are fundamental and appear obvious, the details can be easily overlooked, potentially leading to the selection of an animal model that is not entirely appropriate for the specified research. When defining the pathological progression of infectious disease, basic steps in the progression of the disease process will be identified. From a simple linear view of events, a disease-causing agent must gain exposure to the host, bind to and enter the host, distribute within the host to the target tissue, and exert disease through a specific mode of action. Pathogens may gain access to the systemic circulation by injection (via bites from parasites such as fleas and ticks) or through abrasions. The pathogen may also interact with the mucosa, such as found in the intestines and lungs. In these situations, the pathogen may bind to specific receptors on the host cells. The pathogen then may enter the host through the mucosal cells by commandeering the host's cellular processes to take up the pathogen and enter the systemic circulation. After entering the host, the pathogen may be distributed in the body (such as by the circulatory system). During this distribution, the pathogen can target specific tissues by binding to receptors on those tissues. The pathogen can then enter the cells of the target tissue and cause disease by affecting specific biochemical processes in the target cells. Some pathogens, however, do not invade the host's body or target tissues but produce extracellular factors, such as toxins and tissue-damaging enzymes. These factors may be transported into the body and be subsequently distributed by the circulatory system to target tissues or cells. For example, if the biological phenomenon being investigated is to determine the 50% human lethal dose of an agent such as botulinum toxin, then the anticipated intrinsic factors, such as pathogenic steps in the progression of the intoxication process from absorption into the body to the toxin's effect on the neurons, might be identified as follows:

- Toxin/agent penetration/absorption and biological stability
- Toxin/agent persistence in circulation and transit to target tissues
- Toxin/agent binding and uptake into target tissues
- Toxin/agent mechanism of action in target tissues

Superimposed on this simple linear view of the disease process is a complex interplay between the host and the pathogen. The pathogen will significantly change its physiology and expression of virulence factors in response to interactions with the host, and the host will also change in response to the pathogen. For example, the host cells may produce specific receptors only after exposure to the pathogen [23]. In addition, invasion by the pathogen will prompt the host's innate and acquired immune responses. The pathogen must circumvent the host's resistance, including competitive exclusion by the normal microflora, assault by host factors such as antimicrobial peptides and enzymes, and destruction by the innate and acquired immune response. In some cases, this evasion of the immune response leads to misdirection and deregulation of the immune response, resulting in the host's immune response actually contributing to the pathogenesis of the disease.

As this interaction progresses, the invading organism will typically harness the cellular processes of the host to promote its own replication and may directly cause damage to the host's cells and tissues. The ability of the host to respond to the pathogen in a manner that halts the infection determines the degree of the disease that the host will experience. Thus, virulence is not solely a property of the invading organism but, rather, an expression of the interaction of the pathogen with its host.

A model of disease attempts to mimic the host–pathogen interaction. Therefore, the combination of both the host and the pathogen defines a model for a disease and collectively makes up the intrinsic factors of the model.

4.4.3 Step 3: Define the Extrinsic Factors

Other useful parameters that are not intrinsic to the host–pathogen/agent interaction, but that can affect the process, are known as extrinsic factors. Functionally, extrinsic factors are variables that may be manipulated outside of the host–pathogen/agent relationship. Although extrinsic factors are not routinely considered part of the animal model, they are in fact a critical component. Extrinsic factors can influence the intrinsic factors as they relate to the host–pathogen interaction, which in turn defines the specific animal model. For example, results may be affected by factors affecting the pathogen, such as the means of preparing, handling, and formulating the agent. Extrinsic factors may also influence the response of the host. For instance, the bedding used for the animals, temperature and light cycles provided, and even the time of administering agents may affect the immunological response of the animal or the pharmacokinetics of therapeutic agents that are being studied. Extrinsic factors are an extension of the experimental design. As such, these must be identified and documented to allow comparison of data and to aid in the extrapolation of results to the human disease. The application of this requirement may be complicated by the reality that some of these factors may not be recognized.

The functional definitions of intrinsic and extrinsic factors are not uniformly accepted. Alternate definitions describe the animal as the only intrinsic factor and the pathogen or biological agent as an extrinsic factor that can be manipulated in the experimental design [24]. We assert that the interaction between the host and pathogen must be considered the model for the disease but recognize that there is a philosophically different opinion held by some in the scientific community. Notwithstanding this difference of opinion, it is well accepted that the extrinsic factors influence the intrinsic factors of a model and that, collectively, these factors affect the design of experiments using animal models (Figure 4.1).

Many of the basic steps in pathogenesis may be modeled *ex vivo*. However, for the *ex vivo* models to be predictive of the *in vivo* pathogenic process, the model must account for the potential factors that can influence the interplay of host and pathogen as occurring *in vivo*. A preliminary review of the literature may be necessary to adequately define the distinct features of the biological phenomenon under investigation.

4.4.4 Step 4: Create a Search Strategy

A preliminary, brief review of the literature using freely accessed information may be necessary to confirm the relevant intrinsic and extrinsic factors identified in steps 2 and 3. The preliminary review should include previous studies using animal models and human clinical data. This review will allow for development of a detailed search strategy. If there are no previous data on animal models used for the specific condition being modeled, it is reasonable to search for animal species that have been used for modeling similar conditions. Animals with a close phylogenetic relationship to humans, such as monkeys, should be considered because it is reasonable to assume they may have a higher degree of homology and therefore may respond in a more similar manner. However, caution must be exercised because analogy does not

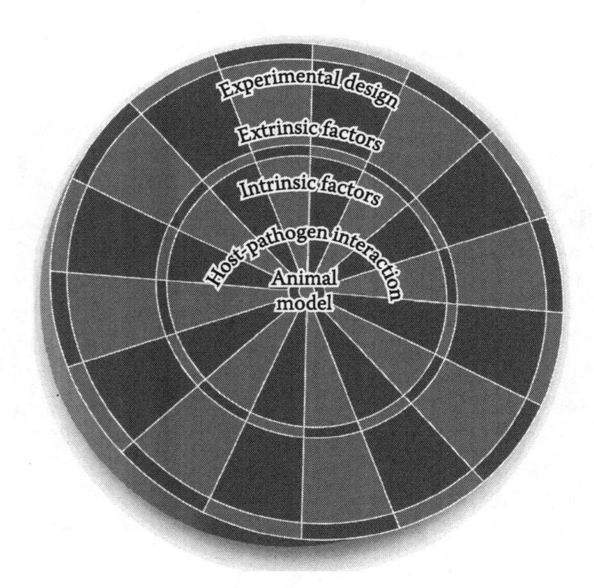

FIGURE 4.1 An animal model of disease is the interaction between the host animal and pathogen or biological agent. This interaction is influenced by intrinsic factors of the host and pathogen that cannot be directly manipulated, as well as by extrinsic factors that can be directly manipulated. Collectively, extrinsic and intrinsic factors are the components of the experimental design for a given animal model and must be defined to gain understanding and control of the model.

always follow homology. This is demonstrated in monkeys, which do not develop acquired immunodeficiency when infected with the human immunodeficiency virus. Instead, the more distantly related feline infected with the feline leukemia virus is considered a more appropriate model for AIDS in humans [25].

A comprehensive literature search strategy can be designed based on the relevant intrinsic and extrinsic factors associated with the biological phenomenon of interest that were identified by the preliminary literature review. Such a review is often an overlooked endeavor, but it is absolutely necessary to acquire a body of knowledge on which to make scientifically informed decisions during the animal model identification and development process. The search strategy should be designed to provide a comprehensive survey of the relevant information from libraries of publications and data. No single database is comprehensive, so the ideal search strategy should include a comprehensive search of all relevant informational resources. However, this may be cost prohibitive, and a tiered search strategy may be more appropriate, starting with the most relevant and free informational resources and expanding to the additional proprietary resources as needed. In addition to the electronic search for information, it is prudent to personally consult clinicians and scientists with experience of the disease or its models. If the comprehensive literature review identifies additional factors, such as pathological features, animal species, or other parameters that were not found in the preliminary literature review, then the review strategy should be changed accordingly.

4.4.5 Step 5: Create a Biological Information Matrix

Following the preliminary assessment of the literature, a biological information matrix of the relevant intrinsic and extrinsic factors for each of the animal species can be prepared. The biological information matrix is an index of the information used to compare the factors of the models to the human disease. The biological information matrix should reveal what animal models are available and which are the most relevant for the proposed research (Figure 4.2). As the matrix is filled in with discrete data, comparative analogies can be made between the different species and the human data. The species that most accurately reflects the human condition of study is then identified on the basis of the current state of scientific knowledge. It may become apparent at this point that more than one species is needed to address the research objective accurately. In addition, the best animal to model a specific component in the disease process may be different than the animal species chosen to model the entire disease process (Figure 4.2). It is paramount that the model be judged by how well it can be applied to the specific research question, rather than how well the animal models the entire array of the disease process in humans. For instance, yeasts may not be used to model central nervous system dysfunction caused by prions because they do not have a central nervous system. However, yeasts are used to model the biology of prion infection and propagation [26].

4.4.6 Step 6: Define Unique Research Resources

In addition to the biological matrix of information, there are many other considerations that must be taken into account when choosing the animal model. Because of animal availability, suitable housing, or other restrictions, some animal models may not be feasible for a particular researcher. For these investigators, only the more distantly related animals, such as mice, rats, guinea pigs, and rabbits, may be available. The researcher should prepare a list of unique resource requirements. It may be helpful to use an integrated team approach when identifying the resource requirements. The primary and secondary investigators in collaboration with the laboratory animal veterinarian and statistician would best be able to address unique requirements [2]. There are many lists cited in the literature for general considerations in choosing the ideal model. The following is a partial list of the general qualities of an ideal model, and these should be considered against the available resources of the researcher when selecting a model [27,28]:

1. Accurately mimic the desired function or disease: This is a fundamental cornerstone for extrapolation of data.
2. Exhibit the investigated phenomenon with relative frequency: The phenomenon must be readily present to lend itself to unhindered scientific study.
3. Be available to multiple investigators: The animals should be handled easily by most investigators. This facilitates leveraging of the scientific community.
4. Be exportable from one laboratory to another: The model should lend itself to widespread usage. This implies that the model must be compatible with

Biological matrix of information

	Mice	Rat	Guinea pig	Rabbit	Nonhuman primate	Human
Biological stability	2	2		1		
Absorption	2		2	3		
Persistence/transit to target						
Uptake at target tissues			1	3	3	
Mech./effects at target	3	3	1	4	4	
Toxicity/lethality	2	2	3	3	3	
Clinical signs		2	1	3	3	
Epidemiology	2	2	2	2	2	
Immune response	3	2	4	3	4	
Therapeutic response						
Deposition	2	2	0	1	3	

FIGURE 4.2 This representation of a biological matrix provides an evaluation of some intrinsic and extrinsic factors of infection and disease relative to the animal used to model the disease. The shades in the sections represent the relative knowledge of the factors, with the darkest shade representing little or no knowledge, the next lighter shade representing a factor that is well defined and understood, and the lightest shade representing knowledge lying somewhere in between. This color coding represents the level of knowledge about a factor in the specific species and does not relate the disease process in that species to that in humans. The number in the sections represents the correlation of the factor in the animal to that of the human, with zero indicating no correlation and scaling up to 4 representing full correlation; a blank indicates there is not enough information to determine a correlation. This is a simplified matrix, and each part may be further subdivided. For example, nonhuman primates may be divided into specific species, and mice may be divided into strains. Likewise, the factors may be subdivided on the basis of the needs of the researcher. For example, absorption may be divided into dermal, pulmonary, or gastrointestinal routes. Further, these subdivisions may be divided into organism, system, organ, tissue, and subcellular absorption based on the questions being asked by the researcher. Thus, a researcher interested in total absorption from an oral exposure would create a single cumulative score for oral exposure from literature, including evaluating subcellular through gross absorption in the specific species, whereas a researcher interested in absorption at the tissue level would create subsections based on organs and different tissues that line the alimentary canal and would create scores based on literature from studies of absorption at the subcellular to tissue levels. In addition, some of the factors listed here may not be relevant to the research question and may be omitted, and other factors may need to be added. This example of a matrix demonstrates several points that are likely to be encountered when evaluating animal models. There will be steps in the disease process that are not understood and that are identified as areas that must be addressed by research to establish the correlation of the model to humans, such as the process of "persistence/transit to target" and "therapeutic response" in the figure. This matrix also demonstrates examples in which one animal may provide a good model for a particular step in the disease process (e.g., guinea pig modeling the immune response) but provide a poor model for the total disease. Conversely, it also presents an example in which a good model of the total disease may have a poor correlation to human disease at one step in the process, as exemplified by the poor correlation of biological stability of the agent in rabbits, even though the rabbit provides an overall good model of the disease. Thus, a researcher

available animal-housing facilities. This facilitates leveraging of the scientific community.

5. Be a polytocous species: The number of offspring produced is a limiting factor for future unrestrained availability. This criterion is especially relevant for spontaneous models for genetic disorders.
6. Be of sufficient size to allow appropriate sampling: The animals must be of sufficient size to allow for appropriate methods of data collection, such as for the sampling of multiple blood collections. This also implies being amenable to investigation with appropriate technological tools.
7. Be of appropriate longevity to be functional: The animal should survive long enough to allow for experimental manipulation and investigation.
8. Be accompanied by readily available background data: The availability of extensive background data may readily contribute to the biological information matrix and enhance interpretation of new data.
9. Be of defined genetic homogeneity or heterogeneity: This has traditionally been relevant for spontaneous and transgenetic models. This criterion is now achieving increased importance with the advances in microarray and proteomic technology.

4.4.7 Step 7: Identify Preliminary Animal Models of Choice

The biological information matrix should provide information to identify potential animal models. The animal models identified at this stage are only preliminary assessments that are meant to help focus the remaining animal model development process. Optimally, at least two species of animals should be selected for modeling to allow for comparison of results between the models as well as with humans. Concordance between animal models increases the level of confidence in the biological response.

4.4.8 Step 8: Conduct Research

Research to fill all the critical gaps of knowledge in the biological matrix of information may be cost prohibitive. Therefore, because of financial constraints, only the gaps viewed to be the most important may be addressed with research. Because many biothreat agents are rare infections in any population, the human condition may not be well documented. This makes the animal model development process much more difficult. However, this may be partially overcome by addressing the gaps in knowledge using a reductionist approach (Figure 4.2). Using the many-to-many animal model methodology, the intrinsic and extrinsic factors of the disease

FIGURE 4.2 Continued. studying a specific step in the disease process could be justifiedfor selecting a different animal model than a researcher studying the entire disease process. This matrix also highlights the fact that animals other than nonhuman primates may be just as effective as models, as indicated for the rabbit in this example. Defining the experimental question that is to be addressed by the model and analyzing the available information for each model should allow the researcher to identify whether a model has been adequately defined and, if so, which model may be best to address the specific question.

process or biological phenomenon under investigation should be identified and characterized with *ex vivo* experimentation. Technology should be explored to determine what *ex vivo* assays are available that may adequately reflect the factors in the disease process or biological phenomenon under investigation. These *ex vivo* experiments should be evaluated with both animal and human tissues or with cell lines. This allows for data to be compared and evaluated for concurrence of data between the animal and human. *In vitro* experimentation may also be necessary to supplement the *ex vivo* studies. The same intrinsic and extrinsic factors in the disease process or biological phenomenon under investigation also should be evaluated using a holistic approach. This approach involves *in vivo* animal experimentation. It is anticipated that the *in vivo* study may differ from the *ex vivo* study because of the unique relationships and interactions of the cells within the intact animal, and these differences will need to be considered when interpreting the data.

Microarrays and proteomics have the potential to lend valuable insight for data interpretation from *ex vivo* and *in vivo* studies. These techniques are obviously limited to species that have been sequenced and for which microarrays have been developed. However, if available, they will rapidly indicate whether the cells have similar or different responses to the agent. Additional approaches to rapidly evaluate the similarity of the mechanism of infection between the species could include using proteomics and electron microscopy to monitor stages of entry and propagation in host cells. Further, comparing the agent's effect on human cells derived from different organs/tissues to its effect on similar cells derived from the species used for the animal models can provide profiles of activity that may be used to evaluate the animal model. These approaches provide relatively rapid means to evaluate the correlation of the agent in animals to the agent in humans. Data obtained from *in silico* models, such as computer assimilation models, can also be evaluated by comparing the appropriateness of data as compared to the *in vivo* and *ex vivo* studies.

Ex vivo modeling may be done at the same time as animal models or may even precede the animal models if observations allow identification of target tissues. The data and conclusions from the animal *ex vivo* experiments should be compared and evaluated for concurrence of data from animal *in vivo* experiments. If there is concurrence of the data between animal *in vivo*, *ex vivo*, and *in silico* studies, as well as human *ex vivo* studies and available human case studies, then the extrapolation of data can be made with increased confidence. This process is an ongoing endeavor and should build on information learned previously.

4.4.9 Step 9: Evaluate the Validity of the Animal Models of Choice

What is required to validate an animal model, and at what point does the model become validated? Simply, a validated model is one in which a significant overlap of analogies for the intrinsic and extrinsic factors exists between the animal model and human disease. The definition of "significant" in this context is a point of contention that must be defined by the individual researcher and accepted by scientific peers. The animal model and human condition being modeled should have similar characteristics in the biological information matrix (Figure 4.2). The experimental design of the research to validate the model is similar to the research done for step 8, except that a more comprehensive approach is taken to further fill gaps of knowledge in the biological matrix of information. If there is not a sufficient amount of overlapping data between the animal model and humans in the biological information matrix, then additional experiments should be conducted to fill in the gaps. If these gaps are filled and the overlap of analogies is determined to be insignificant, then the model must be deemed invalid and another model sought.

The degree of accuracy of the animal model depends on the reliability of methods used to measure the pathological process or biological phenomenon under investigation. The techniques used for evaluation must be sensitive. A failure to accurately identify similarities and differences between the animal model and human can lead to erroneous extrapolations. Hierarchical evaluation of each factor, at the system, organ, tissue, cellular, and subcellular levels, can provide invaluable insight. It is important that this evaluation be done early in the animal model development process. The greater the sensitivity of measurements, the more reliable the validation will become.

Histopathology often provides the initial point of evaluation, but other techniques, such as immunohistochemistry, *in situ* hybridization, microarrays, proteomics, and other technologies, will be required to adequately assess the various hierarchical levels of anatomy and physiology. Validation is an evolving process that is never completed because the models are always subject to further definitive re-examinations and revalidation as new technology becomes available.

4.5 EXTRAPOLATING ANIMAL MODEL RESULTS TO HUMANS

Models are a copy or imitation of the study target. They will never be perfect in every instance. A thorough understanding of the model and an appreciation of its weaknesses will enable the researcher to make more accurate assessments and extrapolate results with a higher degree of confidence. What can be extrapolated and what cannot is one of the challenges of working with models. To extrapolate data directly from the animal to the human without first investigating and evaluating other sources of data such as *ex vivo*, *in vitro*, and *in silico* modeling and clinical case studies would not promote a high degree of confidence in the validity of the extrapolated data (Figure 4.3). The goal of an animal model is to have a high degree of valid extrapolation to the humans.

A consideration that is critical to the extrapolation process is the experimental design and methodology used to collect the data. Ideally, the experimental design and methodology should mirror the conditions being modeled as closely as possible and must consider the relevant intrinsic and extrinsic factors of the model. Biothreat agent research commonly uses animal challenges via aerosol or oral exposure; however, both of these exposure routes may provide misleading data if they are not designed correctly. For example, the pattern of deposition within the animal varies by particle size in the generated aerosol. A particle size of 1 μm provides a similar pattern of pulmonary particle deposition in the guinea pig, nonhuman primate, and human, but at 5 μm, the pulmonary particle deposition is much lower in the guinea pig [29]. Similar considerations are important for oral challenges, such as the effects of stomach pH in the fed and unfed animals and between the various species, or the gastric

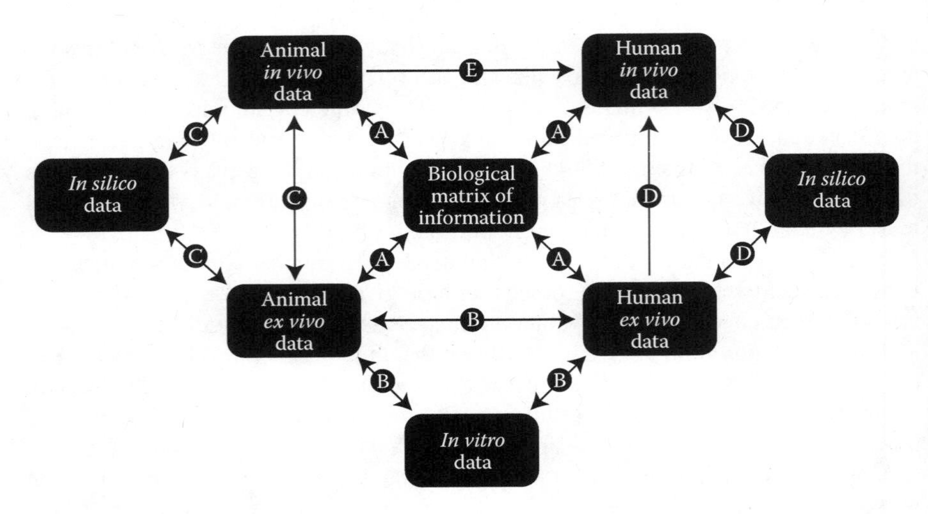

FIGURE 4.3 Based on concurrence of the data between animal *in vivo*, *ex vivo*, and *in silico* studies, as well as human *ex vivo* studies and available human case studies, extrapolation of data can be made with increased confidence. All the sources of data contribute to the biological matrix of information. Green represents data related to animals, and red represents data related to humans. The arrows show the relationships between the various sources of data. Biological matrix of information is collected from all data sources (A). Animal *ex vivo* and *in vitro* data are compared to human *in vitro* data (B). The animal *ex vivo* data are validated with animal *in vivo* data (C) and *in silco* data (C). On the basis of the concurrence of (B) and (C), extrapolation of data from human *in vitro* can be made to the human *in vivo* (D) and *in silico* models (D). Ultimately animal data are extrapolated to the human (E).

emptying time compared to the volume of the challenge dose [30,31]. The different strains or isolates of agents, and the differences in their preparation, must also be considered when comparing data from different research experiments and making extrapolations [32].

The time points for therapeutic intervention in animal models often require much deliberation to accurately reflect the human time course of intervention. Animals do not provide symptoms, and the progression or time course of clinical signs is not always the same. For example, the botulinum intoxication process in the guinea pig differs by time course, and the clinical signs are not always appreciable in these small species as compared to nonhuman primates and humans [33].

In vitro models may provide a more ethical or humane course of research, and the reductionist approach of *in vitro* assays can offer the advantages of controlling the variables in the environment. However, *in vitro* studies are often limited in what can be accurately extrapolated to a more complex biological system. Mathematical and computer models and other *in silico* models are constructed by data already gained from research and are therefore limited by what is already known about the disease. Cell culture studies are limited because they may not behave normally in the *ex vivo* setting when removed from the animal. In addition, cell lines are obtained from individual members of the species and, because of intraspecies variability in the biological response, may not accurately reflect the general population.

Animal models are limited because the conservation of biochemistry among species and the physiological differences between species are not fully defined. Any predictions based on models must be tempered by the realization that the interaction of a disease-causing agent with its host is complex and influenced by numerous intrinsic and extrinsic factors, many of which probably have not been identified. At least some of these factors may be specific to a given species, and others will vary even within a species. Factors such as nutrition, stress, and rest are known to influence animal and human response to infections. Controlling the significant intrinsic and extrinsic factors, as well as proper experimental design and statistical analysis, can normally overcome the biological variability of a model.

Human studies have the potential to provide the most accurate data. However, controlled experimental studies using humans are limited in scope or may not be possible for more virulent or untreatable disease. Human clinical case studies used to model the general population may also be limited because the data are not generated from a controlled environment. The patients may have preexisting conditions, and thus conclusions drawn may be less clear. In addition, the relatively small number of patients typically described in case studies of rare diseases may not reflect the general population, and this may affect the accuracy of the interpretation of the results. The human population is considered to be genetically limited relative to most other animals, with the members of a single tribe of monkeys demonstrating more genetic diversity than that observed for the whole of humankind. Nevertheless, there are known genetic differences in the human population that influence susceptibility to diseases. For instance, susceptibility to infection by *Plasmodium vivax* is dependent on the host expressing the Duffy blood type, which is the receptor in humans for that parasite [34]. Further, for every disease described for humans there have been survivors, but in most cases we do not know what parameters influence survival. Simply, individual responses to an agent are not uniform.

The understanding of the disease process is further complicated by the fact that there are variants for most pathogenic agents, and these variants are associated with different virulence potentials. Realizing the large array of factors that influence disease, and the limited understanding of these factors, models have still been used to determine whether a disease-causing agent follows a similar pathological progression between species and how the steps in this process contribute to the disease. When the disease processes and host's responses are similar to that of humans, the model provides a reference to allow predictions of responses in humans. However, numerous examples exist of accepted models that failed to be predictive at some level. For example, lethality of *Yersinia pestis* to small rodents is considered indicative of the virulence of the bacteria; however, strains of *Y. pestis* have been described that kill mice but do not cause disease in larger animals, including humans [35]. As another example, primates are considered to be predictive of infections with the Ebola virus, yet the Reston strain of the Ebola virus that causes disease and death in primates apparently does not cause disease in humans [36]. Therefore, the degree of accuracy of predictions based on animal models can only be definitively accessed by human exposure.

4.6 CONCLUSION

The significant effect that animal models have had in the study of infectious diseases is exemplified by the application of Koch's postulates early in the history of microbiology [37]. The continued use of animal models has been essential to achieving our present understanding of infectious diseases and has led to the discovery of novel therapies. Animal models have been used to provide the preliminary safety and efficacy testing for nearly all therapeutics in use today and have reduced testing in humans of potentially dangerous or ineffective therapies. The role of animal models in safety and efficacy testing has only increased with time. With the implementation of the "animal rule" by the FDA, the animal model will provide the only premarketing efficacy data available for the evaluation of new therapeutics targeting diseases caused by certain biological threat agents. This underscores the need for well-characterized animal models. Confidence in the correlation of results from a model to the human disease can be achieved only if the relationship of the model to the human disease is well understood. This chapter provided a systematic approach to achieve the required understanding of an animal model so that it may be applied with confidence.

REFERENCES

1. Committee on New and Emerging Models in Biomedical and Behavioral Research, Institute for Laboratory Animal Research (U.S.), Commission on Life Sciences, and National Research Council, *Biomedical Models and Resources: Current Needs and Future Opportunities*, Washington, DC: National Academy Press, 1998, p. 10.
2. Kriesberg, K., *Animals as Models*, http://www4.ncsu.edu/ ~ jherkert/ori/models.htm, accessed on April 1, 2005.
3. Quimby, F., Animal models in biomedical research. In *Laboratory Animal Medicine*, J.G. Fox, L.C. Anderson, F.M. Loew, and F.W. Quimby, eds., 2nd edn., New York: Academic Press, 2002, pp. 1185–1225.
4. National Research Council (U.S.). Committee on Models for Biomedical Research, *Models for Biomedical Research: A New Perspective*. Washington, DC: National Academy Press, 1985, pp. 12–23.
5. Hau, J., Animal models for human disease. In *Source Book of Models for Biomedical Research*, P. Michael Conn, ed., New Jersey: Humana Press Inc., 2008, pp. 4–5.
6. Committee on the Assessment of Future Scientific Needs for Live Variola Virus, Board on Global Health, and Institute of Medicine, *Assessment of Future Scientific Needs for Live Variola Virus*, Washington, DC: National Academy Press, 1999, http://books.nap.edu/html/variola_virus/chg.html, accessed on August 2, 2005.
7. Needham, A.J. et al., *Drosophila melanogaster* as a model host for *Staphylococcus aureus* infection, *Microbiology*, 150, 2347–2355, 2004.
8. Garsin, D.A. et al., A simple model host for identifying gram-positive virulence factors, *Proc. Natl. Acad. Sci. USA*, 98, 10892–10897, 2001.
9. Riley, W.J., Insulin dependent diabetes mellitus, an autoimmune disorder? *Clin. Immunol. Immunopathol.*, 53, S92–S98, 1989.
10. Jun, H.S. and Yoon, J.W., A new look at viruses in type 1 diabetes, *ILAR J.*, 45, 349–374, 2004.
11. DaSilva, L. et al., Humanlike immune response of human leukocyte antigen-DR3 transgenic mice to staphylococcal enterotoxins: A novel model for superantigen vaccines, *J. Infect. Dis.*, 185, 1754–1760, 2002.

12. Pruimboom-Brees, I.M. et al., Cattle lack vascular receptors for *Escherichia coli* O157:H7 Shiga toxins, *Proc. Natl. Acad. Sci. USA*, 97, 10325–10329, 2000.
13. Tesh, V.L. et al., Comparison of the relative toxicities of Shiga-like toxins type I and type II for mice, *Infect. Immun.*, 61, 3392–3402, 1993.
14. Keusch, G.T. et al., Shiga toxin: Intestinal cell receptors and pathophysiology of enterotoxic effects, *Rev. Infect. Dis.*, 13, S304–S310, 1991.
15. Lyons, C.R. et al., Murine model of pulmonary anthrax: Kinetics of dissemination, histopathology, and mouse strain susceptibility, *Infect. Immun.*, 72, 4801–4809, 2004.
16. Chang, T.W. et al., Clindamycin-induced enterocolitis in hamsters as a model of pseudomembranous colitis in patients, *Infect. Immun.*, 20, 526–529, 1978.
17. Hardy, W.D. Jr. and Essex, M., FeLV-induced feline acquired immune deficiency syndrome. A model for human AIDS, *Prog. Allergy*, 37, 353–376, 1986.
18. Tsolis, R.M. et al., Of mice, calves, and men. Comparison of the mouse typhoid model with other *Salmonella* infections, *Adv. Exp. Med. Biol.*, 473, 261–274, 1999.
19. Bray, M. et al., Haematological, biochemical and coagulation changes in mice, guinea-pigs and monkeys infected with a mouse-adapted variant of Ebola Zaire virus, *J. Comp. Pathol.*, 125, 243–253, 2001.
20. Massoud, T.F. et al., Principles and philosophy of modeling in biomedical research, *FASEB J.*, 12, 275–285, 1998.
21. Mylonakis, E., Ausubel, F.M., Tang, R.J., and Calderwood, S.B. The art of serendipity: Killing of *Caenorhabditis elegans* by human pathogens as a model of bacterial and fungal pathogenesis, *Expert. Rev. Anti. Infect. Ther.*, 1, 167–173, 2003.
22. Title 21: Food and Drugs. Part 314.510: Approval based on a surrogate endpoint or on an effect on a clinical endpoint other than survival or irreversible morbidity. In *Code of Federal Regulations*, Washington, DC, Government Printing Office, April 1, 2004; see also Parts 314.610, 601.41, and 601.91.
23. Hooper, L.V. and Gordon, J.I., Glycans as legislators of host-microbial interactions: Spanning the spectrum from symbiosis to pathogenicity, *Glycobiology*, 11, 1R–10R, 2001.
24. Lipman, N.S. and Perkins, S.E., Factors that may influence animal research. In *Laboratory Animal Medicine*. J.G. Fox, L.C. Anderson, F.M. Loew, and F.W. Quimby, eds., New York: Academic Press, 2002, pp. 1143–1184.
25. Gardner, M.B. and Luciw, P.A., Animal models of AIDS, *FASEB J.*, 3, 2593–2606, 1989.
26. Burwinkel, M., Holtkamp, N., and Baier, M., Biology of infectious proteins: Lessons from yeast prions, *Lancet*, 364, 1471–1472, 2004.
27. Leader, R.A. and Padgett, G.A., The genesis and validation of animal models, *Am. J. Pathol.*, 101, s11–s16, 1980.
28. National Research Council (U.S.). Committee on Animal Models for Research on Aging, *Mammalian Models for Research on Aging*, Washington, DC: National Academy Press, 1981, pp. 1–6.
29. Palm, P., McNerney, J., and Hatch, T., Respiratory dust retention in small animals, *AMA Arch. Indust. Health*, 13, 355–365, 1956.
30. Dressman, J.B. and Yamada, K., Animal models for oral drug absorption. In *Pharmaceutical Bioequivalence*, P. Welling, and F.L. Tse, eds., New York: Marcel Dekker, 1991, pp. 235–266.
31. Stevens, C.E., Comparative physiology of the digestive system. In *Dukes Physiology of Domestic Animals*. M.J. Swenson, ed., Ithaca, NY: Comstock, 1993, pp. 216–232.
32. Ohishi, I., Oral toxicities of *Clostridium botulinum* type A and B toxins from different strains, *Infect. Immun.*, 43, 487–490, 1984.
33. Sergeyeva, T., Detection of botulinal toxin and type A microbe in the organism of sick animals and in the organs of cadavers, *Zhurnal. Mikrobiologii* 33, 96–102, 1962.

34. Miller, L.H. et al., The resistance factor to *Plasmodium vivax* in blacks. The Duffy-blood-group genotype, FyFy, *N. Engl. J. Med.*, 295, 302–304, 1976.
35. Zhou, D. et al., DNA microarray analysis of genome dynamics in *Yersinia pestis*: Insights into bacterial genome microevolution and niche adaptation, *J. Bacteriol.*, 186, 5138–5146, 2004.
36. Jahrling, P.B. et al., Experimental infection of cynomolgus macaques with Ebola-Reston filoviruses from the 1989–1990 U.S. epizootic, *Arch. Virol. Suppl.*, 11, 115–34, 1996.
37. Koch, R., Die Aetiologie der Tuberkulose. *Mitt. Kaiserl. Gesundheitsamt*, 2, 1–88, 1884.

5 Infectious Disease Aerobiology
Aerosol Challenge Methods

Chad J. Roy and M. Louise M. Pitt

CONTENTS

5.1 INTRODUCTION

Aerosol exposure is the most probable route in a biological warfare or terrorist attack [1]. In natural aerosol infections, only a few biological agents are considered either obligate or opportunistic airborne pathogens (e.g., *Mycobacterium tuberculosis*, influenza) [2]. Because natural aerosol infection is a poor surrogate for studying airborne infection, modeling these interactions within a controlled experimental setting allows for intensive study of the process of aerosol-acquired disease. The modern history of studying aerosolized infectious disease agents using a homogenous synchronized experimental aerosol dates back to early twentieth-century efforts involving the infection of guinea pigs with aerosolized *M. tuberculosis* [3]. From these early studies to present-day investigations, important distinctions can be drawn between modeling natural infection and experimental infection. Natural infection, or the use of the communicability of an infectious agent to cause disease, is very much a different process from that of experimental infection. Natural infection of a naïve host from the aerosolized secretions (e.g., a cough or sneeze) of an infected host relies on a variety of uncontrollable factors, the majority of which are propagated by the clinical course of disease in the infected host. Characteristics of the infectious agent (infectious stability), amount (concentration), particle size distribution, and form (particle constituents) of a biological aerosol are highly variable when clinically ill animals are used for the purposes of infection. Performing this type of exposure will result in an asynchronous and heterogeneous infection, potentially affecting experimental design and the desired biological outcome. Controlled experimental infection, in contrast, allows the researcher to exert control over a wide range of experimental parameters to generate a biological response in a group of animals exposed to an approximately similar dose. The design and operation of the laboratory apparatus to support this type of experimentation is technically and logistically demanding, and coordination between the biological and engineering sciences is critical to successful aerosol challenge.

5.2 GENERAL

A number of considerations must be taken into account when performing an aerosol challenge with infectious biological agents. Foremost is safety: care must be taken at every juncture of the procedure not to contaminate the laboratory, equipment within the laboratory, or personnel present during the experiment. Successful performance of an animal aerosol challenge consists of generating and effectively delivering an aerosol concentration of an infectious or toxic agent in sufficient quantity to induce a desired biological response, including inducing an infection or toxic response that will produce illness in unprotected (sham-exposed) control animals. To achieve this goal, one must consider a variety of experimental variables, all of which are interdependent and rely heavily on the quality of the aerosol system in use, animal selection, technical capability, veterinary resources, and microbiological support.

5.3 AEROSOL TEST FACILITIES AND SYSTEMS

The basic design for inhalation systems used in an infectious aerosol challenge generally consists of a container or chamber wherein an aerosol is introduced and is allowed to either decay (static) or is pulled through the chamber via an exhaust (dynamic) at a rate approximating the input flow [4]. Static systems allow the exponential decay of an introduced agent as the available oxygen is exhausted and waste gases (e.g., carbon dioxide, ammonia) increase in concentration. Dynamic inhalation systems, in contrast, are desirable in experimental infection because the inhalation unit has a continuous air flow through the chamber, creating a unidirectional flow and a constant introduction of "fresh" aerosol and removal of unwanted waste products. This configuration is first described by Henderson [5] when used with infectious aerosols. Later descriptions [6] punctuate the advantages of dynamic systems for aerosol challenge; most microbial components generally decrease in viability as atmospheric residence time increases [7,8]. The process of performing any type of aerosol exposure demands that all parameters (air flow, humidity, temperature, pressure, etc.) be monitored and controlled throughout the duration of the procedure. The various types of system configurations used for infectious disease aerosol challenges are generally simpler versions of inhalation systems used in commercial inhalation toxicology laboratories.

The basic engineering requirements for the operation of this type of experimentation necessitate use of high-containment biological safety cabinets (BSCs) to protect personnel and other parts of the laboratory from unwanted contamination. All aerosol exposures of live infectious agents and protein toxins are generally performed within class III BSCs [9]. The class III BSC is considered to be the secondary containment; the sealed modular aerosol exposure chamber serves as the primary containment. When used for infectious bioaerosol studies, class III BSCs need continuous mechanical, structural, and leak-seal maintenance. Structurally, the drains, electrical outlets, autoclaves, windows, gloves, and high-efficiency particulate air filters are continuously checked, repaired, and replaced. Leak-seal testing is performed and the cabinet is certified on a semiannual basis. During operation, the cabinet is maintained continuously at least at a negative pressure of 0.5″ water.

Both the cabinets and the aerosol exposure systems within are decontaminated between the usages of biological agents for regular equipment maintenance and to prevent cross contamination from residual genus or strain-differentiated organisms. Gaseous decontamination of the inhalation equipment and the interior of the class III BSC are performed immediately after each challenge experiment. Paraformaldehyde gas has been used historically at the U.S. Army Medical Research Institute of Infectious Diseases (USAMRIID) for decontamination; hydrogen peroxide vapor may also be used to safely decontaminate equipment and safety cabinets. Preceding decontamination, paper strips impregnated with *Bacillus atrophaeus* spores are strategically placed in the cabinet and the exposure equipment. Subsequent to decontamination, the spore strips are removed and cultured for growth. No growth from the paper strips ensures the spores were killed and provides assurance that the safety cabinet was successfully decontaminated. Thereafter, the seal on the class III BSC can be breached and the equipment can be safely removed and replaced.

5.3.1 Recommended Inhalation Exposure Systems for Animal Challenge

At present, all exposure chambers being used at facilities presently performing animal exposures with select biological agents are modular and infrastructure independent. In many of the inhalation systems used for animal exposures, a primary air flow (usually at a particular pressure) passes through an aerosol generator loaded with the inoculum under study. Thereafter, the aerosol, entrained in the primary flow, is usually supplemented and mixed with a secondary air flow before entering into the exposure system at an inlet aerosol chamber. The exposure chamber, ranging from about 15 to 30 L, will then reach a steady-state aerosol concentration. Chamber exhaust is segregated between a primary flow and a sampling flow. The input and exhaust flows are maintained by manipulation of the flow and vacuum control, and chamber pressure is monitored using an attached magnehelic gauge to ensure neutrality (0.0 ± 0.5 mm Hg) during exposure.

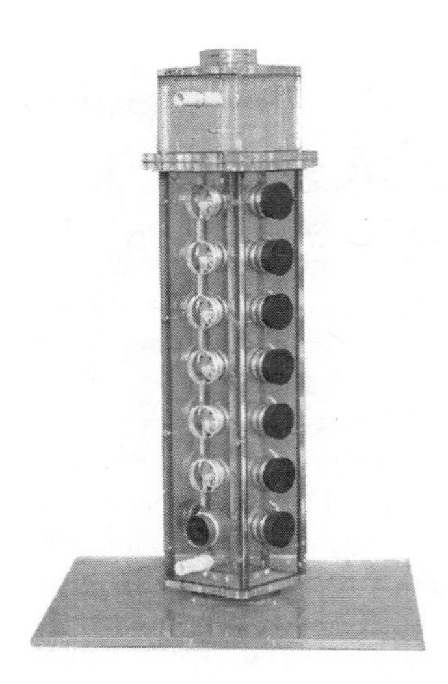

FIGURE 5.1 Nose-only rodent exposure chamber presently used at the U.S. Army Medical Research Institute of Infectious Diseases. The design allows for fresh aerosol to be delivered directly to the snout of the rodent, which is then exhausted via a secondary manifold under slightly negative pressure.

Animals are exposed to aerosols either *en masse* or singly. In general, smaller animals (rodents, guinea pigs) are exposed as a group, whereas larger animals (rabbits, primates) are exposed singly. Small animals are exposed via either nose-only or whole-body inhalation systems. Nose-only inhalation chambers, commonly used in pharmaceutical preclinical laboratories, are commercially available from a number of vendors [10]. The nose-only inhalation chamber, as the name implies, is configured as a vertical tower, such as the in-house model presently in use at the USAMRIID [11] (Figure 5.1). Animals are restrained in stanchion-type holders, which restrain the animal without restricting airflow. The loaded restraints are then attached to the tower in such a way as to only have their noses exposed to aerosol. The whole-body exposure chamber, in contrast, is a sealed box with an access door for loading and unloading animals [12] (Figure 5.2). The animals to be exposed are placed in wire cages that are in turn placed inside the exposure chamber. Although they are much easier to manipulate in the class III BSC, animals exposed in the whole-body chamber are susceptible to fomite formation on the fur during aerosol exposure. Larger animals such as rabbits and nonhuman primates are exposed muzzle or head only, respectively.

In the USAMRIID's configuration, a rectangular head-only exposure chamber with one side containing a circular cutout with a modified fitted latex dam, with a porthole cut in its center, stretched across the opening is used [13] (Figure 5.3). Rabbits are unanesthetized during the aerosol challenge; the rabbit is placed in a nylon restraint bag that allows one to comfortably hold the animal without risk of being scratched from thrashing or kicking. The rabbit's nose/muzzle is then pushed snugly against the rubber dam, forming a seal between the fur and the dam. The rabbit is held in place during the exposure. Primates, in contrast, are anesthetized before

FIGURE 5.2 Whole-body exposure chamber presently used at the U.S. Army Medical Research Institute of Infectious Diseases. This chamber is used for aerosol exposure of rodents *en masse*. Smaller steel reinforced mesh boxes (4) are loaded with the rodents and inserted into the chamber before initiation of exposure. This chamber is capable of exposing up to 40 mice per exposure iteration.

aerosol exposure. The head of the primate is inserted into the chamber through the latex port, fitting snugly around the animal's neck without restricting respiration. The animal's head rests on a stainless-steel mesh, which is integrated into the chamber. The operating characteristics of the aforementioned inhalation configurations are generally operated at low air flows (12–25 L/min) and pressures (15–30 psig).

FIGURE 5.3 Head-only exposure chamber presently used at the U.S. Army Medical Research Institute of Infectious Diseases. This chamber is used for single aerosol challenge of primates (head only) or rabbits (muzzle only) through the use of a modified rubber dam stretched across the circular opening on one of the sides of the chamber.

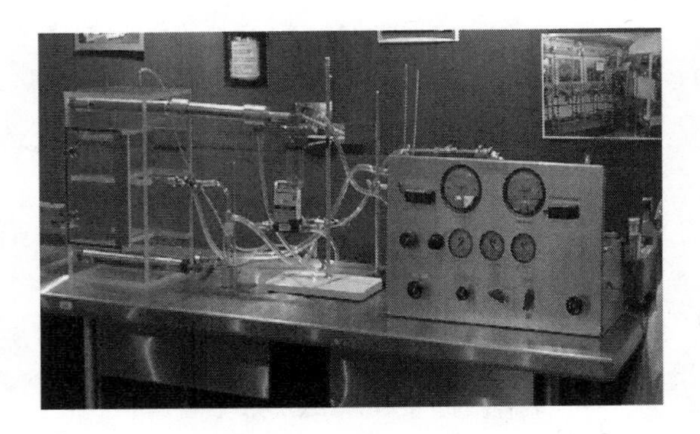

FIGURE 5.4 Instrumentation panel for manual control of aerosol challenge (shown with a whole-body exposure chamber). The panel has been historically used at the U.S. Army Medical Research Institute of Infectious Diseases to monitor and control aerosol generation, air flows, and chamber pressure once modular inhalation systems replace infrastructure-dependent inhalation systems. These panels are interchangeable with many different chamber configurations and can easily be moved in and out of class III biological safety cabinets for repair and servicing.

It is obligatory that one exhibits substantial monitoring and control over aerosol exposures involving infectious aerosols. Manually controlled modular inhalation systems have been used historically and are still in use today. Because of the number of meters and gauges that are involved in the control and data acquisition associated with aerosol studies, it is desirable to bundle instrumentation for ease of use in high containment. An instrument panel containing flow meters, air pressure gauges, vacuum controls, and on–off switches for the generator and aerosol sampler was used at the USAMRIID for several years (Figure 5.4). Recently, an automated aerosol exposure control platform has been developed specifically for use within high containment [14,15]. The automated platform provides a microprocessor-driven inhalation platform that imparts exquisite data acquisition and control over all aspects of the exposure and incorporates a dosimetry function based on the respiratory parameters of the animal during exposure (Figure 5.5). This improves on the precision and accuracy of inhaled dose delivery and calculation. This system has been introduced into the aerosol capability at the USAMRIID; it is routinely used and has been validated for use in regulatory studies.

5.3.2 Aerosol System Characterization

Once monitoring and control of the inhalation system have been ensured, one of the key parameters to the successful performance of experimental infection is the characterization of aerosol behavior during the process of generation and sampling. It is important to determine the effect of the aerosol generation process on the activity and viability of the microbial agent before exposure of the animal. Without explicit knowledge of the activity of the microorganism within the particular to the exposure chamber and control system used, dose estimation and subsequent biological response may be compromised. Aerosols of different forms can exhibit vastly

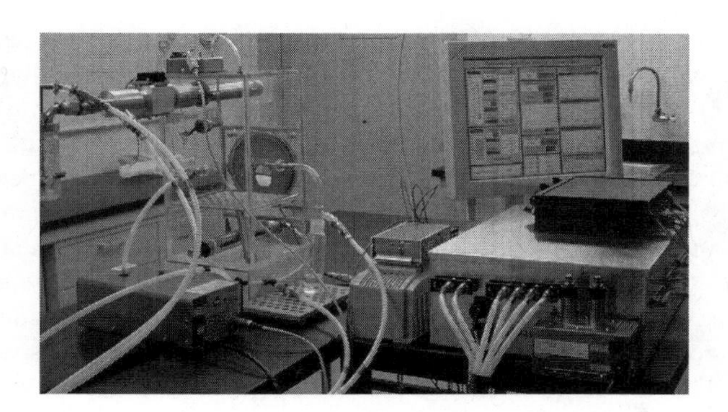

FIGURE 5.5 The automated bioaerosol exposure system for aerosol challenge (shown with a whole-body chamber configuration). The automated bioaerosol exposure system platform is an updated, automated version of the manually controlled instrumentation panel historically used at the U.S. Army Medical Research Institute of Infectious Diseases. In addition to full electronic acquisition and control of generation, air flows, and pressure, the automated bioaerosol exposure system unit has integrated user-defined humidification of the chamber and real-time respiration monitoring in single exposure with larger animals (e.g., primates).

different stability constants; aerosols composed of vegetative bacterial cells are considered especially susceptible to dehydration and death in aerosol form, whereas negative-strand RNA viruses are considered to be quite hardy in aerosol form [7]. Strain can also impact aerosol viability. Mutants lacking capsules from the cell wall can have deleterious effects on survivability in aerosol [C. J. Roy, unpublished data, 2004]. Microbial airborne concentrations are generally established using a viable counting technique such as bacterial culture or viral plaque assay. Because aerosol viability is of paramount importance in this type of inhalation exposure, one must establish the extent of the effect on the microorganism during generation and sampling. Environmental factors such as temperature and relative humidity can have a deleterious effect on the viability of a bacterial or viral aerosol [16]. Similarly, viability may be affected by mechanical disruption from aerosol generation.

It is therefore necessary to establish a reliable viable aerosol concentration using similar experimental parameters before the commencement of any animal exposure. The results of the preparative experiments will ensure that the in-use inhalation system is properly characterized with the specific biological agent, and the agent is properly characterized in aerosol, using the configured inhalation system. A well-characterized exposure chamber is necessary to generate a reproducible atmosphere. This characterization includes leak rates, mixing characteristics and optimum flow rates to achieve equilibrium concentrations rapidly [17].

The most efficient method to empirically determine aerosol stability for the purposes of animal challenge is determination of an aerosol dilution factor, or "spray factor," within a particular inhalation exposure configuration. Before determination of the spray factor, determination of optimal cell culture harvest and isolation of the select agent should be performed using general microbiology techniques. This will ensure that optimal and standardized cultures are being used to determine the spray factor.

To determine the aerosol spray factor, a series of aerosol "runs" are performed in the particular inhalation system that will be used in the animal challenges (see Section 5.3.2.1). During each run, the starting concentration (C_s) within the aerosol generator should be increased in concentration sequentially. The resulting aerosol concentration (C_a) that is produced from each of the aerosol experiments, usually determined by analysis of the substrate from the particular aerosol sampler in use, is expressed in units per liter of aerosol (colony-forming units (CFUs) or plaque-forming units (PFUs) or μg/, aerosol). The ratio of these two values will produce a unitless factor that expresses the dilution that one should expect during an animal challenge experiment with the same inhalation configuration and strain-specific select agent. Determination of the aerosol factor provides a starting point for one to exhibit quality control over subsequent experimentation and to establish a standard for aerosol experiments with that particular agent. In addition, the determination of the spray factor allows one to estimate the aerosol generator starting concentration that is needed to achieve a desired presented dose in a particular species of animal (see Section 5.3.2.2). An agent-specific database that maintains the spray factors for each inhalation system can then be used for accurate calculation of starting concentrations, overall performance of the system, and quality control of the generation process.

5.3.2.1 Part I: Determination of an Aerosol Spray Factor (F_s)

Experiment/assumptions: An aerosol dilution factor, or spray factor, needs to be determined (based upon a particular inhalation system configuration) used for animal challenge. The agent is *Yersinia pestis*, strain CO92, that will be used in aerosol challenge. The maximum achievable liquid concentration (C_s) is 1.0×10^{11} CFU/mL. An integrated aerosol sample, pulling at a continuous 6 L/min (Q_{agi}), will be collected during each of discrete runs of the system. Each sampler is preloaded with 10 mL of collection medium (V_{agi}); an evaporation constant (E_c) of 0.15 mL/min is assumed. The t_d for each run is 10 min. Samples will be cultured immediately (C_{sam}) after each aerosol run. A total of five runs of the system with logarithmic increases in starting concentrations (C_s), originating at 1.0×10^7 CFU/mL are planned for the experiment.

1. Determination of aerosol concentration (C_a) from each aerosol run ($C_s = 1.0 \times 10^7$ CFU/mL shown):

$$C_a = \frac{C_{sam} \times (V_{agi} - (E_c \times t_d))}{Q_{agi} \times t_d}$$

$$= \frac{2.0 \times 10^3 \,\text{CFU/mL}^* \times (10\,\text{mL} - (0.15\,\text{mL/m} \times 10\,\text{m}))}{6\,\text{l/m} \times 10\,\text{m}}$$

$$= 283\,\text{CFU/L}$$

where "*" indicates an average from culture plates performed in triplicate.

The other data for subsequent aerosol runs are given:

$$C_s = 1.0\times10^8; C_a = 2.9\times10^3 \text{ CFU/L}$$
$$C_s = 1.0\times10^9; C_a = 2.6\times10^4 \text{ CFU/L}$$
$$C_s = 1.0\times10^{10}; C_a = 2.7\times10^5 \text{ CFU/L}$$
$$C_s = 1.0\times10^{11}; C_a = 2.7\times10^6 \text{ CFU/L}$$

2. Thereafter, the F_s is calculated for each C_s and C_a ($C_s = 1.0 \times 10^7$ CFU/mL shown)

$$\begin{aligned} F_s &= \frac{C_a}{C_s(\text{CFU/L})} \\ &= \frac{283\text{CFU/L}}{1.0\times10^9\text{CFU/L}} \\ &= 2.8\times10^{-7} \end{aligned}$$

3. Therefore, assuming relative linearity, an average F_s can be used for the presented dose (D_p) calculation:

$$Average = 2.71\times10^{-7}$$

5.3.2.2 Part II: Starting Concentration Calculation for Aerosol Challenge: An Example

Experiment/assumptions: A group of vaccinated guinea pigs weighing an average of 900 g will be challenged by aerosol to *Y. pestis*, CO92. The desired presented dose (D_p) is 4.2×10^5 CFU/animal, equating to 100 median lethal doses (LD_{50}s). A historical (Part I) aerosol spray factor (F_s) of 2.1×10^{-7} will be used. The acute challenge duration (t_d) is 10 min.

4. Determination of predicted respiratory volume (V_e) during exposure:
   a. The minute volume (V_m) is calculated using Guyton's formula:

$$\begin{aligned} V_m &= 2.10\times \text{BW}(g)^{0.75} \\ &= 2.10\times 900^{0.75} \\ &= 345\,\text{mL} \end{aligned}$$

   b. Accounting for the duration of the exposure (t_d):

$$\begin{aligned} V_e &= 345\,\text{mL}\times t_d \\ &= 3.45\,\text{mL}\times 10\,\text{m} \\ &= 3.45\,\text{L} \end{aligned}$$

5. The necessary aerosol concentration (C_a) to achieve D_p is then calculated:

$$V_e C_a = D_p$$
$$3.451(C_a) = 4.2 \times 10^5 \text{ CFU}$$
$$= \frac{4.2 \times 10^5 \text{ CFU}}{3.45 \text{ L}}$$
$$C_a = 1.21 \times 10^5 \text{ CFU/L}$$

6. The starting concentration (C_s) is then calculated by application of the F_s:

$$F_s = \frac{C_a}{C_s}$$
$$2.1 \times 10^{-7} = \frac{1.21 \times 10^5 \text{ CFU/L}}{C_s}$$
$$= \frac{1.21 \times 10^5 \text{ CFU/L}}{2.1 \times 10^{-7}} = 5.79 \times 10^{11} \text{ CFU/L} \times 1000 \text{ mL/1 L}$$
$$C_s = 5.79 \times 10^8 \text{ CFU/mL}$$

7. Therefore, a C_s of 5.79×10^8 CFU/mL for a t_d of 10 min to deliver a D_p of 4.2×10^5 CFU in the experiment.

5.4 PARTICLE SIZE AND GENERATION

Another important component of animal aerosol challenge, in conjunction with microbial stability of the aerosol, is characterization and optimization of the aerosols generated. The ability of a particle to penetrate the respiratory system is solely dependent on the aerodynamic size of the particle generated. Without any control over the size of the particle generated, no assurances can be made on deposition onto susceptible tissues. It is preferable in experimental infection to generate aerosols that will deposit mostly into the lower bronchial and pulmonary space. These tissues are the most susceptible to injury and infection and are considered a target for any engineered threat in an offensive or bioterrorist-type exposure. Particle size in experimental infection is governed by the type of generator used and the residence time within the chamber before exposure. There are a variety of pressure- and flow-driven atomizers and nebulizers that have been developed; the Collison nebulizer has been used far more than any other type of generator in experimental exposures to infectious aerosols [18]. The Collison nebulizer is a simple, low-pressure, improved atomizer that is operated by application of compressed air between 20 and 25 psig and that produces an aerosol flow rate of 7.5 L/min output. Under these conditions, the nebulizer produces a highly respirable aerosol size (1.0 μm mass median aerodynamic diameter; 1.4 geometric standard deviation [σ_g]), as measured by an aerodynamic particle sizer. In addition to its ease of use, relatively small aliquots of the biological agent (10 mL) are needed for generation. The starting concentration (placed in the vessel of the nebulizer) for the exposure is calculated on the basis of aerosol efficiency for each individual agent and the desired delivered dose. Other generation devices can be used to modulate particle size distribution in experimental exposures. These instruments, such as the spinning top aerosol generator (STAG), can be configured to be used with modular inhalation systems described herein. The proper use of particle-modulating generators in this capacity significantly complicates the experimental setup, and working in high containment with such devices should only be pursued if particle modulation is absolutely necessary as a component of the experimental design. In using the STAG for particle modulation, material requirement can increase dramatically (>10–30 mL/single exposure) and manipulation of the solvent–solute ratio can vary according to the liquid density of the inoculum being used.

Recently, a flow-focusing aerosol generator (FFAG) for generating large-particle aerosols has been developed. Liquid is pumped through a capillary needle of defined diameter and drawn through a critical orifice. Upon application of an appropriate pressure from a compressed air supply, the focused liquid splits into particles of a defined size at a specific distance from the orifice of the aerosol generator. The orifice diameter is fixed; however, upon modulation of the liquid flow rate or gas pressure, particle sizes greater than 10 μm can be generated. The aerosols are produced with a geometric standard deviation of 1.1–1.3, indicating particle uniformity [19].

Results of particle modulation in aerosol exposures can dramatically affect pathogenesis and shift the onset and tempo of diseases previously performed using a "standard" particle size distribution similar to the output from a Collison nebulizer [20].

5.5 AEROSOL SAMPLING DEVICES

Either interval or continuous air sampling of the exposure chamber during aerosol challenge is a requisite step for determination of aerosol concentration. Although there are many methods available for aerosol sampling, the sampler of choice for infectious biological aerosols is an impinger. Impingers rely on impaction of the aerosol-laden air onto a wetted surface (Figure 5.6). A vacuum is drawn on the collection vessel, which accelerates the sampled air through a capillary jet. The sampler takes advantage of Stokes diameter to collect particles under 20 μm at a relatively high efficiency (97%). In presently operating systems at the USAMRIID, aerosols are sampled continuously using a low-flow version of the all-glass impinger (AGI, Ace Glass Inc., Vineland, New Jersey). For low-flow collection, the AGI-4 (denoting the distance, in millimeters, of the capillary end to the vessel bottom) is commonly used for bioaerosol sampling. In continuous sampling during aerosol generation and exposure, the AGI collection vessel is filled (10 mL) with the medium appropriate to the microorganism being used in the aerosol challenge. The AGI capillary contains a 6 L/min critical orifice to regulate flow. With this sampler, a sustained vacuum of at least 15 in. of mercury is required to maintain the critical pressure ratio across the orifice. After the aerosol challenge is completed, the liquid collection medium is decanted for further analysis. The AGI is by no means the only sampling methodology that can be used for bioaerosol collection. Sample collection using filter substrates housed within a cassette are commonly used to collect bioaerosols. Similarly, cascade impactors loaded with microbial plates have been used for size selection of aerosol samples.

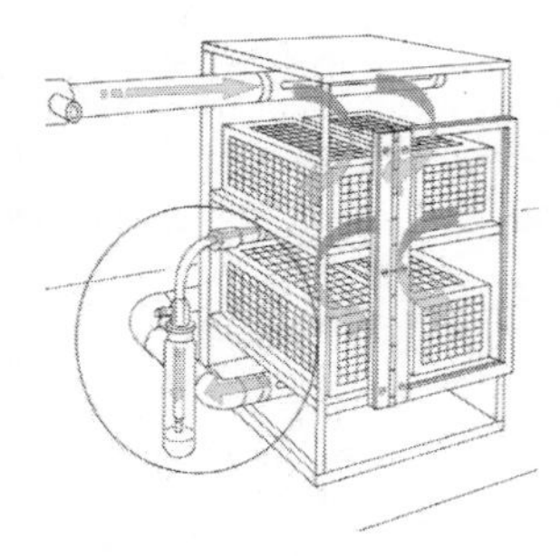

FIGURE 5.6 An all-glass impinger (circled) sampling a whole-body exposure chamber containing rodent restraint caging. The blue arrows indicate aerosol flow through the chamber. The all-glass impinger is operated continuously during the aerosol challenge. The collection fluid is then decanted and assayed for microbial content.

5.6 ASSAY METHODS

Assay of the integrated air sample and the liquid inoculum used for the experimental infection is generally performed immediately after the experimental procedure. Conventional culture is widely used for quantitative analysis in bacterial challenges; cell-based plaque assays are used for most of the viral agents. A simple protein assay is used for samples taken during protein agent aerosols. In addition to determining the starting concentration (C_s) from the aerosol generator, the aerosol concentration (C_a) will be determined. Before assay, it is advantageous to estimate the approximate concentration of the agent in the sample fluid. This will allow adequate dilution of the sampling medium before culturing or testing the agent and avoids the unnecessary reproduction of plates or cell culture plates that will contain too few or too many colonies or plaques to count. In addition to standard microbiological/virological techniques for sample assay, quantitative methods such as polymerase chain reaction (PCR) can be used for agent estimation. A disadvantage of using PCR, however, is the ability to estimate viable agent contained in a particular sample. Viable (or replicating) agent concentrations within aerosol samples remains an essential portion of dose estimation and the use of quantitative methods such as PCR are discounted as an alternative to traditional culture or plaque assay.

5.7 AEROSOL CHALLENGE DOSIMETRY

The calculation of dose in aerosol challenge has historically used a "presented" dose to express the mass of agent administered to the animal. Presented dose, when described as a part of infectious disease studies, refers to the inhaled dose (irrespective of percentage deposition) estimated from the multiplication of the aerosol concentration (in CFU, PFU, μg) and the total volume of air breathed by the animal during the time of exposure. For rodent species, the respiratory rate (expressed as minute volume, V_m) is prospectively calculated using a formula based on body weight ($2.10 \times BW^{0.75}$) [21] or other similar formulae [22]. For larger species (rabbits and nonhuman primates), V_m is determined using whole-body plethysmography immediately preceding the exposure [23]. The total volume of air breathed is then determined by multiplication of exposure time (t_e). Variation in presented dose among group and study cohorts is inherent in both approaches to dose estimation; plethysmography offers a method to specify the margin of interanimal variability among dose.

Newer methods in dosimetry for aerosol exposures with animals in modular inhalation systems have emerged recently [14,15] that utilize computer-based monitoring and control to better estimate dose. These systems utilize the active respiration of a single animal (usually a nonhuman primate) within the chamber to estimate tidal volume via a differential pressure transducer. Cumulative tidal volume is then used to estimate a total volume of aerosol-laden air that would be required to achieve a particular dose. This method is innovative in that it is a departure from set time-based exposures that have been the standard used in the past. The near "real-time" dosimetry concept remains limited by the quantitative methods still in use for agent concentration—one does not know the aerosol sample (C_a) until *post hoc* bacterial culture or viral plaque assay of the collected aerosol sample. This is a limiting factor when using newer techniques for dose estimation such as near "real-time" dosimetry.

5.8 TRANSPORT AND HUSBANDRY OF AEROSOL-CHALLENGED ANIMALS

One of the more potentially hazardous procedures during aerosol challenge is the animal transport immediately after aerosol exposure to a select agent. The potential for reaerosolization of fomites on the fur/hair of the exposed animal is high; certain agents that are generally hardy in the environment (e.g., *Bacillus anthracis* spores, staphylococcal enterotoxin B) can maintain their infectious/toxic potential while in residence on the animal's coat [6]. The modality of exposure and animal species used in the challenge dictates an innate risk for reaerosolization of the agent. The fur of animals with a large portion or all of their body (head-only or whole-body) exposed to the agent generally carry the most agent on their coat. To avert any potential exposures, all animals that undergo aerosol challenge are treated as infectious and the entire coat contaminated with the agent. Subsequently, transport of animals out of the safety cabinet into the observation room is performed with care.

A number of procedures have been implemented at the USAMRIID for transport of exposed animals out of the safety cabinet and back to a designated caging unit in the observation room. Exposed animals are removed from the exposure chamber and placed in an interlocking double-door high-flow pass box within the safety cabinet before delivery back to the observation animal room. A mobile class III cabinet has been designed to transport animals safely to and from the animal-holding room; this cabinet then locks onto the class III biosafety cabinet containing the aerosol exposure system thus acting as the high-flow passbox. The entire observation room should be considered contaminated once exposed animals are placed back into their caging units. Any naïve animals leaving the room to be exposed should be considered, for the purposes of hallway transport, as contaminated, infectious animals. Proper care and husbandry procedures for aerosol-challenged animals by animal care staff is a critical part of the aerosol experiment. After the animals are in their caging units, there remains a potential of exposure to respirable particles reaerosolized from preening activities. In addition, some biological agents will rapidly induce a disease state after exposure (i.e., staphylococcal enterotoxin B) consistent with emesis, diarrhea, and other excretions with potentially high concentrations of the agent.

5.9 SUMMARY

Aerosol challenge of animals is one of the critical subject matter areas in infectious disease research that requires a combination of skill sets to ensure successful experimental results. Many of the pitfalls in aerosol challenge are procedural; the simple process of ensuring unidirectional air flow during the aerosol can be the determinative factor between success and failure of an experimental infection. Working with infectious agents in aerosol is not without potential risk—this type of research does present some of the more potentially hazardous environments in infectious disease research. Strict adherence to safety protocols in every procedural step of aerosol challenge is obligatory, no matter the experience of the researcher. Similarly, performance of aerosol challenge using specially developed standard operating procedures ensures continuity of dose in studies involving animal species exposed singly or *en masse*. A fully functional aerobiology facility that has thoroughly vetted all of its operating procedures is indeed a valuable resource for infectious disease research involving aerosol challenge.

REFERENCES

1. Franz, D.R. (ed.), *Textbook of Military Medicine*, Chapter 30, Part I: Warfare, weaponry, and the casualty: Medical aspects of chemical and biological warfare. Washington, DC: Surgeon General Printing Office, 1997.
2. Roy, C.J. and Milton, D.K., Airborne transmission of communicable infection—The elusive pathway. *New England Journal of Medicine*, 350(17):1710–1712, 2004.
3. Wells, W.F., Airborne contagion and air hygiene: An ecological study of airborne infection, The Commonwealth Fund. Cambridge, MA: Harvard University Press, 1955.
4. Cheng, Y. and Moss, O., Inhalation exposure systems. In: R.O. McCellan and R.E. Henderson (eds.), *Inhalation Toxicology* (pp. 25–66). New York: Pergamon Press, 1997.
5. Henderson, D., An apparatus for the study of airborne infection. *Journal of Hygiene*, 50, 53–68, 1952.
6. Jemski, J.V. and Phillips, G.B., Aerosol challenge of animals. In: W. Gay (ed.), *Methods of Animal Experimentation* (pp. 274–341). New York: Academic Press, 1965.
7. Larson, E., Dominik, J. and Slone, T., Aerosol stability and respiratory infectivity of Japanese B encephalitis virus. *Infection and Immunity*, 30, 397–401, 1980.
8. Dunklin, G.W. and Puck, T.T., The lethal effect of relative humidity on airborne bacteria. *Journal of Experimental Medicine*, 87, 87–101, 1948.
9. Richmond, J.Y. and McKinney, R.W., *Primary Containment for Biohazards: Selection, Installation, and Use of Biological Safety Cabinets*, 2nd edn., Washington, DC: Government Printing Office, 2000.

10. Cannon, W.C., Blanton, E.F. and McDonald, K.E., The flow-past chamber: an improved nose-only exposure system for rodents. *American Industrial Hygiene Association Journal*, 44, 923–928, 1983.
11. Stephenson, E.H., Moeller, R.B., York, C.G. and Young, H.W., Nose-only versus whole-body aerosol exposure for induction of upper respiratory infections of laboratory mice. *American Industrial Hygiene Journal*, 49, 128–135, 1988.
12. Pitt, M.L.M., Fleeman, G., Turner, G.R. and Young, M., Head-only bioaerosol exposure system for large animals. *Proceedings of the American Association of Aerosol Research*, A75.5, 1991.
13. Roy, C.J., Hale, M.L., Hartings, J.M., Duniho, S. and Pitt, M.L.M., Impact of inhalation exposure modality and particle size on the respiratory deposition of ricin in BALB/c mice. *Inhalation Toxicology*, 15(6):619–638, 2003.
14. Roy, C.J. and Hartings, J.M., Automated inhalation toxicology exposure system. U.S. Patent and Trademark Office, Patent Application 09/919,741, 2001.
15. Hartings, J.M. and Roy, C.J., The automated bioaerosol exposure system: development of the platform and a dosimetry application with nonhuman primates. *Journal of Toxicological and Pharmacological Methods*, 49, 39–55, 2004.
16. Ehrlich, R., Miller, S. and Walker, R., Relationship between atmospheric temperature and survival of airborne bacteria. *Applied Microbiology*, 19, 245–249, 1970.
17. Dabisch, P.A., Kline J., Lewis, C., Yeager, J. and Pitt, M.L., Characterization of a head-only aerosol exposure system for nonhuman primates. *Inhalation Toxicology*, 22(3):224–233, 2010.
18. May, K.R., The Collision nebulizer. Description, performance & application. *Journal of Aerosol Science*, 4(3):235, 1973.
19. Thomas, R.J., Webber, D., Sellors, W., Collinge, A., Frost, A., Stagg, A.J., Bailey, S.C. et al., Characterization and deposition of respirable large- and small-particle bioaerosols. *Applied and Environmental Microbiology*, 74(20):6437–6443, 2008.
20. Roy, C.J., Reed, D.S., Wilhelmsen, C.L., Hartings, J.M., Norris, S. and Steele, K.E., Pathogenesis of aerosolized eastern equine encephalitis virus infection in guinea pigs. *Virology Journal*, 6, 170, 2009.
21. Guyton, A.C., Measurement of the respiratory volumes of laboratory animals. *American Journal of Physiology*, 150, 70–77, 1947.
22. Bide, R.W., Armour, S.J. and Yee, E., Allometric respiration/body mass data for animals to be used for estimates of inhalation toxicity to young adult humans. *Journal of Applied Toxicology*, 20, 273–290, 2000.
23. Besch, T.K, Ruble, D.L., Gibbs, P.H. and Pitt, M.L., Steady-state minute volume determination by body-only plethysmography in juvenile rhesus monkeys. *Laboratory Animal Sciences*, 46, 539–544, 1996.

6 Characterization of New and Advancement of Existing Animal Models of *Bacillus anthracis* Infection

Elizabeth K. Leffel and M. Louise M. Pitt

CONTENTS

6.1 BACKGROUND

Anthrax, an ancient disease of animals and humans, has been extensively studied over several decades [1]. *Bacillus anthracis*, a large (approximately 1–10 μm), nonmotile, spore-forming bacillus, is the etiologic agent of anthrax. Details of the life cycle of the bacterium are well described. Spores are released into the soil from carcasses of infected animals, and vegetative spore growth occurs in the soil. Herbivores are commonly infected by grazing on contaminated land or ingesting contaminated feed. Carnivores and scavengers can also be infected by feeding on an infected carcass before putrefaction occurs. Wild animals in captivity are even vulnerable, as demonstrated by an outbreak of anthrax that occurred in cheetahs after they were fed infected baboon meat [2]. Human infection generally occurs via the cutaneous or gastrointestinal route, when people handle infected animals or animal by-products or ingest infected meat. In addition, infection can occur by the inhalational route which is the most lethal. The incubation period and degree of illness will vary, depending on the exposure route and dose, in both humans and animals. The initial symptoms in humans are nonspecific and may include malaise, headache, fever, nausea, and vomiting [3,4]. Respiratory distress and shock ensue and the disease has a mortality rate near 100% of untreated cases [5,6] and 40% in cases treated with antibiotics at varying stages of infection [4]. Inhalation anthrax is the most likely form of disease resulting from a warfare or terrorist attack because of the perceived ease of aerosol distribution of spores. Exposure route and dose are two areas that require careful consideration when developing an animal model and will be discussed.

As early as 1886, Hans Beuchner studied "pulmonary anthrax" under experimental conditions in mice, guinea pigs, and rabbits. The animals were exposed to clouds of spores in small chambers and lung histology was examined. Over 100 years later, the mechanism of infection is still generally agreed upon. If *B. anthracis* spores are inhaled by susceptible animals, a generalized systemic disease occurs when the spores undergo phagocytosis by macrophages and are carried to the draining tracheobronchial lymph nodes. The spores then germinate in the nodes, the bacilli proliferate and spread to mediastinal nodes and surrounding tissue. The lymphatic system allows bacilli to spread into systemic circulation, resulting in septicemia and the seeding of multiple organs. The meninges can become involved and often hemorrhagic meningitis results. In tissue, the bacteria are encapsulated and may be found in chains of two to three organisms.

Specifically in humans, aerosolized spores are deposited, phagocytized, transported, and germinate, as described above. Fatal toxemia leads to hypotension and systemic hemorrhage after an incubation period of approximately 1–6 days. Initial respiratory symptoms last for 2–3 days, and if not treated within 24–48 h after exposure, death occurs in 100% of cases [4–10].

Because of the highly infectious characteristics of anthrax spores by the inhalational route, and the high mortality associated with the respiratory illness, *B. anthracis* is considered a serious military and bioterrorist threat [11]. In 1979, anthrax spores were accidentally released from a military research institute in Sverdlovsk, Russia, resulting in an epidemic of cases of inhalational anthrax reported up to 50 km from the site of release [12]. It was discovered in the late 1990s that Iraq produced and fielded *B. anthracis* spores for use in the Gulf War [13]. This threat moved to the forefront of realistic possibilities when *B. anthracis* spores were distributed in letters, via the U.S. postal service, resulting in 22 cases of anthrax in 2001 with 5 of these being fatal inhalational anthrax [3,4,14]. Despite being studied for decades, the animal model most appropriate for extrapolation to humans remains uncertain for a variety of reasons. The main one is the fact that there has been limited available information on inhalational anthrax in humans and that makes comparison with animal data difficult. Relatively recent bioterrorism events of 2001, surveys of endemic disease in Turkey, and outbreaks in populations of intravenous drug users has resulted in a small database of knowledge of clinical disease and will allow better comparison for animal models [4,15–17].

6.2 DESCRIPTION OF CURRENT ANIMAL MODELS

There are many factors to consider before choosing an animal model. The most obvious question is whether or not the model can be extrapolated to human disease.

How close is the animal model related to humans anatomically, and how similar is the animal and human pathophysiology? Is the mechanism of pathogenesis well understood? Are the immune systems closely related? Are the disease endpoints similar? The answers require some knowledge of the anatomy and physiology of the animal and how that will affect pathogenesis of a given organism. Often, this is unknown and proof-of-concept studies must be designed in animals determined to be the best predictors based on the information available. When one wishes to study an induced disease model, such as anthrax (biothreat agent), the following variables are important and should be addressed on a species-specific basis.

First, one must decide whether the chosen species is genetically and/or physiologically comparable to humans. This requires basic natural history studies to characterize and document the pathogenesis of anthrax in the chosen model. The next studies would evaluate the efficacy of a vaccine or therapeutic/postexposure countermeasure in the animal model because efficacy studies cannot be performed in humans because ethical and/or practical reasons such as the incidence of disease is rare, the disease is deadly with no licensed treatment, or inhalation is not usually a natural route of exposure. In recognition of these facts, the U.S. Food and Drug Administration (FDA) has written regulation (21 CFR 314.601.90-95 and 21 CFR 314.600-650) to support the scientific community's reliance on data from well-characterized animal models to predict human disease processes and treatment efficacy. This regulation is often referred to as the "FDA Animal Rule" and generally states that, after safety has been established in humans, the FDA may license a product based on well-controlled animal studies when the results establish that the product is "reasonable likely to produce clinical benefit in humans."

The second consideration must be the route of exposure. If the model is not based on a route of exposure relevant to expected human exposure, then it may not be worthwhile. If the chosen experimental route is not the same, it must be demonstrated that the resulting disease course is comparable. The dose of the agent that will actually be presented to the animal must be well calculated and based on what a human might be expected to encounter. Thought must be given to the metabolic processes in the chosen animal. For example, will *B. anthracis* be viable after exposure? Is there a species that has an enzyme that is toxic to the capsule of the spores? If delivered to the lung, the deposition and retention of the organism should be considered and how it might affect outcome of predicted disease course.

Lastly, there is an endless list of factors that seem minor but have to be considered when developing an appropriate animal model. Is the animal available and can it be properly housed at a research facility? Will the life span or size of the animal impact housing capabilities? Is there familiarity with species and are reagents readily available to process biological samples? What is the cost? Is gender important; can the endocrine system unduly influence the model design? There are no standard answers to these questions. They may even change depending on whether one is looking at a predictive model (such as vaccine efficacy) or an induced disease model (anthrax).

The choice of an animal model is not trivial. Ideally, the disease pathogenesis or intoxication should mimic the human disease; however, lack of human data makes comparison with the animal data difficult. Animal species and strains differ not only in susceptibility to infectious biological agents or toxins, but also in their qualitative and quantitative responses to vaccines and therapies. The principal animal models used in laboratory investigations of experimental anthrax have been mice, rats, guinea pigs, rabbits, and nonhuman primates. The next sections of this chapter will address the issues associated with each model and update what are currently regarded as the best models.

6.2.1 Mouse

There is a need for well-defined small animal models for screening potential vaccines and therapeutics against anthrax. Mouse models are very desirable due to the number of well-characterized inbred mouse strains, well-understood immune mechanisms, extensively studied genetic polymorphism, differences in disease resistance, availability of reagents, and convenient size that allows for ease of maintenance and facilitates use of statistically adequate numbers.

Abalakin and Cherkasskii [18] provided some evidence for differences in resistance among inbred mice. Several mouse strains were killed by 400 spores of a fully virulent encapsulated strain of *B. anthracis*. However, mouse strain CC57BR survived after a 100-fold-higher challenge. When challenged with a nonencapsulated, toxin-producing vaccine strain (STI), two mouse strains, A/Sn and DBA/2, died whereas other mouse strains were resistant. The authors proposed that susceptibility to infection was directly related to the sensitivity of the animals to the edematogenic and immunosuppressing action of anthrax toxin. The genetic analysis indicated that resistance to anthrax is probably controlled by a dominant gene, not linked with histocompatibility complex H-2, and is probably unrelated to the presence of hemolytic activity in mouse sera, determined by the C5 component of complement.

The suitability of inbred mouse strains as a model for studying anthrax was further investigated by Welkos et al. [19], who tested 10 inbred mouse strains. Welkos et al., in contrast to Abalakin and Cherkasskii, found that all strains had low parenteral LD_{50} (median lethal dose in 50% of the population) values for a virulent *B. anthracis* strain (5–30 spores). However, time-to-death analysis revealed significant differences among the mouse strains. DBA/2J and A/J were killed more rapidly; C3H/HeN, C57BL/6J, C3HHeJ, C57L/J, and C58/J were intermediate; and CBA/J, BALB/cJ, and C57BR/cdJ clearly had prolonged survival times in comparison with the others. In contrast, the mouse strains were either distinctly susceptible (A/J and DBA/2J) or resistant (C3H/HeN, C57BL/6J, C3HHeJ, C57L/J, and C58/J; CBA/J, BALB/cJ, and C57BR/cdJ) to lethal infection by a toxigenic, nonencapsulated *B. anthracis* "vaccine" strain [20]. Both A/J and DBA/J mice are deficient in the C5 component of complement and have defective H*c* genes [21,22]. C5-derived peptides are important anaphylotoxins and chemoattractants for macrophages and neutrophils during inflammation. The absence of these factors results in a delay in the influx of macrophages to the site of infection allowing the bacteria to overwhelm the host before a suitable immune response can be mounted [22]. Nontoxigenic encapsulated strains were shown to be lethal in both A/J and CBA/J mice [23]. Nontoxigenic encapsulated strains that are avirulent in guinea pigs are virulent in mice. Thus, the capsule appears to be the dominant virulence factor in mice.

Studies using fully virulent *B. anthracis* strains have included both outbred and inbred mice. Lyons et al. [24] compared intratracheal and intranasal routes of infection in the inbred BALB/c mouse. The most consistently identified pathological lesions of disseminated anthrax appeared in the spleen, not in the lung. Pulmonary changes at 48 h consisted primarily of diffuse distention of septal capillaries with bacilli, with minimal to no parenchymal inflammation.

In contrast, in BALB/c mice that received an aerosol dose of 50 LD_{50} Ames spores, the most common changes were the presence of bacteria in a variety of organs, most notably in alveolar septal capillaries, in renal glomerular capillaries, and in hepatic sinusoids. Focally extensive areas of mediastinitis usually characterized by mild infiltration of polymorphonucleated leukocytes (PMN) and fewer macrophages and lymphocytes were found occasionally [25] whereas bacilli were frequently found in these sites of inflammation. Only one-third of the mice had splenic lesions. Pulmonary lesions were rare whereas organisms cultured from the lungs were heat shock resistant, indicating they were ungerminated spores, a finding similar for human inhalational anthrax. Bacilli were found in the blood early in infection followed by a rapid decrease and then an increase until the mice succumbed. The lack of pulmonary lesions in the aerosol mouse model differs from the intranasal and intratracheal route [24] and re-emphasizes that the route of exposure should be considered and the resulting disease well understood before choosing a model for a particular application.

Several laboratories have taken advantage of the fact that nonencapsulating toxigenic-producing strains (vaccine strains) are lethal in mice and have developed anthrax models via various infection routes [26–29]. These models have the advantage of being able to be used in a BSL-2 laboratory, increasing the amount of basic research that can be done. However, it must be remembered that the classic pathological changes seen in human anthrax cases caused by virulent strains may not be well represented. After vaccination with various vaccines, mice can be protected against nonvirulent, unencapsulated anthrax strains but not virulent organisms [20]. Advantage has been taken of the susceptibility/resistance of the various mouse strains to the unencapsulated vaccine strains for screening potential vaccines [30,31]. The urgent need for effective countermeasures for inhalational anthrax since 2001 has led to mice being used more extensively for screening various potential therapies [24,31,32] and elucidating virulence factors and their mechanisms of action [24,33–37] and early host responses [38,39].

Care should be taken in the interpretation of these studies particularly when extrapolating the results to human experience. If the goal of model development is to predict the efficacy of a therapeutic or vaccine in humans exposed to virulent strains of *B. anthracis*, then the mouse may not be an appropriate small animal model.

6.2.2 Rat

The rat has a natural resistance to *B. anthracis* infection which increases as the animal ages. It takes about 1 million spores to kill a rat, compared to 10,000 to kill a rabbit, and only 10 spores to kill a mouse, parenterally [40]. Not surprisingly, even in immune rats, this resistance to spore challenge is only faintly improved [41].

However, the rat is extremely sensitive to anthrax toxin and although several species (Fischer 344, NIH black, Wistar and Norvegicus black) have been used to study toxin effects, the Fischer 344 has been shown to be most sensitive [42–44]. Subsequent to toxin exposure by intravenous or intraperitoneal routes, rats will develop clinical signs of illness that lead to respiratory compromise, pulmonary edema, and death [43,45]. Many studies are done *in vitro*, or via a parenteral route; no recent studies are described in the literature in which the aerosol route of exposure is used [46–50]. Therefore, the rat may be a good model to study isolated events of toxin exposure. It is not a practical model to predict effects on humans of a biological warfare event such as an anthrax attack, or to evaluate therapeutics or vaccines.

6.2.3 Guinea Pig

Guinea pigs are a well-established model for anthrax [50–55]. In the 1950s, Ross [56,57] used the guinea pig to compare intratracheal and aerosol exposure to *B. anthracis* and to correlate systemic organ failure with location of spores/bacilli over the time course of infection. These studies were a classic set of experiments that describe what we currently accept as anthrax pathogenesis. Ross [56,57] showed that spores are inhaled, undergo phagocytosis, germinate, and the vegetative cells move through the lymph and blood causing systemic disease. When guinea pigs were exposed to "spore clouds," Ross found a small number of spores in the lungs and recent pathology findings suggest there may be a role for local phagocytic cells in the nasal passages, in the pathogenesis of disease. Histological studies were repeated by Ross after instilling spores by intratracheal implantation to increase the numbers and introduce a more practical way to determine the spore location. In both cases, Ross showed that spores were ingested by alveolar macrophages, moved into the lymph where they germinated, both intracellularly and extracellularly of the macrophages, in the regional lymph nodes. Bacilli were then found in the bloodstream, resulting ultimately in systemic disease [56]. A recent review article describes that some lesions in the guinea pig are consistent with human and rabbit pathology.

Guinea pigs have since been used extensively in anthrax pathogenesis studies because they are sensitive to infection with *B. anthracis* spores, they are a physically manageable animal, and the disease course mimics that seen in rabbits [51,53,55,58–60]. Approximately 50–500 spores are lethal for a guinea pig after parenteral exposure [61]. In comparison, the guinea pig is slightly more resistant after aerosol challenge, the LD_{50} of the Ames strain is in the range of 23,000–79,000 colony-forming units (CFUs) [62,63]. The intranasal LD_{50} for the Vollum strain is approximately 40,000 CFU and for the ATCC 6605 strain is about 80,000 CFU [64]. In contrast to the rat, the guinea pig is resistant to toxin. Only 15 units of toxin per kilogram will kill a rat whereas it takes 1125 units of toxin per kilogram to kill a guinea pig [61].

Vaccines are tested in potency assays to evaluate the consistency of manufacture from lot to lot. Based upon protection seen in guinea pigs vaccinated against parenteral spore challenge, the FDA has approved potency assays for the licensed anthrax vaccine BioThrax® in guinea pigs [65]. This use of an animal model is unique and quite specific. The assay is designed to test lot variation, which occurs much later than routine safety or efficacy studies in which most models are used.

Outside of the specific intended use discussed above, the guinea pig has been used extensively for pathogenesis studies [50,56,57,66–70]. Ivins et al. [50] compared *B. anthracis* strains that varied in virulence, in the guinea pig model, to determine their immunizing potential. In doing so, the group confirmed other studies showing that fully virulent strains must possess both the toxin plasmid and the capsule plasmid, in addition to expressing both proteins in order to induce clinically significant disease. Much of the other work was completed around the 1950s and subsequent pathogenesis studies have been combined with vaccine efficacy experiments.

The Hartley guinea pig is most often used in testing vaccine immunogenicity and efficacy, perhaps because this is one of the few outbred stocks [50,55,60,66,71–73]. It has been demonstrated that passive immunity can be achieved in this model which demonstrated how important anti-protective antigen (PA) antibodies are in imparting protection to guinea pigs exposed to *B. anthracis* [66,74,75].

Obviously, when one endeavors to establish an animal model to be used in vaccine efficacy studies, correlates of immunity must be established. In the guinea-pig model, it has been shown that a good correlate for protection is neutralizing antibodies to PA; a direct correlation between survival and neutralizing-antibody titer has been reported. In the same paper, the researchers determined that there was no such consistent correlation between survival and immunoglobulin G (IgG) anti-PA antibody titers measured by enzyme-linked immunosorbent assay (ELISA) [76]. In fact, the quantity of PA antibodies may not be the determining factor in guinea-pig protection; it may be the quality of the antibodies sustained at a critical level [77]. Immunological memory for PA may also contribute to the level of protection afforded by vaccine [78]. Although the guinea pig is an acceptable model for pathogenesis, there are conflicting data reported in the literature regarding the usefulness of the model for efficacy studies. Protection by vaccination with the licensed human vaccine yields variable survival results in guinea pigs, depending on the vaccine/adjuvant combination [63]. This may be due to the fact that the adjuvant, aluminum hydroxide, may be the least effective choice in the guinea pig [63,72,79]. Interestingly, the guinea pig can be protected when the same antigen is combined with other adjuvants [63,80]. Therefore, the guinea pig is an acceptable model for evaluating vaccines if careful consideration is given to adjuvants used in vaccine efficacy studies. In other words, basic knowledge of the mechanism of protection (mechanism of action of the medical countermeasure) in the model is particularly important when using the guinea pig.

The guinea pig has been used less frequently for therapeutic efficacy studies, but it may be becoming more popular for screening new therapeutics [81]. In an intranasal model, a monoclonal antitoxin antibody was protective and had a synergistic effect on survival when administered with a suboptimal dose of ciprofloxacin [82]. The studies were completed under both prophylactic and therapeutic scenarios and correlates of protection (neutralizing antibodies) were confirmed. This suggests that the guinea-pig model, under a properly designed study, would be an acceptable model for therapeutic efficacy studies.

6.2.4 Rabbit

The rabbit has been shown to be sensitive to *B. anthracis* and to display pathological changes similar to those reported in naturally occurring disease in humans. As early as 1886, the rabbit was used in experiments to investigate inhalational anthrax and, since that time, there have been numerous publications to show that the model is also predictive of outcome in nonhuman primates [67]. Parenteral and aerosol routes of exposure to spores have been studied and ensuing pathology and disease pathogenesis are well documented and have been shown to be similar to the nonhuman primate models or to that observed in the limited human cases available [4,6,67,68, 83–85]. It is preferable to perform early studies in a species lower on the phylogenetic scale than nonhuman primates. These data indicate that the rabbit model is an appropriately susceptible model to be useful in early nonclinical efficacy studies.

Zaucha et al. [83] completed an extensive pathology study to compare inhalation and subcutaneous exposure to *B. anthracis* in New Zealand white rabbits. The rapid occurrence of disease, in both exposure groups, made the comparison of clinical signs rather insignificant. It is suggested by the authors that the short time course of disease observed in naive rabbits may impair the influx of leukocytes in response to the bacilli, resulting in hemorrhage and necrosis. Rabbits died so quickly that although lesions were found in the brain and meninges upon necropsy, the animal rarely showed clinical signs to indicate such involvement. Some cases described in the literature do indicate central nervous system (CNS) involvement, based on clinical signs in infected rabbits [40,85]. Very early work, in 1943, compared lesions of the CNS from rabbits that died from anthrax to known human cases and found the lesions similar [86]. This data was substantiated by the later Zaucha et al. [83] work which found CNS pathology although CNS clinical signs were not obvious.

The overall striking pathological differences [83] between the rabbits and other species was that, in the rabbits, there was a lack of leukocytic response in the brain and meninges, mild mediastinal lesions, and a low incidence of anthrax-related pneumonia. These differences may be due to the susceptibility of the rabbit model that results in the more rapid progression to death. Interestingly, pulmonary lesions were found in the majority of rabbits, regardless of exposure group. The authors conclude that, although anthrax is a disease of rapid progression in rabbits, pathology at the time of death is very similar to that reported for human cases of inhalational anthrax. It is suggested that advantage should be taken of the fact that the disease progresses so rapidly in the rabbit; the model provides a "rigorous test of candidate products" [83].

Various vaccines, to include bacterial supernatants, purified recombinant proteins, and DNA-based vaccines, have been shown to be efficacious in the rabbit model against a lethal challenge of virulent *B. anthracis* spores [72,87–91]. In order to be an effective vaccine efficacy model, an immune biomarker must be identified that correlates with protection, for the model to have predictive relevance in the human. Using PA-based vaccines with an aluminum-based adjuvant, it was demonstrated that anti-PA IgG and toxin-neutralizing antibodies were predictive of protection [88,92,93]. It is widely accepted that correlates of immunity for the rabbit model are toxin-neutralizing antibodies (measured by the toxin neutralization assay) or IgG anti-PA antibodies (measured by ELISA), although the original work [79,88] to establish correlates has not been updated. Unpublished data have been presented at recent scientific meetings [94] to indicate that neutralizing antibodies may be the best predictor of infection, as also published by Weiss et al. [93]. The ability to passively protect rabbits from lethality with anti-PA antibodies lends support to the belief that antibodies play a role in active immunity [82,95].

In conducting vaccine efficacy studies in this model, similar to the guinea pig, the choice of adjuvant must be carefully taken into consideration. Adjuvants other than Alhydrogel, such as CpG, may [96] or may not [80] enhance the immune response and therefore may not be predictive of efficacy in the nonhuman primate and humans.

The rabbit model has also been shown to be useful for therapeutic studies. Biomarkers predictive of active infection have been established as triggers for treatment initiation. Two biological indicators of infection, an increased body temperature and antigenemia (presence of systemic PA), have been correlated with bacteremia, the gold standard of infection verification [97–99]. These triggers-to-treat have been used in therapeutic models to evaluate a monoclonal antibody [97] and an antibiotic [99].

The rabbit is a useful model to study inhalational anthrax. It is an appropriate pathogenesis model that can be used to predict the response of both nonhuman primates and humans infected with *B. anthracis*, therefore it would be useful for evaluating efficacy of lead therapeutics or vaccine candidates. Although there has not yet been a product licensed by the FDA, the New Zealand white rabbit has been presented as one species utilized under the Animal Rule and was not rejected by the FDA as a reasonably predictive model (FDA Anti-Infective Drugs Advisory Committee Meeting, October 27, 2009; Biologic License Application 125349).

6.2.5 Nonhuman Primate

Over the years, scientists have tested different strains of nonhuman primates: some very early work was done in the chimpanzee [100] and then there was concentration on the rhesus [8,61,72,83,101–106]. Research expanded into the cynomolgus macaque [107]. Since the publication of the first edition of this book in 2006, there have been few manuscripts published that report advancements on the basic (natural history and disease pathogenesis) model developments in the nonhuman primate models: the African green monkey [98,108] and, most recently, the marmoset has been introduced [109]. This section will describe recent research to characterize these new models and compare them to what is known about clinical anthrax.

Haas [110] prepared a review article in 2002 in which the risk of inhaled *B. anthracis* spores was analyzed. The article underscores why it has been difficult to extrapolate nonhuman primate data to human risk and why identifying the animal model is critical when considering this risk. The basic lesions of human inhalational anthrax are hemorrhage, edema, and necrosis, with only a relatively mild cellular inflammatory component in the majority of tissues. The spleen, lymph nodes, mediastinum, lungs, gastrointestinal tract, and brain are principal sites of involvement in humans [111–114]. Similar pathogenesis would ideally also be seen in the nonhuman primate model chosen to evaluate a medical countermeasure.

Differences in inhalational anthrax among animal models and humans may be attributed to greater susceptibility of the animal models, resulting in more rapid progression to the fatal outcome, compared with the more protracted time course of the disease in humans. The differences may be associated with differences in knowledge around the time of exposure: in models, researchers know the precise time of exposure and the dose within relative certainty whereas in documented clinical cases [4,14] the exposure dose is unknown and the time of exposure is sometimes uncertain. Researchers monitor exposed animals continually for indication of disease, and humans generally present when they are very ill, as initial symptoms are generally thought to be a nonthreatening virus. Lastly, human cases may have underlying disease such as pneumonia or cardiovascular disease [4] which increases susceptibility.

The *Macaca mulatta* monkey (rhesus macaque) is one nonhuman primate species that has been extensively used in anthrax studies; it has been considered an appropriate model for studying anthrax and testing vaccine efficacy [103,104,115–117]. The robustness of the rhesus model is now somewhat debated, especially in recent years, because it seems to be fairly resistant to inhalation anthrax. Although it is still unclear why this is occurring, it may be related to age [109] or major histocompatibility complex 1 (MHC1) [118,119]. One cannot discount other possibilities that could be responsible for the observation of this apparent resistance to anthrax: (1) Could the origin of animals be a factor? (2) Over the years that research has progressed in this model, have properties of spore preparations changed? (3) Are technical methods sufficiently different at different laboratories to cause this effect? These are questions for which we do not have answers and the relatively recent perceived lack of confidence in the rhesus therapeutic/vaccine model is probably responsible for the paucity of published research documenting refinement of it as a model for inhalation anthrax.

In the event that there is a suspected exposure to *B. anthracis*, people may be treated before there is confirmation of infection [14], referred to as postexposure prophylaxis (PEP). Despite a movement from the rhesus in vaccine research, the rhesus has been used in two recent studies to evaluate the effects of PEP. In one study, ciprofloxacin and BioThrax (anthrax vaccine adsorbed or AVA) were evaluated after a very high exposure (~1600 LD_{50}s) of aerosolized spores. Monkeys were treated beginning ~2 h after exposure with either oral antibiotic alone (twice daily for 14 days) or antibiotic plus three doses of vaccine. There was only 44% survival in the antibiotic group, compared to 100% survival in the combination therapy group [120]. In a second study, a PEP and therapeutic scenario were compared [121]. Monkeys received intravenous ciprofloxacin for 10 days, either ~2 h after aerosol exposure or when they had been bacteremia-positive for two consecutive time points. The therapeutic group was matched to a monkey in the PEP group and each received treatment until the therapeutic group had reached 10 days of therapy. In the PEP group, there was only 20% survival after therapy was stopped, compared to 70% in the therapeutic group (3 died while on treatment). There were no untreated controls in either study, but this body of work is suggestive that the rhesus model might be manipulated for use in PEP scenarios.

Because the rhesus macaques were becoming increasingly less available for anthrax research work at the time of writing the last edition of this book, work to develop comparable nonhuman primate models had begun. Cynomolgus monkeys (*Macaca fascicularis*) and African green monkeys (*Chlorocebus aethiops*) were previously discussed [87,88]. In 2008, the natural history of the common marmoset (*Callithrix jacchus*) was published [109].

The African green monkey has been established as an acceptable model for inhalation anthrax because the disease pathogenesis is similar to what has been reported for the rhesus, rabbit, and humans [108]. Of particular interest is documentation of a widened mediastinum, observable on x-ray—a clinical sign that is frequently observed in human cases of anthrax [3,108,122]. In addition, the African green monkey model can be manipulated to evaluate therapeutic countermeasures, as a biomarker for infection has been identified [98]. The authors showed that systemic PA could be diagnosed with a fast electrochemiluminescence (ECL) assay and that it correlated with the gold standard assay for bacteremia (plate culture) which requires 24 h. The use of the ECL assay allowed the African green monkey to be used as a therapeutic model for medical countermeasures [123].

Research in the cynomolgus monkey model has progressed from basic model development to a therapeutic model to evaluate the efficacy of new medical countermeasures [97,124,125]. In one published study, cynomolgus monkeys were exposed to approximately 200 aerosol LD_{50}s and monitored for signs of anthrax infection. The trigger-to-treat was antigenemia, monitored by the ECL assay [97]. There was also a strong correlation with bacteremia seen in this model.

In addition, the cynomolgus monkey model has also been established for use as a prophylactic model to evaluate medical countermeasures. In another study, drug was administered 2 days before an aerosol challenge of 100 aerosol LD_{50}s, resulting in survival after 28 days of 70–90%, depending on the dose of drug [97]. Interestingly, when survivors were rechallenged a year later, there was a 100% survival rate, indicating that the immune system had generated a memory response.

In 2008, Lever et al. [109] published a study introducing a new nonhuman primate model for inhalation anthrax—the common marmoset (*C. jacchus*). It was reported that not only was the marmoset susceptible to *B. anthracis*, it exhibited pathology that was similar to that for the rhesus, cynomolgus monkey, and humans. The marmoset has an LD_{50} (1.5×10^3 CFU) similar to that of the cynomolgus monkey (5.5×10^4 CFU) and similar time-to-death after aerosol challenge (2–6.5 days). The marmoset is a smaller nonhuman primate and may present an advantage in the ease of housing and handling.

Because anthrax animal model research is now shifting from the characterization of natural history to that of useful manipulations for the evaluation of medical countermeasures, there has been interest in utilizing traditionally "non-biodefense" models. Stearns-Kurosawa et al. [126] have adapted the sepsis model (using baboons) to evaluate the effects of toxin by infusing bacteria to simulate a more controlled bacteremic phase of anthrax, over that of the inhalation models. The authors concede that although the model is enlightening regarding possible implications of toxemia, it does not mimic the kinetics of the disease and they are refining the model to address this issue [126].

Therefore, the nonhuman primate is a reliable model for pathogenesis, therapeutic, or vaccine efficacy studies. It is predictive of human disease, particularly inhalation anthrax. Although the most historical work has still been completed in the rhesus macaque, it appears that the cynomolgus macaque, African green monkey, and marmoset are promising alternatives.

6.3 SUMMARY

When selecting a small animal model for anthrax, the mouse would preferably be used only for screening studies and not for efficacy determination because it is not predictive of human disease. The guinea pig is an acceptable efficacy model, but should be used only after carefully considered study designs. For example, the guinea pig could be valuable in vaccine efficacy studies, but the adjuvant used may affect survival results. Therefore, the study endpoints would determine whether either of these animals would be appropriate.

Perhaps the best smaller animal model for anthrax is the rabbit. Although disease progression is rapid, pathogenesis is similar to that seen in nonhuman primates and in humans. Vaccine efficacy studies in rabbits are also predictive of the outcome in nonhuman primates. As with the guinea pig, careful consideration should be given to the design of a vaccine efficacy study, regarding the adjuvant used and an appropriate study endpoint. This model may be a rigorous test for anthrax therapeutics due to the fact that the disease does progress so rapidly.

The nonhuman primate is still considered the gold standard and best compared to humans. Although the rhesus has been the best characterized, for the longest period of time, the natural history of anthrax in the African green monkey, cynomolgus macaque, and marmoset has been characterized and the body of work in these species is expanding.

REFERENCES

1. Hambleton, P., J.A. Carman, and J. Melling, Anthrax: The disease in relation to vaccines. *Vaccine*, 1984, **2**(2): 125–132.
2. Jager, H.G., H.H. Booker, and O.J. Hubschle, Anthrax in cheetahs (*Acinonyx jubatus*) in Namibia. *J Wildl Dis*, 1990, **26**(3): 423–424.
3. Mayer, T.A. et al., Clinical presentation of inhalational anthrax following bioterrorism exposure: Report of 2 surviving patients. *JAMA*, 2001, **286**(20): 2549–2553.
4. Jernigan, D.B. et al., Investigation of bioterrorism-related anthrax, United States, 2001: Epidemiologic findings. *Emerg Infect Dis*, 2002, **8**(10): 1019–1028.
5. Albrink, W.S., Pathogenesis of inhalation anthrax. *Bacteriol Rev*, 1961, **25**: 268–273.
6. Brachman, P.S., Inhalation anthrax. *Ann N Y Acad Sci*, 1980, **353**: 83–93.
7. Friedlander, A.M., Anthrax: Clinical features, pathogenesis, and potential biological warfare threat. *Curr Clin Top Infect Dis*, 2000, **20**: 335–349.
8. Hail, A.S. et al., Comparison of noninvasive sampling sites for early detection of *Bacillus anthracis* spores from rhesus monkeys after aerosol exposure. *Mil Med*, 1999, **164**(12): 833–837.
9. Lew, D., *Bacillus anthracis* (anthrax), in *Principles and Practices of Infectious Disease*, G. Mandell, J. Bennett, and D. R., editors. New York: Churchill Livingstone Inc., 1995, pp. 1885–1889.
10. Tekin, A., N. Bulut, and T. Unal, Acute abdomen due to anthrax. *Br J Surg*, 1997, **84**(6): 813.

11. Inglesby, T.V. et al., Anthrax as a biological weapon: Medical and public health management. Working Group on Civilian Biodefense. *JAMA*, 1999, **281**(18): 1735–1745.
12. Friedlander, A.M., Anthrax, in *Textbook of Military Medicine: Medical Aspects of Chemical and Biological Warfare*. Washington, DC: Office of the Surgeon General at TMM Publications, 1997, p. 467.
13. Zilinskas, R.A., Iraq's biological weapons. The past as future? *JAMA*, 1997, **278**(5): 418–424.
14. Doolan, D.L. et al., The US capitol bioterrorism anthrax exposures: Clinical epidemiological and immunological characteristics. *J Infect Dis*, 2007, **195**(2): 174–184.
15. Becker, C., 20/20 Hindsight. Months after anthrax claimed the lives of several Americans, hospitals review their reaction to the event—and plan for future crises. *Mod Healthc*, 2002, **32**(8): 8–9, 12.
16. Doganay, M. and G. Metan, Human anthrax in Turkey from 1990 to 2007. *Vector Borne Zoonotic Dis*, 2009, **9**(2): 131–140.
17. Beaumont, G., Anthrax in a Scottish intravenous drug user. *J Forensic Leg Med*, 2010, **17**(8): 443–445.
18. Abalakin, V.A. and B.L. Cherkasskii, Use of inbred mice as a model for the indication and differentiation of *Bacillus anthracis* strains (in Russian). *Zh Mikrobiol Epidemiol Immunobiol*, 1978(2): 146–147.
19. Welkos, S.L., T.J. Keener, and P.H. Gibbs, Differences in susceptibility of inbred mice to *Bacillus anthracis*. *Infect Immun*, 1986, **51**(3): 795–800.
20. Welkos, S.L. and A.M. Friedlander, Pathogenesis and genetic control of resistance to the Sterne strain of *Bacillus anthracis*. *Microb Pathog*, 1988, **4**(1): 53–69.
21. Shibaya, M., M. Kubomichi, and T. Watanabe, The genetic basis of host resistance to *Bacillus anthracis* in inbred mice. *Vet Microbiol*, 1991, **26**(3): 309–312.
22. Welkos, S. et al., The role of antibodies to *Bacillus anthracis* and anthrax toxin components in inhibiting the early stages of infection by anthrax spores. *Microbiology*, 2001, **147**(Pt 6): 1677–1685.
23. Welkos, S.L., N.J. Vietri, and P.H. Gibbs, Non-toxigenic derivatives of the Ames strain of *Bacillus anthracis* are fully virulent for mice: role of plasmid pX02 and chromosome in strain-dependent virulence. *Microb Pathog*, 1993, **14**(5): 381–388.
24. Lyons, C.R. et al., Murine model of pulmonary anthrax: kinetics of dissemination, histopathology, and mouse strain susceptibility. *Infect Immun*, 2004, **72**(8): 4801–4809.
25. Heine, H.S. et al., Determination of antibiotic efficacy against *Bacillus anthracis* in a mouse aerosol challenge model. *Antimicrob Agents Chemother*, 2007, **51**(4): 1373–1379.
26. Glomski, I.J. et al., Inhaled non-capsulated *Bacillus anthracis* in A/J mice: Nasopharynx and alveolar space as dual portals of entry, delayed dissemination, and specific organ targeting. *Microbes Infect*, 2008, **10**(12–13): 1398–1404.
27. Glomski, I.J. et al., Noncapsulated toxinogenic *Bacillus anthracis* presents a specific growth and dissemination pattern in naive and protective antigen-immune mice. *Infect Immun*, 2007, **75**(10): 4754–4761.
28. Loving, C.L. et al., Murine aerosol challenge model of anthrax. *Infect Immun*, 2007, **75**(6): 2689–2698.
29. Duong, S., L. Chiaraviglio, and J.E. Kirby, Histopathology in a murine model of anthrax. *Int J Exp Pathol*, 2006, **87**(2): 131–137.
30. Flick-Smith, H.C. et al., Mouse model characterisation for anthrax vaccine development: Comparison of one inbred and one outbred mouse strain. *Microb Pathog*, 2005, **38**(1): 33–40.
31. Chabot, D.J. et al., Anthrax capsule vaccine protects against experimental infection. *Vaccine*, 2004, **23**(1): 43–47.
32. Heine, H.S. et al., Efficacy of oritavancin in a murine model of *Bacillus anthracis* spore inhalation anthrax. *Antimicrob Agents Chemother*, 2008, **52**(9): 3350–3357.

33. Drysdale, M. et al., Capsule synthesis by *Bacillus anthracis* is required for dissemination in murine inhalation anthrax. *EMBO J*, 2005, **24**(1): 221–227.
34. Pickering, A.K. et al., Cytokine response to infection with Bacillus anthracis spores. *Infect Immun*, 2004, **72**(11): 6382–6389.
35. Pickering, A.K. and T.J. Merkel, Macrophages release tumor necrosis factor alpha and interleukin-12 in response to intracellular *Bacillus anthracis* spores. *Infect Immun*, 2004, **72**(5): 3069–3072.
36. Steward, J. et al., Post-exposure prophylaxis of systemic anthrax in mice and treatment with fluoroquinolones. *J Antimicrob Chemother*, 2004, **54**(1): 95–99.
37. Chand, H.S. et al., Discriminating virulence mechanisms among *Bacillus anthracis* strains by using a murine subcutaneous infection model. *Infect Immun*, 2009, **77**(1): 429–435.
38. Cote, C.K., N. Van Rooijen, and S.L. Welkos, Roles of macrophages and neutrophils in the early host response to *Bacillus anthracis* spores in a mouse model of infection. *Infect Immun*, 2006, **74**(1): 469–480.
39. Drysdale, M. et al., Murine innate immune response to virulent toxigenic and nontoxigenic *Bacillus anthracis* strains. *Infect Immun*, 2007, **75**(4): 1757–1764.
40. Lincoln, R.E., *Advances in Veterinary Science*. New York: Academic Press, 1964.
41. Jones, W.I., Jr. et al., *In vivo* growth and distribution of anthrax bacilli in resistant, susceptible, and immunized hosts. *J Bacteriol*, 1967, **94**(3): 600–608.
42. Beall, F.A., M.J. Taylor, and C.B. Thorne, Rapid lethal effect in rats of a third component found upon fractionating the toxin of *Bacillus anthracis. J Bacteriol*, 1962, **83**: 1274–1280.
43. Beall, F.A. and F.G. Dalldorf, The pathogenesis of the lethal effect of anthrax toxin in the rat. *J Infect Dis*, 1966, **116**(3): 377–389.
44. Haines, B.W., F. Klein, and R.E. Lincoln, Quantitative assay for crude anthrax toxins. *J Bacteriol*, 1965, **89**: 74–83.
45. Fish, D.C. et al., Pathophysiological changes in the rat associated with anthrax toxin. *J Infect Dis*, 1968, **118**(1): 114–124.
46. Mourez, M. et al., Mapping dominant-negative mutations of anthrax protective antigen by scanning mutagenesis. *Proc Natl Acad Sci USA*, 2003, **100**(24): 13803–13808.
47. Sarac, M.S. et al., Protection against anthrax toxemia by hexa-D-arginine *in vitro* and *in vivo. Infect Immun*, 2004, **72**(1): 602–605.
48. Sawada-Hirai, R. et al., Human anti-anthrax protective antigen neutralizing monoclonal antibodies derived from donors vaccinated with anthrax vaccine adsorbed. *J Immune Based Ther Vaccines*, 2004, **2**(1): 5.
49. Webster, J.I., M. Moayeri, and E.M. Sternberg, Novel repression of the glucocorticoid receptor by anthrax lethal toxin. *Ann N Y Acad Sci*, 2004, **1024**: 9–23.
50. Ivins, B.E. et al., Immunization studies with attenuated strains of *Bacillus anthracis. Infect Immun*, 1986, **52**(2): 454–458.
51. Turnbull, P.C. et al., Development of antibodies to protective antigen and lethal factor components of anthrax toxin in humans and guinea pigs and their relevance to protective immunity. *Infect Immun*, 1986, **52**(2): 356–363.
52. Ivins, B.E. et al., Transposon Tn916 mutagenesis in *Bacillus anthracis. Infect Immun*, 1988, **56**(1): 176–181.
53. Ivins, B.E. et al., Immunization against anthrax with aromatic compound-dependent (Aro-) mutants of *Bacillus anthracis* and with recombinant strains of *Bacillus subtilis* that produce anthrax protective antigen. *Infect Immun*, 1990, **58**(2): 303–308.
54. Ivins, B.E. et al., Immunization against anthrax with *Bacillus anthracis* protective antigen combined with adjuvants. *Infect Immun*, 1992, **60**(2): 662–668.
55. Little, S.F. and G.B. Knudson, Comparative efficacy of *Bacillus anthracis* live spore vaccine and protective antigen vaccine against anthrax in the guinea pig. *Infect Immun*, 1986, **52**(2): 509–512.

56. Ross, J.M., On the histopathology of experimental anthrax in the guinea-pig. *Br J Exp Pathol*, 1955, **36**(3): 336–339.
57. Ross, J.M., The pathogenesis of anthrax following the administration of spores by the respiratory route. *J Pathol Bacteriol*, 1957, **73**: 485–495.
58. Puziss, M. and G.G. Wright, Studies on immunity in anthrax. X. Gel-adsorbed protective antigen for immunization of man. *J Bacteriol*, 1963, **85**: 230–236.
59. Ivins, B.E. and S.L. Welkos, Cloning and expression of the *Bacillus anthracis* protective antigen gene in *Bacillus subtilis. Infect Immun*, 1986, **54**(2): 537–542.
60. Ivins, B.E. and S.L. Welkos, Recent advances in the development of an improved, human anthrax vaccine. *Eur J Epidemiol*, 1988, **4**(1): 12–19.
61. Lincoln, R.E. et al., Value of field data for extrapolation in anthrax. *Fed Proc*, 1967, **26**(5): 1558–1562.
62. Benjamin, E. et al., LD50 of aerosolized *B. anthracis* and deposition of spores in the Hartley guinea pig, in *ASM Biodefense Research Meeting*, 2005, Baltimore, MD, USA. p. Abstract # 96(R).
63. Ivins, B. et al., Experimental anthrax vaccines: efficacy of adjuvants combined with protective antigen against an aerosol *Bacillus anthracis* spore challenge in guinea pigs. *Vaccine*, 1995, **13**(18): 1779–1784.
64. Altboum, Z. et al., Postexposure prophylaxis against anthrax: evaluation of various treatment regimens in intranasally infected guinea pigs. *Infect Immun*, 2002, **70**(11): 6231–6241.
65. Little, S.F. et al., Development of an in vitro-based potency assay for anthrax vaccine. *Vaccine*, 2004, **22**(21–22): 2843–2852.
66. Kobiler, D. et al., Efficiency of protection of guinea pigs against infection with *Bacillus anthracis* spores by passive immunization. *Infect Immun*, 2002, **70**(2): 544–560.
67. Barnes, J.M., The development of anthrax following the administration of spore by inhalation. *Br J Exp Pathol*, 1947, **28**: 385–394.
68. Dalldorf, F.G. and F.A. Beall, Capillary thrombosis as a cause of death in experimental anthrax. *Arch Pathol*, 1967, **83**(2): 154–161.
69. Fasanella, A. et al., Detection of anthrax vaccine virulence factors by polymerase chain reaction. *Vaccine*, 2001, **19**(30): 4214–4218.
70. Kolesnik, V.S. et al., Experimental anthrax infection in laboratory animals with differing species susceptibility to the causative agent (in Russian). *Zh Mikrobiol Epidemiol Immunobiol*, 1990(6): 3–7.
71. Terril, L. and D. Clemons, *The Laboratory Guinea Pig*. Boca Raton: CRC Press, 1998.
72. Fellows, P.F. et al., Efficacy of a human anthrax vaccine in guinea pigs, rabbits, and rhesus macaques against challenge by *Bacillus anthracis* isolates of diverse geographical origin. *Vaccine*, 2001, **19**(23–24): 3241–3247.
73. Gauthier, Y.P. et al., Efficacy of a vaccine based on protective antigen and killed spores against experimental inhalational anthrax. *Infect Immun*, 2009, **77**(3): 1197–1207.
74. Little, S.F. et al., Passive protection by polyclonal antibodies against *Bacillus anthracis* infection in guinea pigs. *Infect Immun*, 1997, **65**(12): 5171–5175.
75. Mabry, R. et al., Passive protection against anthrax by using a high-affinity antitoxin antibody fragment lacking an Fc region. *Infect Immun*, 2005, **73**(12): 8362–8368.
76. Reuveny, S. et al., Search for correlates of protective immunity conferred by anthrax vaccine. *Infect Immun*, 2001, **69**(5): 2888–2893.
77. Hambleton, P. and P.C. Turnbull, Anthrax vaccine development: A continuing story. *Adv Biotechnol Processes*, 1990, **13**: 105–122.
78. Marcus, H. et al., Contribution of immunological memory to protective immunity conferred by a *Bacillus anthracis* protective antigen-based vaccine. *Infect Immun*, 2004, **72**(6): 3471–3477.

79. Pitt, M.L. et al., In vitro correlate of immunity in an animal model of inhalational anthrax. *J Appl Microbiol*, 1999, **87**(2): 304.
80. Ionin, B., Development of a guinea pig model for post-exposure prophylaxis of inhalation anthrax, in *Biodefense Vaccines & Therapeutics Conference—Animal Models Forum*, June 15, 2010, Washington, D.C.
81. Richter, S. et al., Capsule anchoring in *Bacillus anthracis* occurs by a transpeptidation reaction that is inhibited by capsidin. *Mol Microbiol*, 2009, **71**(2): 404–420.
82. Peterson, J.W. et al., Human monoclonal anti-protective antigen antibody completely protects rabbits and is synergistic with ciprofloxacin in protecting mice and guinea pigs against inhalation anthrax. *Infect Immun*, 2006, **74**(2): 1016–1024.
83. Zaucha, G.M. et al., The pathology of experimental anthrax in rabbits exposed by inhalation and subcutaneous inoculation. *Arch Pathol Lab Med*, 1998, **122**(11): 982–992.
84. Nordberg, B.K., C.G. Schmiterlow, and H.J. Hansen, Pathophysiological investigations into the terminal course of experimental anthrax in the rabbit. *Acta Pathol Microbiol Scand*, 1961, **53**: 295–318.
85. Klein, F. et al., Pathophysiology of anthrax. *J Infect Dis*, 1966, **116**(2): 123–138.
86. Lebowich, R.J., B.G. McKillip, and J.R. Convoy, Cutaneous anthrax—A pathologic study with clinical correlation. *Am J Clin Pathol*, 1943, **13**: 505–515.
87. Aleksandrov, N.I. et al., Further experimental study of the efficacy of chemical anthrax vaccine (in Russian). *Zh Mikrobiol Epidemiol Immunobiol*, 1964, **41**: 45–50.
88. Pitt, M.L. et al., In vitro correlate of immunity in a rabbit model of inhalational anthrax. *Vaccine*, 2001, **19**(32): 4768–4773.
89. Galloway, D.R. and L. Baillie, DNA vaccines against anthrax. *Expert Opin Biol Ther*, 2004, **4**(10): 1661–1667.
90. Little, S.F. et al., Duration of protection of rabbits after vaccination with *Bacillus anthracis* recombinant protective antigen vaccine. *Vaccine*, 2006, **24**(14): 2530–2536.
91. Mikszta, J.A. et al., Protective immunization against inhalational anthrax: A comparison of minimally invasive delivery platforms. *J Infect Dis*, 2005, **191**(2): 278–288.
92. Little, S.F. et al., Defining a serological correlate of protection in rabbits for a recombinant anthrax vaccine. *Vaccine*, 2004, **22**(3–4): 422–430.
93. Weiss, S. et al., Immunological correlates for protection against intranasal challenge of *Bacillus anthracis* spores conferred by a protective antigen-based vaccine in rabbits. *Infect Immun*, 2006, **74**(1): 394–398.
94. Nuzum, E. National Institute for Allergic and Infectious Diseases (NIAID) studies, in *FDA Vaccines and Related Biological Products Advisory Committee*, Maryland, USA, 2010.
95. Mohamed, N. et al., A high-affinity monoclonal antibody to anthrax protective antigen passively protects rabbits before and after aerosolized *Bacillus anthracis* spore challenge. *Infect Immun*, 2005, **73**(2): 795–802.
96. Thomas, L.J. et al., Co-administration of a CpG adjuvant (VaxImmune, CPG 7909) with CETP vaccines increased immunogenicity in rabbits and mice. *Hum Vaccin*, 2009, **5**(2): 79–84.
97. Migone, T.S. et al., Raxibacumab for the treatment of inhalational anthrax. *N Engl J Med*, 2009, **361**(2): 135–144.
98. Rossi, C.A. et al., Identification of a surrogate marker for infection in the African green monkey model of inhalation anthrax. *Infect Immun*, 2008, **76**(12): 5790–5801.
99. Yee, S. et al., Aerosolized *Bacillus anthracis* infection in New Zealand white rabbits: Natural history and intravenous levofloxacin treatment. *Comp Med*, 2010, **60**(6): 461–468.
100. Albrink, W.S. and R.J. Goodlow, Experimental inhalation anthrax in the chimpanzee. *Am J Pathol*, 1959, **35**: 1055–1065.
101. Klein, F. et al., Anthrax toxin: Causative agent in the death of rhesus monkeys. *Science*, 1962, **138**: 1331–1333.

102. Kelly, D.J. et al., Serum concentrations of penicillin, doxycycline, and ciprofloxacin during prolonged therapy in rhesus monkeys. *J Infect Dis*, 1992, **166**(5): 1184–1187.
103. Fritz, D.L. et al., Pathology of experimental inhalation anthrax in the rhesus monkey. *Lab Invest*, 1995, **73**(5): 691–702.
104. Ivins, B.E. et al., Comparative efficacy of experimental anthrax vaccine candidates against inhalation anthrax in rhesus macaques. *Vaccine*, 1998, **16**(11–12): 1141–1148.
105. Kihira, T., J. Sato, and T. Shibata, Pharmacokinetic–pharmacodynamic analysis of fluoroquinolones against *Bacillus anthracis*. *J Infect Chemother*, 2004, **10**(2): 97–100.
106. Klinman, D.M. et al., CpG oligonucleotides improve the protective immune response induced by the anthrax vaccination of rhesus macaques. *Vaccine*, 2004, **22**(21–22): 2881–2886.
107. Vasconcelos, D. et al., Pathology of inhalation anthrax in cynomolgus monkeys (*Macaca fascicularis*). *Lab Invest*, 2003, **83**(8): 1201–1209.
108. Twenhafel, N.A., E. Leffel, and M.L. Pitt, Pathology of inhalational anthrax infection in the African green monkey. *Vet Pathol*, 2007, **44**(5): 716–721.
109. Lever, M.S. et al., Experimental respiratory anthrax infection in the common marmoset (*Callithrix jacchus*). *Int J Exp Pathol*, 2008, **89**(3): 171–179.
110. Haas, C.N., On the risk of mortality to primates exposed to anthrax spores. *Risk Anal*, 2002, **22**(2): 189–193.
111. Borio, L. et al., Death due to bioterrorism-related inhalational anthrax: Report of 2 patients. *JAMA*, 2001, **286**(20): 2554–2559.
112. Grinberg, L.M. et al., Quantitative pathology of inhalational anthrax I: Quantitative microscopic findings. *Mod Pathol*, 2001, **14**(5): 482–495.
113. Gleiser, C.A., Pathology of anthrax infection in animal hosts. *Fed Proc*, 1967, **26**(5): 1518–1521.
114. Guarner, J. et al., Pathology and pathogenesis of bioterrorism-related inhalational anthrax. *Am J Pathol*, 2003, **163**(2): 701–709.
115. Berdjis, C.C. et al., Pathogenesis of respiratory anthrax in *Macaca mulatta*. *Br J Exp Pathol*, 1962, **43**: 515–524.
116. Gleiser, C.A. et al., Pathology of experimental respiratory anthrax in *Macaca mulatta*. *Br J Exp Pathol*, 1963, **44**: 416–426.
117. Ivins, B.E., P.F. Fellows, and G.O. Nelson, Efficacy of a standard human anthrax vaccine against *Bacillus anthracis* spore challenge in guinea-pigs. *Vaccine*, 1994, **12**(10): 872–874.
118. Blasky, A.J. et al., Rapid high-resolution MHC class I genotyping of Chinese rhesus macaques by capillary reference strand-mediated conformational analysis. *Immunogenetics*, 2008, **60**(10): 575–584.
119. Karl, J.A. et al., Identification of MHC class I sequences in Chinese-origin rhesus macaques. *Immunogenetics*, 2008, **60**(1): 37–46.
120. Vietri, N.J. et al., Short-course postexposure antibiotic prophylaxis combined with vaccination protects against experimental inhalational anthrax. *Proc Natl Acad Sci USA*, 2006, **103**(20): 7813–7816.
121. Vietri, N.J. et al., A short course of antibiotic treatment is effective in preventing death from experimental inhalational anthrax after discontinuing antibiotics. *J Infect Dis*, 2009, **199**(3): 336–341.
122. Gill, J.R. and J. Melinek, Inhalation of anthrax: Gross autopsy findings. *Arch Pathol Lab Med*, 2002, **126**(8): 993–994.
123. Dyer, D. et al., Development of a therapeutic African green monkey model for inhalational anthrax to demonstrate added value of adjunct therapies, in *BacillusACT 2009: The International Bacillus anthracis, B. cereus, and B. thuringiensis Conference*, New Mexico, USA, 2009.

124. Vitale, L. et al., Prophylaxis and therapy of inhalational anthrax by a novel monoclonal antibody to protective antigen that mimics vaccine-induced immunity. *Infect Immun*, 2006, **74**(10): 5840–5847.
125. Whipple, E.C. et al., Low doses of antigen coupled to anti-CR2 mAbs induce rapid and enduring IgG immune responses in mice and in cynomolgus monkeys. *Mol Immunol*, 2007, **44**(4): 377–388.
126. Stearns-Kurosawa, D.J. et al., Sepsis and pathophysiology of anthrax in a nonhuman primate model. *Am J Pathol*, 2006, **169**(2): 433–444.

7 Glanders

David L. Fritz and David M. Waag

CONTENTS

7.1 BACKGROUND

Glanders is one of two forms of clinical disease caused by *Burkholderia mallei* in a host animal. It is a zoonotic disease primarily of solipeds: horses, donkeys, and mules [1]. However, nearly all mammals are susceptible to the *B. mallei*, which is a Gram-negative aerobic bacillus and is likely an obligate mammalian pathogen [2–4]. Glanders is endemic in the Far and Middle East, northern Africa, eastern Mediterranean, and southeastern Europe [5,6]. Although historically the incidence of glanders in human populations has been low, the possibility of its emerging as a human pathogen cannot be discounted. The need for preemptive study of this disease is underscored by past failures in protective vaccine development, resistance of the organism to antibiotic therapy, and by the protean manifestations of the disease within the host.

At one time, *B. mallei* had worldwide occurrence, but with modernization of transportation and the implementation of strict controls, the incidence of glanders has decreased significantly during the last century. By 1939 glanders had been eradicated from most of Western Europe, Canada, and the United States, and it was eradicated from the remainder of Western Europe by 1960. As a side note, some horses exported from Mexico into the United States are still found to be reactors to *B. mallei* by the complement-fixation test, although investigators in Mexico can find no active glanders in that country. The reason has yet to be explained [7]. At present, sporadic infections are still being reported from some countries in the Far East where the disease in animals is seasonal, and from southeastern Europe, North Africa, the eastern Mediterranean (Turkey), and the Middle East (Iran, Syria, and Iraq) where enzootic areas are believed to exist [5,8]. In these endemic areas infected animals serve as reservoirs for human disease [9]. At present, there are no accurate epidemiological data from any of these countries because of the difficulties in the identification of glanders. Serological data are misinterpreted because of the cross-reactivity with antibodies to closely related *Burkholderia pseudomallei*, the etiologic agent of melioidosis, which is also found in similar environments [10].

Glanders is one of the oldest diseases known, described by the Greeks in 450–425 BC and the Romans in AD 400–500 [7,10–13]. The disease was probably carried to Europe from Central Asia by the Mongols or Tartars [11,14]. In the seventeenth century, glanders was recognized as a contagious disease, but the etiologic agent was not known until 1882, when the bacillus was isolated and identified by Loeffler and Schutz in Berlin, Germany. In the nineteenth century, transmission of glanders among horses via water troughs was identified, and it was shown that humans in contact with diseased horses could also develop infection. The modern species name for the agent is derived from the Greek *malis* and *melis*, terms used for diseases of Equidae, and from the Latin *malleus*, meaning "severe malignant disease" [3,10,11,13,14].

Historically, glanders was a serious disease, as horses were the mainstay of civilian and military transport [11]. Because the disease thrived in war times, glanders was the scourge of military horses, and postwar transfer of Army horses to civilian services contributed to the spread of glanders far and wide. In the United States, the American Civil War caused the disease to spread over the eastern coast and to flourish in the cities with large concentrations of horses. During the early 1900s, diagnostic tests were discovered, and glanders was rapidly brought under control. The advent of the automobile diminished the horse population and helped to stamp out the disease in the United States [7,15].

Although natural glanders infectious occur primarily in horses, donkeys, and mules [16], natural disease is occasionally seen in goats, sheep, dogs, and cats [17,18], and it has been rarely reported in carnivores that have eaten infected horse meat [17,18]. Cattle, pigs [18], and birds [2] are very resistant to natural disease. Experimentally, most domesticated animals can be infected, with the exception of cattle, pigs, and rats [2]. Guinea pigs and hamsters are the laboratory rodents most susceptible to glanders infection [2]. Miller et al. [1] found that, although there appears to be a variance in susceptibility among individual guinea pigs, hamsters are more uniformly susceptible to infection with *B. mallei*. Vyshelesskii [19] stated his studies showed that, of the animal species tested, cats were most susceptible, followed by guinea pigs, rabbits, field mice, and moles. Laboratory mice are only slightly susceptible to infection unless the organisms are given in very high doses [3].

The human form of glanders is primarily an occupational disease of veterinarians, horse caretakers, and slaughterhouse employees [7,14,20]. Infections occur through mucous membranes (eyes, mouth, nose), respiratory tract, and abraded or cut skin [7,9,15,20,21]. Sources of infections can be contaminated animal nasal discharges, secretions from pustules [4,9,13], or droplets from an infected animal coughing or snorting on a person's face [7,10]. Infections of aerosol origin are suspected to occur in natural settings because laboratory infections with aerosolized agents have been reported among laboratory workers [5,10,15,22]. The disease is not generally contagious among humans, although a few cases of human-to-human transmission have been reported [5,7,9,10]. The potential threat of aerosolized

B. mallei is best illustrated by the number of laboratory infections and personnel exposed to the disease during animal or human autopsies [4,10,20].

The number of laboratory infections is rather surprising in view of the comparatively low human infectivity when glanders was a common disease among horses [7,20]. At present, human susceptibility to glanders has not been determined. However, it has been stated that there are few organisms as dangerous to work with as the glanders bacillus [10,17]. Although human susceptibility to *B. mallei* has not been studied in detail, the organism has proven to be highly infectious in laboratory settings [4,10,17]. In spite of high laboratory infectivity, however, natural infections of humans have been sporadic and are usually subclinical. In the 2000 years since glanders was first described, no documented epidemics have occurred in humans [5,12,13]. The documented sporadic cases have been primarily occupational, occurring mostly in veterinarians, horse caretakers, and slaughterhouse employees [6,7,14,17]. However, autopsy findings of glanders-associated nodules in large numbers of humans with equine contact suggest that the prevalence of mild subclinical infections may be higher than previously suspected [10,11]. This is supported by studies of glanders infections in laboratory workers, demonstrating the high infectivity of the glanders bacillus [10,17].

7.2 ETIOLOGIC AGENT

Burkholderia mallei, the glanders bacillus, is a small rod ranging from 0.3 to 0.7 by 0.5 to 5 μm [9,20]. Bacilli are nonmotile and do not form spores; they are encapsulated and Gram-negative [5,9,15,23,24]. Glanders bacilli are reputedly difficult to find in tissue sections because of their low numbers and because they do not stain well with many conventional stains [4,22,25]. In the experience of some, bacilli stain best with Giemsa stain but are often obscured by heavily stained background material; likewise, in immunolabeled sections, the abundance of bacterial antigen obscures bacilli [26]. *Burkholderia mallei* grows well aerobically but slowly in ordinary laboratory media [9,15]. Primary isolation requires 48 h at 37°C to develop colonies of 0.5–1.0 mm in diameter [9]. The organism tolerates moderate variations in pH and grows equally well on mildly acid or alkaline substrates [27]. The glanders bacillus is only slightly resistant to desiccation, heat, and chemicals. The organism is killed by short-time desiccation and sunlight in 24 h [10,27], temperatures of 55°C for 10 min, and disinfectants such as hypochlorites and iodine [9,10]. Although *B. mallei* is an obligate animal parasite, it can survive under favorable conditions up to 3 days outside the host [15]. Laboratory experiments have shown that this organism can survive for 1 month in tap water, and that viability/virulence of cultures are well preserved by lyophilization for 3–6 months [7].

7.3 ANIMAL MODELS

Various animal models of human *B. mallei* infection have been previously reported, including monkeys [1,28–30], guinea pigs [1], hamsters [1,26,31,32], and mice [1,24,32–35]. Similar to humans, nonhuman primates are a nonsusceptible host for glanders, but there is little information in the Western scientific literature reporting work with glanders in monkeys. In one study, six rhesus monkeys were given graded doses of a virulent strain; the monkey receiving the largest dose developed a cutaneous abscess that resolved completely after 3 weeks [1]. In the Russian literature, some investigators have worked with baboons [30], but details of the disease in monkeys from this and other works are vague [28,29]. Of the laboratory rodents tested, guinea pigs and hamsters are the most susceptible to infection with *B. mallei*. However, guinea pigs vary from individual to individual in their susceptibility to infection [1], whereas Syrian hamsters are uniformly infected; for this reason, they have been used in more recent studies than have guinea pigs [26,32]. Although laboratory mice vary from strain to strain somewhat in their susceptibility to glanders infection, they are all moderately resistant to infection, as are humans. When high doses of the organisms are given, mice become uniformly infected. Because of the widespread use of BALB/c mice in scientific research, the availability of knockouts, and so forth for the BALB/c, these mice have been used preferentially in several recent studies [24,35].

7.4 CLINICAL DISEASE

Two major presentations of the disease caused by *B. mallei* occur: the nasal-pulmonary form (glanders) and the cutaneous form (farcy). These two forms may be present simultaneously and are usually accompanied by systemic disease [11,14,21]. The route of infection, dose, and virulence of *B. mallei* determine the severity of the disease. Clinically evident disease may be acute or chronic, but subclinical and even latent infections may occur [13,14,17]. Severity of the disease may also depend on the susceptibility of the host. With equids, the chronic form is typically seen in horses, characterized by a more gradual onset and fewer systemic signs. Conversely, the acute form is more commonly observed in donkeys and mules, with death occurring in 3–4 weeks [7,20,27]. Humans are most often afflicted with the acute form of the disease, which is characterized by a rapid onset of pneumonia, bacteremia, pustules, and death occurring within days [9]. In contrast, the chronic form of the disease is characterized by intermittent recrudescence, and milder signs and symptoms, and may last up to 25 years. The course of the disease in humans is intensely painful and is invariably fatal if not treated effectively [2,4,7,10,12,13,17].

Humans cases may exhibit one of several manifestations: acute fulminating septicemia, with a sudden onset, chill, pyrexia and a degree of prostration out of all proportion to the clinical signs; acute pulmonary infection with nasal mucopurulent discharge, bronchopneumonia and lobar pneumonia with or without bacteremia; acute suppurative infection with generalized pyemia and multiple cutaneous eruptions; chronic suppurative syndrome with remission and exacerbation for up to 15 years; extended latent (dormant, carrier state) infection with prolonged incubation period and crudescence of the characteristic clinical picture; and occult (nonclinical) glanders, expressed only by encapsulated nodules in internal organs and usually discovered during autopsies [4,7,10,20,36].

Disease can result from an extremely low infectious dose inoculated by either aerosol, oral, or parenteral routes. The incubation period is short, and the definitive diagnosis of glanders is confounded by nonspecific signs and symptoms. Although glanders is a serious, life-threatening zoonotic disease, relatively little is known about the pathogenesis, virulence factors, strain differences, and host immunopathological responses to infection. The acute form of glanders has a violent onset with the earliest sign being high fever, followed by chills and prostration.

7.5 GROSS LESIONS

In natural cases of glanders in equids, there are, typically, lesions of the nasal passages consisting of deep crateriform ulcers in the septum and nasal turbinates. Nasal lesions begin as submucosal nodules that soon rupture to form ulcers that exude a thick, sticky purulent exudate from both nostrils. Nasal involvement is often accompanied by swelling of the lymph nodes of the neck. These nodes tend to ulcerate or form draining sinuses. There is almost always lung involvement regardless of the route of infection. Lesions of the lungs are small granulomas with liquid (caseous) or calcified centers; these may eventually become expansive to yield a diffuse pneumonic process. Similarly, by the time infection has reached the lungs, granulomas may be present in the spleen, liver, and other internal organs. In nonclinical cases, lesions are predominantly pulmonary granulomas, and occasionally abscesses. Interestingly, the gastrointestinal tract is never involved, indicating it has a high degree of immunity from infection.

In experimental studies with laboratory rodents, the guinea pig, Syrian hamster, and white inbred mouse have most often been used. When given a lethal intraperitoneal dose of *B. mallei*, typical gross pathological findings in the hamster consist of splenomegaly with multiple splenic white foci and, in the later stages of the disease, scattered white foci in the lungs [26]. In mice given a lethal dose, gross pathological findings are similar to hamsters, except that lung involvement is extremely rare [24]. Following aerosol exposure to a lethal dose of glanders, gross pathological changes are similar to those following intraperitoneal injection, except that there are striking pneumonic changes in both species (D.L. Fritz, D. DeShazer, and D.M. Waag, unpublished data).

There are only scant autopsy data from humans who have succumbed to glanders infection. Inoculation of a cutaneous abrasion, scratch, or wound would result in painful nodules, erythematous swelling of the face or limbs, and lymphangitis. Infection of the nasal mucosa might result in a mucopurulent discharge, similar to what is found in horses. In the acute pulmonary form, pulmonary abscesses would be present early in the disease course [22], and later there would be lobar or bronchopneumonia. In the chronic form, pathological lesions again would be similar to those in horses. There would be multiple subcutaneous and intramuscular abscesses, usually on arms and legs, with associated enlargement of lymphatics and local lymph nodes [20]. In addition, there might be deep ulcerative lesions of the skin, osteomyelitis, and rarely, meningitis [13]. Nodules may form in the nasal mucosa, and these may rupture and ulcerate [20].

7.6 MICROSCOPIC LESIONS

Histologically, the principal lesions of acute glanders infection consist of an inflammatory cell infiltrate of equal numbers of macrophages and polymorphonuclear leukocytes (PMNs; pyogranulomatous inflammation) in multiple organs [24,26,37,38]. The results of several experimental glanders studies indicate that organs rich in reticuloendothelial tissue are particularly susceptible to localization of the glanders bacillus and genesis of lesions [23,24,26,38]. These organs include, but are not limited to, the spleen, lymph nodes, liver, and bone marrow.

In the susceptible host (hamster, guinea pig), the inflammatory cell infiltrate is organized into compact nodules (pyogranulomas), typically with a necrotic center. As is also typical in the susceptible host, there is widespread necrosis of both leukocytes and parenchymatous tissue within the lesions. Lesions secondary to vascular changes occur in the later stages of the disease process [26]. These included infarcts in the spleen and possibly bone, plus variably sized areas of ischemia in the liver and brain. Vascular thrombi consisted of fibrin, PMN, fewer macrophages, and glanders bacillus antigen (confirmed by immunohistochemistry). Some authors believe these septic thrombi are the precursors of expansive foci of pyogranulomatous inflammation [2,26].

Another histological feature in the susceptible host is the persistence of karyorrhectic debris in necrotic pyogranulomas [2,10,26,39]. In the older literature, this feature has been termed chromatotexis [22,39] or chromatin masses [37]. The presence of multinucleated giant cells in glanders nodules appears to depend on both host species and bacterial strain. Most authors [2,17,22,37,39,40] report their frequent occurrence in a variety of host species, whereas some report their absence [24,26,41]. Duval and White [37] and Howe [2] state that multinucleated giant cells are usually present in glanders lesions caused by infective strains of low virulence but are absent when more highly virulent bacteria are involved. M'Fadyean [39] and Howe [2] report that these cells are common in older lesions but may be absent early in the disease. In chronic lesions, inflammatory foci become heavily walled-off by collagen; the inflammatory cell infiltrate consists largely of macrophages and multinucleated giant cells; and the central zone is largely liquefied cellular debris, often with foci of calcification. Viable glanders bacilli are scant. These chronic lesions have many histological similarities with tubercles resulting from chronic infection with *Mycobacterium tuberculosis*.

Inflammation of the testes in male hamsters inoculated intraperitoneally, termed the Straus reaction [1], is seen in glanders, as well as several other pathogenic organisms. It is believed that this results from an extension of the infectious material from the peritoneal cavity to the testicles via the tunic vaginalis, which is typically patent in male rodents.

Also, in susceptible hosts such as the hamster, the organisms can become widely disseminated and can involve almost any organ in the body. In addition to organs mentioned above, extension to skin, nasal cavity, tooth pulp, brain, and female reproductive organs were also seen. Inflammation in the bone marrow of long bones, and even the marrow of the calvaria, was so severe as to cause pathological fractures and extension of the inflammation into surrounding tissues [26]. Infection of the nonsusceptible host, such as the BALB/c mouse, yields many histological similarities, as with the susceptible host. First, the character of the inflammatory cell infiltrate is similar in that it typically consists of equal numbers of macrophage and PMN, as in the hamster. These cells are rarely aggregated into a true nodule, or pyogranuloma, but rather, they are more loosely aggregated. Necrosis is only rarely seen in the mouse, unlike with the hamster, and thrombosis is only present if the disease process is prolonged. Hemorrhage is rarely present. In addition, there is a difference in distribution of glanders lesions.

Unlike the extension of infection to most organs in the intraperitoneally infected hamster, lesions in the mouse (following intraperitoneal inoculation) are confined to the spleen, liver, and various lymph nodes, and there are occasional inflammatory foci in the marrow of the femur. However, in the latter, these lesions are always localized and never extensive/destructive, as in the hamster [26]. Following aerosol exposure of the BALB/c mouse to a lethal dose of *B. mallei*, Lever et al. [35] described finding foci of inflammation and necrosis in the lungs as early as 24 h after exposure. In time, these foci coalesced to produce areas of extensive consolidation, and, in time, the character of the inflammatory cell infiltrate changed from acute (predominance of PMN) to chronic (predominance of mononuclear cells such as macrophages and lymphocytes). Foci of inflammation, with some accompanying hepatocellular degeneration and necrosis of hepatocytes in the livers of infected mice, were noted at 24 h. The first lesions in the spleen were found at 48 h after exposure, with an inflammatory infiltrate similar to that in the liver. In time, the numbers of megakaryocytes in the red pulp increased, indicating increased extramedullary hematopoiesis.

Our studies of aerosol-exposed BALB/c mice (unpublished data) corroborate those of Lever et al. [35], with one noteworthy addition: aerosol-exposed mice all developed acute inflammation of the nasal passages, and, in time, this inflammatory process extended back to involve the nasal sinuses. The acute inflammation developed into pyogranulomas, there was frequent erosion and ulceration of the mucosal lining (both respiratory and olfactory epithelia) of these passages, and eventually the passages were blocked by necrotic cellular debris. More interesting, however, is the fact that the inflammatory process eventually extended from the innermost portions of the sinuses into the braincase. We have hypothesized that macrophages carrying phagocytosed glanders bacilli move along olfactory nerves that extend from the olfactory epithelium (lining the deep sinus passages) through small holes in the cribriform plate, into aggregations of the nerves within the brain case called the olfactory tracts. The olfactory tracts feed into (and out of) the cranial-most portions of the brain, the olfactory lobes.

We have also hypothesized that macrophages move as they do because of the pressure that has built up in the sinuses passages (necrotic cellular debris plugging up passages). The resultant changes are very uniform in these mice in the later stages of infection: inflammation becomes pronounced in the olfactory tract and, to a lesser degree, in the meninges and neuropil of the olfactory lobes of the mouse brain. It is also noteworthy from our other unpublished studies that mice given low levels of antibiotics and that survive the acute phase of infection still, in many cases, harbor sites of active infection in the innermost sinus passages. We believe these findings underscore concerns we have on recrudescing infection in any animal that has been aerosol-exposed to *B. mallei*.

7.7 IMMUNITY

Currently, there is no evidence for immunity to glanders by virtue of previous infection or vaccination [5,42]. Because of the severe effect of glanders enterprises involving horses up through the early twentieth century, there were many attempts to develop efficacious vaccines. These vaccine preparations were made by chemically treating [5] or drying [42] *B. mallei* whole cells. Trials using these vaccine preparations were uniformly unsuccessful in protecting horses, although isolated and inconsistent resistance to infection was observed [19]. In addition, glanderous horses that seemed to symptomatically recover from glanders would recrudesce upon challenge with *B. mallei*. The results of studies indicated that control and eradication of glanders was dependent on the elimination of infected horses and on the prevention of infected horses from entering stables free from disease.

Because of the bioterrorist threat posed by *B. mallei*, a category B agent, there is a renewed interest in developing an efficacious vaccine against glanders. In a recent study, Amemiya et al. [43] found that nonviable *B. mallei* whole cells failed to protect mice from a parenteral live challenge. This vaccine was found to induce interleukin (IL)-2; γ-interferon; measurable amounts of IL-4 and IL-5, and IL-10; and a much higher level of IgG1 than IgG2a. Taken together, the data suggested that nonviable *B. mallei* cell preparations did not protect mice in the study because they induced a mixed T-cell helper (Th)1- and Th2-like immune response to all of the nonviable cell preparations. More recent experiments have shown that an irradiation-killed cellular vaccine can protect approximately 50% of mice challenged with a lethal aerosol dose of *B. mallei*, although the survivors remain infected (unpublished data). Results of a study by Trevino et al. [44] indicate humoral antibodies may be important for immune protection. In that study, mouse monoclonal antibodies, presumably against *B. mallei* lipopolysaccaride, protected mice against lethal aerosol challenge. Another study by Goodyear et al. [45] indicates that monocyte chemotactic protein-1 plays a key role in regulating both cellular immunity and γ-interferon production following pneumonic infection. Efforts are ongoing to identify virulence factors as vaccine components so that a recombinant or subcellular vaccine can be developed.

7.8 LABORATORY DIAGNOSIS

Diagnostic tools for glanders were developed in the early twentieth century so that infected animals could be identified and culled. These tools were especially needed for the diagnosis of chronic glanders because isolating the etiologic agent was more difficult. The first diagnostic reagent was mallein, a skin-test antigen that was developed in Russia in 1891 and was composed of a filtrate of bacteria cultured for 4–8 months [5]. Mallein is injected into the eyelid of suspected horses, where it causes an inflammatory, purulent reaction within 48 h of injection if the horse is positive. In the United States, where glanders has been eradicated, the complement fixation (CF) test is used for glanders screening [46]. The mallein test is performed only on those animals that are equivocal for CF antibodies.

There are no specific serodiagnostic tests for glanders in humans. The indirect hemagglutination and CF tests have been used [47,48]. However, the CF test may not detect chronic cases of glanders [19]. Although unpublished, an enzyme-linked immunosorbent assay for human glanders that uses irradiation-killed *B. mallei* whole cells has been developed. This test is able to differentiate glanders from anthrax, brucellosis, tularemia, Q fever, and spotted fever. However, because of the antigenic similarity between *B. mallei* and *B. pseudomallei*, the assay is currently unable to distinguish between cases of melioidosis and glanders. However, polymerase chain reaction-based methodologies have been developed to provide specific detection of *B. mallei* in laboratory mouse tissues [49,50].

Small Gram-negative bacteria may be seen in Gram stains of lesion exudates, but microorganisms are generally very difficult to find, even in acute abscesses [22]. Blood cultures are frequently negative until terminal stages of the disease [22].

7.9 TREATMENT

Burkholderia mallei is susceptible to a wide range of antibiotics *in vitro* [51,52]. In general, most *B. mallei* strains are susceptible to gentamicin, streptomycin, tobramycin, azithromycin, imipenem, ceftazidime, tetracycline, doxycycline, ciprofloxacin, erythromycin, sulfadiazine, and amoxicillin-clavulanate. Aminoglycosides will likely not be effective *in vivo* [32,51]. Most *B. mallei* strains exhibit resistance to amoxicillin, ampicillin, penicillin G, cephalexin, ceftriaxone, metronidazole, and polymyxin B. A class A β-lactamase gene (*penA*) has recently been identified in *B. mallei* ATCC 23344, and the encoded β-lactamase is probably responsible for resistance to penicillins and cephalosporins [53]. Experimental animals, including hamsters, guinea pigs, and monkeys, have been used to determine the *in vivo* efficacy of chemotherapeutic agents [32,54–58]. Sodium sulfadiazine was effective for treating acute glanders in hamsters [59]. Penicillin and streptomycin, however, were not useful chemotherapeutic agents in hamsters. Doxycycline and ciprofloxacin were also examined in hamsters [32]. Doxycycline therapy was superior to ciprofloxacin therapy, but relapse did occur in some of the treated animals 4–5 weeks after challenge.

In a separate study, hamsters infected subcutaneously or by aerosol with *B. mallei* were treated with ofloxacin, biseptol, doxycycline, and minocycline [56]. Whereas all of the antimicrobials exhibited some activity in animals challenged subcutaneously, ofloxacin was superior. None of the antimicrobials demonstrated appreciable activity against a high aerosol dose of *B. mallei* delivered, but doxycycline provided 70% protection against a low dose [56]. The results of other studies demonstrated that a combination of antimicrobials were therapeutically useful in *B. mallei*-infected hamsters [57,58]. It is difficult to directly compare the results of different experimental chemotherapy studies due to the number of variables involved (animal model, route of infection, challenge dose, antibiotic, treatment dose, duration of treatment, and length of follow-up). However, these studies indicated that a prolonged course of therapy with a combination of antimicrobials (doxycycline, ciprofloxacin, and ofloxacin) may provide the best chance of recovery from experimental glanders.

However, caution is warranted with antibiotic therapy against *B. mallei* due to the organism's ability to survive treatment and "hide out" in various tissues. This has been demonstrated in our own studies (unpublished data), and also in a study by Judy et al. [60]. In that study, although both ceftazidime and levofloxacin were able to clear organisms *in vitro* (mouse macrophages), and insure survival of mice infected with a lethal dose of *B. mallei*, presence of bacteria was found in lungs and spleens of survivors 34 days after infection.

Most human glanders cases occurred before the antibiotic era, and the mortality rate was above 90% [61]. There have been several cases of human glanders, primarily in laboratory workers, that have been successfully treated with antibiotics [62–65]. Eight cases were successfully treated with sulfadiazine [62,65]. Streptomycin was used to treat a patient infected with *B. mallei* and *M. tuberculosis* [64]. Treatment with streptomycin was apparently successful against glanders, but had little effect on the tuberculosis in the bones in this patient. In a recent case of laboratory-acquired glanders, the patient received imipenem and doxycycline intravenously for 1 month, followed by oral azithromycin and doxycycline for 6 months [63]. This treatment regimen was successful and there was no relapse of disease.

REFERENCES

1. Miller, W.R. et al., Studies on certain biological characteristics of *Malleomyces mallei* and *Malleomyces pseudomallei*. II. Virulence and infectivity for animals. *J Bacteriol*, 1948. **55**: 127–135.
2. Howe, C., Glanders, in *The Oxford Medicine*, H.A. Christian, editor. New York: Oxford University Press, 1949, pp. 185–201.
3. Pitt, L.L., Pseudomonas, in *Topley & Wilson's Principles of Bacteriology, Virology, and Immunity*, M.T. Parker and L. Collier, editors. Philadelphia: H.B.C. Decker, 1990, pp. 255–273.
4. Sanford, J.P., Pseudomonas species (including melioidosis and glanders), in *Principles and Practice of Infectious Diseases*, G.L. Mandell, R.G. Douglas, and J.E. Bennett, editors. New York: Churchill Livingstone, 1990, pp. 1692–1696.
5. Kovalev, G.K., Glanders (Review). *Zh. Mikrobiol Epidemiol Immunobiol*, 1971, **48**: 63–70.
6. Benenson, A.S., *Control of Communicable Diseases Manual*. Washington, DC: American Public Health Association, 1995.
7. Steele, J.H., Glanders, in *CRC Handbook Series in Zoonoses*, J.H. Steele, editor. Boca Raton: CRC Press, 1979, pp. 339–362.
8. Benenson, A.S., Glanders, in *Control of Communicable Diseases in Man*. Washington, DC: American Public Health Association, 1990, p. 77.
9. Freeman, B.A., Pseudomonas and Legionella, in *Burrow's Textbook of Microbiology*. Philadelphia: W.B. Saunders, 1985, pp. 544–557.
10. Redfearn, M.S. and N.J. Palleroni, Glanders and melioidosis, in *Diseases Transmitted from Animals to Man*, W.T. Hubbert, W.F. McCulloch, and P.R. Schnurrenberger, editors. Springfield, IL: Charles C. Thomas, 1975, pp. 110–128.
11. Howe, C., A. Sampath, and M. Spotnitz, The pseudomallei group: A review. *J Infect Dis*, 1971, **124**(6): 598–606.
12. Wilkinson, L., Glanders: Medicine and veterinary medicine in common pursuit of a contagious disease. *Med Hist*, 1981, **25**: 363–384.
13. Hornick, R.B., Diseases due to *Pseudomonas mallei* and *Pseudomonas pseudomallei*, in *Infections in Children*, R.J. Wedgewood, editor. Philadelphia: Harper & Row, 1982, pp. 910–913.
14. Steele, J.H., The zoonoses: An epidemiologist's viewpoint. *Prog Clin Pathol*, 1973, **5**: 239–286.
15. Gillespie, J.H. and J.F. Timoney, The genus *Pseudomonas*, in *Hagan and Bruner's Microbiology and Infectious Diseases of Domestic Animals*. Ithaca, NY: Cornell University Press, 1981, pp. 51–60.
16. Miller, W.R., L. Pannell, L. Cravitz, W.A. Tanner, and M.S. Ingalls, Studies on certain biological characteristics of *Malleomyces mallei and Malleomyces pseudomallei*.

I. Morphology, cultivation, viability, and isolation from contaminated specimens. *J Bacteriol*, 1948, **55**: 115–126.
17. Parker, M., Glanders and melioidosis, in *Principles of Bacteriology, Virology, and Immunity*, M.T. Parker and L.H. Collier, editors. Philadelphia: H.B.C. Decker, 1990, pp. 392–394.
18. Dungworth, D.L., The respiratory system, in *Pathology of Domestic Animals*, K.V.F. Jubb, P.C. Kennedy, and N. Palmer, editors. New York: Academic Press, 1993, pp. 553–555.
19. Vyshelesskii, S.N., Glanders (Equina). *Trudy Vsessoiuznyi Institut Eksperimental'noi Veterinarii (in Russian)*, 1974, **42**: 67–92.
20. Smith, G.R., A.D. Pearson, and M.T. Parker, Pasteurella infections, tularemia, glanders, and melioidosis, in *Topley & Wilson's Principles of Bacteriology, Virology, and Immunity*, G.R. Smith and C.S. Easmon, editors. Philadelphia: H.B.C. Decker, 1990, pp. 392–394.
21. von Graevenitz, A., Clinical microbiology of unusual *Pseudomonas* species. *Prog Clin Pathol*, 1973, **5**: 185–218.
22. Sanford, J.P., Melioidosis and glanders, in *Harrison's Principles of Internal Medicine*, J.D. Wilson, E. Braunwald, A.S. Fauci, K.J. Isselbacher, J.B. Martin, R.B. Petersdorf, and R.K. Root, editors. New York: McGraw-Hill, 1990, pp. 606–609.
23. Popov, S.F., V. Kurilov, and A.T. Iakovlev, *Pseudomonas pseudomallei* and *Pseudomonas mallei*—Capsule-forming bacteria (in Russian). *Zh Mikrobiol Epidemiol Immunobiol*, 1995(5): 32–36.
24. Fritz, D.L. et al., Mouse model of sublethal and lethal intraperitoneal glanders (Burkholderia mallei). *Vet Pathol*, 2000, **37**(6): 626–636.
25. Bartlett, J.G., Glanders, in *Infectious Diseases*, S.L. Gorbach, J.G. Bartlett, and N.R. Blacklow, editors. Philadelphia: W.B. Saunders, 1982, pp. 1293–1295.
26. Fritz, D.L. et al., The hamster model of intraperitoneal *Burkholderia mallei* (glanders). *Vet Pathol*, 1999, **36**(4): 276–291.
27. Smith, D.T., N.F. Conant, and H.P. Willett, *Actinobacillus mallei* and glanders; melioidosis and actinobacillus, in *Zinsser Microbiology*. New York: Appleton-Century Crofts, 1968, pp. 744–750.
28. Manzenyuk, I.N. et al., Some indices of infectious process in therapy of malleus in monkeys. *Antibiot Khimioter*, 1996, **41**: 13–18.
29. Manzenyuk, I.N. et al. Homeostatic changes in monkeys in a model of glanders. *Antibiot Khimioter*, 1997, **42**: 29–34.
30. Khomiakov Iu.N. et al., The principles of the therapy of glanders in monkeys (in Russian). *Zh Mikrobiol Epidemiol Immunobiol*, 1998(1): 70–74.
31. Dyadishchev, N.R., A.A. Vorobyev, and S.B. Zakharov, Pathomorphology and pathogenesis of glanders in laboratory animals. *Zh Mikrobiol Epidemiol Immunobiol*, 1997, **2**: 60–64.
32. Russell, P. et al., Comparison of efficacy of ciprofloxacin and doxycycline against experimental melioidosis and glanders. *J Antimicrob Chemother*, 2000, **45**(6): 813–818.
33. Alekseev, V.V. et al., The early laboratory diagnosis of the pulmonary form of glanders and melioidosis by using rapid methods of immunochemical analysis (in Russian). *Zh Mikrobiol Epidemiol Immunobiol*, 1994(5): 59–63.
34. Manzenyuk, I.N. et al., *Burkholderia mallei* and *Burkholderia pseudomallei*. Study of immuno- and pathogenesis of glanders and melioidosis. *Antibiot Khimioter*, 1999, **44**: 21–26.
35. Lever, M.S. et al., Experimental aerogenic *Burkholderia mallei* (glanders) infection in the BALB/c mouse. *J Med Microbiol*, 2003, **52**(Pt 12): 1109–1115.
36. Sonnenwirth, A.C., Pseudomonads and other nonfermenting bacilli, in *Microbiology*, B.D. Davis, R. Dulbecco, H.N. Eisen, and H.S. Ginsberg, editors. Hagerstown, MD: Harper & Row, 1980, pp. 674–677.

37. Duval, C.W. and P.G. White, The histological lesions of experimental glanders. *J Exp Med*, 1907, **9**(4): 352–380.
38. Ferster, L.N. and V. Kurilov, Characteristics of the infectious process in animals susceptible and resistant to glanders (in Russian). *Arkh Patol*, 1982, **44**(11): 24–30.
39. M'Fadyean, J., Glanders. *J Comp Pathol*, 1904, **17**: 295–317.
40. Coleman, W. and J. Ewing, A case of septicemic glanders in the human subject. *J Med Res*, 1903, **9**(3): 223–240.
41. Galati, P., V. Puccini, and F. Contento, An outbreak of glanders in lions. Histopathological findings. *Acta Med Vet*, 1973, **19**: 261–277.
42. Mohler, J.R. and A. Eichhorn, Immunization tests with glanders vaccine. *J Comp Path*, 1914, **27**: 183–185.
43. Amemiya, K. et al., Nonviable *Burkholderia mallei* induces a mixed Th1- and Th2-like cytokine response in BALB/c mice. *Infect Immun*, 2002, **70**(5): 2319–2325.
44. Trevino, S.R. et al., Monoclonal antibodies passively protect BALB/c mice against *Burkholderia mallei* aerosol challenge. *Infect Immun*, 2006, **74**(3): 1958–1961.
45. Goodyear, A. et al., Critical protective role for MCP-1 in pneumonic *Burkholderia mallei* infection. *J Immunol*, 2010, **184**(3): 1445–1454.
46. Hagebock, J.M. et al., Serologic responses to the mallein test for glanders in solipeds. *J Vet Diagn Invest*, 1993, **5**(1): 97–99.
47. Gangulee, P.C., G.P. Sen, and G.L. Sharma, Serological diagnosis of glanders by haemagglutination test. *Indian Vet J*, 1966, **43**: 386–391.
48. Sen, G.P., G. Singh, and T.P. Joshi, Comparative efficacy of serological tests in the diagnosis of glanders. *Indian Vet J*, 1968, **45**: 286–293.
49. Lee, M.A., D. Wang, and E.H. Yap, Detection and differentiation of Burkholderia pseudomallei, *Burkholderia mallei* and *Burkholderia thailandensis* by multiplex PCR. *FEMS Immunol Med Microbiol*, 2005. **43**(3): 413–417.
50. Ulrich, M.P. et al., Using real-time PCR to specifically detect *Burkholderia mallei*. *J Med Microbiol*, 2006, **55**(Pt 5): 551–559.
51. Heine, H.S. et al., *In vitro* antibiotic susceptibilities of *Burkholderia mallei* (causative agent of glanders) determined by broth microdilution and E-test. *Antimicrob Agents Chemother*, 2001, **45**(7): 2119–2121.
52. Kenny, D.J. et al., *In vitro* susceptibilities of *Burkholderia mallei* in comparison to those of other pathogenic Burkholderia spp. *Antimicrob Agents Chemother*, 1999, **43**(11): 2773–2775.
53. Tribuddharat, C. et al., *Burkholderia pseudomallei* class a beta-lactamase mutations that confer selective resistance against ceftazidime or clavulanic acid inhibition. *Antimicrob Agents Chemother*, 2003, **47**(7): 2082–2087.
54. Batmanov, V.P., Sensitivity of *Pseudomonas mallei* to fluoroquinolones and their efficacy in experimental glanders (in Russian). *Antibiot Khimioter*, 1991, **36**(9): 31–34.
55. Batmanov, V.P., Sensitivity of *Pseudomonas mallei* to tetracyclines and their effectiveness in experimental glanders (in Russian). *Antibiot Khimioter*, 1994, **39**(5): 33–37.
56. Iliukhin, V.I. et al., Effectiveness of treatment of experimental glanders after aerogenic infection (in Russian). *Antibiot Khimioter*, 1994, **39**(9–10): 45–48.
57. Manzenyuk, I.N., V.V. Dorokhin, and E.A. Svetoch, The efficacy of antibacterial preparations against *Pseudomonas mallei* in *in vitro* and *in vivo* experiments (in Russian). *Antibiot Khimioter*, 1994, **39**: 26–30.
58. Manzenyuk, I.N. et. al., Resistance of *Pseudomallei mallei* to tetracyclines; assessment of the feasibility of chemotherapy (in Russian). *Antibiot Khimioter*, 1995, **40**: 40–44.
59. Miller, W.R., L. Pannell, and M.S. Ingalls, Experimental chemotherapy in glanders and melioidosis. *Am J Hyg*, 1948, **47**: 205–213.
60. Judy, B.M. et al., Comparison of the *in vitro* and *in vivo* susceptibilities of *Burkholderia mallei* to Ceftazidime and Levofloxacin. *BMC Microbiol*, 2009, **9**: 88.

61. Howe, C., Glanders, in *The Oxford Medicine*, H.A. Christian, editor. New York: Oxford University Press, 1950, pp. 185–202.
62. Howe, C. and W.R. Miller, Human glanders: Report of six cases. *Ann Intern Med*, 1947, **26**: 93–115.
63. Srinivasan, A. et al., Glanders in a military research microbiologist. *N Engl J Med*, 2001, **345**(4): 256–258.
64. Womack, C.R. and E.B. Wells, Co-existent chronic glanders and multiple cystic osseous tuberculosis treated with streptomycin. *American J Med*, 1949, **6**: 267–271.
65. Anasbi, M. and M. Minou, Two cases of chronic human glanders treated with sulfamides (in French). *Ann Inst Pasteur (Paris)*, 1951, **81**: 98–102.

8 Plague

Jeffrey J. Adamovicz and Patricia L. Worsham

It appears that the nature of the experimental animal was far more essential to the results than the nature of the vaccine used.

Otten 1936

CONTENTS

8.1 BACKGROUND

Plague, a severe febrile illness caused by the Gram-negative bacterium *Yersinia pestis*, is a zoonosis usually transmitted by flea bites. It is foremost a disease of rodents; over 200 species are reservoirs of *Y. pestis* [1,2]. When fleas feed on a bacteremic animal, the organism is taken with the blood meal into the midgut of the flea where it multiplies, eventually forming a mass of aggregated bacteria that blocks the proventriculus, a valve-like structure leading to the midgut. This blockage starves the flea, which then makes repeated, desperate attempts to feed. Because of the blockage, blood carrying *Y. pestis* is regurgitated into the bite wounds, thus spreading the disease to new hosts. The blocked flea, also a victim of the disease, eventually starves to death [3]. Most often, humans become infected by flea bite during an epizootic event. Less frequently, human disease is a result of contact with blood or tissues of infected animals (including ingestion of raw or undercooked meat), or exposure to aerosol droplets containing the organism [2,4]. Infectious aerosols can be generated by humans or animals with plague pneumonia, particularly cats [5,6].

There have been three pandemics of plague recorded in modern times. The first, known as the Justinian plague, began in the busy port of Pelusium in Egypt in the sixth century, ultimately spreading to Mediterranean Europe and Asia Minor. There have been estimates of 15%–40% mortality for a particular plague epidemic during this period [7]. The second pandemic, originating in central Asia in the early fourteenth century, spread along trade routes from China, eventually encompassing the Mediterranean basin, the Middle East, and most of Europe. The first European epidemic, which began in Messina in 1347, is thought to have killed approximately 30–40% of the European population and eventually became known as the Black Death [8]. For hundreds of years, this second pandemic ravaged Europe, with epidemics continuing late into the seventeenth century. The current (Modern or Third) pandemic most likely began in China, reaching Hong Kong and other Asian ports in 1894. In just a few years, plague had been disseminated via rat-infested steamships to ports worldwide, leading to an estimated 26 million plague cases and 12 million deaths during the first 35 years of the pandemic [9]. It was during the early years of the Modern pandemic that *Y. pestis* was introduced to new locations in North and South America, Southern Africa, Australia, the Philippines, and Japan [9].

8.1.1 TAXONOMY

Yersinia pestis, the causative agent of plague, is a Gram-negative coccobacillus belonging to the family Enterobacteriaceae. The genus was named in honor of Alexandre Yersin, the scientist who originally isolated *Y. pestis* during a plague outbreak in Hong Kong in 1894; the species name *pestis* is derived from the Latin for plague or pestilence. Previous designations for this species have included *Bacterium pestis*, *Bacillus pestis*, *Pasteurella pestis*, and *Pesticella pestis* [10]. This species is closely related to two other pathogens of the genus *Yersinia*: *Yersinia pseudotuberculosis* and *Yersinia enterocolitica*. The extensive genetic similarity (>90%) between *Y. pseudotuberculosis* and *Y. pestis* led to a recommendation that *Y. pestis* be reclassified as a subspecies of *Y. pseudotuberculosis* [11]. This proposal was not well received, primarily due to fear that this change in nomenclature would increase the potential for laboratory-acquired infections. The most recent molecular fingerprinting analysis of *Y. pestis* suggests that this pathogen arose from *Y. pseudotuberculosis* through microevolution over millennia, during which the enzootic Pestoides isolates evolved (see Section 8.1.4). The Pestoides strains appear to have split from *Y. pseudotuberculosis* over 10,000 years ago. This was followed by a binary split approximately 3500 years later that led to populations of *Y. pestis* more frequently associated with human disease. The isolation of *Y. pestis* "Pestoides" from both Africa and Asia suggests that *Y. pestis* spread globally long before the first documented plague of Justinian in AD 784 [12].

8.1.2 Morphology

The characteristic "safety pin" bipolar staining of this short bacillus (0.5–0.8 μm by 1.0–3.0 μm) is best seen with Wayson's or Giemsa stain. Depending on growth conditions, *Y. pestis* can exhibit marked pleomorphism with rods, ovoid cells, and short chains present. A gelatinous envelope, known as the F1 capsular antigen, is produced by the vast majority of strains at a growth temperature of 37°C. Unlike the other mammalian pathogens of the genus, which produce peritricous flagella at growth temperatures less than 30°C, *Y. pestis* is nonmotile [10,13].

8.1.3 Growth Characteristics

In the laboratory, *Y. pestis* is capable of growth at a broad range of temperatures (4–40°C), with an optimal growth temperature of 28°C. Although it grows well on standard laboratory media such as sheep blood agar, MacConkey agar, or heart infusion agar, growth is slower than that of *Y. pseudotuberculosis* or *Y. enterocolitica*; more than 24 h of incubation are required to visualize even pinpoint colonies. Appearance of colonies can be hastened by growth in an environment containing 5% CO_2. The round, moist, translucent, or opaque colonies are nonhemolytic on sheep blood agar and exhibit an irregular edge. A fried-egg appearance is common in older colonies and is more pronounced in certain strains. Long-term laboratory passage of *Y. pestis* or short-term growth under less-than-optimal conditions is associated with irreversible genetic changes leading to attenuation. These changes include the deletion of a large chromosomal pathogenicity island that encodes factors necessary for growth in both the flea and the mammalian host and the loss of one or more virulence plasmids [7,10,13].

8.1.4 Biochemistry and Genetics

Yersinia pestis is a facultative anaerobe, fermenting glucose with the production of acid. An obligate pathogen, it is incapable of a long-term saprophytic existence, due in part to complex nutritional requirements, including a number of amino acids and vitamins. It also lacks certain enzymes of intermediary metabolism that are functional in the closely related but more rapidly growing species such as *Y. enterocolitica* or *Y. pseudotuberculosis*. *Yersinia pestis* strains have traditionally been separated into three biovars, based on the ability to reduce nitrate and ferment glycerol [7]. Some molecular methods of typing, such as ribotyping and restriction fragment length polymorphisms of insertion sequence locations, support this division of strains [14,15]. Biovar orientalis (Gly^-, Nit^+) is distributed worldwide and is responsible for the third (Modern) plague pandemic. It is the only biovar present in North and South America. Biovar antiqua (Gly^+, Nit^+) is found in Central Asia and Africa and may represent the most ancient of the biovars [7,12]. Biovar mediaevalis (Gly^+/Nit^-) is geographically limited to the region surrounding the Caspian Sea. There are no apparent differences in pathogenicity among the biovars [7,16]. Using a more modern approach, the microevolution of *Y. pestis* was investigated by three different multilocus molecular methods. Eight populations were recognized by the three methods and an evolutionary tree for these populations, rooted on *Y. pseudotuberculosis*, was proposed. The eight population groups do not correspond directly to the biovars; thus, it was suggested that future strain groupings be rooted in molecular typing. Four of the groups were made up of transitional strains of *Y. pestis*, "Pestoides," that exhibit biochemical characteristics of both *Y. pestis* and *Y. pseudotuberculosis* [17]. These isolates represent the most ancient of the *Y. pestis* strains characterized to date [12]. Recently, a large and diverse *Y. pestis* collection was assayed for the presence of 933 distinct single-nucleotide polymorphisms; this study suggests that the *Y. pestis* strains in the United States and Madagascar each arose from a single radiation from China, whereas there appear to have been multiple radiations from China to Europe, South America, Africa, and Southeast Asia [18].

8.1.5 Isolation and Identification

Procedures for the isolation and presumptive identification of *Y. pestis* by Level A laboratories can be downloaded from the Centers for Disease Control and Prevention (CDC) website http://www.bt.cdc.gov/agent/plague/index.asp [19]. The World Health Organization offers their Plague Manual online at http://www.who.int/csr/resources/publications/plague/WHO_CDS_CSR_EDC_99_2_EN/en/index.html [20]. A recent review of the methodology for isolating and indentifying *Y. pestis* from clinical samples and animals is available [13].

Standard bacterial methodologies include staining and microscopic analysis of the organism, isolation on culture medium, and biochemical tests. Laboratories experienced in the identification of *Y. pestis* with the appropriate containment facilities should perform diagnostic tests for plague. Care should be taken to avoid aerosols; in this regard, fixing slides with methanol rather than heat-fixing is preferred.

A rapid and accurate presumptive diagnosis of plague can be made by using a fluorescent antibody test (FA) to detect the plague-specific capsular antigen. Because F1 is produced only at temperatures >33°C, this method requires a relatively fresh sample from the animal or from a laboratory culture incubated at the appropriate temperature. Flea samples will be negative, as will be samples refrigerated for more than 30 h [13]. The test is performed at some public health and veterinary diagnostic laboratories and by the Centers for Disease Control; air-dried slides should be submitted for FA testing. Confirmatory testing includes lysis by a species-specific bacteriophage [2]. The standard method of serodiagnosis for humans is the passive hemagglutination assay or the enzyme-linked immunosorbent assay using F1 antigen. Paired sera from the acute and convalescent periods are compared [13]. A rapid test kit based on detection of immunoglobulin G antibodies to the F1 capsular antigen has been used in surveillance of plague in human and rat populations, although it is not yet commercially available [21]. However, F1 antigen capture-based immunodiagnostic kits are available through New Horizons Diagnostics (Columbia, Maryland) and Tetracore (Rockville, Maryland).

Although they are generally not yet validated, genetic methods such as the polymerase chain reaction (PCR) may become more common for identifying *Y. pestis* [22–24]. Such assays have been promising in identifying experimentally infected fleas and animals and can be used to detect *Y. pestis* in cases where cultures and serum are not available [25,26]. The use of more than one plague-specific primer set in diagnostic PCRs will allow the identification of rare genetic variants such as F1- and Pla-strains, as well as wild-type strains. Recently, a real-time PCR assay targeting the *Y. pestis*-specific virulence plasmids was validated using human clinical samples. This assay identified more of the patients with a clinical diagnosis of plague than either culture or an F1-based immunoassay [27]. A rabbit polyclonal antiserum has also been developed that identifies both F1-positive and F1-negative strains of *Y. pestis*, although there is some cross reactivity with *Y. pseudotuberculosis* (John W. Ezzell, personal communication).

8.2 HUMAN DISEASE

In humans, plague is generally classified as bubonic, septicemic, or pneumonic. *Yersinia pestis* is a lymphotrophic pathogen. Thus, in bubonic plague, it migrates from the site of entry to the regional lymph nodes, where it multiplies and forms a bubo, the exquisitely painful enlarged node that is, the hallmark of the disease. The bubo is packed with bacilli and is often accompanied by an overlying edema. At times, bubonic plague leads to bacteremia and hematogenous spread to other organs including the liver, spleen, lungs, and, less commonly, the meninges [2,28]. Cases of plague bacteremia without obvious lymphadenopathy are termed septicemic plague [4,28,29]. A small percentage of plague patients develop pneumonic plague secondary to bubonic or septicemic plague and these individuals are capable of spreading the disease directly to other humans. Primary pneumonic plague, acquired by inhaling infectious aerosols generated by the coughing of a plague-infected person or animal, is rare but rapidly fatal [28,29]. It appears that pharyngeal plague can be acquired by ingesting or inhaling the organism. In some cases, this form of the disease appears to be asymptomatic [2,4,29,30].

The clinical symptoms and pathogenic lesions that develop after plague infection are not universal and are highly dependent on the age and health of the individual, the relative amount of the inoculum, and the duration of disease. Generally, there is an increase in symptoms and lesion severity with time [31]. We have attempted to note the most prevalent clinical symptoms and pathogenic lesions in human plague infections, believing that certain symptoms and lesions would eventually be noted in most patients in the absence of treatment. However, these observations are intended to serve only as a guide to select the most appropriate animal model for plague pathogenesis, vaccine, or therapeutic development studies, not to serve as absolute criteria for assessment of animal models.

8.2.1 Bubonic and Septicemic Plague

Bubonic plague is characterized by a suppurative lymphadenitis [28]. Clinically, most patients manifest with fever and lymphadenopathy 2–7 days after a flea bite. The lymphadenopathy may progress to ulceration and cutaneous fistulae. However, this presentation is not universal and additional or different clinical symptoms may be observed (Table 8.1). The presence of vesicles, pustules, papules, eschars, or necrotic lesions indicating the site of flea bites or initial entry through the skin may or may not be noted. In cases of untreated bubonic plague, septicemia, secondary pneumonic plague, or plague meningitis may develop [2,28,31]. It is estimated that untreated bubonic plague is fatal in approximately 60% of patients [2]. Primary septicemic plague is difficult to diagnose as the bubo is absent and the clinical signs are general in nature: fever, chills, nausea, vomiting, diarrhea, and abdominal pain. Due to difficulties in diagnosis of septicemic plague, the fatality rate for this form of the disease is 28% even if treatment is initiated [32]. Gastrointestinal symptoms are most often

TABLE 8.1
Clinical Signs of Peripheral Infection (Bubonic/Septicemic Plague)

Symptom	Human	African Green Monkey	Cynomolgus Macaque	Rhesus Macaque	Langur Monkey	Domestic Cat
Lymphadenopathy (bubonic)	+	++	+	+	++	++
Fever	+++	+++	+++	++	+++	++
Myalgia	+	+	+	+	+	+
Malaise	+	+	+	ND	ND	ND
Diarrhea	++	ND	ND	ND	ND	++
Vomiting	++	ND	ND	ND	ND	++
Nausea	++	ND	ND	ND	ND	ND
Anorexia	++	+	+	+	ND	++
Lethargy	++	++	++	++	++	++
Ataxia	ND	ND	ND	ND	ND	++
Chills	++	ND	ND	ND	ND	ND
Elevated pulse	++	++	++	+	++	ND
Cyanosis	+ (Late)	+	+	+ (Late)	+	+
Headache	+	ND	ND	ND	ND	ND
Pharyngitis	+	ND	ND	ND	ND	++
Neutrophilic leukocytosis	++	++	++	++	ND	++
Thrombocytopenia	+	ND	ND	++	ND	ND

Note: ND, not determined.

observed without noted lymphadenopathy and are believed to be associated with the septicemic form of the disease. Patients may also develop a blackening or purpura of the distal skin or cervical skin region [33]. This purpura is the basis of the term "black death" used to describe late-stage plague victims. Cervical lymphadenitis has been noted in several fatal human cases of plague and is associated with the septicemic form of the disease. However, it is possible that these patients were exposed by the oral/aerosol route and developed pharyngeal plague that then progressed to a systemic infection [2,29–31]. This mode of infection has not been extensively studied. Secondary pneumonic plague can be associated with either bubonic or septicemic plague and occurs after an episode of bacteremia. The secondary plague pneumonia patient is likely to have exhibited signs of sepsis 1–2 days before pneumonia is observed. These individuals tend to produce a thick, mucopurulent sputum rather than the watery, serosanguinous sputum associated with primary pneumonic disease [2].

The pathogenesis of bubonic plague most often follows inoculation by the bite of an infected flea or, in rare cases, introduction into a lesion after contact with infected animal tissues. The organisms are phagocytosed and transported to the lymphangioles and the local draining lymph node. In humans, this is often the inguinal nodes in the groin or the auxiliary nodes in the armpit [31]. Bacteria that escape the macrophages establish a primary lesion in the lymph node, generally with some involvement of all nodes in the drainage [7]. There is marked swelling and the development of a painful node; in late-course disease the node can hemorrhage and/or become necrotic. This process is the basis of the development of the observed bubo. In some cases, the draining lymph node is not noticeably involved but the organisms gain access to the peripheral circulation, resulting in septicemic plague and the seeding of other organs, including the liver, spleen, and lung [31,33,34].

The most frequently noted pathology in humans who succumb to bubonic or septicemic plague is summarized in Table 8.2. In addition to the characteristic bubo, most terminal patients will also develop severe general vascular dilation and engorgement with interstitial hemorrhage. These hemorrhages are frequently noted in the epicardium, the pleura, and the peritoneal tissue associated with the digestive tract, although these lesions are almost always sterile. The most frequently noted infected organ is the spleen. The level of infection is generally low although there is usually marked swelling. If the disease progresses, a fatty degeneration of the liver is observed, and a secondary pneumonia can occur. Other major organs are not routinely involved.

Some human histopathology observations have been made (Table 8.3). The absence or presence of a particular lesion may be temporally related to the time of death or affected by treatment. Additional pathology associated with bubonic plague includes kidney damage due to disseminated intravascular coagulation and bacterial or bacterial/fibrin clots [35–37]. Since the advent and use of antibiotics to treat plague, the notation of the former lesion has decreased. Humans make a robust immune response to plague antigens even after antibiotic treatment [38]. Although it is likely that antibiotics may have blunted the development of antibody responses to some plague antigens, those that were uniformly recognized include F1, V antigen, lipopolysaccharide, plasminogen activator (Pla) and *Yersinia* outer proteins (Yops) M and H. This seroreactivity can be used as a comparator when selecting an animal model for vaccine studies.

TABLE 8.2
Gross Pathology Bubonic Plague

Lesion	Human	African Green Monkey	Cynomolgus Macaque	Rhesus Macaque	Langur Monkey	Domestic Cat
Hemorrhage/ suppuration at inoculation site	+	++	+	+	++	++
Formation of primary bubo	+	++	+	+	++	+
Hemorrhagic necrosis of primary lymph node	++	++	++	++	++	++
Petechial lesions intestinal tissue	+	+	+	ND	++	+
Petechial skin lesions	++	++	++	ND	+++	++
Enlarged spleen	++	++	++	++	++	+
Pulmonary edema/ hemorrhage	+ (Late)	+ (Late)	+ (Late)	ND	+++	+
Fluid in pericardium	++	++	++	ND	++	ND
Liver congestion	++	++	++	+	++	ND

Note: ND, not determined.

TABLE 8.3
Histopathology Bubonic Plague

Lesion	Human	African Green Monkey	Cynomolgus Macaque	Rhesus Macaque	Langur Monkey	Domestic Cat
Bacteria in lymph node (ground glass)	++	++	++	++	++	++
Lympholysis in lymph node	++	++	++	+	+	+
Hemorrhage/necrosis in lymph node	++	+++	++	++	+++	++
Degeneration of liver parenchyma	+	+	+	++	ND	+
Fibrin thrombi (liver/ kidney)	+	+	+	+	+	ND
Bacteria in splenic red pulp	+ (Late)	+	+	++	++	+

Note: ND, not determined.

8.2.2 Primary Pneumonic Plague

Primary pneumonic plague is much less common than bubonic or septicemic disease. Humans infected by the aerosol route will usually begin to show symptoms of primary pneumonic disease within 24–48 h of exposure, although it may take up to 7 days [31]. The principal clinical features are fever with severe headache and a nonproductive cough that may progress to hemoptysis with rales over one or more lobes [2,20,28,29]. Additional clinical symptoms are listed in Table 8.4. Without treatment, most patients will die within 1–3 days of disease onset. Bacteremia is an ominous indicator of death, even for patients treated with appropriate antibiotics.

The pathology of pneumonic plague is somewhat dependent on the depth of penetration into the lung. The primary lesions may be in the bronchia and lead to bronchiolitis or, if the organisms settle in the alveolar spaces, an alveolitis may occur. Both conditions eventually lead to lobular pneumonia and lobar consolidation. The appearance of hemorrhage on the pleural surface and bronchial mucosa increases with the duration of disease. The lung volume becomes progressively less due to extensive edema and congestion due to the presence of degenerate neurophils and numerous bacilli. The organisms may also be found in the cervical, hilar, and/or mediastinal lymph nodes. The lymph nodes can progress to necrosis and hemorrhage

TABLE 8.4
Clinical Signs of Aerosol Infection

Symptom	Human	African Green Monkey	Cynomolgus Macaque	Rhesus Macaque	Langur Monkey	Domestic Cat[a]
Lymphadenopathy	++	++	++	++	++	++
Fever	+++	+++	+++	+	+++	++
Myalgia	+	ND	ND	ND	ND	ND
Malaise	++	++	++	++	ND	++
Anorexia		ND	+	ND	ND	+
Lethargy	++	++	++	++	++	+
Ataxia		ND	ND	ND	ND	+
Chills	++	ND	ND	ND	ND	ND
Elevated pulse	+	+	+	+ (Late)	+	ND
Cyanosis	+ (Late)	+	+	+ (Late)	++	+
Headache	++	ND	ND	ND	ND	ND
Pharyngitis	+	+	+	ND	+	+
Chest pain	+	ND	ND	ND	ND	ND
Cough	++	ND	ND	+	ND	++
Rales	++	+	+	+	ND	ND

Note: ND, not determined.
[a] Cat pathology consistent with incidental aerosol exposure.

TABLE 8.5
Gross Pathology Primary Pneumonic Plague

Lesion	Human	African Green Monkey	Cynomolgus Macaque	Rhesus Macaque	Langur Monkey	Domestic Cat[a]
Fibrinous pleuritis	++	++	ND	+	ND	ND
Multilobar pneumonia	+++	+++	+++	+ (Challenge-dependent)	++	+ (Focal)
Hemorrhagic mediastinum	+	++	+	+	++	ND
Congestion of trachea/bronchi	++	++	+	+	+++	ND

Note: ND, not determined.

[a] Cat pathology consistent with incidental aerosol exposure.

if the patient survives long enough. The gross pathology and histopathology of pneumonic plague are summarized in Tables 8.5 and 8.6. It must be mentioned that histopathology of human plague has only been studied for a relatively low number of cases with obvious temporal differences between patients; therefore, caution should be used in interpreting these data.

TABLE 8.6
Histopathology Primary Pneumonic Plague

Lesion	Human	African Green Monkey	Cynomolgus Macaque	Rhesus Macaque	Langur Monkey	Domestic Cat[a]
Pulmonary alveolar flooding	+++	+++	++	++	++	++
Necrohemorrhagic foci	++	+++	+	++	++	++
Fibrinous pleuritis	++	++	++	ND	ND	++
Mediastinitis	+	++	+	++	+	+
Fibrin thrombi (liver/kidney)	+	+	+	++	ND	ND
Disseminated intravascular coagulopathy	+	+	+	++	ND	ND
Fibrin in lung	–	–	–	–	ND	ND
Neutrophil infiltration of lung	+++	+++	+++	+++	ND	++
Bacteria in lung	+++	+++	+++	+++	+++	++
Bacteria in splenic red pulp	++	+++	+	++	++	++

Note: ND, not determined.

[a] Cat pathology consistent with incidental aerosol exposure.

8.2.3 Pharyngeal Plague

Although *Y. pestis* is not generally classified as an oral pathogen, the closely related *Y. pseudotuberculosis* typically infects hosts orally. In nature, it appears that some hosts, including humans, may acquire plague as a result of ingesting infected animals [2]. The ubiquitous nature of flea vectors and the associated possibility of vector-borne disease make it difficult to assess the significance of oral infections in endemic plague, although cannibalism is thought to play a role in some rodent species [39]. It has been reported that humans can be infected by ingestion of fleas or by consuming undercooked meat from infected animals such as rodents, camels, or goats. At times, this takes the form of pharyngeal or tonsillar plague, which may be followed by the septicemic form of the disease [2,29–31,40–43].

8.3 SMALL ANIMAL MODELS OF PLAGUE

8.3.1 Mouse

Perhaps the most widely used animal model for plague in recent years has been the mouse (*Mus musculus*). This model offers a number of advantages for investigators interested in pathogenesis, vaccine development, or evaluation of therapeutics. Practical considerations include the low cost of the animal, the obvious advantages in terms of space requirements, and ease of handling. U.S. Food and Drug Administration (FDA)-approved vaccines have previously been evaluated in the mouse model [44] and the immune response to *Y. pestis* in infected mice is similar to that of humans [38]. The mouse responds vigorously to antigens known to be important in human immunity such as F1, whereas some other models, such as the guinea pig, may require use of adjuvants not approved for human use to obtain a significant response [45]. Mice can also be used in models of passive protection [46]. The existence of numerous inbred strains and "knockout" mice enable investigators to dissect the immune response to vaccines and infection [47–49]. Both outbred strains of mice, such as the Swiss–Webster, and the BALB/c inbred strain have been used in recent vaccine efficacy testing and characterization of the immune response [46,50–56]. For testing therapeutics, outbred Swiss–Webster, OF1, and Porton strains of mice have been used along with the inbred BALB/c strain [57–62]. Some subtle variation between inbred strains in their response to a plague vaccine has been reported [63] and it has been suggested that some inbred mice may exhibit exaggerated responses to DNA-based vaccines [64]. In addition, live vaccines appear to be more virulent in mice of certain haplotypes; murine MHC classes H-2k and H-2a tolerate the live vaccine EV76 better than mice of haplotypes H-2 and H-2b [65]. Differences in subcutaneous (s.c.), intraperitoneal (i.p.), and intranasal median lethal dose (LD_{50}) values among BALB/c, NIH/s, (inbred), and Porton outbred mice have been reported, but the statistical significance of this is unclear, as confidence limits for the LD_{50} were not given [66]. Recently, inbred and knockout mouse strains have been useful in dissecting the immune response to plague in mice [47–49,67,68]. Females are generally used because males can be more aggressive and infighting among male mice has complicated interpretation of some experiments [63].

The mouse is highly susceptible to infection by *Y. pestis* by parenteral, intravenous (i.v.), and aerosol routes [7,46,59]. Thus, bubonic, septicemic, and pneumonic plague can be modeled. Bacterial strain selection is not complex in this model; susceptibility of outbred Swiss–Webster mice to a panel of genetically and geographically diverse strains of *Y. pestis* has been reported [1]. The mouse, like the nonhuman primate (NHP), is sensitive to infection by strains expressing the F1 capsular antigen and to F1-negative strains [16,69–71].

The LD_{50} of wild-type *Y. pestis* in mice when administered by the s.c. route (mimicking the flea bite) is generally between 1 and 10 colony-forming units (CFUs) [7]. Meyer [72] found that the pathogenesis of plague infection in mice, rats, guinea pigs, and monkeys was similar when the animals were inoculated s.c. There are also a number of similarities of these models to the pathology of disease in humans, although gross changes in the size of regional nodes in mice may not be obvious. When mice are challenged s.c., the organism is rapidly carried to the regional lymph nodes and transferred to the thoracic duct and bloodstream [73]. The initial bacteremia ensues within 6–12 h, seeding the liver, spleen, and bone marrow. After growth in these organs, a terminal bacteremia generally appears within 3–5 days. Because of the low LD_{50}, mice are often used for surveillance of naturally occurring plague. Samples containing triturated fleas or other potentially contaminated materials are injected s.c. and the organism is then isolated from the spleen. Alternatively, tissues can be examined by FA for expression of the *Y. pestis* F1 capsular antigen [13,26]. Historically, the s.c. mouse model was also used in the FDA potency assay for the formalin-killed plague vaccine, Plague USP, which is no longer manufactured [44].

Aerosol models have been used for vaccine and therapeutic efficacy testing, as well as pathogenesis studies [17,50,57,58,69,71,74]. Small-particle aerosols of *Y. pestis* produce primary pneumonic plague in the mouse [50,57]. As reviewed by Meyer [75], the lesions observed in mice after inhaling *Y. pestis* are quite similar to those of human primary pneumonic plague. Early in the infection, a cellular infiltration occurs in the alveolar septa followed by massive hemorrhage and edema, multiplication of the bacteria within the blood vessels, and spread to adjacent lobes of the lung. Deaths occur between 72 and 96 h. The investigators found little evidence of cross infection between cagemates. Meyer [75] hypothesized that the physical structure of the mouse respiratory system prevents particles exhaled by the infected animal from reaching the lung of the contact animal and is responsible for lack of pneumonic spread between mice. In this respect, the mouse is not a good model for examining the spread of pneumonic plague between animals. Aerosols with variable particle sizes may yield disease characterized by cervical buboes and septicemia rather than pneumonic disease; this could mirror reports of pharyngeal plague in humans [75–77]. The whole-body aerosol LD_{50} of *Y. pestis* strain CO92 (one of the best characterized and most commonly used strains in current plague research) in Swiss Webster mice has been reported to be 2.1×10^3 [78] and 6×10^4 [79]. Differences in aerosol generation systems and in the calculation of inhaled dose may be responsible for the observed variation in LD_{50}.

Intranasal models of respiratory infection rather than aerosol exposures have been used extensively in recent years [54,80–83]. Details of the resulting disease pathology and descriptions of the model itself have not yet reached the literature; however, it is clear that the mouse is susceptible to infection by this route and that pneumonia results from installation of the pathogen in the nares. Estimated LD_{50} values for *Y. pestis* strain CO92 by intranasal infection are routinely one to two logs lower than the values reported when an aerosol challenge is used. It is not yet clear whether this reflects damage to the pathogen during the aerosolization process or whether it is indicative of different disease processes. Meyer et al. [84] reported that 10% of the inoculum reaches deeper respiratory passages when anesthetized mice are exposed intranasally. It was also noted that, after intranasal installation, the early lesions were observed in the bronchi; peribronchial masses of bacteria were observed before cellular infiltration in the alveoli. This is in contrast to the pathology of disease produced by small-particle aerosols. Time to death was similar to that observed with aerosol-induced disease. As the disease progressed, however, it came to more closely resemble human pneumonic plague [75,85]. It seems logical that intranasal installation might, in some respects, mimic a polydisperse aerosol rather than the small-particle aerosols generated by a Collison nebulizer (see the discussion in Section 8.3.2). It was recently shown that there are clear pathological differences in mice exposed to large-particle and small-particle aerosols [86]. This pathological difference was seen with nose-associated lymphoid tissue (NALT) damage and secondary pneumonia in large-particle exposure whereas primary pneumonia was the result of small-particle aerosols. This indicates that the method of aerosol exposure is a critical parameter that should be properly selected to support experimental aims. For instance, large-particle exposure may be appropriate for prediction of person-to-person spread whereas small-particle aerosols would be recommended for testing protection against a deliberate aerosol release. A side-by-side comparison of disease kinetics in mice exposed to both types of aerosols and intranasally at various doses would be very useful in comparing the intranasal small-particle aerosol and the polydisperse aerosol models. Ideally, the studies would be performed simultaneously using the same inoculum preparation and source of mice. This would minimize the large number of variables that currently plague the available data.

Septicemic plague can be induced in the mouse by i.v. challenge by fully virulent organisms and some therapeutics have been assessed in this model [59,73]. More importantly, the i.v. virulence of pigmentation-deficient (Pgm^-) strains of *Y. pestis*, which are highly attenuated by the s.c. route, has enabled numerous investigators to safely assess the importance of several *Y. pestis* virulence factors under biosafety level-2 conditions [7]. Retro-orbital challenges are often used in lieu of injections into the mouse tail vein [87]. Because the organism does not normally have direct access to the bloodstream, these models do not reflect naturally acquired disease. However, they do encompass an important part of the disease progression (septicemia followed by seeding of the spleen and liver) and are invaluable to scientists lacking higher containment facilities. Furthermore, certain Pgm^- strains with a defined chromosomal deletion are exempt from CDC select agent regulations, an advantage for laboratories without CDC registration for select agents. For a description of this exclusion, the following website may be consulted: http://www.cdc.gov/od/sap/exclusion.htm.

Butler et al. [88] described mouse, guinea pig, and rat models of oral infection by *Y. pestis*, including both intragastric installation and incorporation of bacteria into drinking water. In mice, there was evidence of both fatal systemic infections and self-limited disease with subsequent seroconversion. *Yersinia pestis* did not appear to produce a true intestinal infection; instead, the organism appeared to invade through the gastrointestinal epithelium and multiply in the blood, liver, and spleen. Thus, the pathology was similar to that of mice infected parenterally rather than resembling disease caused by enteric pathogens. In more recent studies, Kokushkin et al. [89] presented a mouse model based on a more natural ingestion process in which mice were challenged by feeding organisms imbedded in agar pellets. Very early in the infection, the authors observed organisms within regional nodes along the gastrointestinal tract, particularly at the ileal–caecal junction. Systemic infection was observed at later time points [89].

8.3.2 Guinea Pig

Like the mouse, the guinea pig is an attractive model in terms of expense, space, and ease of handling. Historically, however, there have been some problems with the use of this model. Numerous investigators have reported difficulties in successfully vaccinating guinea pigs with killed, whole-cell plague vaccines or with antigenic extracts, including F1 capsular antigen, although these preparations were highly effective in the mouse, the rat, and the NHP [45,90–95]. In some cases, incorporation of oil-based adjuvants was necessary to achieve protection from *Y. pestis* challenge in the guinea pig [45,90]. This makes the guinea pig a less attractive model as these adjuvants are not approved for human use. Passive protection models, in which immune sera were administered to guinea pigs before challenge, have generally not been useful [90,96], although Meyer [72] reported some success when relatively large amounts of specific antibody were administered i.p. Live attenuated vaccines have generally been more successful in protecting guinea pigs [90]. However, serious concern was raised regarding the use of this animal as a model for live plague vaccines when some vaccine strains that were essentially avirulent in the guinea pig proved fatal when tested in NHPs [97]. Meyer et al. [93] felt that the inability to protect guinea pigs with F1 antigen made this model unsuitable for vaccine efficacy testing.

Although many western investigators prefer the mouse model, the small animal model of choice for plague in the former Soviet Union (FSU) is the guinea pig rather than the mouse. Anisimov et al. [98] have stated that guinea-pig virulence is the best predictor of likely virulence for humans. Part of the rationale for this assertion is that certain subspecies of *Y. pestis* found in the FSU are only rarely associated with human infections. These isolates are attenuated for guinea pigs when they are infected by parenteral routes, but retain virulence for other animal models [98,99]. However, this phenomenon might be a result of limited contact between humans and the hosts that maintain these endemic foci. Furthermore, attenuation by parenteral routes of infection does not necessarily predict aerosol virulence [100,101]. The emphasis on live bacterial vaccines in the FSU might also have made the guinea-pig model appealing, as it is known to respond well to this type of vaccination. The diverse groups of bacterial and animal strains used in western laboratories and the FSU make direct comparisons of many research studies difficult. Because of the obvious differences between the mouse and guinea pig, investigators should take care in extrapolating findings between these animal models.

In many cases, it is unclear what strains of guinea pigs were used in a particular study. Recently, the outbred Dunkin Hartley and "breedless" guinea pigs have been used as models of plague [69,92,94,102]. Due to the variability of guinea pigs in their response to vaccines, Smith and Packman [103] suggested the use of inbred and/or specific pathogen-free animals. However, Meyer et al. [97] reported that the use of specific pathogen-free animals did not address the issue of this variability in response to vaccination and challenge.

Like the mouse, the guinea pig is considered to be highly susceptible to infection by *Y. pestis* [69,74], with an s.c. $LD_{50} < 10$. However, there are striking differences in the sensitivity of these rodents to certain strains of *Y. pestis*. For example, unlike the mouse and NHP, the guinea pig is relatively resistant to infection by nonencapsulated strains (F1); the capsule is an essential virulence factor in this animal [69,104]. Another case of a unique pattern of resistance in the guinea pig was noted during a Brazilian outbreak. Nearly all of 200 isolates from a plague focus exhibited high pathogenicity for mice but were avirulent in guinea pigs [105]. These strains of *Y. pestis* were F1-postitive but were subsequently found to be asparagine auxotrophs. Apparently, an asparaginase present in the serum of guinea pigs, but not in mice (or humans), depleted the guinea pig host environment of an essential nutrient for the auxotrophic strains. In this case, extrapolating virulence data acquired in guinea pig studies to humans would be problematic. Some F1-positive isolates of *Y. pestis* from the Caucus region are virulent for mice, voles, susliks, and gerbils, but attenuated in the guinea pig [99]. The genetic basis for this host specificity is unknown. Finally, there are reports of some seasonal variation in sensitivity to *Y. pestis* in guinea pigs; the animals appeared to be more sensitive in the summer than in the winter [72,85,90]. Spivak et al. [90] suggested that the climate of laboratories located in the tropics might have promoted the lack of seasonal variation in some early studies.

Bubonic plague models, using intradermal or s.c. injection, as well as a flea bite model have been developed; the guinea pig is highly susceptible to infection by these routes [31,90,94,106]. As reviewed by Pollitzer [31], rubbing infectious material onto shaved, depiliated, or even intact guinea-pig skin was enough to initiate an infection. Papules are observed when the animal is infected by the intradermal route or by the flea. Lymphadenopathy, septicemia, and death follow. When the guinea pig is infected parenterally, the course of the disease is protracted when compared to many other models and is not always dose related [30,31,69,106].

The aerosol LD_{50} for the Hartley guinea pig with *Y. pestis* strain CO92, the most commonly used isolate in the United States at this time, is ~40,000 CFU, similar to that of the mouse [69]. In aerosol studies of mixed particle size, infection of the guinea pig initiated a disease characterized by cervical and laryngeal edema, lymphadenopathy, hemorrhagic nodes, septicemia, and hemorrhage of the intestinal wall. Miliary abcesses of the spleen were present. Approximately one quarter of animals had evidence of pneumonia; however, this appeared to be a secondary lung infection rather than a primary one [75,76,107–109]. Monkeys exposed under the same conditions developed a primary pneumonic plague. Based on these results, Strong and Teague [109] concluded that the infection in guinea pigs originates in the mucous membranes of the mouth and throat. Invasion of local lymph nodes is then followed by septicemia and, in some cases, secondary pneumonia. They attributed this disease progression both to shallow respiration and the relatively small size of the guinea pig airway [109]. Others have suggested that the thick cluster of hair present in the guinea-pig nose might affect deposition of particles [31]. Druett et al. [110] reported that the guinea-pig respiratory tract does not allow particles >4 μm to reach the lungs. Particles <1 μm initiated primary bronchopneumonia whereas those 10–12 μm deposited in the upper airway [110]. The latter group of animals exhibited pathology similar to that described by Strong and Teague [109]. In another model, intratrachael installation of $5 \times 10E^7$ bacteria produced pneumonic plague in some groups of guinea pigs [75,111] Success was also reported using an intranasal installation in animals anesthetized with barbiturates [84,85]. According to Meyer [75], this technique flushes approximately 10% of the inoculum into deeper respiratory passages. In contrast to the mice in this study, the guinea pigs did not appear visibly ill; rather, they died suddenly. Some guinea pigs infected by these methods have transmitted the infection to control cagemates, although this was not typical [75].

8.3.3 Multimammate Mouse (*Mastomys natalensis* and *Mastomys coucha*)

Two species of *Mastomys* (previously *Praomys*) have been used as laboratory models for plague. These sibling species are ubiquitous agricultural pests in Southern Africa and reside in close proximity to humans. *Mastomys coucha* is a known reservoir for *Y. pestis* whereas *Mastomys natalensis* carries the arenavirus responsible for Lassa fever [112,113]. Recently, the identification of isozyme and allotype markers distinguishing these species was reported; however, it is not yet clear why they do not act as reservoirs for the same pathogens [112]. In the laboratory, *M. natalensis* is orders of magnitude less sensitive to infection by *Y. pestis* than *M. coucha* [114,115]. Naïve *M. natalensis* but not *M. coucha* react to *Y. pestis* mitogens nonspecifically and this may explain the greater innate immunity of the former species [116]. Laboratory colonies of both species have been established [116,117]. It has been suggested that the *M. coucha* model may more closely mimic the susceptibility of *Cercopithicus aethiops* to live vaccines than mice or guinea pigs and would be a more appropriate screening model for this type of vaccine [117].

8.3.4 Rat

Plague models of a number of rat species (*Rattus norvegicus*, *Rattus rattus*, *Rattus alexandrinas*, as well as Sprague–Dawley and Wistar laboratory rats) [93,118,119] have been described. This genus is more resistant to infection by *Y. pestis* than either mice or guinea pigs, with the s.c. lethal dose approximately 1000-fold higher, depending on the strain of *Y. pestis* and the type of rat. Resistance to plague was noted in laboratory rats and in rats captured from both endemic and nonendemic areas [119]. Various challenge routes have been used with the rat, including s.c., intradermal, aerosol, and intranasal. Williams and Cavanaugh [120] found that the intranasal route was not as reliable as an aerosol in establishing pneumonic plague, as the intranasal infection often involved the tonsils and larynx rather than primary pneumonic disease.

They felt, however, that the intranasal challenge route was a more stringent test of vaccines than the s.c. challenge and was, therefore, suitable for efficacy testing [120]. A rat oral infection model was developed by feeding rats infected tissues; approximately 22% of the rats that succumbed to plague became bacteremic [118].

Otten [121] demonstrated that both wild-caught *R. rattus* and laboratory rats could be protected with live attenuated strains of *Y. pestis*. However, he noted that wild rats showed visible signs of stress in captivity and was concerned about using this model for assessing duration of immunity [121]. Rats were protected by the same cell fractions (predominantly F1) as mice and monkeys. They responded to live attenuated, whole-cell (killed), and F1-based vaccines with significant F1 titers; these titers appeared to correlate with protection from *Y. pestis* challenge [93,120]. It has been demonstrated that protective antibody is passed to newborn rats by their immune dams *in utero* [122].

A recent study reported that 500 CFU of *Y. pestis* strain C092 was lethal to 100% of Brown Norway, Sprague–Dawley, and Wistar rats when administered intrademally in the lower back. Some animals survived, however, when the pathogen was injected intradermally into the ear [123]. Currently, the Brown Norway inbred rat is the most utilized for models of both bubonic and pneumonic plague; the genome of this strain has been completely sequenced, which makes it particularly attractive as a model. Kinetics of disease progression, identification of virulence factors, and transcriptomic analyses have all been reported in the rat [124–127]. The course of disease appears to be similar to that of the human. The intranasal LD_{50} for *Y. pestis* strain CO92 was reported to be approximately 200 CFU whereas the aerosol LD_{50} was calculated to be 1.6×10^3 CFU [125,126]. A strain of *R. norvegicus* "WR," derived from the Wistar strain, was reportedly highly susceptible to plague regardless of age, sex, or season of the year. The LD_{50} of *Y. pestis* in this rat strain was <20 CFU by s.c. challenge. It is not clear, however, whether the WR strain is currently available [120] and the genetic basis for this enhanced sensitivity is unknown.

8.3.5 Rabbit

Rabbits (*Oryctolagus cuniculus*) are known to acquire plague in the wild and have been associated with human disease. Recently, a human case of plague associated with skinning an infected rabbit was reported in the United States [128]. Historically, rabbits were not thought to be a reliable laboratory model, as there is significant variability in the response to challenge doses [31]. The rabbit responds well to *Y. pestis* antigens, however, and has been used extensively in the production of plague antiserum. The antiserum has been demonstrated to passively protect mice [72,96]. As reviewed by Pollitzer [31], some investigators found that rabbits that inhaled *Y. pestis* developed septicemic disease rather than pneumonic plague. Regardless of the concerns cited above, the rabbit model has been revisited in recent years. These studies include vaccine efficacy trials, the identification of essential virulence genes, and an experimental model to evaluate the human body louse as a vector of plague [95,129–131]. In artificial flea-feeding models, fleas fed on rabbit blood "spiked" with *Y. pestis* had lower bacterial burdens and were less likely to maintain the infection than fleas fed rat blood spiked with the same inoculum [132]. It is not clear whether this reflects differences in the way the fleas digest different sources of blood, nutritional differences in the blood sources, or perhaps the presence of substances inhibitory to fleas in the blood of certain mammalian species. Likewise, it is uncertain whether this phenomenon might affect transmission to fleas (and eventually mammalian hosts) in nature. However, it does appear that results of such *in vitro* feeding models are dependent on the type of blood used and this should be kept in mind when designing such experiments. Future studies with inbred and outbred strains of rabbits may eventually produce a more accepted model for plague.

8.3.6 Cotton Rat

The cotton rat (*Sigmodon hispidus*) is relatively resistant to plague [72]. Meyer et al. [85] described an intranasal model of infection for cotton rats along with guinea pigs and mice. The mice and cotton rats succumbed to infection 3–4 days after the intranasal installation, having exhibited clear signs of illness [85]. The resulting primary pneumonia was said to resemble that of humans. The cotton rat appeared to be more difficult to vaccinate than mice or monkeys when a cell fraction containing F1 was used [93].

8.3.7 Ground and Rock Squirrels (*Ostospermophilus beecheyi* and *Ostospermophilus variegatus*)

Many cases of human plague have arisen from contact with the California ground squirrel. The LD_{50} value for *Y. pestis* in ground squirrels collected in California was approximately 25-fold higher than the ID_{50}. Some animals survived challenge without seroconversion, whereas others developed antibody to the capsular antigen F1. Although there was heterogeneity in their response to infection with *Y. pestis*, some animals did develop nasal bleeding, petechial hemorrhage, splenomegaly, and pneumonitis. This animal is useful for surveillance, but has rarely been used for laboratory research on plague [133]. Quan et al. [134] noted that the rock squirrel developed coagulopathy and pneumonia at a frequency similar to that of humans and suggested that this animal might be appropriate as a laboratory model of plague. Gross pathology of disease in the rock squirrel fell into three categories: a rapidly lethal form with hemorrhagic buboes and splenomegaly but without necrotic lesions in the liver and spleen; a subacute form with a longer time to death that was characterized by buboes and necrotic lesions in the spleen and liver; and a nonfatal lymphadenopathy [135].

8.3.8 Vole (*Microtus californicus*)

Studies involving laboratory-bred *Microtus californicus* crossing plague-resistant and plague-susceptible animals suggest that the nature of the innate resistance in this species is multigenic and may be due to differences in phagocytic activity [136]. This model might prove useful in exploring the nature of innate immunity.

8.3.9 Other

There are numerous small animals in nature that serve as reservoirs or contribute to outbreaks of endemic plague. Most have been of interest primarily from the standpoint of epidemiology. However, although extensive laboratory work has not been performed to date with these models, some may prove useful in the future. For example, it was demonstrated that *Y. pestis* can remain latent in hibernating little susliks (*Citellus pygmaeus* Pallas) that had been experimentally infected; such models may be useful in the study of natural plague cycles [137].

Recent reviews of interest include References [9,138–140].

8.4 NHP Models of Plague

NHPs have been used for over a century as a model for plague infection, pathogenesis, and vaccine efficacy. Primates, like humans, are an incidental host for plague. However, there are some differences in plague susceptibility between NHPs and humans and between different species of NHPs. The extent of and basis for these differences are not completely understood. Comparing results from the literature is difficult as often the source of the monkeys used was not reported, and other variables such as the strain of plague, route, and dose of *Y. pestis* confound interpretation. Recently, we have begun to identify subtle differences in human/NHP physiology, anatomy, and immunity that can be traced to differences in the genetic code. The comparability of humans and NHPs at the genetic level is an unfinished story; however, understanding this information will eventually assist in selecting the most appropriate NHP model for the plague researcher. Several species of NHPs have been described in the literature as models to study plague pathogenesis and vaccine efficacy. We will briefly describe the known similarities and differences in plague infection in NHPs and how this compares to the human infection.

8.4.1 Rhesus Macaque

Rhesus macaques (*Macaca mulatta*) have been used extensively in older plague vaccine studies. They have been described as more resistant to s.c. plague infection than other primates, including humans. The s.c. LD_{50} for the rhesus monkey is apparently several million organisms, far above the predicted infective dose of several hundred to several thousand organisms in humans [141,142]. The rhesus monkey is also somewhat resistant to aerosolized *Y. pestis*. The calculated aerosol LD_{50} for the rhesus monkey is 20,000 inhaled organisms, whereas the LD_{50} for humans is estimated to be about 3000 organisms [143–145]. However, a lowered resistance to plague has been observed in the rhesus monkey after intratracheal challenge with a calculated LD_{50} of 100 CFU [141]. Although infection by the intratracheal route does lead to the development of pneumonic plague, there are some differences in the pathology when compared to aerosol-exposed animals. Most notably, the nature of the pneumonia after intratracheal instillation is often more confined than that seen with aerosol delivery, and there is evidence in some primates that the pneumonia may be coincident with a primary septicemia [145]. Although the rhesus monkey can develop acute plague pneumonia, a large number of monkeys exposed to aerosolized plague will develop a protracted disease with unique lesions referred to as chronic pneumonic plague [146]. Because chronic plague is a very infrequent finding in humans, we will not discuss this syndrome at length. However, an important observation regarding the two forms of pneumonic plague in the rhesus monkey is the notation that the numbers of viable bacilli were "controlled" in the chronic form of the disease. Thus, the study of chronic plague in the rhesus monkey may lead to a better understanding of important host factors required for an enhanced protective response to aerosolized *Y. pestis*.

Conversely, the observed lesions in acute pneumonic plague in rhesus monkeys closely match those in humans (Tables 8.5 and 8.6). Fever has been reported to manifest early in the disease and 2–4 days after the animal becomes febrile, the blood pressure and hematocrit begin to fall, with death occurring within 24 h [145]. The levels of circulating eosinophils decline during the course of the disease. Animals develop tachypnea and are rapidly prostrate. In the acute form of the disease, rhesus monkeys develop lobar pneumonia similar to the human disease (Table 8.5). Pneumonic plague, but not bubonic plague, in the rhesus monkey has been reported to cause an early disruption of liver function with rapid colonization of the liver [145]. This observation appears to be the opposite of what happens in humans; however, interpretation is complicated by the use of the intratracheal challenge route in this rhesus study, as intratracheal infection may result in direct plague septicemia.

Finegold et al. [36] examined plague-infected rhesus monkeys for evidence of disseminated intravascular coagulation. Monkeys exposed to aerosols demonstrated a time-dependent increase in clotting times, partial thromboplastin times, mean prothrombin times, and circulating fibrinogen with a concomitant decrease in platelet counts [36]. This phenomenon has been described to variably occur in humans and is likely tempered by the use of antimicrobials in the human cases [37,147]. Collectively, the data suggest that rhesus macaques may not be the best NHP model for plague vaccine studies although their use in this context has historical precedent. Rhesus macaques may be more useful for pathogenesis and innate immunity studies. However, they have been used in vaccine immunity studies [148].

8.4.2 Cynomolgus Macaque

The cynomolgus monkey (*Macaca fascicularis*) has been used in plague vaccine trials since the beginning of the twentieth century. Importantly, in the last 5 years the cynomolgus macaque has become the most published NHP model for plague pathogenesis and vaccine studies [149–158]. The susceptibility to plague has been described to be similar for both the cynomolgus and the rhesus macaques by the s.c. and intratracheal routes. However, a review of the literature suggests a large variance in susceptibility [97]. Whether this difference is real or an artifact of experimental variables remains to be determined. Our own work suggests that the cynomolgus macaque is highly susceptible to aerosolized plague with an LD_{50} of approximately 300 inhaled organisms [159]. This value was determined via a staircase method. More recent studies suggest that the LD_{50} for small-particle aerosols is even lower with values of 24 [158] or 66 [151] CFU as calculated by probit analysis or linear regression, respectively.

The clinical and pathological responses of the cynomolgus macaque are similar to the human disease, although subtle differences can be discerned. As with human pneumonic plague, cynomolgus macaques infected by the aerosol route manifest a fever generally 2 days after infection. Tachycardia and tachypnea are observed and the animals become lethargic. The animals develop detectable rales and lobular and lobar consolidation [97]. Bacteremia can only be detected in the peripheral circulation within 24 h of death and death usually occurs from day 3 to 5 after exposure [159]. The gross pathology of the lungs appears to be similar to the human disease with the exception that the development of fibrinous pleuritis is notably reduced (Table 8.5). This may be a temporal phenomenon as closer examination of lung lesions reveals damage consistent with fibrinous pleuritis (Table 8.6). The dissemination of bacteria outside of the lung appears to be reduced in cynomolgus macaques with lower levels in the spleen and peripheral circulation although this may reflect a dose phenomenon. The necrosis and hemorrhage of the lung, as well as that of the mediastinal lymph nodes, is reduced compared to noted human lesions, again this may be a dose effect or temporal phenomenon. The most recent published studies on the pathology of primary pneumonic plague confirm earlier observations and extend the findings such that previous pathopneumonic findings such as fever and or difficulty in breathing may not always manifest in infected animals [151,158], and there is variability in histopathology [157]. However, these studies all conclude that clinical symptoms, gross lesions, and histopathology were all consistent with end-stage human primary pneumonic plague.

Cynomolgus macaques make a robust, though variable, response to plague antigens. F1-V vaccinated macaques make an antibody response to fraction one (F1), V antigen, LPS, Yop B, Yop D, and Yop M after aerosol plague challenge [160]. This model has also been used to test killed whole cell, live attenuated, and F1 + V recombinant plague vaccines [149,150] or the recombinant fusion protein vaccine rF1-V [152,154,161]. The killed vaccines induced protection against parenteral but not aerosol challenge, whereas the live attenuated and recombinant vaccines protected the majority of animals against significant pneumonic plague morbidity and mortality [97]. We, and others, have attempted to correlate this level of protection with what would be expected to occur in humans. Measures of antibodies against F1 and V antigens seem to be the most promising correlate of immunity [160]. Although absolute titer seems to correlate with protection, the production of antibody to specific epitopes appears critical. The clinical and pathological results collected to date indicate that the cynomolgus macaque is an excellent model for plague pathogenesis and vaccine studies. Additional genetic data for comparative genomic studies are required to determine the ability of this macaque to reflect the human response to infection and treatment.

8.4.3 Vervet

The African green monkey (*Cercopithecus aethiops*) species consists of several subspecies with varied innate resistance to plague. Those described as originating from Kenya were susceptible to infection and mortality by an EV76 vaccine strain whereas Ethiopian green monkeys were not killed by the vaccine strain [97,162,163]. The response of a third subspecies currently located on the Caribbean island of

St. Kitts, originally derived from South Africa, has not been determined. The ability to infect and kill certain subspecies of monkeys with a human vaccine strain calls into question the use of African green monkeys for plague studies. Although susceptibility to wild-type organisms is desirable, the ability to resist attenuated plague organisms is equally important. Note, however, that there are significant differences between EV76 strains and it is not clear whether the aforementioned studies used the same challenge organism.

Fortunately, vaccine studies on non-Kenyon species of African green monkeys have been extremely productive. African green monkeys, like cynomolgus macaques, have an inhaled aerosol LD_{50} [159] of about 300 organisms of wild-type *Y. pestis*. The clinical manifestation of plague in St. Kitts-derived monkeys is similar to that in humans and other NHPs that develop acute pneumonic plague (Table 8.4). The pathology has been reported to be similar to that in humans; however, there did not appear to be a correlation with pathology and challenge dose (Tables 8.5 and 8.6). In contrast, animals challenged s.c. exhibited longer survival and protracted pathology with lower challenge doses [162]. An LD_{50} for the s.c. route was not calculated; however, it was estimated to fall between several hundred and several thousand organisms—similar to the predicted LD_{50} for humans. Conversely, the LD_{50} by the intradermal route has been reported to be between 5 and 50 CFU, which may be significantly lower than in humans [164].

African green monkeys originating in South Africa or Ethiopia have been used in vaccine studies with mixed results. Although the animals exhibited a robust anti-F1 antibody response, an oral live attenuated plague vaccine protected 3 of 6 vaccinated monkeys from pneumonic plague, whereas a recombinant plague vaccine protected only 7 of 28 animals from aerosolized plague [159,164]. Similar mixed partial protection results were also observed with various live vaccine trials in African green monkeys challenged by aerosol or s.c. [97]. The variable response to all plague vaccines tested to date, as well as an obvious "susceptible" plague phenotype, make interpretation of these vaccine studies difficult. Collectively, these data raise concerns about the utility of using African green monkeys for plague vaccine trials. Conversely, these monkeys are useful for plague pathogenesis and pathology studies. In addition, the African green monkey has been found to be suitable for evaluation of plague therapeutics [159].

8.4.4 Langur

Langur monkeys (*Semnopithecus* or *Presbytis entellus*) were some of the first animals to be chosen for the study of bubonic plague. The langur monkeys were described to be "variably susceptible" to plague [162]. Like the African green monkey, they have been noted to be much more susceptible to plague s.c. than rhesus macaques. The s.c. LD_{50} for langurs was reported to be 210 CFU [162]. Langur monkeys have been noted to succumb to fatal infection with the live attenuated human plague vaccine strain; this is also similar to observations with the vervet [97].

The incubation period for bubonic plague in this model is 2–4 days after infection, with most deaths occurring between days 3 and 10. Clinically, most infected langurs initially manifest with fever, and septicemia as well as other symptoms of bubonic plague (Table 8.1). The symptoms rapidly progress, with death immediately preceded by prostration and tachypnea. Organisms can be recovered from the spleen and heartblood. Animals that survive infection make a robust antibody response and survivors appear to be resistant to reinfection [165].

Langur monkeys develop pathological lesions after s.c. inoculation that are similar to human bubonic plague [165]. Most notably, like humans, the formation of primary bubos is variable; however, there is a tendency to form petechial hemorrhages in the skin and intestinal tract at a frequency higher than that noted in humans (Table 8.6). Infection of langurs s.c. seems to most resemble what has been described as septicemic plague in humans; thus, the langur may make an appropriate model to study plague septicemia. Only limited aerosol/transtracheal challenges of langurs have been described in the literature and it is unclear whether these data represent primary pneumonic plague in humans [162,165]. The few clinical and pathological lesions that have been described are noted in Tables 8.4 through 8.6.

8.4.5 Sacred Baboons (*Papio hamadryas*)

The baboon has been used extensively in the FSU for evaluation of plague therapeutics and vaccines. Both s.c. and aerosol models have been described [166,167]. The s.c. LD_{50} in this model is approximately 2000 CFU and this route of administration causes a glandular plague that is similar to the bubonic form. Approximately 18% of these infected baboons developed a secondary plague pneumonia. The baboon and guinea pig have been used as models for assessment of FSU live and subunit vaccine candidates [168].

8.4.6 Marmoset (*Callithrix* sp.)

There are currently no published plague vaccine or pathogenesis studies using marmosets. However, the size, reproductive capacity, and overall relatedness to humans make the marmoset an attractive animal model for future plague pathogenesis and vaccine studies. The marmoset has recently been successfully used to study pathogenesis of several infectious diseases, including human herpes viruses, malaria, *Legionella* and *Chlamydia*.

8.5 OTHER MODELS OF PLAGUE

Domestic cats and dogs can become intermediate hosts for plague in endemic areas. However, outside of epidemiological and diagnostic studies, neither dogs nor cats have been used extensively as a model for studying plague pathogenesis or plague vaccines. Dogs, like most carnivores, are relatively resistant to plague morbidity/mortality but may become transiently infected and transmit the disease. On rare occasions, dogs have been described to succumb to plague infection, especially by the aerosol route [108]. Conversely, the cat is more susceptible to plague infection and manifests many clinical symptoms and lesions similar to those observed in humans

[169]. Dogs make a rapid and sustained antibody response to plague; this makes dogs useful epidemiological sentinels and producers of diagnostic reagents [170].

Cats and dogs frequently come into contact with plague through ingestion of infected animals or the acquisition of infected fleas during forays into the habitat of infected rodents/mammals. In the case of dogs, transmission to humans can be mediated by dog fleas that previously fed on a bacteremic dog followed by mechanical transmission of dog or rodent fleas into human habitat. Dogs that become infected orally may also carry culturable organisms in their throat and tonsils for several days after oral infection [169,171]. There is recent evidence that the dog can transmit an infectious dose to humans by the respiratory route. A recent case of primary pneumonic plague in China was likely due to a terminally infected dog that transmitted infection to its owner. The owner transmitted pneumonic plague to 12 other close contacts. The owner and two contacts subsequently died [172].

Conversely, there is ample evidence that cats are the most prevalent source of primary pneumonic plague for humans in the United States. Several cases of cat-transmitted primary pneumonic plague in humans have been documented [5,173]. Cats can be infected by the aerosol route and develop primary pneumonic plague similar to humans (Table 8.4). They may also become acutely infected with plague after ingesting infected animals [174]. After ingesting an infected rodent, cats become acutely ill and bacteremic. Most infected cats will succumb to infection within 7 days; however, cats can also become chronically infected with a disease course lasting several weeks. Like chronically infected dogs, cats carry plague in their throat and pharyngeal tissue [174,175]. Most, but not all, infected cats will eventually succumb to plague infection without treatment. This may make the cat suitable to investigate innate immunity to plague and late-course plague pathogenesis, in particular the development of secondary plague pneumonia.

Cats develop lesions in their lymphoid tissue and internal organs consistent with systemic plague infection in humans after a flea bite (Tables 8.2 and 8.3). However, they tend to form partially encapsulated abscesses containing some fibrin around infected lymph nodes, whereas humans generally fail to encapsulate infected nodes. In addition, when infected by the oral or aerosol route, cats develop necrotic abscesses, particularly about the head and neck. These lesions are associated with bilaterally infected submandibular or cervical lymph nodes [174,175]. Similar lesions have been observed in both humans and NHPs. Feline lesions eventually rupture and can infect humans who come in contact with the purulent material. The importance of this mode of transmission in a human-to-human context is unknown, but is likely to be restricted. As opposed to cats, fistulating lymph nodes are only occasionally observed in humans; however, this difference could be due to the protracted length of infection in some cats. Likewise, the importance of oral infection in humans is unclear. Humans have been noted to develop buboes in cervical lymph nodes associated with plague septicemia and it is possible that the oral route was the portal of exposure [176]. Cats, therefore, may also be an appropriate model to study oral infection. Cats also make a robust antibody response to F1 and other plague antigens. This may indicate that they could be considered as animal models for plague vaccine and pathogenesis studies.

8.6 SUMMARY

Recent years have brought a resurgence of effort into plague research. We have tried in this review to draw from both these modern approaches and the classic studies of pathogenesis and immunology conducted by the pioneers of the field. Their work, some of it over a century old, is still relevant for investigators interested in *Y. pestis*, the pathogenesis of plague, and animal models of disease.

In particular, we recommend the World Health Organization monograph by Pollitzer [31] and the extensive reviews on plague by Meyer and his colleagues, including the supplementary issue of the *Journal of Infectious Disease* published in 1974 that celebrated the long and distinguished career of Dr. Meyer [73,74,177]. These scientists left a wealth of insight into research on plague pathogenesis, epidemiology, immunology, and vaccine development—beginning with Yersin himself.

Combining the astute and painstaking observations of our predecessors with our modern facilities and knowledge of genomics, bacterial physiology, and animal husbandry can only accelerate our progress. Likewise, a better understanding of the models chosen by investigators outside the western scientific community will be a tremendous advantage in choosing the most appropriate mirrors for human disease and immunity.

REFERENCES

1. Smith C.R. et al., Plague studies in California: A review of long-term disease activity, flea-host relationships and plague ecology in the coniferous forests of the Southern Cascades and northern Sierra Nevada mountains. *J Vector Ecol.* 2010, **35**(1):1–12.
2. Poland, J.D. and D.T. Dennis, Plague, in *Infectious Diseases of Humans: Epidemiology and Control*, A.S. Evans and P.S. Brachman, editors. New York: Plenum, 1998, pp. 545–558.
3. Hinnebusch, B.J., Bubonic plague: A molecular genetic case history of the emergence of an infectious disease. *J Mol Med*, 1997, **75**(9): 645–652.
4. Tigertt, W.D., Plague, in *Bacterial Infections of Humans*, A.S. Evans and P.S. Brachman, editors. New York: Plenum, 1991, pp. 513–523.
5. Doll, J.M. et al., Cat-transmitted fatal pneumonic plague in a person who traveled from Colorado to Arizona. *Am J Trop Med Hyg*, 1994, **51**(1): pp. 109–114.
6. Gage, K.L. et al., Cases of cat-associated human plague in the Western US, 1977–1998. *Clin Infect Dis*, 2000, **30**(6): 893–900.
7. Perry, R.D. and J.D. Fetherston, *Yersinia pestis*—Etiologic agent of plague. *Clin Microbiol Rev*, 1997, **10**(1): 35–66.
8. McEvedy, C., The bubonic plague. *Sci Am*, 1988, **258**: 118–123.
9. Dennis, D.T., Plague as an emerging disease, in *Emerging Infections*, Vol. 2, W.M. Scheld, W.A. Craig, and J.M. Hughes, editors. Washington, DC: ASM Press, 1998.
10. Bercovier, H. and H.M. Mollaret, *Yersinia*, in *Bergey's Manual of Systematic Bacteriology*, N.R. Kreig and J.G. Holt, editors. Baltimore, MD: Williams & Wilkens, 1984, pp. 498–503.
11. Bercovier, H. et al., Intra-and interspecies relatedness of *Yersinia pestis* by DNA hybridization and its relationship to *Yersinia pseudotuberculosis. Curr Microbiol*, 1980, **4**: 225–229.
12. Achtman, M. et al., Microevolution and history of the plague bacillus, *Yersinia pestis. Proc Natl Acad Sci USA*, 2004, **101**(51): 17837–17842.

13. Bockemuhl, J. and J.D. Wong, Yersinia, in *Manual of Clinical Microbiology*, P.R. Murray, editor. Washington, DC: ASM Press, 2003, pp. 672–683.
14. Achtman, M. et al., *Yersinia pestis*, the cause of plague, is a recently emerged clone of *Yersinia pseudotuberculosis. Proc Natl Acad Sci USA*, 1999, **96**(24): 14043–14048.
15. Guiyoule, A. et al., Plague pandemics investigated by ribotyping of *Yersinia pestis* strains. *J Clin Microbiol*, 1994. **32**(3): 634–641.
16. Worsham, P.L. and M. Hunter. Genetic Diversity of Yersinia pestis: Selection of a Panel of Strains for Vaccine Testing. *Abstract, Annual Meeting of the American Society for Microbiology*, New Orleans Louisiana, 24–27 May 2004.
17. Worsham, P.L. and C. Roy, Pestoides F, a *Yersinia pestis* strain lacking plasminogen activator, is virulent by the aerosol route. *Adv Exp Med Biol*, 2003, **529**: 129–131.
18. Morelli, G. et al., *Yersinia pestis* genome sequencing identifies patterns of global phylogenetic diversity. *Nat Genet*, 2010, **42**(12): 1140–1143.
19. CDC, ASM, and APHL. *Basic Protocols for Level A Laboratories for the Presumptive Identification of* Yersinia pestis, 2002 [cited; Available from: www.ype.asm.cp.042202].
20. Dennis, D.T. et al. *WHO/CDS/CSR/EDC/99.2 Plague Manual: Epidemiology, Distribution, Surveillance, and Control.* 1999 [cited; Available from: http://www.who.int/csr/resources/publications/plague/WHO_CDS_CSR_EDC_99_2_EN/en/].
21. Thullier, P. et al., Short report: Serodiagnosis of plague in humans and rats using a rapid test. *Am J Trop Med Hyg*, 2003, **69**(4): 450–451.
22. Radnedge, L. et al., Identification of nucleotide sequences for the specific and rapid detection of *Yersinia pestis. Appl Environ Microbiol*, 2001, **67**(8): 3759–3762.
23. Zhou, D. et al., Identification of signature genes for rapid and specific characterization of *Yersinia pestis. Microbiol Immunol*, 2004, **48**(4): 263–269.
24. Stewart, A. et al., A quadruplex real-time PCR assay for the detection of *Yersinia pestis* and its plasmids. *J Med Microbiol*, 2008, **57**(Pt 3): 324–331.
25. Higgins, J.A. et al., 5′ uclease PCR assay to detect *Yersinia pestis. J Clin Microbiol*, 1998, **36**(8): 2284–2288.
26. Engelthaler, D.M. et al., PCR detection of *Yersinia pestis* in fleas: Comparison with mouse inoculation. *J Clin Microbiol*, 1999, **37**(6): 1980–1984.
27. Riehm, J.M. et al., Detection of *Yersinia pestis* using real-time PCR in patients with suspected bubonic plague. *Mol Cell Probes*, 2011, **25**(1): 8–12.
28. Worsham et al., Plague, in *Textbook of Military Medicine: Medical Aspects of Chemical and Biological Warfare*, Z.F. Dembeck, editor. Washington DC: Office of the Surgeon General, 2007, pp. 91–119.
29. Butler, T., Plague, in *Plague and other Yersinia Infections*, W.B. Greenough and T.C. Harigan, editors. New York: Plenum Press, 1983, pp. 73–108.
30. Poland, J.D. and A.M. Barnes, Plague, in *CRC Handbook Series in Zoonoses. Section A. Bacterial, Rickettsial, Chlamydial, and Mycotic Diseases*, J.H. Steele, editor. Boca Raton, FL: CRC Press, 1979, pp. 515–559.
31. Pollitzer, R., Plague. *WHO Monogr Ser*, 1954, **22**: 1–698.
32. Dennis, D.T. and F.A. Meier, Plague, in *Pathology of Emerging Infections*, C.R. Horsburgh and A.M. Nelson, editors. Washington, DC: ASM Press, 1997, pp. 21–48.
33. Smith, J.H. and B.S. Reisner, Plague, in *Pathology of Infectious Diseases*, D.H. Conner, editor. New Jersey: Prentice Hall, 1997, pp. 729–738.
34. Rollins, S.E., S.M. Rollins, and E.T. Ryan, *Yersinia pestis* and the plague. *Am J Clin Pathol*, 2003, **119**(Suppl): S78–S85.
35. Inglesby, T.V. et al., Plague as a biological weapon: Medical and public health management. Working Group on Civilian Biodefense. *JAMA*, 2000, **283**(17): 2281–2290.

36. Finegold, M.J. et al., Studies on the pathogenesis of plague. Blood coagulation and tissue responses of *Macaca mulatta* following exposure to aerosols of *Pasteurella pestis*. *Am J Pathol*, 1968, **53**(1): 99–114.
37. Crook, L.D. and B. Tempest, Plague. A clinical review of 27 cases. *Arch Intern Med*, 1992, **152**(6): 1253–1256.
38. Benner, G.E. et al., Immune response to Yersinia outer proteins and other *Yersinia pestis* antigens after experimental plague infection in mice. *Infect Immun*, 1999, **67**(4): 1922–1928.
39. Rust, J.H., Jr. et al., Susceptibility of rodents to oral plague infection: a mechanism for the persistence of plague in inter-epidemic periods. *J Wildl Dis*, 1972, **8**(2): 127–133.
40. Tieh, T.H., Primary pneumonic plague in Mukden, 1946, and report of 39 cases with 3 recoveries. *J Infect Dis*, 1948, **82**: 52–58.
41. Christie, A.B., T.H. Chen, and S.S. Elberg, Plague in camels and goats: Their role in human epidemics. *J Infect Dis*, 1980, **141**(6): 724–726.
42. Meyer, K.F., Pasturella and Francisella, in *Bacterial and Mycotic Infections in Man*, R.J. Dubos and J.G. Hirsch, editors. Philadelphia: Lippincott, 1965, pp. 659–697.
43. Cleri, D.J. et al., Plague pneumonia disease caused by *Yersinia pestis*. *Semin Respir Infect*, 1997, **12**(1): 12–23.
44. Williams, J.E. et al., Potency of killed plague vaccines prepared from avirulent *Yersinia pestis*. *Bull World Health Organ*, 1980, **58**(5): 753–756.
45. von Metz, E., D.M. Eisler, and G.A. Hottle, Immunogenicity of plague vaccines in mice and guinea pigs. *Appl Microbiol*, 1971, **22**(1): 84–88.
46. Anderson, G.W., Jr. et al., Protection of mice from fatal bubonic and pneumonic plague by passive immunization with monoclonal antibodies against the F1 protein of *Yersinia pestis*. *Am J Trop Med Hyg*, 1997, **56**(4): 471–473.
47. Elvin, S.J. and E.D. Williamson, Stat 4 but not Stat 6 mediated immune mechanisms are essential in protection against plague. *Microb Pathog*, 2004, **37**(4): 177–184.
48. Elvin, S.J. et al., Evolutionary genetics: Ambiguous role of CCR5 in *Y. pestis* infection. *Nature*, 2004, **430**(6998): 417–418.
49. Green, M. et al., The SCID/Beige mouse as a model to investigate protection against *Yersinia pestis*. *FEMS Immunol Med Microbiol*, 1999, **23**(2): 107–113.
50. Anderson, G.W., Jr. et al., Recombinant V antigen protects mice against pneumonic and bubonic plague caused by F1-capsule-positive and -negative strains of *Yersinia pestis*. *Infect Immun*, 1996, **64**(11): 4580–4585.
51. Simpson, W.J., R.E. Thomas, and T.G. Schwan, Recombinant capsular antigen (fraction 1) from *Yersinia pestis* induces a protective antibody response in BALB/c mice. *Am J Trop Med Hyg*, 1990, **43**(4): 389–396.
52. Eyles, J.E. et al., Intra nasal administration of poly-lactic acid microsphere co-encapsulated *Yersinia pestis* subunits confers protection from pneumonic plague in the mouse. *Vaccine*, 1998, **16**(7): 698–707.
53. Garmory, H.S. et al., Protection against plague afforded by immunisation with DNA vaccines optimised for expression of the *Yersinia pestis* V antigen. *Vaccine*, 2004, **22**(8): 947–957.
54. Torres-Escobar, A. et al., Fine-tuning synthesis of *Yersinia pestis* LcrV from runaway-like replication balanced-lethal plasmid in a *Salmonella enterica* serovar typhimurium vaccine induces protection against a lethal Y. pestis challenge in mice. *Infect Immun*, 2010, **78**(6): 2529–2543.
55. Little, S.F. et al., Quantitative anti-F1 and anti-V IgG ELISAs as serological correlates of protection against plague in female Swiss Webster mice. 2010, *Vaccine*, **28**(4): 934–939.

56. Lin, J.S. et al., TNFalpha and IFNgamma contribute to F1/LcrV-targeted immune defense in mouse models of fully virulent pneumonic plague. 2010, *Vaccine*, **29**(2): 357–362.
57. Russell, P. et al., Efficacy of doxycycline and ciprofloxacin against experimental *Yersinia pestis* infection. *J Antimicrob Chemother*, 1998, **41**(2): 301–305.
58. Byrne, W.R. et al., Antibiotic treatment of experimental pneumonic plague in mice. *Antimicrob Agents Chemother*, 1998, **42**(3): 675–681.
59. Rahalison, L. et al., Failure of oily chloramphenicol depot injection to treat plague in a murine model. *J Antimicrob Chemother*, 2000, **45**(4): 541–545.
60. Steward, J. et al., Efficacy of the latest fluoroquinolones against experimental *Yersinia pestis*. *Int J Antimicrob Agents*, 2004, **24**(6): 609–612.
61. Xiao, X. et al., Human anti-plague monoclonal antibodies protect mice from *Yersinia pestis* in a bubonic plague model. *PLoS One*, 2010, **5**(10): e13047.
62. Ayyadurai, S. et al., Lovastatin protects against experimental plague in mice. *PLoS One*, 2010, **5**(6): e10928.
63. Jones, S.M. et al., Protection conferred by a fully recombinant sub-unit vaccine against *Yersinia pestis* in male and female mice of four inbred strains. *Vaccine*, 2001, **19**(2–3): 358–366.
64. Brandler, P. et al., Weak anamnestic responses of inbred mice to Yersinia F1 genetic vaccine are overcome by boosting with F1 polypeptide while outbred mice remain nonresponsive. *J Immunol*, 1998, **161**(8): 4195–4200.
65. Nazarova, L.S. et al., Morphological study of the damaging effect of an EB vaccine strain of the plague microbe in inbred mice (in Russian). *Biull Eksp Biol Med*, 1988, **105**(6): 761–764.
66. Russell, P. et al., A comparison of plague vaccine, USP and EV76 vaccine induced protection against *Yersinia pestis* in a murine model. *Vaccine*, 1995, **13**(16): 1551–1556.
67. Turner, J.K. et al., The resistance of BALB/cJ mice to *Yersinia pestis* maps to the major histocompatibility complex of chromosome 17. *Infect Immun*, 2008, **76**(9): 4092–4099.
68. Turner, J.K., J.L. Xu, and R.I. Tapping, Substrains of 129 mice are resistant to *Yersinia pestis* KIM5: Implications for interleukin-10-deficient mice. *Infect Immun*, 2009, **77**(1): 367–373.
69. Welkos, S.L. et al., Studies on the contribution of the F1 capsule-associated plasmid pFra to the virulence of *Yersinia pestis*. *Contrib Microbiol Immunol*, 1995, **13**: 299–305.
70. Davis, K.J. et al., Pathology of experimental pneumonic plague produced by fraction 1-positive and fraction 1-negative *Yersinia pestis* in African green monkeys (*Cercopithecus aethiops*). *Arch Pathol Lab Med*, 1996, **120**(2): 156–163.
71. Worsham, P.L., M.P. Stein, and S.L. Welkos, Construction of defined F1 negative mutants of virulent *Yersinia pestis*. *Contrib Microbiol Immunol*, 1995, **13**: 325–328.
72. Meyer, K.F., Immunity in plague: a critical consideration of some recent studies. *J Immunol*, 1950, **64**: 139–163.
73. Walker, D.L. et al., Studies on immunization against plague. V. Multiplication and persistence of virulent and avirulent *Pasteurella pestis* in mice and guinea pigs. *J Immunol*, 1953, **70**(3): 245–252.
74. Titball, R.W. and E.D. Williamson, Vaccination against bubonic and pneumonic plague. *Vaccine*, 2001, **19**(30): 4175–4184.
75. Meyer, K.F., Pneumonic plague. *Bacteriol Rev*, 1961, **25**: 249–261.
76. Martini, E., Ueber inhalationspest der ratten. *Z Hyg Infektionskrankh*, 1901, **38**: 332–342.
77. Meyer, K.F. and A. Larson. The pathogenesis of cervical septicemic plague developing after exposure to pneumonic plague prdouced by intratracheal infection in primates, in *Proceedings of the Symposium of the Diamond Jubilee of Haffkine Institute*, January 10–14, 1960, Bombay, India, pp. 1–12.

78. Agar, S.L. et al., Characterization of a mouse model of plague after aerosolization of *Yersinia pestis* CO92. *Microbiology*, 2008, **154**(Pt 7): 1939–1948.
79. Glynn, A. et al., Protection against aerosolized *Yersinia pestis* challenge following homologous and heterologous prime-boost with recombinant plague antigens. *Infect Immun*, 2005, **73**(8): 5256–5261.
80. Wang, S. et al., A DNA vaccine producing LcrV antigen in oligomers is effective in protecting mice from lethal mucosal challenge of plague. *Vaccine*, 2004, **22**(25–26): 3348–3357.
81. Sha, J. et al., Braun lipoprotein (Lpp) contributes to virulence of yersiniae: Potential role of Lpp in inducing bubonic and pneumonic plague. *Infect Immun*, 2008, **76**(4): 1390–1409.
82. Bubeck, S.S., A.M. Cantwell, and P.H. Dube, Delayed inflammatory response to primary pneumonic plague occurs in both outbred and inbred mice. *Infect Immun*, 2007, **75**(2): 697–705.
83. Lathem, W.W. et al., A plasminogen-activating protease specifically controls the development of primary pneumonic plague. *Science*, 2007, **315**(5811): 509–513.
84. Meyer, K.F., S.F. Quan, and A. Larson. Prophylactic immunization and specific therapy of experimental pneumonic plague, in *43rd Annual Meeting of the National Tuberculosis Foundation*, 1947, San Francisco, California.
85. Meyer, K.F., S.F. Quan, and A. Larson, Prophylactic immunization and specific therapy of experimental pneumonic plague. *Am Rev Tuberc*, 1948, **57**: 312–321.
86. Thomas, R.J. et al., Different pathologies but equal levels of responsiveness to the recombinant F1 and V antigen vaccine and ciprofloxacin in a murine model of plague caused by small- and large-particle aerosols. *Infect Immun*, 2009, **77**(4): 1315–1323.
87. Straley, S.C. and W.S. Bowmer, Virulence genes regulated at the transcriptional level by Ca^{2+} in *Yersinia pestis* include structural genes for outer membrane proteins. *Infect Immun*, 1986, **51**(2): 445–454.
88. Butler, T. et al., Experimental *Yersinia pestis* infection in rodents after intragastric inoculation and ingestion of bacteria. *Infect Immun*, 1982, **36**(3): 1160–1167.
89. Kokushkin, A.M. et al., The results of the measured oral infection of white mice with strains of the causative agent of plague differing in their plasmid profile (in Russian). *Zh Mikrobiol Epidemiol Immunobiol*, 1994(2): 15–20.
90. Spivak, M.L. et al., The immune response of the guinea pig to the antigens of *Pasteurella pestis*. *J Immunol* 1958, **80**(2): 132–141.
91. Lawton, W.D., G.M. Fukui, and M.J. Surgalla, Studies on the antigens of *Pastuerella pestis* and *Pastueurella pseudotuberculosis*. *J Immunol*, 1960, **84**: 475–479.
92. Chen, T.H., L.E. Foster, and K.F. Meyer, Comparison of the immune response to three different *Yersinia pestis* vaccines in guinea pigs and langurs. *J Infect Dis*, 1974, **129**(Suppl): S53–S61.
93. Meyer, K.F., Experimental appraisal of antiplague vaccination with dead virulent and living avirulent plague bacilli. *Abstr Int Congr Trop Med Malar*, 1948, **56**(4th Congr):16.
94. Jones, S.M. et al., Protective efficacy of a fully recombinant plague vaccine in the guinea pig. *Vaccine*, 2003, **21**(25–26): 3912–3918.
95. Qi, Z. et al., Comparison of mouse, guinea pig and rabbit models for evaluation of plague subunit vaccine F1 + rV270. *Vaccine*, 2010, **28**(6): 1655–1660.
96. Jawetz, E. and K.F. Meyer, The behavior of virulent and avirulent *P. pestis* in normal and immune experimental animals. *J Infect Dis*, 1944, **74**: 1–13.
97. Meyer, K.F. et al., Live, attenuated *Yersinia pestis* vaccine: virulent in nonhuman primates, harmless to guinea pigs. *J Infect Dis*, 1974, **129**(Suppl): S85–S120.
98. Anisimov, A.P., L.E. Lindler, and G.B. Pier, Intraspecific diversity of *Yersinia pestis*. *Clin Microbiol Rev*, 2004, **17**(2): 434–464.
99. Kovaleva, R.V., Certain characteristics of *Pasteurella pestis* strains isolated from *Microtus brandii* and other rodents. *Zh Mikrobiol Epidemiol Immunobiol*, 1958, **8**: 30–36.

100. Welkos, S. et al., Determination of the virulence of the pigmentation-deficient and pigmentation-/plasminogen activator-deficient strains of *Yersinia pestis* in non-human primate and mouse models of pneumonic plague. *Vaccine*, 2002, **20**(17–18): 2206–2214.
101. Worsham, P., Hunter, M., Mou, S, Bassett, A., and Fritz, D., Virulence of the *Yersinia pestis* strain Pestoides B in the mouse and guinea pig models. Abstract, *International Symposium on Yersinia*, Lexington Kentucky, 2006.
102. Lebedinskii, V.A. et al., Experience using fraction I of the plague microbe for revaccinating experimental animals (in Russian). *Zh Mikrobiol Epidemiol Immunobiol*, 1982(5): 60–63.
103. Smith, H. and L.P. Packman, A filtered non-toxic plague vaccine which protects guinea-pigs and mice. *Br J Exp Pathol*, 1966, **47**(1): 25–34.
104. Burrows, T.W. and G.A. Bacon, The effects of loss of different virulence determinants on the virulence and immunogenicity of strains of *Yersinia pestis*. *Br J Exp Pathol*, 1958, **39**: 278–291.
105. Burrows, T.W. and W.A. Gillett, Host specificity of Brazilian strains of Pasteurella pestis. *Nature*, 1971, **229**(5279): 51–52.
106. Wayson, N.E., C. McMahon, and F.M. Prince, An evaluation of three plague vaccines against infection in guinea pigs induced by natural and artificial methods. *Public Health Rep*, 1946, **61**: 1511–1518.
107. Martini, E., Ueber die wirkung des pestserums bei experimenteller pestpneumonie an ratten, mausen, meerschweinchen, und kaninchen. *Klin Jahrb*, 1902, **10**: 137–176.
108. Strong, R.P. and O. Teague, Studies on pneumonic plague and plague immunization. VIII. Susceptibility of animals to pneumonic plague. *Phillip J Sci*, 1912, **7B**: 223–228.
109. Strong, R.P. and O. Teague, Studies on pneumonic plague and plague immunization. *Phillip J Sci*, 1912, **7B**: 173–180.
110. Druett, H.A. et al., Studies on respiratory infection. II. The influence of aerosol particle size on infection of the guinea pig with Pasteurella pestis. *J Hyg (Lond)*, 1956, **54**: 37–48.
111. Bablet, J. and G. Girard, Lesions histologiques dans le peste pulmonaire primitive experimentale du cobaye. *Ann Inst Pasteur (Paris)*, 1934, **52**: 155–165.
112. Smit, A.A. and H.F. Van der Bank, Isozyme and allozyme markers distinguishing two morphologically similar, medically important Mastomys species (Rodentia: Muridae). *BMC Genet*, 2001, **2**(1): 15.
113. Green, C.A., D.H. Gordon, and N.F. Lyons, Biological species in Praomys (Mastomys) natalensis (Smith), a rodent carrier of Lassa virus and bubonic plague in Africa. *Am J Trop Med Hyg*, 1978, **27**(3): 627–629.
114. Shepherd, A.J. et al., Comparative tests for detection of plague antigen and antibody in experimentally infected wild rodents. *J Clin Microbiol*, 1986, **24**(6): 1075–1078.
115. Shepherd, A.J., P.A. Leman, and D.E. Hummitzsch, Experimental plague infection in South African wild rodents. *J Hyg (Lond)*, 1986, **96**(2): 171–183.
116. Arntzen, L., A.A. Wadee, and M. Isaacson, Immune responses of two Mastomys sibling species to *Yersinia pestis*. *Infect Immun*, 1991, **59**(6): 1966–1971.
117. Hallett, A.F., Evaluation of live attenuated plague vaccines in Praomys (Mastomys) natalensis. *Infect Immun*, 1977, **18**(1): 8–13.
118. Williams, J.E. and D.C. Cavanaugh, Potential for rat plague from nonencapsulated variants of the plague bacillus (*Yersinia pestis*). *Experientia*, 1984, **40**(7): 739–740.
119. Chen, T.H. and K.F. Meyer, Susceptibility and antibody response of *Rattus* species to experimental plague. *J Infect Dis*, 1974, **129**(Suppl): S62–S71.
120. Williams, J.E. and D.C. Cavanaugh, Measuring the efficacy of vaccination in affording protection against plague. *Bull World Health Organ*, 1979, **57**(2): 309–313.
121. Otten, L., Immunization against plague with live vaccine. *Indian J Med Res*, 1936, **24**(1): 73–101.

122. Williams, J.E. et al., Antibody and resistance to infection with *Yersinia pestis* in the progeny of immunized rats. *J Infect Dis*, 1974, **129**(Suppl): S72—S77.
123. Sebbane, F. et al., Kinetics of disease progression and host response in a rat model of bubonic plague. *Am J Pathol*, 2005, **166**(5): 1427–1439.
124. Lemaitre, N. et al., *Yersinia pestis* YopJ suppresses tumor necrosis factor alpha induction and contributes to apoptosis of immune cells in the lymph node but is not required for virulence in a rat model of bubonic plague. *Infect Immun*, 2006, **74**(9): 5126–5131.
125. Agar, S.L. et al., Characterization of the rat pneumonic plague model: Infection kinetics following aerosolization of *Yersinia pestis* CO92. *Microbes Infect*, 2009, **11**(2): 205–214.
126. Anderson, D.M. et al., Pneumonic plague pathogenesis and immunity in Brown Norway rats. *Am J Pathol*, 2009, **174**(3): 910–921.
127. Comer, J.E. et al., Transcriptomic and innate immune responses to *Yersinia pestis* in the lymph node during bubonic plague. *Infect Immun*, 2010, **78**(12): 5086–5098.
128. Lowell, J.L. et al., Identifying sources of human exposure to plague. *J Clin Microbiol*, 2005, **43**(2): 650–656.
129. Houhamdi, L. et al., Experimental model to evaluate the human body louse as a vector of plague. *J Infect Dis*, 2006, **194**(11): 1589–1596.
130. Zhang, Q.W. et al., Evaluation of immunization protection efficacy of plague subunit vaccine (in Chinese). *Zhonghua Yu Fang Yi Xue Za Zhi*, 2009, **43**(9): 785–788.
131. Andrews, G.P. et al., Identification of *in vivo*-induced conserved sequences from *Yersinia pestis* during experimental plague infection in the rabbit. *Vector Borne Zoonotic Dis*, 2010, **10**(8): 749–756.
132. Eisen, R.J. et al., Source of host blood affects prevalence of infection and bacterial loads of *Yersinia pestis* in fleas. *J Med Entomol*, 2008, **45**(5): 933–938.
133. Williams, J.E., M.A. Moussa, and D.C. Cavanaugh, Experimental plague in the California ground squirrel. *J Infect Dis*, 1979, **140**(4): 618–621.
134. Quan, T.J. et al., Experimental plague in rock squirrels, Spermophilus variegatus (Erxleben). *J Wildl Dis*, 1985, **21**(3): 205–210.
135. McCoy, G.W., Studies on plague in ground squirrels. *Public Health Bull*, 1911, **43**: 1–51.
136. Hubbert, W.T. and M.I. Goldenberg, Natural resistance to plague: Genetic basis in the vole (*Microtus californicus*). *Am J Trop Med Hyg*, 1970, **19**(6): 1015–1019.
137. Bizanov, G. and N.D. Dobrokhotova, Experimental infection of ground squirrels (*Citellus pygmaeus* Pallas) with *Yersinia pestis* during hibernation. *J Infect*, 2007, **54**(2): 198–203.
138. Swearengen, J.R. and P.L. Worsham, Plague, in *Emerging Diseases of Animals*, C. Brown and C. Bolin, editors. Washington, DC: ASM Press, 2000, pp. 259–279.
139. Anisimov, A.P., Factors of *Yersinia pestis* providing circulation and persistence of plague pathogen in ecosystems of natural foci. Communication 2 (in Russian). *Mol Gen Mikrobiol Virusol*, 2002(4): 3–11.
140. Anisimov, A.P., *Yersinia pestis* factors, assuring circulation and maintenance of the plague pathogen in natural foci ecosystems. Report 1 (in Russian). *Mol Gen Mikrobiol Virusol*, 2002(3): 3–23.
141. Ehrenkranz, N.J. and K.F. Meyer, Studies on immunization against plague. VIII. Study of three immunizing preparations in protecting primates against pneumonic plague. *J Infect Dis*, 1955, **96**(2): 138–144.
142. Hinnebusch, B.J., K.L. Gage, and T.G. Schwan, Estimation of vector infectivity rates for plague by means of a standard curve-based competitive polymerase chain reaction method to quantify *Yersinia pestis* in fleas. *Am J Trop Med Hyg*, 1998, **58**(5): 562–569.
143. SIPRI, The problem of chemical and biological warfare, in *Chemical Biological Weapons Today*, S. Institute, editor. New York: Humanities, 1973, pp. 3–6.

144. Speck, R.S. and H. Wolochow, Studies on the experimental epidemiology of respiratory infections. VIII. Experimental pneumonic plague in Macacus rhesus. *J Infect Dis*, 1957, **100**(1): 58–69.
145. Ehrenkranz, N.J. and L.P. White, Hepatic function and other physiologic studies in monkeys with experimental pneumonic plague. *J Infect Dis*, 1954, **95**(3): 226–231.
146. Ransom, J.P. and A.P. Krueger, Chronic pneumonic plague in *Macaca mulatta. Am J Trop Med Hyg*, 1954, **3**(6): 1040–1054.
147. Finegold, M.J., Pathogenesis of plague. A review of plague deaths in the United States during the last decade. *Am J Med*, 1968, **45**(4): 549–554.
148. Qiu, Y. et al., Comparison of immunological responses of plague vaccines F1 + rV270 and EV76 in Chinese-origin rhesus macaque, *Macaca mulatta. Scand J Immunol*, 2010, **72**(5): 425–433.
149. Williamson, E.D. et al., Kinetics of the immune response to the (F1 + V) vaccine in models of bubonic and pneumonic plague. *Vaccine*, 2007, **25**(6): 1142–1148.
150. Williamson, E.D. et al., Immunogenicity of the rF1 + rV vaccine for plague with identification of potential immune correlates. *Microb Pathog*, 2007, **42**(1): 11–21.
151. Van Andel, R. et al., Clinical and pathologic features of cynomolgus macaques (*Macaca fascicularis*) infected with aerosolized *Yersinia pestis. Comp Med*, 2008, **58**(1): 68–75.
152. Mizel, S.B. et al., Flagellin-F1-V fusion protein is an effective plague vaccine in mice and two species of nonhuman primates. *Clin Vaccine Immunol*, 2009, **16**(1): 21–28.
153. Mett, V. et al., A plant-produced plague vaccine candidate confers protection to monkeys. *Vaccine*, 2007, **25**(16): 3014–3017.
154. Honko, A.N. et al., Flagellin is an effective adjuvant for immunization against lethal respiratory challenge with *Yersinia pestis. Infect Immun*, 2006, **74**(2): 1113–1120.
155. Cornelius, C.A. et al., Immunization with recombinant V10 protects cynomolgus macaques from lethal pneumonic plague. *Infect Immun*, 2008, **76**(12): 5588–5597.
156. Chichester, J.A. et al., A single component two-valent LcrV-F1 vaccine protects non-human primates against pneumonic plague. *Vaccine*, 2009, **27**(25–26): 3471–3474.
157. Koster, F. et al., Milestones in progression of primary pneumonic plague in cynomolgus macaques. *Infect Immun*, 2010, **78**(7): 2946–2955.
158. Warren, R. et al., Cynomolgus macaque model for pneumonic plague. *Microb Pathog*, 2011, **50**(1): 12–22.
159. Pitt, M.L. et al., unpublished data.
160. Adamovicz, J., unpublished data.
161. Fellows, P. et al., Protection in mice passively immunized with serum from cynomolgus macaques and humans vaccinated with recombinant plague vaccine (rF1V). *Vaccine*, 2010, **28**(49): 7748–7756.
162. Chen, T.H. and K.F. Meyer, Susceptibility and immune response to experimental plague in two species of langurs and in African green (grivet) monkeys. *J Infect Dis*, 1974, **129**(Suppl): S46–S52.
163. Hallett, A.F., M. Isaacson, and K.F. Meyer, Pathogenicity and immunogenic efficacy of a live attentuated plaque vaccine in vervet monkeys. *Infect Immun*, 1973, **8**(6): 876–881.
164. Chen, T.H., S.S. Elberg, and D.M. Eisler, Immunity in plague: protection of the vervet (*Cercopithecus aethips*) against pneumonic plague by the oral administration of live attenuated *Yersinia pestis. J Infect Dis*, 1977, **135**(2): 289–293.
165. Chen, T.H. and K.F. Meyer, Susceptibility of the langur monkey (*Semnopithecus entellus*) to experimental plague: pathology and immunity. *J Infect Dis*, 1965, **115**(5): 456–464.
166. Romanov, V.E. et al., Standardization of conditions for the evaluation of effectiveness of antibacterial drugs in pneumonic plague in sacred baboons (in Russian). *Antibiot Khimioter*, 1995, **40**(6): 23–30.

167. Romanov, V.E. et al., Effect of antibacterial therapy on the epidemic threat of experimental pneumonic plague in monkeys (in Russian). *Antibiot Khimioter*, 2001, **46**(4): 16–18.
168. Byvalov, A.A. et al., Effectiveness of revaccinating hamadryas baboons with NISS live dried plague vaccine and fraction I of the plague microbe (in Russian). *Zh Mikrobiol Epidemiol Immunobiol*, **1984**(4): 74–76.
169. Rust, J.H., Jr. et al., The role of domestic animals in the epidemiology of plague. I. Experimental infection of dogs and cats. *J Infect Dis*, 1971, **124**(5): 522–526.
170. Rust, J.H., Jr. et al., The role of domestic animals in the epidemiology of plague. II. Antibody to *Yersinia pestis* in sera of dogs and cats. *J Infect Dis*, 1971, **124**(5): 527–531.
171. Gage, K.L., J.A. Montenieri, and R.E. Thomas. The role of predators in the ecology, epidemiology, and surveillance of plague in the United States, in *Proceedings of the 16th Vertebrate Pest Conference*, 1994, University of California, Davis, pp. 200–206.
172. Wang, H. et al., A dog-associated primary pneumonic plague in Qinghai Province, China. *Clin Infect Dis*, 2011, **52**(2): 185–190.
173. Carlson, M.E., *Yersinia pestis* infection in cats. *Feline Practice*, 1996, **24**(6): 22–24.
174. Gasper, P.W. et al., Plague (*Yersinia pestis*) in cats: Description of experimentally induced disease. *J Med Entomol*, 1993, **30**(1): 20–26.
175. Eidson, M., J.P. Thilsted, and O.J. Rollag, Clinical, clinicopathologic, and pathologic features of plague in cats: 119 cases (1977–1988). *J Am Vet Med Assoc*, 1991, **199**(9): 1191–1197.
176. Herzog, M., *The Plague: Bacteriology, Morbid Anatomy, and Histopathology*, D.O.T. Interior, editor. Manila: Bureau of Public Printing, 1904. pp. 3–149.
177. Multiple, *J Infect Dis*, 1974. **129**(Suppl. 1): 1–120 [Available from: http://jid.oxfordjournals.org/content/129/Supplement_1.toc].

9 Tularemia

Jeffrey J. Adamovicz and David M. Waag

CONTENTS

9.1 HISTORY/BACKGROUND

Tularemia is an infectious disease caused by the Gram-negative bacterium, *Francisella tularensis*. The first observed clinical cases were seen in Japan in 1837, where people eating meat from rabbits developed fever, chills, and glandular tumors [1]. In 1911, the disease was first described in the United States as a plague-like illness in ground squirrels by Dr. Edward Francis, Tulare County, California. Three years later, the first human case of tularemia in the United States occurred in a butcher [2]. Dr. Francis, after whom the microorganism was named, characterized the clinical signs, symptoms, and the mode of transmission after infection.

Francisella tularensis is a Category A biothreat agent because of its high infectivity, ease of dissemination, and capacity to cause illness and death. During

World War II, *F. tularensis* was developed and possibly used as a biowarfare agent [3,4]. An outbreak of tularemia among German troops during the 1942 Battle of Stalingrad may have resulted from the deliberate spraying of the agent by Soviet defenders or the outbreak might have been caused by natural infection [5]. This microorganism was subsequently weaponized by the United States and the Soviets during the Cold War. The causative microorganism elicits one of the most pathogenic diseases known to humans, causing disease after inoculation or inhalation of as few as 10 microorganisms. This organism is infectious by many routes, including ingestion of contaminated food or water, through microabrasions or bites from various arthropods, or inhalation of contaminated air. Insect vectors commonly associated with causing tularemia are ticks, especially *Dermacentor* and *Amblyomma* species, deer flies, and mosquitoes. *Francisella tularensis* is also maintained in a number of mammals including rabbits, squirrels, and other rodents, and persists in soils and water, possibly in association with amoebae [6].

9.2 MICROORGANISM

9.2.1 Taxonomy

The genus *Francisella* currently consists of two species, *F. philomiragia* and *F. tularensis. Francisella tularensis* has several subspecies including *tularensis, holarctica, novidia*, and *mediaasiatica*. Type A *F. tularensis* subspecies *tularensis* is highly virulent in humans and is the dominant species in North America. Type B *F. tularensis* subspecies *holarctica* (also called *palearctica*) is usually found in European countries and is less virulent [7]. Type A strains are associated with tick-borne tularemia in rabbits, whereas type B strains cause water-borne disease of rodents. *Francisella tularensis* subspecies *novidia* and *Francisella philomiragia* are of low virulence, posing a danger primarily to immunocompromised hosts [8]. Genetic analysis reveals significant differences between type A and type B strains. DNA microarray analysis revealed that *F. tularensis* subspecies *tularensis* possess segments that are missing from *F. tularensis* subspecies *holartica*, and an additional three DNA segments were found missing in the live vaccine strain (LVS) as compared to the parent *holartica* [9]. There may be an association with the attenuation in virulence and the absence of certain regions of the chromosome. The best information on the genetic relatedness of *F. tularensis* subspecies *tularensis* has come from Multiple-locus Variable-Number Tandem repeat Analysis (MLVA). An analysis of 192 isolates from all four subspecies revealed that 120 different genotypes could be identified [10]. Type A strains were found to be the most variable whereas type B strains were found to be the most geographically diverse. Although type A strains are associated with the most severe infections in humans, type B infections are the most prevalent. A number of other genetic techniques have been used to speciate *Francisella* including single-nucleotide polymorphism, ribotyping, pulsed-field gel electrophoresis, and amplified fragment-length polymorphism, and all essentially substantiate the taxonomic designations of MLVA [10–14]. The relative virulence of *F. tularensis* in humans and the currently used animal models to study the human pathogenic strains are depicted in Figure 9.1. Whereas Figure 9.1 is a recapitulation

Bio[A]type	MLVA clusters/ clade [10]	Species	Subspecies	Representative strains	Human pathology[B]	Animal models[C]				
A	A1	*F. tularensis*	*Tularensis*	SCHU S4 (North America)			X	X	X	X
	A2	*F. tularensis*	*Tularensis*	ATCC 6223 (North America)						
	A2	*F. tularensis*	*Tularensis*	(North America)				X	X	
	NA	*F. tularensis*	*Mediaasiatic*	GIEM 543 (Central Asia)		X				
	NA	*F. tularensis*	*Novicida*	ATCC 15452 (North America)		X		X		X
B	B1	*F. tularensis*	*Holartica*	Eurasia		X		X		
	B2	*F. tularensis*	*Holartica*	Scandinavia/ North America		X		X		
	B3	*F. tularensis*	*Holartica* LVS	Eurasia	NA	X	X	X	X	X
	B3	*F. tularensis*	*Holartica*	GIEM 503 Eurasia/ North America		X		X		
	B4	*F. tularensis*	*Holartica*	North America/ Sweden		X		X		
	B5	*F. tularensis*	*Holartica*	Japan		X		X	X	
NA	NA	*F. philomiragia*	NA	Eurasia						

A. Biotype as originally determined from virulence in the rabbit and ferment glycerol [149].
B. Color of figure indicates relative virulence for humans. Red being the most harmful. Green being the least harmful; hospital bed indicates virulence for immune-compromised only.
C. Denotes frequently used experimental animal models including the mouse, rat, nonhuman primate and the hamster; the hamster symbol is representative of several small rodents that include hamsters, voles, guinea pigs, and Sprague Dawley or Fischer 344 rats.

FIGURE 9.1 **(See color insert.)** Common animal models for *F. tularensis.*

of animal models used in previous studies with *Francisella*, it is important to understand that these studies may not have been performed with the optimum species or strain especially for studies involving the mouse. The focus of our detailed species discussions is on those animal models and endpoints that were appropriately matched to the experimental aims.

9.2.2 Virulence Factors

Francisella tularensis is a Gram-negative, encapsulated, nonmotile, highly virulent microorganism that requires cysteine for growth [15]. The bacteria can survive in water, soil, or decaying organic material for long periods of time [16]. Because the infectious dose is low, the organism should be propagated in a biosafety level three (BSL-3) environment. Relatively little is known about *Francisella* pathogenesis. No secreted toxins have been detected. No proteins for invasion have been discovered. No type III or type IV secretion systems have been identified; however, some type III effectors and type IV components remain in the genome [17]. Virulent *F. tularensis* has a thin lipid capsule, but whether the capsule contributes to virulence is uncertain.

For example, a strain of bacteria without capsule remained virulent for Porton mice and Hartley guinea pigs [18], and although the LVS can also persist in mice, injecting capsule-deficient LVS results in bacterial clearance in the spleen [19]. The presence of the capsule appears to protect the organism from complement-mediated killing [19], but it is not required for intracellular survival in polymorphonuclear macrophages.

The role of type IV pili in *Francisella* virulence is also unclear. The genes for the expression of pili are present in both the virulent SCHU S4 and the attenuated LVS strain. Both strains were shown to express pili *in vitro* [20]. However, in a separate proteomics analysis of 13 *F. tularensis* subspecies only virulent type A strains and not *mediaasiatica*, *holaractica*, or *novicida* strains expressed PilP protein [21]. This finding was complicated by the observation that *holartica* strains from North America but not LVS possessed genomic DNA for the genes encoding type IV pili [9]. Therefore, although type IV pili are known to be a virulence determinant in other pathogens by facilitating bacterial adhesion, their role in tularemia pathogenesis remains to be determined. A more direct comparison of the genomic sequences of type strains may clarify the distribution of type IV pili and the roles it may play in pathogenesis.

Francisella tularensis lipopolysaccharide (LPS) is less toxic than typical Gram-negative endotoxin [22]. In addition, LPS does not induce tumor necrosis factor alpha (TNF-α) secretion, nitric oxide production (at low LPS doses), interleukin (IL)-12, IL-6, IL-4, or gamma interferon (IFN-γ) production [23], and, therefore, does not elicit the stress response typical of other intracellular pathogens. *Francisella tularensis* LPS is different from other Gram-negative LPS in that it does not appear to be required for pathogenesis. Toll-like receptor (TLR4) knockout mice were no different from their wild-type littermates in susceptibility to virulent aerosol type A challenge [24]. However, these data are difficult to interpret given the extreme virulence of type A strains in the mouse model [25]. Whether these characteristics aid in evading host immune responses is unclear. LVS LPS can be used as a vaccine to protect mice from intraperitoneal (i.p.) or intradermal (i.d.) LVS or type B but not type A challenge, respectively [26,27]. *Francisella novicida* LPS can also induce protection against an i.p. or i.d. challenge but cannot induce protection against *F. novicida* [28]. Because *F. novicida* LPS but not LPS from LVS is noted to induce the proinflammatory cytokines TNA-α and IL-12, the authors concluded that TNA-α contributes to virulence. It is not clear whether human macrophages would respond in the same manner nor is it clear whether the increased virulence of *F. novicida* for the mouse skewed interpretation of the results. Interestingly, the failure to produce inflammatory cytokines in response to LVS LPS may be due to the blockage of TLR2 mRNA expression. LVS was shown to block J774 production of TNA-α and IL-1 in response to cocultured *Escherichia coli* LPS or bacterial lipopeptides [29,30]. This blockage was associated with the presence of an unknown 23 kDa protein. The identity and distribution of this protein in other species of tularemia is unknown.

Additionally, survival in macrophages appears particularly important for *F. tularensis* persistence and dissemination *in vivo* [31]. Microorganisms are taken into phagosomes and prevent phagolysosome fusion [32]. Subsequently, Francisella escapes into the cysosol, where it multiplies [33]. Apoptosis might be involved in final release from the host cell [34]. Phagocytized bacteria (LVS) can escape the phagosome into the host cell cytoplasm [32,35], and acidification of the phagosome is essential for *F. tularensis* growth and iron acquisition. Nano et al. [36] recently discovered a 30-kb low G + C content pathogenicity island required for growth in macrophages. Although the pathogenicity island was shown to be duplicated in LVS, the pathogenicity determinant proteins *pdpABCD* were absent. In particular, when *pdpD* was mutated in the parent SCHU S4 strain, the ability to replicate within macrophages was greatly attenuated and is required for full virulence [36,37]. Contrary to the hypothesis that *F. tularensis* resides primarily within host cells, recent studies in the mouse have demonstrated that the microorganism can also be found freely in the plasma [38]. However, it is unclear whether this extracellular phase occurs in other animals including humans.

9.3 GENETICS

The complete genome of *F. tularensis* was first published in January 2005 [39], available at http://www.ncbi.nlm.nih.gov/genomes/framik.cgi?db=genome&gi=563. Genes encoding type IV pili, a surface polysaccharide, and iron-acquisition systems and a small pathogenicity island are indicated. Interestingly, genes encoding transport/binding, gene regulation, energy metabolism, and cellular processes were found under-represented in the portions of the *F. tularensis* genome studied thus far. The virulent SCHU S4 strain has a 34% G + C overall content for its <2 Mbp genome. Efforts to sequence the LVS strain are completed and available at http://www.ncbi.nlm.nih.gov/genomeprj/16421. The LVS genome G + C content is slightly lower at 32%. The LVS strain was determined to possess two cryptic plasmids (pFMN10 and pOM1); however, these plasmids are absent in the fully virulent SCHU S4 strain [40]. On the other hand, eight "regions of difference" (0.6–11.5 kb) were present in virulent *F. tularensis* and not *F. tularensis* subspecies *holarctica*. Several additional virulent and avirulent strains of *F. tularensis* have been sequenced or are in process of being sequenced (see http://www.ncbi.nlm.nih.gov/sites/entrez?Db=genome&Cmd=Search&Term=txid263[orgn]) including all or parts of other *F. tularensis* strains including the *novicida* and *mediaasiatic* subspecies and *F. philomiragia*. Comparative genomics using these available sequences will allow a better understanding of the evolution of *F. tularensis* as well as elucidation of its pathogenic mechanisms.

A prerequisite for genetic manipulation of *F. tularensis* like any organism is the development of a genetic toolbox. Successful development of genetic techniques for *Francisella* has been limited to the last few years. Lauriano et al. [41] successfully developed and tested an allelic exchange mechanism using polymerase chain reaction products with an erythromycin resistance cassette insertion in *F. tularensis* subspecies *novicida*. Two independent groups reported successful construction and testing of transposon tools for creating insertional mutants. Kawula et al. [42] created a TN5-based transposon encoding kanymycin resistance but lacking transposase. When coupled with a transposase enzyme the complex was shown to establish stable insertions in an LVS background. The second strategy used a λTNphoA construct to create insertional mutants in genes encoding exported extra-cytoplasmic proteins of the virulent SCHU S4 strain [43]. Maier et al. [44]

recently constructed a shuttle vector for LVS, *F. tularensis novicida*, and *E. coli* DH5α. The plasmid is stably maintained in the absence of antibiotic *in vitro* and *in vivo* and contains an expanded multiple cloning site and a temperature-sensitive mutation. This vector and other recently developed tools will greatly facilitate future research. A more recent study examined the effects of in-frame or transposon insertional mutagenesis and found seven essential virulence genes (*iglB*, *iglD*, *purl*, *purF*, *ggt*, *kdtA*, *glpX*) in SCHU S4 [45]. Although some of these candidates were previously identified by other means, the *in vivo* screening methodology allowed for the rapid identification of the most potentially useful mutations for additional vaccine testing. The development of additional genetic tools and the characterization of portions of the genetic code remain as important understudied areas of knowledge.

9.4 VACCINES

A type B, *F. tularensis* subspecies *holarctica*, attenuated vaccine was developed in the Soviet Union [46] and brought to the United States in 1956. From this preparation, a strain suitable for vaccination, designated the *F. tularensis* LVS, was isolated and characterized [47]. Culturing LVS on glucose cysteine blood agar produced two colony variants, a gray colony variant and a blue variant [47]. The blue variant was protective when used as a vaccine, but the gray colony variant was not. The blue colony variant was also more virulent in mice and guinea pigs. Different lots of LVS vaccine are noted to contain varying percentages of blue and gray colonies although all lots tested are principally of the blue phenotype [48]. Immunization with a single dose of LVS induces measurable anti-LVS antibody and lymphocyte proliferation within 14 days in the majority of vaccinates [48–50]. LVS was found to be effective in reducing the incidence of typhoidal tularemia but not ulceroglandular tularemia [51,52]. LVS has been used extensively in murine studies to define the innate and adaptive immune responses [53]. LVS use in the mouse is discussed in greater detail below (Section 9.6.1). The U.S. Food and Drug Administration's (FDA) approval of this vaccine for public use has been hindered by the fact that the exact parental strain from which it was derived is not known (and, therefore, the mechanism of attenuation is not known). A comparison of the historical i.p. median lethal dose (LD_{50}) values for mice reveals a dramatic increase in virulence for LVS from 10^5 blue colony CFUs (colony-forming units) in the early 1960s to 1 CFU today [47]. Strain stability/reversal to virulence has not been fully studied, but is likely to be problematic since it has been shown that LVS cultured *in vitro* increases capsule expression and subsequently virulence in the mouse [54]. Consequently, LVS remains as an investigational new drug product with informed consent use for researchers with an occupational exposure hazard. However, newly developed vaccines will likely need to be compared to LVS.

One approach to a newly developed vaccine is the improved LVS or iLVS vaccine. This iLVS material was derived from original LVS material and found to consist of 100% blue colonies [55]. The vaccine was produced under current good manufacturing practice conditions and characterized for safety and immunogenicity in rabbits. The iLVS is being evaluated in human clinical trials.

9.5 HUMAN DISEASE

Francisella tularensis is found worldwide in over 100 species of wild animals, birds, and insects. Hares tend to be the reservoirs for *holoarctica* infections in central Europe, voles in Eurasia, and beavers and muskrats in North America [11]. Rabbits tend to be the primary reservoir for subpecies *tularensis* infections in North America [11]. Humans can become infected from bites of infected arthropods, such as ticks or mosquitoes, or from contact with small animals, such as voles or muskrats [56–60]. Other modes of infection include handling infected tissues or fluids, ingesting contaminated food or water, or breathing infectious aerosols [61–64]. The minimal human aerosol infectious dose of type A *F. tularensis* is around 10 microorganisms [65]. Infected humans develop an acute febrile illness.

Tularemia can be found worldwide. A few hundred cases of tularemia are reported annually in the United States. As with most such diseases, the majority of cases are likely unreported or misdiagnosed. Symptoms of all forms of tularemia vary directly with the virulence of the infecting microorganism. Since no toxins have been associated with the microorganism, pathogeneis is likely due to the ability to survive and replicate within the host [66]. The incubation period is between 3 days and 1 week, when the patients develop high fever, chills, fatigue, body aches, headache, lymph node enlargement, and nausea [67]. Symptoms can progress to weakness, malaise, weight loss, and anorexia. If the disease is not treated, symptoms get progressively worse. This disease can be difficult to diagnose due to nonspecific and influenza-like symptoms. Human to human transmission has not been reported [68].

Tularemia can be found in different forms: ulceroglandular, glandular, oculoglandular, oropharyngeal, pneumonic, and typhoidal, depending on the bacterial strain, dose, and route of infection [51,52,69]. The ulceroglandular form is similar to the glandular form, except for the presence of an ulcer at the site of infection. Usually, manifestations of tularemia can be broken down into two forms: the more common ulceroglandular form, where local or regional symptoms and signs predominate, and the more lethal typhoidal form, characterized by systemic symptoms without a skin lesion [68]. After inoculation into skin or mucous membranes, by inhalation or by oral ingestion, the infecting microorganisms spread hematogenously, resulting in secondary pleuropneumonia, sepsis, and even meningitis [70]. The majority of naturally occurring tularemia cases are ulceroglandular, with only a small percentage of cases as pneumonic or typhoidal. *Francisella tularensis* is a facultative intracellular pathogen and multiplies in macrophages. Target organs are associated with the reticuloendothelial system including the lymph nodes, lungs and pleura, spleen, liver, and kidneys.

9.5.1 Ulceroglandular Tularemia

This most common form of tularemia usually occurs as a result of the bite of an infected athropod and from handling infected animals. In ulceroglandular tularemia, fever is a typical symptom, and the presence of cutaneous ulcers is a frequent clinical sign, usually developing within 3 days of infection [59]. The duration of illness is from 3 to 110 days, but usually the patient recovers within 2 weeks. Regional lymph nodes become enlarged and the patient develops fever, chills, and muscle pain [67,68]. Ulceroglandular tularemia can cause death in approximately 3% of untreated patients [67]. Many patients experience malaise and weakness for weeks after recovering from acute disease. In the outbreak referenced above, antibody titers, measured by the direct fluorescent antibody test, ranged from 160 to 10,240, with a mean of 1297. Titers peaked 3–5 weeks after exposure. This particular outbreak resulted from people handling infected muskrats. In an experimental setting, 10 CFU of the virulent type A *F. tularensis* strain SCHU S4 was administered to humans i.d. [52,71]. Patients became infected and experienced fever, headache, myalgia, skin lesions, and anorexia.

9.5.2 Oropharyngeal Tularemia

Humans can also develop oropharyngeal tularemia after ingesting contaminated water or food [16,72]. Only about 5% of tularemia cases are oropharyngeal. Bacteria infect mucous membranes in the pharynx, and exudative pharyngitis develops in 4–5 days [67]. Cervical adenopathy may occur. The pharynx, tonsils, and soft palate frequently ulcerate. The patients become febrile and develop chills and malaise. Outbreaks have occurred in Norway, Kosovo, and the Black Sea region of Turkey. The suspected cause of the outbreak in Norway was infected hares [73]. Patients developed serological enzyme-linked immunosorbent assay titers in the range of 80–640. The source of the outbreak in Kosovo was rodent-contaminated food [16], whereas the presumed source of infection in Turkey was an infected rat carcass in a water reservoir [74]. The mechanism of infection in oropharyngeal tularemia is uncertain, as *F. tularensis* does not appear to colonize the gut and can be lysed by bile [75,76].

There is evidence, albeit somewhat controversial, that *F. tularensis* is able to penetrate unbroken skin. Dr. Ohara's wife was infected after the hearts of dead hares infected with *F. tularensis* caused her to develop symptoms of tularemia [1]. Guinea pigs and rabbits have also been infected after unbroken skin came into contact with infected tissues [77]. Finally, baby albino mice became infected when *F. tularensis* in saline was applied to their abdomens [78]. These observations may indicate direct penetration, penetration through unseen microabrasions, or subsequent transfer from the infected skin surface to other portals.

9.5.3 Pneumonic Tularemia

Human pneumonic tularemia is the most acute form of the disease where infection progresses systemically from the lungs to other organs and causes pathological changes, primarily to the liver and spleen [51,67,68]. The initial presentation may not include an indication of respiratory involvement. Typical signs include abrupt onset of fever 5–7 days after exposure, a nonproductive cough, and pain in the lower back. Peribronchial infiltrates, bronchial pneumonia, pleural effusions, and hilar lymphadenopathy may be noted in radiographs. Granulomatous lesions of the pleura or parenchyma of the lungs may also be observed. Humans infected by aerosol exposure also show hemorrhagic inflammation of the airways, which may proceed to bronchopneumonia [79]. Alveolar spaces become filled with a mononuclear cell infiltrate. Pleuritis with adhesions and effusion and hilar lymphadenopathy are commonly found [80]. When used as experimental models of human disease, monkeys exhibit signs and symptoms similar to those of humans. Pathological changes and patterns of bacterial dissemination in mice, rabbits, and monkeys are similar to those found in humans [71,81]. The fatality rate for untreated pneumonic infections with type A *F. tularensis* is approximately 30% and for infections with type B *F. tularensis* is approximately 10% [68,82]. Fatalities can usually be reduced to around 2% of cases with antibiotic therapy [68].

9.5.4 Immunity

The mechanism of immunity to tularemia is not well characterized. In humans, recovery from a previous infection usually induces protection against a subsequent infection. In mice, survival of a sublethal i.d. challenge always leads to the development of protective immunity from a subsequent i.p. challenge [53]. Collectively, these results indicate the induction of a long-lived immunological memory to one or more protective tularemia antigens. Vaccination with the LVS vaccine does provide excellent protection against i.d. or aerosol challenge in humans; however, the basis for protection is unclear. Because the organism is a facultative intracellular pathogen we hypothesize that cellular immunity plays an important role in controlling infection. Patients that recover from tularemia have been noted to have long-lived antibody titers indicating that humoral responses may also be important in immunity [83]. It is likely that cellular immunity is suppressed during the acute phase of infection but is important in the resolution of infection and the development of immunological memory. Antibodies likely play a secondary role to control the spread of extracellular organisms during the bacteremic phase of acute infection and/or to help prevent reinfection. Antibodies may still prove useful as correlates of immunity as suggested in recent rabbit studies with the iLVS vaccine [55]. In a second unrelated study with a different improved LVS vaccine in the mouse, antibodies were again used to measure vaccine take [84]. The cellular and molecular aspects of immunity to tularemia have been studied most intensely in the mouse model and these findings are discussed in Section 9.6.1.

9.5.5 Treatment

Interestingly, many effective antibiotics for the treatment of tularemia are used off-label. The efficacy of licensed antibiotics was determined from *in vitro* susceptibility testing and human use. Proper antibiotic treatment can lower tularemia fatalities to about 2%. The duration of antibiotic therapy is generally prolonged, and the disease can recrudesce if therapy ends early. Streptomycin given intramuscularly (i.m.) or intravenously (i.v.) is the licensed drug of choice for treating the disease in adults, children, and pregnant women [85]. Gentamicin is more widely available and can be used i.v. but is not label-indicated. Doxycycline is effective by the i.v. route and is licensed for use in all three treatment groups. Tetracycline and chloramphenicol are also treatment options, but patients sometimes relapse [85]. Chloramphenicol is label-indicated for use in adults and children. Ciprofloxacin is also effective and has a low treatment failure rate [68]. Ciprofloxacin was also shown to be effective in preventing type A-induced death in a mouse model [86]. However, other studies in mice have shown that if treatment were delayed for 24 h or 48 h after exposure, gatifloxicin and moxifloxacin were more effective in enhancing the survival rate than was ciprofloxacin [87]. Type B strains have been noted to have natural resistance to azithromycin, ceftazidime, meropenum, or imipenem. Type A strains were constructed with resistance to chloramphenicol and tetracycline by both the United States and the former Soviet Union [68]. Past studies indicated that the LVS challenge of either BALB/c or C57BL/6 mice would be a good therapeutic model to study the effectiveness of additional antibiotics. In particular, the mouse is a good choice to test the concept of aerosol delivery of antibiotics [88]. In addition, the guinea pig and rabbit could also be used to test chemoprophylaxis or treatment protocols with more virulent strains.

9.6 ANIMAL MODELS

Francisella tularensis can cause natural infections in over 250 host species, including rabbits, cats, prairie dogs, voles, raccoons, squirrels, rats, lemmings, and wild mice, and other species, including birds, fish, and protozoa [16,89–97]. Birds and swine appear to be relatively resistant to infection [98,99]. Animal models have been developed in several species to study the pathogenesis of disease and the efficacy of pre- and post-exposure treatments.

9.6.1 Mice (*Mus musculus*)

The majority of previously published studies on tularemia pathogeneisis and vaccine efficacy were conducted in the mouse model. However, this does not mean that the mouse is the best model for vaccine studies or comparative studies to humans. For instance, LVS is lethal in mice but not in humans and unlike the human model LVS vaccination fails to protect mice against a physiologically relevant challenge against type A strains. There have also been noted differences in disease pathology between mice and humans. The differences between mice and humans are especially important in the area of vaccine studies. In the last 5 years, a number of novel vaccine approaches have been tested in mice with statistical increases in survival, mean survival time, and/or reduced bacterial burdens following challenge between vaccine and control groups of mice. Although this vaccine-induced protection is statistically significant, it is not likely to be biologically significant (at least for the mouse). Biological significance as defined here is the ability of a vaccinate to resist a biologically relevant challenge such as one or more tick bites, inhalation of an infected large droplet nuclei, infection of an open cut or wound, or ingestion of infected water or food. Although there is a paucity of data on what the biologically relevant challenges may be (in number of infectious particles or CFUs), what is certain is that an infection or biologically relevant challenge is most likely to be greater than 10–88 CFU used in recent studies for challenge of vaccinated animals) [100–102]. One can posit that the problem may not be in the vaccine construct itself (as in the case of LVS) but in the use of the mouse to measure vaccine effects. Conversely, the mouse is useful for a number of reasons including the wide availability of inbred strains, genetic knockout strains, and well-characterized immune-deficient strains. Naïve BALB/c and C57BL/6 mice were equally susceptible to type A or type B *F. tularensis* aerosol challenge. The LD_{50}s for aerosol exposure of both strains of mice to type A or type B *F. tularensis* were between 10 and 20 organisms, respectively [81]. However, the mice died sooner after type A (day 6) than after type B (day 8) infection. The aerosol LD_{50} for mice is similar to the infectious dose for humans by the same route. Due to the extreme virulence of this microorganism, most experimental infections in rodent model systems have used the LVS, although many recent studies used the more virulent type B and type A strains. The use of LVS minimizes the risk to laboratory workers since it is extremely attenuated in virulence, but not necessarily infectivity, in humans [103]. LVS was lethal in naïve mice and caused pathological reactions similar to those seen with virulent strains in humans [104]. However, using an LVS model of infection is not without potential problems. First, dose lethality varied significantly depending on route of LVS infection; naïve mice were most sensitive to i.p. challenge followed by i.v., i.n., aerosol, and i.d. challenge. Whereas the LD_{50} of LVS given i.p. to BALB/c or C57BL/6 mice was 4 and 1 CFU, respectively [105], the LD_{50} given i.d. was much higher ($>10^6$ CFU) [103]. Curiously, C57BL/6J and C3H/HeJ mice were more resistant to i.p. challenge than other inbred mouse strains despite their lowered resistance to subcutaneous (s.c.) or i.v. challenge [103,106]. Unlike LVS, clinical type A or type B strains have i.d. LD_{50}s < 20 CFU [81]. Furthermore, the LD_{50} for LVS administered by aerosol was $>10^3$, whereas the LD_{50} for virulent strains was <10 CFU [104]. The LD_{50} for LVS administered intranasally (i.n.) to BALB/c and C3H/HeN mice was determined to be 3.2×10^2 and 1.6×10^2 organisms, respectively [103]. When different mouse strains were compared for susceptibility (by LD_{50} determination) to s.c. or i.v. challenge with LVS, the most susceptible strains were C3H/HeJ, CBA/J, C57BL/6J, and A/J, whereas the most resistant strains were SWR/J, SJL/J, BALB/cJ, AKR/J, and the outbred CD-1 (Swiss) [103,106–108]. These LVS studies may indicate an important role for host genetics in immunity to tularemia. LVS vaccination of most strains of mice imparted a solid protection against a subsequent type A challenge by the i.d. route but not the aerosol route [109]. Therefore, it is important to note that contributions of host genetics were not readily apparent with challenge with more virulent type A strains, which seemed to be uniformly virulent in all naïve mouse strains. Although the mouse is a fundamental tool in studying *F. tularensis* pathogenesis, careful thought should be given to the limitations of murine data in comparative human vaccine studies.

The most useful contribution of the mouse to tularemia research has been in the study of the murine immune response to LVS. There is a relatively large volume of literature on the mouse model; a great portion of this is challenge of LVS-immunized mice. Early studies in the late 1940s demonstrated that LVS but not killed bacteria could protect mice from an i.d. challenge with virulent SCHU S4 [110]. An Intranasal vaccination with LVS could partially protect BALB/c mice, but not C57BL/c mice, against an i.n. or s.c. challenge against a type A strain *F. tularensis* isolate [111]. Microorganisms could be recovered from over half of the challenged mice.

Because LVS could not fully protect against a virulent aerosol challenge, later studies used LVS as the vaccine and as the challenge strain. LVS vaccination of mice was shown to prevent lethal infection following an i.d., s.c., or aerosol challenge with LVS [103,104,112]. More recent studies involving challenge by type A and/or type B strains of LVS-vaccinated inbred strains of mice have been used to demonstrate innate differences in vaccine efficacy and to define components of the immune response. LVS-vaccinated BALB/c mice, but not C57BL/6 mice, were protected from a low-dose aerosol type B challenge [108]. Neither mouse strain was protected against a type A challenge. One possible explanation for this observation is because LVS is derived from a type B strain, it may lack crucial protective antigens that are present in type A strains. It has been shown that LVS-immunized mice boosted i.d. with SCHU S4 were more resistant to i.d. or aerosol challenge than mice boosted with LVS [106,108]. The use of *F. novicida* as a vaccine did not improve protection against either a type A or type B aerosol challenge in BALB/C mice [113]. This was an interesting finding since *F. novicida* is more closely related to SCHU S4 than LVS.

LVS vaccination was used to differentiate vaccine efficacy against type A i.d. challenge in mice of different genetic backgrounds [109]. This study was used to differentiate two classes of vaccinated mice based on mean time to death following challenge. The low vaccine responders consisted of A/J, SW, DBA/2, and CF-1 mice whereas the high responders consisted of BALB/c, BDF1, C3H/HeN, CD1, CDF1, and strain 129 mice. These distinctions were not seen in i.d. infected naïve mice indicating that there is no difference in innate resistance to type A challenge, unlike LVS challenge. They may also indicate that host genetics may play a determinant role in the productive response to LVS vaccination. The relevance of these data remains to be determined as LVS vaccination failed to differentiate a high/low vaccine responder subset to type A aerosol challenge. The authors conclude that BALB/c and or C3H/HeN mice are the best choices for future LVS vaccine studies [109]. This finding was substantiated by demonstration of LVS-dependent protection against type B aerosol challenge of BALB/c but not C57BL/6 mice [108]. We concur with the author's recommendations to use BALB/c and/or C3H/HeN mice for future LVS or LVS-like vaccine studies. C57BL/6 mice are also interesting because of their increased resistance to i.p. infection and their usefulness to differentiate LVS protection against i.d. challenge with type A and type B strains.

LVS or LVS- and SCHU S4-immune spleen cells have been shown to passively protect naïve mice against LVS or SCHU S5 i.p. challenge, respectively [103,106]. Subsequent studies in the mouse demonstrated that both CD4+ and CD8+ T cells were required to resolve an infection with LVS [114]. The role of these T cells may be to produce IFN-γ and activate macrophages. Gamma interferon was shown to be stimulated after i.n. infection with LVS [115]. In BALB/c and C57BL/6 mice, NK cells were responsible for early secretion of IFN-γ. Alveolar macrophages may play a role in killing bacteria directly when activated by IFN-γ [116]. Studies in mice have shown that the primary macrophage response to *F. tularensis* LVS is via TLR2 activation, and requires *de novo* bacterial protein synthesis, but occurs independently of intracellular LVS replication [117]. However, since fully virulent bacteria regularly infect and replicate within macrophages, the role macrophages play in protection from disease is unclear. The role of γδ T cells is also unclear, although it is likely they also contribute to immunity. The γδ T cells were noted to be increased in number and duration in tularemia patients and these cells have noted protective roles in other intracellular infections such as listeriosis [118]. Neutrophils are probably critical to controlling infection since neutrophil-depleted mice but not their control littermates were unable to control a sublethal i.v. or i.d. challenge with LVS [119]. B-cells may also play a direct-early nonantibody role in protective immunity. It is possible that B-cells serve to bridge innate and adaptive immunity in tularemia infections possibly through a TLR mechanism. Treatment of mice with CpG oligonucleotides was shown to protect mice against a lethal LVS i.d. challenge [53,120]. TNF-α, IL-1, and IFN-γ also have been shown to play important protective roles in LVS infection. Anti-IFN-γ antibodies were shown to block LVS-dependent protection of BALB/c mice against an i.d. challenge of virulent type A tularemia [108]. As mentioned earlier, LVS (and presumably other strains) secretes a 23 kDa protein that blocks NF-κB signaling leading to an inhibition of LPS-dependent secretion of TNF-α and IL-1β. This deliberate suppression of cytokine expression in pathogenesis is likely an important virulence mechanism. TNF-α receptor and IFN-γ knockout mice are exceptionally susceptible to i.d. or aerosol challenge with LVS [53]. However, similar studies utilizing a type B challenge did not demonstrate an increased susceptibility to lethality in TNF-α receptor, IFN-γ, or B-cell null mice [25]. The authors of this study concluded that host defenses against LVS are not protective against more virulent strains and that LVS only behaves like a virulent strain in the absence of specific host immune mechanisms. Perhaps differences between pathogenesis of LVS in mice versus humans can be explained by expansion of circulating Vγ9/Vδ2 T cells in humans after vaccination/infection [121]. Mice do not have a corresponding cell receptor.

An important caveat is that conclusions regarding the murine LVS model have not been substantiated in other animal models. We predict that these observations in mice are not likely to be substantiated in other infection models such as the rabbit, nonhuman primates (NHPs), or humans. Collectively, these data indicate that cellular immunity likely plays a predominant role in controlling and clearing tularemia infections. However, a role for humoral immunity has also been demonstrated. The relative contribution of humoral immunity is unclear and may be masked by the predominant role of cellular immunity. The evidence that antibodies may play a role in combating infection has largely come from murine studies. For instance, the passive transfer of polyclonal murine antibodies to B-cell null mice significantly reduced bacterial tissue burdens and conferred survival from a sublethal type B i.d. challenge, but littermates that were given normal serum died [122]. Antibodies to LPS have also been shown to passively protect mice against an LVS challenge [26]. However, it is not clear that anti-LPS antibodies play a role in a type A challenge. Importantly, type A tularemia strains have been noted to phase shift their LPS expression; this likely thwarts any protective role anti-LPS antibodies play during acute infection although they may contribute to prevention of reinfection [123].

The lungs, liver, and spleen are organs that support bacterial replication after aerosol challenge with either virulent type A *F. tularensis* or LVS. C3H/HeJ and C57BL/10J mice were exposed to approximately 10 CFU of a type A *F. tularensis* strain by aerosol [24]. The second day after challenge, bacterial counts in the spleen, liver, and lungs were 8.9×10^3, 4.5×10^3, and 3.2×10^3 per gram tissue, respectively. Two days later, counts in these tissues increased to 4×10^9, 3×10^9, and 6×10^8, respectively. Results were similar after an aerosol challenge with LVS. When

BALB/c and C57BL/6 mice were exposed to approximately 10^3 CFU of LVS, bacteria were found in the lungs in 48 h, and then disseminated to the liver and spleen via the blood [81]. On the day of death, mice had 10 times the number of microorganisms in the liver as in the spleen. The histopathogenesis observed in mice infected with *F. tularensis* was similar to that seen in human patients with tularemia pneumonia. Aerosol exposure led to infection in the lungs, which progressed to a systemic infection in other organs. There was deterioration of the spleen and liver accompanied by lesions in the lungs [31]. Also, the BALB/c mice displayed clinical signs such as lethargy, hunched back, and anorexia at least 48 h before death, whereas C57BL/6 mice did not exhibit clinical signs until only a few hours before death. The pathogenesis of infection was also different between mice challenged with type A or type B *F. tularensis*. Mice challenged with type B bacteria developed a moderate leucopenia, whereas mice given the type A bacteria developed severe leucopenia in both mouse strains 4 days after challenge. The histopathological reactions were more severe in mice given the type A microorganism. Gross changes in internal organs were not noted until 4 days after challenge, and inflammatory necrosis could be seen in the liver, with hepatocytes containing microorganisms. Changes in the liver were less severe in mice given the type B microorganism. Changes could also be seen in the lungs at 4 days after infection. Lungs contained severe pulmonary necrosis and pleuritis. Pulmonary tissue surrounding small and medium blood vessels was infiltrated by macrophages and neutrophils. Histopathological changes were of lower magnitude in the spleen than the liver. Within 2 days of infection, the population of neutrophils and macrophages in the spleen increased. The splenic white pulp contained basophilic granules, necrotic debris, and bacteria. This histopathogenic presentation was similar to humans infected with type A *F. tularensis*.

When BALB/c and C3H/HeN mice were infected i.n. with 1000 CFU of LVS, bacteria could be found in the lungs within 24 h after infection [103]. Numbers of bacteria in the lungs increased 100-fold between the second and fifth day after challenge. Within 48 h after infection, bacteria could be found in the spleen and liver. Bacterial numbers in the spleen and liver increased approximately 1000-fold by day 5 after infection. The mice exhibited symptoms of illness by the third day after infection and were dead by day 7. LVS-infected BALB/c and C3H/HeN mice have also been examined histopathologically. The mice displayed dyspnea and signs of bronchopneumonia by day 3 and pneumonia with necrotic lesions by day 6. In the liver, acute inflammation with necrosis progressed to granuloma formation by day 8 of challenge, when the mice died. Thus, the pathological responses in mice after LVS infection were similar to responses in humans infected with virulent *F. tularensis*.

Mice are extremely susceptible by the s.c. route to type A, but not LVS, challenge. After s.c. challenge with 1 CFU of type A *F. tularensis*, bacteria migrated to the lymph nodes within 24 h, to the internal organs within 48 h, and death followed 6 days after challenge [124]. Over 10^{10} CFU of *F. tularensis* per gram of tissue could be found in spleens collected 6 days after infection. The basis for the relative differences in s.c. pathogenicity between LVS and type A strains is unclear. However, it is unlikely to be due to the inability of LVS to escape the site of inoculation. Mice infected with LVS by the s.c. route were noted to have culturable organisms in the lungs, liver, and spleen 3 days after infection [103].

When BALB/c and C3H/HeN mice were given LVS i.d., bacteria were found in the liver within 1 day, and numbers increased from less than 1×10^4 CFU to 3×10^5 CFU by day 5 [103]. Bacteria were present in the spleen on day 2 at 1×10^5 CFU per organ and numbers increased to 3×10^5 CFU on day 5. Bacteria were not found in the lungs until the fourth day after infection, at which time 1×10^3 CFU were present.

When BALB/c and C57BL/6 mice were infected i.d. with type A microorganisms, histopathological changes in the lungs were less severe than when mice were infected by aerosol [81]. However, lesions in the spleen developed 1 day earlier than in mice challenged by aerosol. These lesions were similar to ulceroglandular disease in humans. When mice were challenged i.d. with type B bacteria, there were increased numbers of neutrophils in the spleen on day 4, and by day 5, necrosis of splenic lymphocytes was noted. Overall, LVS and type A strains were clearly different in their pathogenesis, which is problematic for using LVS as a model of virulent tularemia. Therefore, the development of countermeasures against virulent (type A) strains of *F. tularensis* may require the use of a well-characterized, virulent challenge instead of the more popular LVS challenge.

Mice have been used to study the ability of ticks to acquire and transmit virulent strains of *F. tularensis* [125]. Bacteremia were noted 2 days after infection with type A and type B *F. tularensis* and blood concentrations reached 10^8 CFU/mL within 3 days of infection for type A strains and within 4 days of infection with type B *F. tularensis.*

Mice were less susceptible to becoming infected via the oral route, with an LD_{50} of around 10^6 microorganisms, than via parenteral routes [126]. After oral infection, microorganisms appeared first in the mesenteric lymph nodes, and progressed to the liver and spleen where they grew to high numbers and induced neutrophilic infiltration and tissue necrosis. Growth in tissues was accompanied by rapid increases in the levels of IFN-γ, IL-1b, IL-6, and TNF-α. Immunization with LVS i.d. did not protect against oral challenge with type A *F. tularensis.*

9.6.2 Rats (*Rattus* spp.)

Rats are less susceptible to *F. tularensis* infection than mice and there are also susceptibility differences between species of rats. Male Fischer 344 rats were found to be much more susceptible to a lethal aerosol dose of type A *F. tularensis* than Sprague-Dawley rats [127]. Whereas greater than 90% of the Fischer rats died after an infectious dose of 2×10^5 organisms, none of the Sprague-Dawley rats died. This indicates that Fischer rats may be suitable for aerosol pathogenesis studies as compared to Sprague-Dawley rats. Unlike mice, Fischer 344 rats exhibit a range of susceptibilities that better correlated with human susceptibility to *F. tularensis* infection than mice [128] as they were found to be more resistant to pulmonary challenge with LVS or subspecies *novicida*. Furthermore, Fischer 344 rats were sensitive to intratrachial infection with type A *F. tularensis* and exhibited some protection after LVS vaccination and respiratory challenge with type A *F. tularensis* [129]. An important aspect of challenge studies in rats is the noted differences in disease pathology dependent on the method of inhalation challenge. When Fischer 344 rats were exposed to LVS i.n., bacteria could be cultured from the lungs in 3 days. However, when rats were exposed to LVS by aerosol, bacteria could be detected in the lungs within 24 h [127]. A more complete analysis of the route-dependent pathophysiology of the disease is needed in the rat.

Challenging rats s.c. with 10 CFU of type B microorganisms did not kill them [124]. However, organs were infected and the microorganisms reached a peak titer at day 5 in the spleen and lymph nodes of approximately 10^5 CFU per gram of tissue. Although mice and guinea pigs die after this challenge dose, rats tended to recover. Rats, therefore, may be a more appropriate surrogate model for human typoidal/glandular/pneumonic tularensis than either the guinea pig or the mouse.

9.6.3 Guinea Pigs (*Cavia porcellus*)

The susceptibility of guinea pigs to *F. tularensis* infection is less than that of rats but greater than that of mice. However, like humans guinea pigs are protected by LVS vaccination against SCHU S4 challenge. The respiratory LD_{50} of Hartley guinea pigs exposed to type A bacteria was less than 10 microorganisms, whereas the corresponding LD_{50} for type B *F. tularensis* was 38 microorganisms. Like in the mouse and rabbit, the infection progressed systemically from the lungs to the blood and spleen after aerosol infection [69]. When guinea pigs were exposed to type B *F. tularensis* (approximately 3×10^3 CFU), bacterial counts in the spleen increased after 3 days until the time of death, which was usually on day 5. Bacteremia was noted the second day after infection.

The LD_{50} for guinea pigs challenged s.c. with type A or type B *F. tularensis* was 1 CFU [124]. When guinea pigs were challenged s.c. with 10 CFU of type B *F. tularensis*, ulceration and swelling developed at the inoculation site and microorganisms were found in the regional lymph nodes within 2 days [124,130]. By the third day, the animals were febrile and bacteria were detected in the regional lymph nodes and spleen. Granulomas appeared at the site of injection and spread to deep layers of the dermis. These lesions were similar to those observed in humans with the ulceroglandular form of the disease. In 4 days, microorganisms were found in the internal organs and the local lymph nodes, spleen, and liver were enlarged. By day 6, granulomas in the spleen appeared necrotic and the animals were dead by day 10. When guinea pigs were given a higher dose (10^7 CFU) of the type B bacterium, a fever developed within 24 h and ulcerations could be seen at the site of injection. Although the lymph nodes, spleen, and liver were infected, there were no apparent changes in the appearance of these organs. By the third day, the injection site and regional lymph nodes exhibited necrosis and the animal developed splenomegaly and hepatomegaly. The animals died between 4 and 8 days after infection. Because guinea pigs appear to be more sensitive to tularemia infection than most mouse strains their utility for additional research is questionable. An additional hurdle to the use of the guinea pig for surrogate studies for human disease is the undeveloped immunological tools to make comparable pathophysiological observations.

9.6.4 Voles (*Microtus arvalis*)

The LD_{50} of voles challenged s.c. with type B *F. tularensis* was less than 10 microorganisms [92]. Importantly, voles have been used to demonstrate that infection can be passed to cage-mates [131]. Cage-mates were infected by allowing the animals to feed on infected cadavers of voles or white mice. Approximately one-third of infected voles shed bacteria in their urine, whereas bacteria were found in the feces of one vole [92]. Bacterial titers in the urine of infected voles were 10^5 CFU/mL and voles were shown to shed organisms in the urine for 80 days. The kidneys exhibited histopathological changes with focal necrosis in the vascular loops of most glomeruli and within tubules and granulomas in interstitial tissue. Granulomas were also found in the lymph nodes, spleen, and liver. Therefore, voles can be chronically infected and pose a risk of infection to other animals and humans. Voles may be an appropriate model for vaccine studies specifically with regard to the development of veterinary vaccines that target rodents that are known to maintain enzootic foci. An additional aspect of voles is that as a small animal model they offer an alternative or adjunct to the mouse. This was recently demonstrated in a comparative biochemical response study between voles and mice [132]. Interesting differences in the liver enzymes and lipid metabolism were noted between the mouse and the vole models. This may allow for the direct comparison of human biochemical markers in the vole.

9.6.5 Rabbits (Family *Leporidae*)

Most, but not all, species of wild and laboratory rabbits are susceptible to type A tularemia infection and some are susceptible to type B strains. The various hares in Europe are noted to be infected with *F. tularensis holarctica* [133] whereas snowshoe and cottontail rabbits in North America are most frequently infected with *F. tularensis tularensis* and *holarctica* and less frequently with *F. tularensis novicida* [95]. Handling infected rabbits is often associated with zoonotic infections of humans. During the winter months, handling infected rabbit tissue is the principal source of tularemia infection in the United States. Rabbits serve as a useful sentinel animal to study the distribution of strains of *Francisella* and their prevalence. Antitularemia antibodies and occasionally viable organisms can be readily detected in native rabbit populations [95,134–136]. Rabbits are also useful for *in vitro* and *in vivo* immunological studies, in particular the macrophages of New Zealand rabbits are readily infected by both type A (SCHU S4) and type B (LVS) bacteria [137]. Rabbit macrophages were susceptible to SCHU S4 killing but were able to control LVS infection at the same multiplicity of infection. LPS from LVS is a known protective antigen in mice against LVS but not SCHU S4 challenge. Rabbit monoclonal and polyclonal antibody is useful for defining species-specific similarities and differences in LPS [138,139].

In the laboratory, rabbits have been used as models of tularemia pathogenesis. Rabbits can be infected i.d., i.p., s.c., and by aerosol. Rabbits, like humans, are extremely susceptible to type A (SCHU S4) *F. tularensis* with an s.c. LD_{50} value of less than 10 CFU. By comparison, the LD_{50} for a type B strain (425) is more than 1×10^6 CFU [140]. Hornick and Eigelsbach [141] concluded that the infectivity of type A strains in the domestic rabbit (*Oryctolagus*) coupled with glycerol fermentation is comparable to and predictive of virulence in both rhesus macaques and humans. The s.c. route of infection in the rabbit is useful to differentiate strains of lowered virulence. Bell et al. [142] performed a quantitative analysis of infection with various type A and type B strains in rabbits, mice, and guinea pigs [142]. They noted that though the most virulent type A strains were equally virulent in all three animals (<10 CFU), strains of diminished virulence were most readily seen by the comparison of infection in mice, guinea pigs, and rabbits; mice were fatally infected with <10 CFU, guinea pigs were of intermediate susceptibility, and rabbits were highly refractory surviving challenge doses $>10^6$ CFU.

Rabbits can be used for pathological studies. Domestic rabbits infected i.d. into the hind footpad are readily infected by SCHU S4 [143]. The 1000 CFU inoculum reached the popliteal lymph nodes 8 h after infection. From the popliteal nodes, the infection rapidly became systemic with bacteria detected in the heart blood, bone marrow, and lungs by 36 h after infection. The rabbits were septicemic on day 3 and died 4–6 days after infection. Necrotic lesions were noted at the site of inoculation and in the lymph nodes, liver, bone marrow, and lungs. The pathological lesions resembled those in fatal human cases of ulceroglandular tularemia.

Importantly, rabbits are highly susceptible to infection by the aerosol route. The pathology of rabbits exposed to small particle size (1–2 μm) aerosols of SCHU S4 was described by Baskerville and Hambleton [144]. In this study, the rabbits were exposed to retained doses of $1–2 \times 10^5$ or $2–4 \times 10^8$ CFU of viable SCHU S4 and the kinetics of pathological lesions were recorded. The earliest sites of infection are the alveolar ducts and alveoli reflecting the nature of small particle penetration of the deep lungs. An influx of polymorphonuclear macrophages was seen within 24 h of infection in the alveoli. After 24 h bacteria were detectable in the lungs. There was a time-dependent increase in the lung lesions involving the alveoli, alveolar ducts, the bronchial and bronchiolar epithelium, the nasal mucosa, and the trachea eventually manifesting as necrotizing bronchopneumonia. As in human cases of tularemia pneumonia, the cervical and bronchial lymph nodes become enlarged with small to large necrotic lesions that eventually manifest as necrotizing lymphadenitis. Rabbits exposed to infectious aerosols also become systemically infected by day 1–4 after exposure. Necrotic lesions are noted in the spleen and liver that coalesced over time. The pathology of rabbits exposed to SCHU S4 is similar to that of mice exposed to infectious aerosols of either SCHU S4 or LVS or of rhesus monkeys exposed to SCHU S4 [81,108,145]. Importantly, the pattern of bacterial spread and the pathological lesions are similar to observations in humans, although lesions of the upper respiratory tract are not noted in humans.

The similarity of disease pathology in the rabbit and human make the rabbit an excellent model for vaccine studies. Rabbits, like humans, demonstrate a similar pattern of resistance to tularemia infection. The wide availability and relatively low cost of rabbits make them an excellent choice to study tularemia infection and treatment. Results of these studies are likely to be representative of and predictive for clinical outcomes in humans.

Rabbits have been used to evaluate the safety and immunogenicity of an improved LVS vaccine [55]. After vaccination, the development of specific anti-*F. tularensis* antibodies was seen in a dose-dependent manner and no adverse clinical signs or histopathological responses were noted.

9.6.6 Nonhuman Primates

The majority of information available for experimental infection of NHPs with *F. tularensis* comes from studies conducted during the 1960s and 1970s. Many of these efforts focused on comparing disease outcome and bacterial dissemination in LVS-vaccinated and nonvaccinated animals, and results were similar to the disease progression seen in other animals including humans [146]. In cynomolgus (*Macaca fascicularis*) monkeys (male and female; 2–5 kg) infected with LVS intracutaneously or by aerosol (10^5 or 7.6×10^5 organisms, respectively), microorganisms were isolated from the inoculation site (or lungs), regional lymph nodes, liver, and spleen within 24 h. LVS was not isolated from blood or bone marrow, and clearance was apparent by day 14. Agglutination titers peaked at 21–28 days after exposure, and mean titers were 1:940 and 1:1980 (dermal and aerosol, respectively) 2 months after exposure. Subsequent to aerosol challenge with virulent *F. tularensis* (10^3 SCHU S4), bacteria were recovered from lungs of all animals within 3 days whether or not they had been previously vaccinated with LVS (10^2–10^5 organisms recovered per gram of lung tissue). Bacteria were isolated from the spleen and liver of several nonvaccinated animals at this time. SCHU S4 was also recovered from the axillary, inguinal, and coeliac but not cervical lymph nodes of vaccinated animals by day 14. However, higher concentrations of SCHU S4 were obtained at earlier time points from the spleen and liver in nonvaccinated animals. Histopathological changes were evident by days 2, 3, and 5 for nonvaccinated, dermally vaccinated, and aerosol-vaccinated monkeys, respectively. For example, monocyte and neutrophil accumulation was visible in the rudimentary alveoli of bronchioles and alveolar ducts. In addition, whereas vaccinated animals exhibited a self-limiting disease, nonvaccinated animals showed a pattern of infection characterized by rapid and persistent multiplication, systemic involvement, and mortality within 7 days. There was an acute, progressive inflammatory response associated with necrosis and granuloma formation. SCHU S4 was easily isolated from all tissues of nonvaccinated animals by day 5 after challenge. Importantly, those exposed by aerosol to LVS were more resistant to virulent SCHU S4 respiratory challenge than dermally vaccinated monkeys as evidenced by the less extensive bacterial dissemination, earlier clearance, and delayed lymph node involvement in these animals. This may indicate the utility of NHPs in particular for tularemia vaccine development, especially if different vaccination routes (s.c., i.n., mucosal, etc.) for future vaccine candidates are investigated.

An issue in these earlier studies was that the type A SCHU S4 strain infection caused mortality too quickly in unvaccinated monkeys (usually within 96 h and preceding granuloma formation) so only the acute phase of pneumonia could be evaluated. And LVS inoculation resulted in minimal detectable pathology in NHPs. With either strain, stages of tularemic pneumonia could not be resolved. Therefore, *F. tularensis* strain 425, a type B isolate with a moderate virulence intermediate between SCHU S4 and LVS, was used [147]. After whole-body exposure of 20 rhesus monkeys (*Macaca mulatta*) with a mean inhaled dose of 5×10^5 organisms, the course of infection was followed for 35 days until the animals recovered without intervention. The earliest lesion in the lungs, seen by day 3, was a general reddening of all lobes. Histology revealed purulent terminal and respiratory bronchiolitis with numerous erosions of the bronchiolar epithelium at that time. Also, bacteria were readily isolated from copious oculonasal discharges. By day 6, inflammation had spread, and a fibrinous, macrophage infiltrate filled alveolar spaces. Neutrophil proliferation in the spleen was also seen early. At day 9, fibrinous exudation, a significant feature of the pneumonia also seen in humans, had contracted to form plugs in the alveoli at the periphery of purulent foci. These were vascularized by day 15 and cleared by day 35. Over half the number of monkeys developed a mild hepatitis, and pathological alterations were also evident in the tubules and glomeruli of the kidneys. *Francisella tularensis* antigen was detectable by immunofluorescence in the respiratory bronchioles at 24 h and up to 21 days after challenge [140]. The dose response data from one study for strain 425 are summarized in Table 9.1, and the mortality rate did not exceed 18%, even at a dose of 10^6 organisms.

The LD_{50} after aerosol exposure for monkeys was directly related to particle size [148]. For SCHU S4 in *M. mulatta* (males, 4–5 kg), the LD_{50} was 14 bacteria if delivered by aerosol in a particle with a mean diameter of 2.1 μm. LD_{50} increased with particle size: 378 bacteria at 7.5 μm, 874 cells at 12.5 μm, and 4447 cells at 24.0 μm. After exposure to 2.1–7.5 μm aerosols, an increase in rectal temperature from 40°C to 41.1°C was the initial indicator of infection, followed by a subsequent drop to 35°C –36.7°C [149]. X-rays revealed a severe lobar infiltration on days 2 and 3. These animals were also anorexic, photophobic, and sensitive to abdominal palpitation. Those exposed to 12.5–24.0 μm aerosols did not display symptoms until later (6–10 days), and disease was less severe. However, upper respiratory lesions, mucosal congestion, and enlargement of cervical lymph nodes were still observed.

Several other studies have used vervet monkeys (*Cercopithecus aethiops*). If untreated, these animals also developed pyogranulomatous lesions in the liver, spleen, respiratory tract, and lymph nodes, with death 5–7 days after infection with SCHU S4 [150]. Kanamycin treatment, while allowing survival and recovery of the animals, did not prevent development of persistent lesions. Further, to better stimulate clinical conditions, treatment was not initiated until pyrexia was evident (third day after infection). However, if treatment was delayed by 24 h, day 4 lesions were

TABLE 9.1
***Francisella tularensis* 425 Infection in Monkeys After Respiratory Exposure**

Mean Number of Organisms Inhaled	Incubation (Days)	Acute Illness Duration (Days)	Percent Animals Showing Illness[a]				
			None	Mild	Moderate	Severe	Fatal
10^0	6 (4–10)	5 (2–9)	30	40	30	0	0
10^2	5.5 (4–7)	6.5 (1–17)	0	50	45	5	0
10^3	4.5 (2–8)	9 (4–22)	0	57	40	0	3
10^4	4 (2–7)	10 (4–18)	0	3	77	13	7
10^5	3 (3–4)	11 (7–15)	0	0	80	12	8
10^6	3 (2–5)	9 (3–19)	0	0	55	27	18

[a] Values based on groups of 20–30 monkeys, except that 10 animals only were administered 10 cells.

larger and more widespread after a much greater systemic dissemination of bacteria and tissue damage. After i.n. inoculation, infection in vervets was consistently fatal. In general, the disease course and outcome were more severe than tularemia in humans, despite the similar pathology in target organs described above.

For rhesus monkeys, the LD_{50} for SCHU S4 given i.d. was 10 CFU [71]. By comparison, in mice the LD_{50}s (i.d.) of type B *Francisella* (strain not specified) and LVS were 1–9 and ~1.95×10^4–1.06×10^6 CFUs (depending on mouse strain and injection site), respectively. Without LVS vaccination or antibiotic treatment, type A infection of NHPs is consistently fatal. Importantly, the disease progression and pathology of pneumonic tularemia in NHPs is very similar to the described course of illness in humans and other animals (mice, rabbits, guinea pigs). The development of bronchopneumonia, hemorrhagic inflammation in the airways, mononuclear cell infiltration and exudate filling of alveolar spaces is seen in both humans and monkeys. Costs and logistical concerns may preclude extensive use of monkeys in research; however, given the resistance of these animals to LVS (virulent in mice), sensitivity to virulent strains like SCHU S4, and anatomical similarity of the lung to that of humans, use of this animal model should be integral to future research for tularemia.

Marmosets (*Callithrix jacchus*) have also been evaluated as models of inhalational tularemia [151,152]. The minimum lethal dose was less than 10 microorganisms. Animals challenged with 100 CFU of SCHU S4 strain *F. tularensis* developed fever 2.5 days after exposure. Clinical signs were seen within 18 h of fever onset and decreased activity was seen after 3 days. All animals died with 1 week of challenge. Postmortem examination noted gross pathological changes in the lungs, spleen, and liver accompanied by high bacterial numbers in tissues and blood. Histopathological changes included severe suppurative bronchopneumonia, severe multifocal pyogranulomatous hepatitis, splenitis, and lymphadenitis. Many elements of clinical disease in marmosets are also found in human disease and in other animal models.

In another NHP model, African green monkeys (*Chlorocebus aethiops*) given 729 CFU of type A *F. tularensis* by infectious aerosol died or were euthanized 7–11 days after infection [153]. Clinical changes were seen within 2 days of challenge. Gross pathological changes consisted of necrotic foci in the lungs, mediastinal lymph nodes, spleen, heart, mediastinum, diaphragm, liver, urinary bladder, urethra, and mesentery. Lungs, mediastinal lymph nodes, and the spleen were most severely affected.

The relative susceptibility of humans, mice, guinea pigs, rabbits, and monkeys to infection with type A, type B, and LVS strains through aerosol or s.c. exposure is shown in Table 9.2.

9.6.7 Other Animal Models

An outbreak of tularemia in wild prairie dogs (*Cynomys* spp.) occurred at a commercial exotic pet distributor in Texas [90]. Prairie dogs may have been infected by cannibalizing the carcasses of infected animals and subsequently passed the disease to humans. When the animals were necropsied, the animals had enlarged submandibular lymph nodes, a sign of oropharyngeal tularemia. This species has not been developed as an animal model of tularemia but may represent an opportunity to study disease in endemic foci.

TABLE 9.2
The LD_{50} of *Francisella tularensis*

	Aerosol			Subcutaneous		
	Type A	Type B	LVS	Type A	Type B	LVS
Humans[a]	<10 [71]	>10^6 [71]	NV	<10 [71]	>10^6 [71]	NV
Mice	<10–20 [81]	10–20 [81]	>10^3 [81,104]	1 [124]	<10 [131]	>10^6 [104,160]
Guinea pigs	<10 [69]	38 [69]	NR	1 [124]	1 [124]	NR
Rabbits	<10 [144]	<10 [144]	NV	<10 [142]	>10^6 [143]	NV
Monkeys	<10 [71]	>10^6 [71]	NV	<10 [71]	>10^6 [71]	NV

Note: LD_{50}, CFU (Reference); LVS, live vaccine strain; NR, not reported; NV, not virulent.
[a] Infectious dose.

Cats (*Felis catus*) have also acquired tularemia by eating infected rabbits, and may also pass the infection to humans [96,154]. Signs of infection are splenomegaly, hepatomegaly, lymphadenopathy, lethargy, fever, and anorexia. Cats are not currently used as animal models of tularemia. Their utility for studying tularemia remains to be determined.

Hamsters (*Cricetus cricetus*) have been proposed as a useful model for pathogenesis studies. The hamster is considered a reservoir of tularemia infection in Eastern Europe and countries of the former Soviet Union. A number of older manuscripts from the former Soviet Union used the hamster as an animal model; however, these studies did not always clarify the type of hamster used. The hamster has been shown to be susceptible to type A and type B (Gyuranecz) strains [155]. Like the prairie dog, the hamster is a relatively uncharacterized animal model that may be useful in understanding transmission of tularemia.

9.7 SUMMARY

Francisella tularensis is not yet a well-studied pathogen; however, the numbers and types of published research papers have increased since 2003. Continued basic research efforts are required to more completely understand the pathogenesis, virulence, and immune response to this host-adapted bacterium. Fortunately, a number of new genomic sequences have recently been published and may serve as the portal to understanding the many facets of the pathogen and the disease. However, additional tools for genetic manipulation of the organism are required to facilitate the prerequisite molecular biology. In addition, several large and small animal models have been used to study the pathogenesis of both type A and type B strains of *Francisella*. An increasing number of vaccine studies have been conducted in animals. The rat is beginning to emerge as the most useful animal model for comparability to the human. Studies with the LVS vaccine have shown that the mouse is useful but with the caveat that the vaccine does not protect against a virulent type A challenge. This is in contrast to observations in both the NHP and human vaccinates.

None of the animal models described in this chapter behaves in an identical fashion to human infection with respect to their sensitivity of type A or LVS and an ability of LVS vaccination to protect against a type A (SCHU S4) challenge with the possible exception of the Fischer rat. This suggests that significant work toward finding suitable animal models remains [156]. Better animal models that more closely represent human responses to infection are needed to develop succeeding generations of tularemia vaccines.

Mice are not good models of human tularemia. They are exquisitely susceptible to type A and type B tularemia infections whereas humans are not as susceptible. Killed tularemia vaccines generally did not protect mice against a type A infection, whereas the use of those vaccines in humans reduced the severity of disease [47]. It was also noted that inflammatory responses after vaccination occurred sooner in humans than in mice [157,158].

Due to the relatively few number of cases of pneumonic tularemia, tularemia vaccine candidates would need to be licensed by the FDA under the Animal Rule [http://www.fda.gov/OHRMS/DOCKETS/98fr/053102a.htm]. Use of the Animal Rule requires that mechanisms of pathogenesis are understood in that animal model. As mentioned previously, mice are more susceptible to all subspecies of *F. tularensis* than humans [103]. In addition, vaccines given to humans and primates can be protective against virulent *F. tularensis*, but mice are not protected. This should be considered when reviewing published studies in which a statistically significant level of protection is demonstrated in the mouse but without consideration of the biological relevance of this protection. Again, this may be more indicative of a problem with the use of the mouse as a vaccine model versus an inherent problem with the science or candidate vaccine. Thus, mice may not be a good experimental model for evaluating the efficacy of tularemia vaccine candidates. Rabbits and guinea pigs are also more sensitive to *F. tularensis* than humans [159], and the susceptibility of rats differs between strains. In primates, on the other hand, the pathogenesis after inhalation of *F. tularensis* and protection afforded by LVS was similar to humans. Therefore, it is likely that well-characterized primate models will be required for eventual licensure of a tularemia vaccine.

One interesting aspect of tularemia model development is that there is a relative wealth of data for the human model, in particular direct challenge of naïve and immunized humans. It is unlikely that these human studies could be replicated today. Other animals should be investigated more thoroughly for utility in vaccine studies. In particular, the vole and rabbit may be better than the mouse. The rat and or the NHP are likely to be the best models for future pivotal efficacy studies. If we listen to nature the choice of animal model can be narrowed to best match the questions we pose.

DEDICATION

This chapter is dedicated to the brave human volunteers of the Seventh Day Adventists. These medical volunteers participated in the early LVS vaccine trials and in aerosol and cutaneous challenge with type A virulent organisms.

REFERENCES

1. Ohara, S., Studies on yato-byo (Ohara's disease, tularemia in Japan). I. *Jpn J Exp Med*, 1954, **24**(2): 69–79.
2. Wherry, W.B. and B.H. Lamb, Infection of man with *Bacterium tularense*. 1914. *J Infect Dis*, 2004, **189**(7): 1321–1329.
3. Harris, S., Japanese biological warfare research on humans: A case study of microbiology and ethics. *Ann N Y Acad Sci*, 1992, **666**: 21–52.
4. Alibek, K., *Biohazzard*. New York: Random House, 1999.
5. Christopher, G.W. et al., Biological warfare. A historical perspective. *JAMA*, 1997, **278**: 412–417.
6. Greub, G. and D. Raoult, Microorganisms resistant to free-living amoebae. *Clin Microbiol Rev*, 2004, **17**(2): 413–433.
7. Tarnvik, A. et al., *Francisella tularensis*—A model for studies of the immune response to intracellular bacteria in man. *Immunology*, 1992, **76**(3): 349–354.
8. Whipp, M.J. et al., Characterization of a novicida-like subspecies of *Francisella tularensis* isolated in Australia. *J Med Microbiol*, 2003, **52**(Pt 9): 839–842.
9. Samrakandi, M.M. et al., Genome diversity among regional populations of *Francisella tularensis* subspecies *tularensis* and *Francisella tularensis* subspecies *holarctica* isolated from the US. *FEMS Microbiol Lett*, 2004, **237**(1): 9–17.
10. Johansson, A. et al., Worldwide genetic relationships among *Francisella tularensis* isolates determined by multiple-locus variable-number tandem repeat analysis. *J Bacteriol*, 2004, **186**(17): 5808–5818.
11. Keim, P., A. Johansson, and D.M. Wagner, Molecular epidemiology, evolution, and ecology of Francisella. *Ann N Y Acad Sci*, 2007, **1105**: 30–66.
12. Fey, P.D. et al., Molecular analysis of *Francisella tularensis* subspecies *tularensis* and *holarctica. Am J Clin Pathol*, 2007, **128**(6): 926–935.
13. Farlow, J. et al., *Francisella tularensis* in the United States. *Emerg Infect Dis*, 2005, **11**(12): 1835–1841.
14. Svensson, K. et al., Evolution of subspecies of *Francisella tularensis. J Bacteriol*, 2005, **187**(11): 3903–3908.
15. Bernard, K. et al., Early recognition of atypical *Francisella tularensis* strains lacking a cysteine requirement. *J Clin Microbiol*, 1994, **32**(2): 551–553.
16. Reintjes, R. et al., Tularemia outbreak investigation in Kosovo: Case control and environmental studies. *Emerg Infect Dis*, 2002, **8**(1): 69–73.
17. Champion, M.D. et al., Comparative genomic characterization of Francisella tularensis strains belonging to low and high virulence subspecies. *PLoS Pathog*, 2009, **5**(5): e1000459.
18. Hood, A.M., Virulence factors of *Francisella tularensis. J Hyg (Lond)*, 1977, **79**(1): 47–60.
19. Sandstrom, G., S. Lofgren, and A. Tarnvik, A capsule-deficient mutant of *Francisella tularensis* LVS exhibits enhanced sensitivity to killing by serum but diminished sensitivity to killing by polymorphonuclear leukocytes. *Infect Immun*, 1988, **56**(5): 1194–1202.
20. Gil, H., J.L. Benach, and D.G. Thanassi, Presence of pili on the surface of *Francisella tularensis. Infect Immun*, 2004, **72**(5): 3042–3047.
21. Hubalek, M. et al., Comparative proteome analysis of cellular proteins extracted from highly virulent *Francisella tularensis* ssp. *tularensis* and less virulent *F. tularensis* ssp. *holarctica* and *F. tularensis* ssp. *mediaasiatica. Proteomics*, 2004, **4**(10): 3048–3060.
22. Sandstrom, G. et al., Immunogenicity and toxicity of lipopolysaccharide from *Francisella tularensis* LVS. *FEMS Microbiol Immunol*, 1992, **5**(4): 201–210.

23. Ancuta, P. et al., Inability of the *Francisella tularensis* lipopolysaccharide to mimic or to antagonize the induction of cell activation by endotoxins. *Infect Immun*, 1996, **64**(6): 2041–2046.
24. Chen, W. et al., Toll-like receptor 4 (TLR4) does not confer a resistance advantage on mice against low-dose aerosol infection with virulent type A *Francisella tularensis. Microb Pathog*, 2004, **37**(4): 185–191.
25. Chen, W. et al., Susceptibility of immunodeficient mice to aerosol and systemic infection with virulent strains of *Francisella tularensis. Microb Pathog*, 2004, **36**(6): 311–318.
26. Dreisbach, V.C., S. Cowley, and K.L. Elkins, Purified lipopolysaccharide from *Francisella tularensis* live vaccine strain (LVS) induces protective immunity against LVS infection that requires B cells and gamma interferon. *Infect Immun*, 2000, **68**(4): 1988–1996.
27. Conlan, J.W. et al., Mice vaccinated with the O-antigen of *Francisella tularensis* LVS lipopolysaccharide conjugated to bovine serum albumin develop varying degrees of protective immunity against systemic or aerosol challenge with virulent type A and type B strains of the pathogen. *Vaccine*, 2002, **20**(29–30): 3465–3471.
28. Kieffer, T.L. et al., *Francisella novicida* LPS has greater immunobiological activity in mice than *F. tularensis* LPS, and contributes to F. novicida murine pathogenesis. *Microbes Infect*, 2003, **5**(5): 397–403.
29. Lindgren, H. et al., Factors affecting the escape of *Francisella tularensis* from the phagolysosome. *J Med Microbiol*, 2004, **53**(Pt 10): 953–958.
30. Telepnev, M. et al., *Francisella tularensis* inhibits Toll-like receptor-mediated activation of intracellular signalling and secretion of TNF-alpha and IL-1 from murine macrophages. *Cell Microbiol*, 2003, **5**(1): 41–51.
31. Tarnvik, A., Nature of protective immunity to *Francisella tularensis. Rev Infect Dis*, 1989, **11**(3): 440–451.
32. Clemens, D.L., B.Y. Lee, and M.A. Horwitz, Virulent and avirulent strains of *Francisella tularensis* prevent acidification and maturation of their phagosomes and escape into the cytoplasm in human macrophages. *Infect Immun*, 2004, **72**(6): 3204–3217.
33. Chong, A. et al., The early phagosomal stage of *Francisella tularensis* determines optimal phagosomal escape and Francisella pathogenicity island protein expression. *Infect Immun*, 2008, **76**(12): 5488–5499.
34. Lai, X.H., I. Golovliov, and A. Sjostedt, *Francisella tularensis* induces cytopathogenicity and apoptosis in murine macrophages via a mechanism that requires intracellular bacterial multiplication. *Infect Immun*, 2001, **69**(7): 4691–4694.
35. Golovliov, I. et al., An attenuated strain of the facultative intracellular bacterium *Francisella tularensis* can escape the phagosome of monocytic cells. *Infect Immun*, 2003, **71**(10): 5940–5950.
36. Nano, F.E. et al., A *Francisella tularensis* pathogenicity island required for intramacrophage growth. *J Bacteriol*, 2004, **186**(19): 6430–6436.
37. Ludu, J.S. et al., The Francisella pathogenicity island protein PdpD is required for full virulence and associates with homologues of the type VI secretion system. *J Bacteriol*, 2008, **190**(13): 4584–4595.
38. Forestal, C.A. et al., *Francisella tularensis* has a significant extracellular phase in infected mice. *J Infect Dis*, 2007 **196**(1): 134–137.
39. Larsson, P. et al., The complete genome sequence of *Francisella tularensis*, the causative agent of tularemia. *Nat Genet*, 2005, **37**(2): 153–159.
40. Karlsson, J. et al., Sequencing of the *Francisella tularensis* strain Schu 4 genome reveals the shikimate and purine metabolic pathways, targets for the construction of a rationally attenuated auxotrophic vaccine. *Microb Comp Genomics*, 2000, **5**(1): 25–39.

41. Lauriano, C.M. et al., Allelic exchange in *Francisella tularensis* using PCR products. *FEMS Microbiol Lett*, 2003, **229**(2): 195–202.
42. Kawula, T.H. et al., Use of transposon-transposase complexes to create stable insertion mutant strains of *Francisella tularensis* LVS. *Appl Environ Microbiol*, 2004, **70**(11): 6901–6904.
43. Gilmore, R.D., Jr. et al., Identification of *Francisella tularensis* genes encoding exported membrane-associated proteins using TnphoA mutagenesis of a genomic library. *Microb Pathog*, 2004, **37**(4): 205–213.
44. Maier, T.M. et al., Construction and characterization of a highly efficient francisella shuttle plasmid. *Appl Environ Microbiol*, 2004, **70**(12): 7511–7519.
45. Kadzhaev, K. et al., Identification of genes contributing to the virulence of *Francisella tularensis* SCHU S4 in a mouse intradermal infection model. *PLoS One*, 2009, **4**(5): e5463.
46. Tigertt, W.D., Soviet viable Pasteurella tularensis vaccines. A review of selected articles. *Bacteriol Rev*, 1962, **26**: 354–373.
47. Eigelsbach, H.T. and C.M. Downs, Prophylactic effectiveness of live and killed tularemia vaccines. I. Production of vaccine and evaluation in the white mouse and guinea pig. *J Immunol*, 1961, **87**: 415–425.
48. Waag, D.M. et al., Immunogenicity of a new lot of *Francisella tularensis* live vaccine strain in human volunteers. *FEMS Immunol Med Microbiol*, 1996, **13**(3): 205–209.
49. Waag, D.M. et al., Cell-mediated and humoral immune responses induced by scarification vaccination of human volunteers with a new lot of the live vaccine strain of *Francisella tularensis*. *J Clin Microbiol*, 1992, **30**(9): 2256–2264.
50. Waag, D.M. et al., Cell-mediated and humoral immune responses after vaccination of human volunteers with the live vaccine strain of *Francisella tularensis*. *Clin Diagn Lab Immunol*, 1995, **2**(2): 143–148.
51. Saslaw, S. et al., Tularemia vaccine study. II. Respiratory challenge. *Arch Intern Med*, 1961, **107**: 702–714.
52. Saslaw, S. et al., Tularemia vaccine study. I. Intracutaneous challenge. *Arch Intern Med*, 1961, **107**: 689–701.
53. Elkins, K.L., S.C. Cowley, and C.M. Bosio, Innate and adaptive immune responses to an intracellular bacterium, *Francisella tularensis* live vaccine strain. *Microbes Infect*, 2003, **5**(2): 135–142.
54. Cherwonogrodzky, J.W., M.H. Knodel, and M.R. Spence, Increased encapsulation and virulence of *Francisella tularensis* live vaccine strain (LVS) by subculturing on synthetic medium. *Vaccine*, 1994, **12**(9): 773–775.
55. Pasetti, M.F. et al., An improved *Francisella tularensis* live vaccine strain (LVS) is well tolerated and highly immunogenic when administered to rabbits in escalating doses using various immunization routes. *Vaccine*, 2008, **26**(14): 1773–1785.
56. Meka-Mechenko, T. et al., Clinical and epidemiological characteristic of tularemia in Kazakhstan. *Przegl Epidemiol*, 2003, **57**(4): 587–591.
57. Centers for Disease Control and Prevention (CDC), Tularemia—Oklahoma, 2000. *MMWR Morb Mortal Wkly Rep*, 2001, **50**(33): 704–706.
58. Bell, J.F. and S.J. Stewart, Chronic shedding tularemia nephritis in rodents: Possible relation to occurrence of *Francisella tularensis* in lotic waters. *J Wildl Dis*, 1975, **11**(3): 421–430.
59. Young, L.S. et al., Tularemia epidemia: Vermont, 1968. Forty-seven cases linked to contact with muskrats. *N Engl J Med*, 1969, **280**(23): 1253–1260.
60. Morner, T., The ecology of tularaemia. *Rev Sci Tech*, 1992, **11**(4): 1123–1130.
61. Berdal, B.P. et al., Field detection of *Francisella tularensis*. *Scand J Infect Dis*, 2000, **32**(3): 287–291.
62. Greco, D. and E. Ninu, A family outbreak of tularemia. *Eur J Epidemiol*, 1985, **1**(3): 232–233.

63. Hoel, T. et al., Water- and airborne *Francisella tularensis* biovar palaearctica isolated from human blood. *Infection*, 1991, **19**(5): 348–350.
64. Feldman, K.A. et al., Tularemia on Martha's vineyard: Seroprevalence and occupational risk. *Emerg Infect Dis*, 2003, **9**(3): 350–354.
65. McCrumb, F.R., Aerosol infection of man with *Pasteurella tularensis. Bacteriol Rev*, 1961, **25**: 262–267.
66. Hager, A.J. et al., Type IV pili-mediated secretion modulates *Francisella* virulence. *Mol Microbiol*, 2006, **62**(1): 227–237.
67. Evans, M.E. et al., Tularemia: A 30-year experience with 88 cases. *Medicine (Baltimore)*, 1985, **64**(4): 251–269.
68. Dennis, D.T. et al., Tularemia as a biological weapon: Medical and public health management. *JAMA*, 2001, **285**(21): 2763–2773.
69. Samoilova, L.V. et al., Experimental study of the pulmonary form of plague, tularemia and pseudotuberculosis (in Russian). *Zh Mikrobiol Epidemiol Immunobiol*, 1977(3): 110–114.
70. Stuart, B.M. and R.L. Pullen, Tularemia meningitis: Review of the literature and report of a case with post-mortem observations. *Arch Intern Med*, 1945, **76**: 163–166.
71. Eigelsbach, H.T. et al., Tularemia: The monkey as a model for man, in *Use of Nonhuman Primates in Drug Evaluation, a Symposium*, H. Vagtborg, editor. Austin & London: University of Texas Press, 1968, pp. 230–248.
72. Anda, P. et al., Waterborne outbreak of tularemia associated with crayfish fishing. *Emerg Infect Dis*, 2001, **7**(3 Suppl): 575–582.
73. Bevanger, L., J.A. Maeland, and A.I. Naess, Agglutinins and antibodies to *Francisella tularensis* outer membrane antigens in the early diagnosis of disease during an outbreak of tularemia. *J Clin Microbiol*, 1988, **26**(3): 433–437.
74. Gurcan, S. et al., An outbreak of tularemia in Western Black Sea region of Turkey. *Yonsei Med J*, 2004, **45**(1): 17–22.
75. Quan, S.F., Quantitative oral infectivity of tularemia for laboratory animals. *Am J Hyg*, 1954, **59**(3): 282–290.
76. Helvaci, S. et al., Tularemia in Bursa, Turkey: 205 cases in ten years. *Eur J Epidemiol*, 2000, **16**(3): 271–276.
77. Francis, E., A summary of present knowledge of tularaemia. *Medicine*, 1928, **7**: 411–432.
78. Quan, S.F., A.G. McManus, and H. Von Fintel, Infectivity of tularemia applied to intact skin and ingested in drinking water. *Science*, 1956, **123**(3204): 942–943.
79. Syrjala, H. et al., Bronchial changes in airborne tularemia. *J Laryngol Otol*, 1986, **100**(10): 1169–1176.
80. Pullen, R.L. and B.M. Stuart, Tularemia: Analysis of 225 cases. *JAMA*, 1945, **129**: 495–500.
81. Conlan, J.W. et al., Experimental tularemia in mice challenged by aerosol or intradermally with virulent strains of *Francisella tularensis*: Bacteriologic and histopathologic studies. *Microb Pathog*, 2003, **34**(5): 239–248.
82. Stuart, B.M. and R.L. Pullen, Tularemic pneumonia: Review of American literature and report of 15 additional cases. *Am J Med Sci*, 1945, **210**: 223–236.
83. Koskela, P. and A. Salminen, Humoral immunity against *Francisella tularensis* after natural infection. *J Clin Microbiol*, 1985, **22**(6): 973–979.
84. Twine, S.M. et al., Immunoproteomics analysis of the murine antibody response to vaccination with an improved *Francisella tularensis* live vaccine strain (LVS). *PLoS One*, 2010, **5**(4): e10000.
85. Enderlin, G. et al., Streptomycin and alternative agents for the treatment of tularemia: Review of the literature. *Clin Infect Dis*, 1994, **19**(1): 42–47.
86. Russell, P. et al., The efficacy of ciprofloxacin and doxycycline against experimental tularaemia. *J Antimicrob Chemother*, 1998, **41**(4): 461–465.

87. Steward, J. et al., Treatment of murine pneumonic *Francisella tularensis* infection with gatifloxacin, moxifloxacin or ciprofloxacin. *Int J Antimicrob Agents*, 2006, **27**(5): 439–443.
88. Conley, J. et al., Aerosol delivery of liposome-encapsulated ciprofloxacin: Aerosol characterization and efficacy against *Francisella tularensis* infection in mice. *Antimicrob Agents Chemother*, 1997, **41**(6): 1288–1292.
89. Hopla, C.E., The ecology of tularemia. *Adv Vet Sci Comp Med*, 1974, **18**(0): 25–53.
90. Avashia, S.B. et al., First reported prairie dog-to-human tularemia transmission, Texas, 2002. *Emerg Infect Dis*, 2004, **10**(3): 483–486.
91. Petersen, J.M. et al., Laboratory analysis of tularemia in wild-trapped, commercially traded prairie dogs, Texas, 2002. *Emerg Infect Dis*, 2004, **10**(3): 419–425.
92. Bell, J.F. and S.J. Stewart, Quantum differences in oral susceptibility of voles, Microtus pennsylvanicus, to virulent *Francisella tularensis* type B, in drinking water: Implications to epidemiology. *Ecol Dis*, 1983, **2**(2): 151–155.
93. Bigler, W.J. et al., Wildlife and environmental health: Raccoons as indicators of zoonoses and pollutants in southeastern United States. *J Am Vet Med Assoc*, 1975, **167**(7): 592–597.
94. Evans, M.E. et al., Tularemia and the tomcat. *JAMA*, 1981, **246**(12): 1343.
95. Shoemaker, D. et al., Humoral immune response of cottontail rabbits naturally infected with *Francisella tularensis* in southern Illinois. *J Wildl Dis*, 1997, **33**(4): 733–737.
96. Woods, J.P. et al., Tularemia in two cats. *J Am Vet Med Assoc*, 1998, **212**(1): 81–83.
97. Sjostedt, A., Tularemia: History, epidemiology, pathogen physiology, and clinical manifestations. *Ann N Y Acad Sci*, 2007, **1105**: 1–29.
98. Morner, T. and R. Mattsson, Experimental infection of five species of raptors and of hooded crows with *Francisella tularensis* biovar palaearctica. *J Wildl Dis*, 1988, **24**(1): 15–21.
99. Bivin, W.S. and A.L. Hogge, Jr., Quantitation of susceptibility of swine to infection with Pasteurella tularensis. *Am J Vet Res*, 1967, **28**(126): 1619–1621.
100. Huntley, J.F. et al., Native outer membrane proteins protect mice against pulmonary challenge with virulent type A *Francisella tularensis*. *Infect Immun*, 2008, **76**(8): 3664–3671.
101. Duplantis, B.N. et al., Essential genes from Arctic bacteria used to construct stable, temperature-sensitive bacterial vaccines. *Proc Natl Acad Sci U S A*, 2010, **107**(30): 13456–13460.
102. Bakshi, C.S. et al., An improved vaccine for prevention of respiratory tularemia caused by *Francisella tularensis* SchuS4 strain. *Vaccine*, 2008, **26**(41): 5276–5288.
103. Fortier, A.H. et al., Live vaccine strain of *Francisella tularensis*: Infection and immunity in mice. *Infect Immun*, 1991, **59**(9): 2922–2928.
104. Conlan, J.W. et al., Different host defences are required to protect mice from primary systemic vs pulmonary infection with the facultative intracellular bacterial pathogen, *Francisella tularensis* LVS. *Microb Pathog*, 2002, **32**(3): 127–134.
105. Green, M. et al., Efficacy of the live attenuated *Francisella tularensis* vaccine (LVS) in a murine model of disease. *Vaccine*, 2005, **23**(20): 2680–2686.
106. Eigelsbach, H.T. et al., Murine model for study of cell-mediated immunity: Protection against death from fully virulent *Francisella tularensis* infection. *Infect Immun*, 1975, **12**(5): 999–1005.
107. Anthony, L.S., E. Skamene, and P.A. Kongshavn, Influence of genetic background on host resistance to experimental murine tularemia. *Infect Immun*, 1988, **56**(8): 2089–2093.
108. Chen, W. et al., Tularemia in BALB/c and C57BL/6 mice vaccinated with *Francisella tularensis* LVS and challenged intradermally, or by aerosol with virulent isolates of the pathogen: Protection varies depending on pathogen virulence, route of exposure, and host genetic background. *Vaccine*, 2003, **21**(25–26): 3690–3700.

109. Shen, H., W. Chen, and J.W. Conlan, Susceptibility of various mouse strains to systemically- or aerosol-initiated tularemia by virulent type A *Francisella tularensis* before and after immunization with the attenuated live vaccine strain of the pathogen. *Vaccine*, 2004, **22**(17–18): 2116–2121.
110. Downs, C.M. and J.M. Woodward, Studies on pathogenesis and immunity in tularemia; immunogenic properties for the white mouse of various strains of *Bacterium tularense*. *J Immunol*, 1949, **63**(2): 147–163.
111. Wu, T.H. et al., Intranasal vaccination induces protective immunity against intranasal infection with virulent *Francisella tularensis* biovar A. *Infect Immun*, 2005, **73**(5): 2644–2654.
112. Anthony, L.S. and P.A. Kongshavn, Experimental murine tularemia caused by *Francisella tularensis*, live vaccine strain: A model of acquired cellular resistance. *Microb Pathog*, 1987, **2**(1): 3–14.
113. Shen, H., W. Chen, and J.W. Conlan, Mice sublethally infected with Francisella novicida U112 develop only marginal protective immunity against systemic or aerosol challenge with virulent type A or B strains of *F. tularensis. Microb Pathog*, 2004, **37**(2): 107–110.
114. Conlan, J.W., A. Sjostedt, and R.J. North, CD4+ and CD8+ T-cell-dependent and -independent host defense mechanisms can operate to control and resolve primary and secondary *Francisella tularensis* LVS infection in mice. *Infect Immun*, 1994, **62**(12): 5603–5607.
115. Lopez, M.C. et al., Early activation of NK cells after lung infection with the intracellular bacterium, *Francisella tularensis* LVS. *Cell Immunol*, 2004, **232**(1–2): 75–85.
116. Polsinelli, T., M.S. Meltzer, and A.H. Fortier, Nitric oxide-independent killing of *Francisella tularensis* by IFN-gamma-stimulated murine alveolar macrophages. *J Immunol*, 1994, **153**(3): 1238–1245.
117. Cole, L.E. et al., Toll-like receptor 2-mediated signaling requirements for *Francisella tularensis* live vaccine strain infection of murine macrophages. *Infect Immun*, 2007, **75**(8): 4127–4137.
118. Sumida, T. et al., Predominant expansion of V gamma 9/V delta 2 T cells in a tularemia patient. *Infect Immun*, 1992, **60**(6): 2554–2558.
119. Sjostedt, A., J.W. Conlan, and R.J. North, Neutrophils are critical for host defense against primary infection with the facultative intracellular bacterium *Francisella tularensis* in mice and participate in defense against reinfection. *Infect Immun*, 1994, **62**(7): 2779–2783.
120. Elkins, K.L. et al., Bacterial DNA containing CpG motifs stimulates lymphocyte-dependent protection of mice against lethal infection with intracellular bacteria. *J Immunol*, 1999, **162**(4): 2291–2298.
121. Poquet, Y. et al., Expansion of Vgamma9 Vdelta2 T cells is triggered by *Francisella tularensis*-derived phosphoantigens in tularemia but not after tularemia vaccination. *Infect Immun*, 1998, **66**(5): 2107–2114.
122. Stenmark, S. et al., Specific antibodies contribute to the host protection against strains of *Francisella tularensis* subspecies *holarctica. Microb Pathog*, 2003, **35**(2): 73–80.
123. Cowley, S.C., S.V. Myltseva, and F.E. Nano, Phase variation in *Francisella tularensis* affecting intracellular growth, lipopolysaccharide antigenicity and nitric oxide production. *Mol Microbiol*, 1996, **20**(4): 867–874.
124. Olsuf'Ev, N.G. and T.N. Dunayeva, Study of the pathogenesis of experimental tularemia. *J Hyg Epidemiol Microbiol Immunol*, 1961, **5**: 409–422.
125. Eisen, R.J. et al., Short report: Time course of hematogenous dissemination of *Francisella tularensis* A1, A2, and Type B in laboratory mice. *Am J Trop Med Hyg*, 2009, **80**(2): 259–262.
126. KuoLee, R. et al., Mouse model of oral infection with virulent type A *Francisella tularensis. Infect Immun*, 2007, **75**(4): 1651–1660.

127. Jemski, J.V., Respiratory tularemia: Comparison of selected routes of vaccination in Fischer 344 rats. *Infect Immun*, 1981, **34**(3): 766–772.
128. Ray, H.J. et al., The Fischer 344 rat reflects human susceptibility to francisella pulmonary challenge and provides a new platform for virulence and protection studies. *PLoS One*, 2010, **5**(4): e9952.
129. Wu, T.H. et al., Vaccination of Fischer 344 rats against pulmonary infections by *Francisella tularensis* type A strains. *Vaccine*, 2009, **27**(34): 4684–4693.
130. Savel'eva, R.A. and A.P. Gindin, Concerning the pathogenesis of tularemia infection in the nonimmune and immune organism (in Russian). *Zh Mikrobiol Epidemiol Immunobiol*, 1965, **42**(8): 43–50.
131. Olsufjev, N.G., K.N. Shlygina, and E.V. Ananova, Persistence of *Francisella tularensis* McCoy et Chapin tularemia agent in the organism of highly sensitive rodents after oral infection. *J Hyg Epidemiol Microbiol Immunol*, 1984, **28**(4): 441–454.
132. Bandouchova, H. et al., Tularemia induces different biochemical responses in BALB/c mice and common voles. *BMC Infect Dis*, 2009, **9**: 101.
133. Morner, T. et al., Infections with *Francisella tularensis* biovar palaearctica in hares (*Lepus timidus*, *Lepus europaeus*) from Sweden. *J Wildl Dis*, 1988, **24**(3): 422–433.
134. Morner, T. and K. Sandstedt, A serological survey of antibodies against *Francisella tularensis* in some Swedish mammals. *Nord Vet Med*, 1983, **35**(2): 82–85.
135. Morner, T., G. Sandstrom, and R. Mattsson, Comparison of serum and lung extracts for surveys of wild animals for antibodies to *Francisella tularensis* biovar palaearctica. *J Wildl Dis*, 1988, **24**(1): 10–14.
136. Morner, T. et al., Surveillance and monitoring of wildlife diseases. *Rev Sci Tech*, 2002, **21**: 67–76.
137. Nutter, J.E. and Q.N. Myrvik, *In vitro* interactions between rabbit alveolar macrophages and *Pasteurella tularensis*. *J Bacteriol*, 1966, **92**: 645–651.
138. Aronova, N.V. and N.V. Pavlovich, Comparative analysis of the immune response of a rabbit to antigens to live and killed *Francisella* species bacteria (in Russian). *Mol Gen Mikrobiol Virusol*, 2001(2): 26–30.
139. Pavlovich, N.V. et al., Species- and genus-specific antigenic epitopes of *Francisella tularensis* lipopolysaccharides (in Russian). *Mol Gen Mikrobiol Virusol*, 2000(3): 7–12.
140. Schricker, R.L. et al., Pathogenesis of tularemia in monkeys aerogenically exposed to *Francisella tularensis* 425. *Infect Immun*, 1972, **5**(5): 734–744.
141. Hornick, R.B. and H.T. Eigelsbach, Tularemia epidemic—Vermont, 1968. *N Engl J Med*, 1969, **281**(23): 1310.
142. Bell, J.F., C.R. Owen, and C.L. Larson, Virulence of *Bacterium tularense*. I. A study of the virulence of *Bacterium tularense* in mice, guinea pigs, and rabbits. *J Infect Dis*, 1955, **97**(2): 162–166.
143. Schricker, R.L., Pathogenesis of acute tularemia in the rabbit. *Technical Manuscript*, 1964. **178**: 1–13.
144. Baskerville, A. and P. Hambleton, Pathogenesis and pathology of respiratory tularaemia in the rabbit. *Br J Exp Pathol*, 1976. **57**(3): 339–347.
145. Eigelsbach, H.T. and R.B. Hornick. Characteristics of two major types of tularemia in North America: Influence of etiologic agent on severity of illness, in *Twentieth Annual Southwestern Conference on Diseases in Nature Transmissable to Man, Austin, Texas, May 7–8,* 1970.
146. Tulis, J.J., H.T. Eigelsbach, and R.W. Kerpsack, Host-parasite relationship in monkeys administered live tularemia vaccine. *Am J Pathol*, 1970, **58**(2): 329–336.
147. Hall, W.C., R.M. Kovatch, and R.L. Schricker, Tularaemic pneumonia: Pathogenesis of the aerosol-induced disease in monkeys. *J Pathol*, 1973, **110**(3): 193–201.

148. Day, W.C. and R.F. Berendt, Experimental tularemia in *Macaca mulatta*: Relationship of aerosol particle size to the infectivity of airborne *Pasteurella tularensis*. *Infect Immun*, 1972, **5**(1): 77–82.
149. Hambleton, P. et al., Changes in whole blood and serum components during *Francisella tularensis* and rabbit pox infections of rabbits. *Br J Exp Pathol*, 1977, **58**(6): 644–652.
150. Baskerville, A., P. Hambleton, and A.B. Dowsett, The pathology of untreated and antibiotic-treated experimental tularaemia in monkeys. *Br J Exp Pathol*, 1978, **59**(6): 615–623.
151. Nelson, M. et al., Characterization of lethal inhalational infection with *Francisella tularensis* in the common marmoset (*Callithrix jacchus*). *J Med Microbiol*, 2010, **59**(Pt 9): 1107–1113.
152. Nelson, M. et al., Establishment of lethal inhalational infection with *Francisella tularensis* (tularaemia) in the common marmoset (*Callithrix jacchus*). *Int J Exp Pathol*, 2009, **90**(2): 109–118.
153. Twenhafel, N.A., D.A. Alves, and B.K. Purcell, Pathology of inhalational *Francisella tularensis* spp. *tularensis* SCHU S4 infection in African green monkeys (*Chlorocebus aethiops*). *Vet Pathol*, 2009, **46**(4): 698–706.
154. Baldwin, C.J. et al., Acute tularemia in three domestic cats. *J Am Vet Med Assoc*, 1991, **199**(11): 1602–1605.
155. Gyuranecz, M. et al., Susceptibility of the common hamster (*Cricetus cricetus*) to *Francisella tularensis* and its effect on the epizootiology of tularemia in an area where both are endemic. *J Wildl Dis*, 2010, **46**(4): 1316–1320.
156. Rick Lyons, C. and T.H. Wu, Animal models of *Francisella tularensis* infection. *Ann N Y Acad Sci*, 2007, **1105**: 238–265.
157. Fuller, C.L. et al., Transcriptome analysis of human immune responses following live vaccine strain (LVS) *Francisella tularensis* vaccination. *Mol Immunol*, 2007, **44**(12): 3173–3184.
158. Andersson, H. et al., Transcriptional profiling of host responses in mouse lungs following aerosol infection with type A *Francisella tularensis*. *J Med Microbiol*, 2006, **55**(Pt 3): 263–271.
159. Lyons, C.R. and J. Wu, Animal models of *Francisella tularensis* infection. *Ann NY Acad Sci*, 2007, **1105**: 238–265.
160. Elkins, K.L. et al., Introduction of *Francisella tularensis* at skin sites induces resistance to infection and generation of protective immunity. *Microb Pathog*, 1992, **13**(5): 417–421.

10 Q Fever

David M. Waag and David L. Fritz

CONTENTS

10.1 HISTORY

In 1933, Dr. Edward Derrick, the director of Health and Medical Services for Queensland, Australia, was sent to investigate a previously undescribed disease among abattoir workers in Queensland, Australia. Patients presented with fever, headache, and malaise. Because the disease etiology was unknown, this syndrome was called Q (query) fever. Blood and urine from patients were injected into guinea pigs in an attempt to discover the cause of the disease. Infection caused a febrile response that could be passed to successive animals [1]. However, the etiologic agent could not be isolated and Dr. Derrick assumed that the causative agent was a virus. About the same time, in Montana, ticks collected in an investigation into Rocky Mountain spotted fever were injected into guinea pigs. One of these animals became febrile and the infection could be passed to successive animals. Although the infectious organism was isolated, the identity of the resulting disease remained unknown. A breakthrough in studying this microorganism occurred in 1938, when Dr. Cox was able to cultivate *Coxiella burnetii* in large numbers in yolk sacs of fertilized hen eggs [2]. Also in 1938, a researcher in Montana was infected with the tick isolate, and guinea pigs were infected by an injection of a sample of the patient's blood. Ultimately, the agent causing the unidentified disease in Australia was shown to be the same as the one isolated from ticks in Montana by demonstrating that guinea pigs previously challenged with the Montana isolate were resistant to challenge with the Q fever agent [3].

10.2 THE ORGANISM

Coxiella burnetii is an obligate phagolysosomal parasite of the host cell and, until recently [4], could not be cultured on artificial media. That is why laboratory animals were critical for the original isolation of this agent. Laboratory animals also play a role today in understanding disease pathogenesis, in evaluating the safety, immunogenicity, and efficacy of vaccines, and in attempts to isolate unknown organisms from environmental samples. In the laboratory, *C. burnetii* strains are routinely cultured in chicken embryo yolk sacs and in cell cultures [5]. However, the microorganism is a slow grower, with a generation time of approximately 8 h [6].

The organism usually grows as a small bacillus that has a cellular architecture similar to other Gram-negative bacteria [7]. At least two different cell types are found within mature populations in animal hosts [8]. Small-cell variants are distinct from large-cell variants in the population. Small-cell variants are thought to be the infectious particles that are resistant to heat and desiccation [9] and killing by chemicals [10], whereas large-cell variants represent metabolically active organisms. *Coxiella burnetii* displays lipopolysaccharide (LPS) variations similar to the smooth-rough LPS variation in *Escherichia coli* [11]. The organism normally has a smooth (phase I) LPS. As it is passed in a nonimmunocompetent host, such as the yolk sack of embryonated eggs, or in cell culture, the microorganism gradually acquires a rough (phase II) LPS.

10.3 TRANSMISSION

Q fever is generally an acute and self-limited febrile illness that rarely causes a chronic debilitating disease. Fatalities are rare, unless the patient develops chronic Q fever [12]. This zoonotic disease occurs worldwide. Although domestic ungulates, including cattle, sheep, and goats, usually acquire and transmit *C. burnetii*, domestic pets can also be a source of human infection [13,14]. Heavy concentrations of this organism are secreted in milk, urine, feces, and especially in parturient products of infected pregnant animals [15]. During natural infections, the organism grows to high titer (>10^9 microorganisms per gram of tissue) in placental tissues of goats, sheep, and possibly cows [16]. However, natural hosts rarely display clinical symptoms [17], although infection is sometimes accompanied by an increased abortion rate and an increase in the numbers of weak offspring [18]. Q fever during human pregnancy can lead to adverse pregnancy outcomes, including spontaneous abortion and premature delivery [19]. Infection in humans and domestic livestock is most commonly acquired by inhaling infectious aerosols [20]. Less frequent portals of entry include ingestion of infected milk [21] and parenteral acquisition caused by the bite of an infected tick [1]. The infectious dose for humans is estimated to be 10 microorganisms or fewer [22]. The route of infection may influence clinical presentation of the disease [23]. In regions of Europe, where ingestion of raw milk is the more common transmission mode, acute Q fever is found predominantly as a granulomatous hepatitis [24]. However, in Nova Scotia, where infection is predominantly by the aerosol route, Q fever pneumonia is more common [25].

10.4 SYMPTOMS

By definition, the disease presentation in animal models must bear similarities to human disease to be useful surrogates for evaluating products intended for humans. Therefore, a thorough understanding of human symptoms of the disease of interest is needed prior to developing an animal model of any disease. Although no clinical feature is diagnostic for Q fever, certain signs and symptoms tend to be prevalent in acute Q fever cases. Fever, headache, and chills are most commonly seen. The temperature rises rapidly for 2–4 days and generally peaks at 40°C and, in most patients, the usual duration of fever is approximately 13 days [26]. Fatigue and sweats also frequently occur [27]. Other symptoms reported in human cases of acute Q fever include cough, nausea, vomiting, and chest pain. A common clinical manifestation of Q fever is pneumonia. Atypical pneumonia is most frequent, but asymptomatic patients can also exhibit radiological changes [28]. These changes are usually nonspecific and can include rounded opacities and pleural effusions [13]. Both homogenous infiltrates [28] and lobar consolidation are observed [29]. Q fever may also cause a syndrome resembling acute hepatitis with elevations of asparate transaminase (AST) and/or alanine transaminase (ALT) [27]. Elevations in levels of alkaline phosphatase and total bilirubin are seen less commonly. Lactose dehydrogenase (LDH) serum levels are elevated in 33–40% of Q fever patients. The white blood cell count in patients with acute Q fever is usually normal, but mild anemia or thrombocytopenia may be present [30].

Chronic infection is a rare, but often fatal, complication of acute Q fever. This disease is usually seen in patients with prior coronary disease or those who are immunocompromised [31,32]. The usual clinical presentation is endocarditis, but chronic hepatitis has also been seen [33]. Patients with chronic Q fever lack appropriate T-cell responses [34] and the resulting immunosuppression of host cellular immunity is a result of a cell-associated immunosuppressive complex [35]. Suppression of host immunity may allow persistence of the microorganism in host cells during the development of chronic Q fever.

Both acute and chronic Q fever are diagnosed based on clinical signs and symptoms and the patient's history. Although host resistance is not dependent on humoral immunity [36], antibody titers can be useful in supporting a diagnosis of acute or chronic Q fever. Serological assays measure antibodies that react with phase I and phase II *C. burnetii* cellular antigens. Antibody titers that exceed the diagnostic cutoffs can support a clinical diagnosis of acute Q fever where the antiphase II titer exceeds the antiphase I titer. Conversely, when the antiphase I titer exceeds the antiphase II titer, the patient likely has chronic Q fever [37].

Although recovery from acute Q fever is usually routine, patients occasionally have developed chronic fatigue, characterized by prolonged recovery. Antibody levels remained high and *C. burnetii* genomic DNA could be detected by polymerase chain reaction in bone marrow [38]. The authors speculated that contributing to this syndrome was a noninfective, nonbiodegraded complex of *C. burnetii* cellular antigens and specific LPS that inhibited ability of the host to clear the microorganisms [39].

10.5 ANIMAL MODELS OF ACUTE Q FEVER

Not surprisingly, considering its wide host range, *C. burnetii* can infect a wide range of laboratory animals, including mice, rats, rabbits, guinea pigs, and monkeys. Mouse, guinea pig, and monkey disease models have been developed to study the pathogenesis of disease and to evaluate the efficacy of vaccines against Q fever [40,41]. The consequences of infection can range from asymptomatic to lethal, the severity of the symptoms varying in direct proportion to the infectious dose. Because of their susceptibility to infection and ease and economy of use as laboratory animals, mice and guinea pigs are the favored animal models of acute Q fever. A single microorganism is sufficient to cause infection [42].

10.5.1 MICE

Because human illness generally occurs as a result of infection by aerosol, the susceptibility of inbred and outbred strains of mice infected by this route was examined [43]. Investigators determined that all animals evaluated in this study displayed the symptoms of roughened fur, lethargy, and coryza after exposure. Histopathological examination revealed intrastitial pneumonia, multifocal hepatitis, splenitis, and lymphadenitis. Lesions observed in the infected animals were similar to those previously described in infected humans. Pneumonia observed in an inbred strain of mouse (DBA/2J) was greater than that found in the outbred Swiss-ICR mice. DBA/2J mice also had a greater mortality after infection than other mouse strains tested.

In an expansion and refinement of the previous study, Scott et al. [44] tested the susceptibility of inbred strains of female mice to intraperitoneal infection. When the ability of *C. burnetii* infection to cause sickness and death in 47 mouse strains was determined, 33 mouse strains were found to be resistant, 10 were of intermediate sensitivity, and 4 were sensitive. Susceptible mice developed lethargy and a roughened fur coat 3–6 days after infection and were ill for 3–17 days. Infection was inapparent in most resistant mouse strains and there was no discernable mouse strain haplotype that correlated with mortality. The most susceptible mouse strain was the A/J strain, with survivors yielding high concentrations of microorganisms from all tissues examined, including heart, liver, lung, spleen, kidney, and brain. The sensitivity of the A/J strain to infection might be related to low interferon production and deficiencies in the complement pathway and macrophage functions [34,45–47]. Induction of gross pathological responses and antibody production were similar in sensitive (strain A/J) and resistant mice (strain C57BL/6J). The dose of phase I *C. burnetii* required to kill 50% of challenged A/J mice (or the median lethal dose, LD_{50}) was $10^{7.1}$ microorganisms, whereas the LD_{50} for C57BL/6J mice was $10^{9.9}$ microorganisms. The LD_{100} in A/J mice was $10^{7.7}$ microorganisms, and few deaths occurred with doses less than $10^{6.5}$ microorganisms. Mice of both strains developed antibody titers against phase I cells, phase II cells, and phase I LPS, and the magnitude of titers was proportional to the infecting dose. In analysis of cell-mediated immunity, the proliferative responses of splenocytes from C57BL/6J mice to specific recall antigen *in vitro* was greater than that in A/J mice. One interesting feature of the proliferative response is that stimulation responses declined with increases in the infecting dose [44].

The C57BL/10 ScN endotoxin-nonresponder mouse model of infection was used to show that injecting phase I Q fever whole cell (WCV) vaccine could suppress the proliferative response of spleen cells to mitogens and *C. burnetii* phase I cellular antigen [48]. The phase I WCV was not directly cytotoxic to spleen cells from normal or vaccinated mice. Phase II WCV did not induce significant mitogenic hyporesponsiveness or negative modulation of spleen cells. Phase I cell-associated constituents responsible for this negative immune modulation were called the immunosuppressive complex. This complex could be dissociated by chloroform–methanol (CM) (4:1) extraction of phase I cells, yielding CM residue (CMRI) and CM extract (CME) [35]. The suppressive components in either CMRI or CME did not induce immunosuppressive activity in the mouse model when injected separately. Reconstituting the CMRI with CME before injection produced the same pathological reactions characteristic of phase I cells. The suppressive complex was expressed by phase I strains with smooth LPS, but not by phase II strains.

The first observation of histopathological consequences of infection was made by Burnet and Freeman [49], who found that the mouse liver and spleen were the organs most affected. In subsequent studies, mice infected with *C. burnetii* developed granulomatous lesions and a mononuclear cell infusion could be found in the spleen, liver, kidneys, and adrenal glands [50]. Numerous bacteria could be isolated from the spleen and liver. Mice infected intranasally developed purulent bronchopneumonia. Pneumonic histopathological changes were only noted in mice infected intranasally. Changes were characterized by exudates containing large and small mononuclear cells. Microorganisms were intracellular in alveolar epithelial cells and histiocytes, but not in leukocytes. Intraperitoneal infection led to granulomatous lesions, primarily containing mononuclear cells, in the spleen, liver, kidney, adrenal glands, and peritoneal and mediastinal lymph nodes. Similar changes were seen in bone marrow. Pathological changes responsible for death of challenged mice have not been defined.

10.5.2 Guinea Pigs

Guinea pigs offer the benefit of a predictable febrile response (>40°C) when infected, whereas infected mice can be asymptomatic [12]. Mature animals are generally used because they have a more uniform response to *C. burnetii* infection [17]. After infection, guinea pigs develop a febrile response within 1–2 weeks, and they can be bacteremic for 5–7 days. The duration of fever depends on the infectious dose and the virulence of the challenge microorganism. Animals may excrete microorganisms in the urine for months [51]. The aerosol LD_{50} of *C. burnetii* in guinea pigs was approximately 10^8 microorganisms using the California AD strain [22]. Antibodies against phase II microorganisms were detected within 2 weeks after infection, and antibodies against phase I and phase II microorganisms were found 2 months after infection. The minimum infectious dose for guinea pigs or mice infected intraperitoneally or by aerosol was fewer than 10 microorganisms [22,44].

Histological changes, though not diagnostic of *C. burnetii* infection, are most noticeable in organs of the reticuloendothelial system. The natural course of infection in guinea pigs corresponds to an acute illness with formation of granulomas in the liver, spleen, bone marrow, and other organs, with rapid regression of clinical signs and clearance of the granulomas. Lung infiltrations were predominate when the microorganisms were delivered by aerosol [52]. Histopathological changes in the lungs of aerosol-infected guinea pigs included a coalescing panleukocytic bronchointerstitial pneumonia that resolved to multifocal lymphohistiocytic interstitial pneumonia [53]. This type of disease evolution closely mimics Q fever in humans. Liver granulomas were found more often after an intraperitoneal challenge. Pathological changes regressed during convalescence. However, animals could remain latently infected with the disease re-emerging after immunosuppression with irradiation or steroids [54,55]. Unlike other rickettsial diseases, infection was not accompanied by scrotal reaction [17]. The infectious dose and route of infection greatly influenced the pathological changes observed. *Coxiella burnetii* strains have a range of disease potentials after infection. When injected into guinea pigs, bacterial isolates within the same genomic group caused similar pathological responses [56].

10.5.3 Nonhuman Primates

During the mid-1970s, cynomolgus monkeys (*Macaca fascicularis*) were developed as models of human acute Q fever [57,58]. These animals, when challenged by aerosol, developed clinical signs and histopathological changes characteristic of human acute Q fever. Cynomolgus monkeys challenged by aerosol with 10^5 microorganisms of phase I *C. burnetii* developed anorexia, depression, and increased rectal temperatures beginning between 4 and 7 days after challenge and lasting for 4–6 days. Monkeys developed severe interstitial pneumonia and bacteremia between 4 and 11 days after challenge. Radiographs showed granular infiltration in the hilar area on day 5 with resolution of the chest radiograph abnormalities beginning by day 16. In monkeys necropsied 21 days after challenge, histopathogenic evaluation revealed moderate to severe interstitial pneumonia and mild to moderate multifocal granulomatous hepatitis. In addition, plasma fibrinogen levels rose by day 6 and remained elevated through day 14. Neutrophilia and lymphopenia were also noted. There were increases in serum alkaline phosphatase levels from day 3 to day 13. Total bilirubin was significantly higher than baseline. Antibodies to phase II and phase I *C. burnetii* were detected by day 7 and day 14, respectively.

In a more recent comparative study, rhesus (*Macaca mulatta*) and cynomolgus monkeys were evaluated for suitability as models of human acute Q fever [59]. After monkeys were challenged by aerosol with 10^5 microorganisms of phase I *C. burnetii*, the mean temperatures were approximately 3°F above normal for approximately 1 week, a fever profile similar to that found in human Q fever patients [59]. However, unlike findings in the Canadian studies of human Q fever [13], the presence of rounded opacities and pleural effusions were not found. In fact, radiographic changes noted in the nonhuman primates were similar to those described in human patients during the West Midlands outbreak in the United Kingdom [28]. Those cases had poorly defined, largely homogeneous shadowing without distinguishing features. Radiographic changes worsened within 2 weeks of the initial film before improving. Pleural effusions and linear atelectasis were rare.

When changes in serum chemistry were evaluated, cynomolgus monkeys exhibited significant increases from baseline in AST between days 14 and 42 after aerosol challenge [59]. AST levels were increased in rhesus monkeys on days 14 and 42. Significantly higher levels of LDH were observed in cynomolgus, compared to rhesus monkeys (days 10 and 42), and levels were higher than baseline (days 5, 10, 14, and 42 after challenge). Elevations in AST and LDH can suggest liver involvement after infection. Cynomolgus monkeys were also noted to have normal white blood cell counts in blood collected on days 10 through 21 after challenge. Rhesus monkeys exhibited lymphopenia throughout the observation period. A decrease in the percentage of platelets in cynomolgus monkeys, occurring 1 week after infection, was noted. Thrombocytopenia is reported in approximately 25% of Q fever patients. Several parameters indicated that infected cynomolgus and rhesus monkeys developed anemia during infection. These included decreases in hematocrit, hemoglobin, and red blood cell levels and an increase in MCH. In one study, approximately half the number of patients with acute Q fever pneumonia had a drop in hemoglobin levels [27]. Hemolytic anemia due to Q fever has also been reported [60].

The serological profile in infected cynomolgus and rhesus monkeys was similar to that described in acute Q fever patients [59,61]. Within 30 days of disease onset, antiphase II antibodies rose to higher levels than antiphase I or anti-LPS responses. Investigators concluded that rhesus and cynomolgus monkeys provide good models of acute Q fever for vaccine efficacy testing. Ultimately, the cynomolgus monkey model was chosen for vaccine efficacy testing because radiographic changes after aerosol challenge were more pronounced. There was also a strong correlation between effects of *C. burnetii* infection on clinical laboratory results in infected cynomolgus monkeys and in humans with acute Q fever.

10.5.4 Other Animal Models

Hamsters have not been well studied as models of *Coxiella* infection. However, they have been reported to be more susceptible to infection than mice and guinea pigs, and had higher antibody titers after infection [62]. Rabbits have not been developed as an animal model, although they can be the source of infection for humans [63]. Rabbits might be a good model to study the effects of *C. burnetii* infection on the fetus, as infected rabbits were shown to deliver dead fetuses at term [64].

10.6 ANIMAL MODELS OF CHRONIC Q FEVER

There are no natural models of chronic Q fever endocarditis. However, chronic infection can be seen in animals that have been immunocompromised genetically, or by drugs or irradiation [54,55,65]. Four models of Q fever endocarditis have been developed. This disease has been established in immunocompromised [66,67] and pregnant [68] mice, and in guinea pigs with cardiac valves damaged by electrocoagulation [69]. In mice immunosuppressed by cyclophosphamide treatment and infected intraperitoneally, endocarditis of the atrioventricular and semilunar valves, characterized by infiltration by macrophages and neutrophils, was present [66]. *Coxiella burnetii* antigens were found in most organs and in cardiac valves, the aorta, and the pulmonary artery, mainly within macrophages, neutrophils, and endothelial cells. In another study [67], severe combined immunodeficiency (SCID) mice infected with *C. burnetii* showed persistent clinical symptoms and died, whereas immunocompetent mice, similarly infected, became asymptomatic and survived. The infected SCID mice had severe chronic lesions in internal organs, including the heart, lungs, spleen, liver, and kidney. The heart lesions were similar to those detected in humans with chronic Q fever endocarditis. Finally, transgenic mice constitutively expressing interleukin-10 in macrophages developed persistent infection with an inability to kill *C. burnetii* with few granulomas in the spleen and liver, similar to findings in human chronic Q fever patients [70].

Approximately one decade after description of the disease and identification of the causative microorganism, parturient animals, particularly ruminants, were known to be an important risk factor for acquiring Q fever [71]. A murine animal model was developed to study the association between infection, pregnancy, and abortion [68]. Intraperitoneal infection of female BALB/c mice with *C. burnetii*, followed by repeated pregnancies over 2 years, resulted in persistent infection associated with abortion and perinatal death, with a statistically significant decrease in viable offspring. In addition, endocarditis occurred in some of the adult animals, and *C. burnetii* antigens and DNA were detected in their heart valves. The development of endocarditis in these animals could be a result of suppression in cellular immunity during pregnancy.

Another experimental model of endocarditis was developed in guinea pigs using electrocoagulation of native aortic valves [69]. One-half of the treated animals developed infective endocarditis. Those with endocarditis had blood cultures positive for *C. burnetii*, and *C. burnetii* antigens were found in their cardiac valves. This model could provide an additional tool for the investigation of the pathophysiology and antibiotic therapy for Q fever endocarditis.

10.7 ANIMAL MODELS OF VACCINE SAFETY

Although vaccination with the *C. burnetii* phase I WCV protected humans against clinical illness [72], the use of early phase I cellular vaccines was frequently accompanied by adverse reactions, including induration at the vaccination site or the formation of sterile abscesses or granulomas [73,74]. This cellular vaccine could also induce hepatomegaly, splenomegaly, liver necrosis, and death in a dose-dependent manner in laboratory animals [75]. Early experience with the vaccine showed that persons who were previously immune by virtue of prior infection or vaccination were at risk of developing severe local reactions if vaccinated [73]. In 1962, a positive skin test was found to correlate with preexisting immunity to Q fever [76]. After skin testing, those persons exhibiting an induration of at least 5 mm diameter in 7 days were not vaccinated. When skin-test-positive individuals were excluded from vaccination, the incidence of adverse reactions after vaccination decreased dramatically. Currently, potential for adverse vaccination reactions is usually assessed by skin tests, although some laboratories also measure the level of specific antibodies against *C. burnetii* [77]. Only individuals testing negative are vaccinated. However, previously sensitized vaccinees run the risk of developing abscesses and granulomas upon vaccination if the skin test is misread or improperly applied. Evidence suggests that cellular *C. burnetii* vaccines are safe and efficacious if the recipients are not immune due to prior *C. burnetii* infection.

Animal models were developed that assessed the risk of developing adverse vaccination reactions in previously immune recipients. One such model involved administering candidate vaccines intradermally to sensitized hairless Hartley guinea pigs [78,79]. The use of hairless guinea pigs eliminated the necessity of removing the animal's hair in preparation for vaccination or observing the vaccination site. Guinea pigs were sensitized by administering WCV or CMR vaccine subcutaneously in complete Freunds adjuvant [79]. Six weeks after antigen sensitization, animals were injected intraderamally with different doses of WCV or CMR [78]. WCV vaccine caused greater induration of injection sites than where an equivalent amount of CMR was applied. In addition, animals sensitized with WCV developed larger areas of induration at the injection sites than animals sensitized with CMR, irrespective of the vaccine given. The host inflammatory response was greater at the WCV injection sites than at the CMR injection sites. Abscess formation was also observed in the group given WCV. Results of this study indicated that the CMR vaccine might cause fewer adverse reactions than WCV if administered to persons previously immune.

To more directly measure the potential of vaccines to develop adverse reactions after vaccination, an abscess/hypersensitivity animal model was developed in guinea pigs [80]. Guinea pigs were sensitized to *C. burnetii* by intraperitoneal infection or by subcutaneous vaccination and aerosol challenge. The latter animals were survivors of a vaccine efficacy study. Sensitized animals were then vaccinated subcutaneously with WCV or CMR. The ability of each vaccine to cause adverse reactions was evaluated histopathologically and by increases in erythema and induration. In this study, the vaccine was administered at the same dose (30 μg) and route used for human vaccination. Between 2 and 3 weeks after vaccination, the animals were euthanized and the skin and subcutaneous tissue of the vaccination sites were excised and processed for histopathological examination. CMR-injected presensitized guinea pigs were found to have less erythema at the injection sites through 1 week after vaccination. Zones of induration surrounding vaccination sites in CMR-vaccinated guinea pigs were smaller and resolved more quickly than for WCV-vaccinated animals.

Draining abscesses were noted in WCV-vaccinated guinea pigs that were sensitized by vaccination and aerosol challenge. Abscesses were characterized histologically by a deep dermal or subcutaneous pocket of fragmented, necrotic polymorphonuclear leukocytes, rimmed with macrophages. Multinucleated giant cells were noted at the site. At 13 days after vaccination, a peripheral mixed cellular infiltrate composed of lymphocytes, lymphoblasts, immunoblasts, macrophages, and polymorphonuclear leukocytes were noted, and at 22 days after vaccination, plasma cells and fibroblasts were also observed. These observations suggested that both WCV and CMR vaccines could induce erythema and induration at the vaccination sites. However, the magnitude of those reactions was greater with WCV, which also caused abscesses with fistulous draining tracts.

This method of subcutaneously vaccinating sensitized guinea pigs to assess the safety of Q fever vaccines is preferable to skin testing. The route and dose are identical to human vaccination. In addition, proper administration of subcutaneous injections is technically less difficult than giving intradermal injections. Finally, similar adverse vaccination reactions as described in humans (erythema, induration, granulomas, and abscesses) can readily be observed in this animal model. This study also showed how secondary studies can be performed after completion of the primary experiment, which can reduce the overall numbers of experimental animals required. Although most animals used in this study had participated in a vaccine efficacy trial, they were reused to generate additional valuable data on the safety of Q fever vaccines.

Animals can also be used to test the safety of attenuated vaccines. The live attenuated human M-44 vaccine was found to persist in three generations of mice after vaccination [81].

10.8 VACCINE EFFICACY TESTING

An efficacious Q fever vaccine was developed in 1948, 12 years after discovery of the etiologic agent. This preparation, consisting of formalin-killed and ether-extracted *C. burnetii* containing 10% egg yolk sac, was effective in protecting human volunteers from aerosol challenge [82]. Purification methods were improved over the years and vaccine efficacy of these more highly purified preparations was also demonstrated in human volunteers [83].

Because ethical considerations preclude humans from intentionally being infected with *C. burnetii* to evaluate vaccine efficacy, reduction in the incidence of Q fever in vaccinated people at risk for natural disease has been used as an indicator of vaccine efficacy [84]. Guinea pigs and mice were developed as surrogate models of human disease to evaluate old- and new-generation vaccines against Q fever before human trials [72,75]. As mentioned above, phase I cellular Q fever vaccines have been associated with the risk of adverse vaccination reactions if given to individuals previously immune. Therefore, safer vaccines are desirable and an efficacious Q fever vaccine is under development (CMR) that offers increased safety by decreasing the risk of adverse reactions [75]. This vaccine was initially tested for safety and efficacy in rodent models [40]. After demonstrating safety and efficacy in a rodent model, vaccines are usually tested for safety and efficacy in nonhuman primates before similar testing in populations of susceptible human volunteers.

10.8.1 Mice and Guinea Pigs

The A/J mouse and guinea pig animal models were used to compare the protective efficacy of the CMR Q fever vaccine with Q-Vax, a licensed vaccine from Australia [40]. Animals were challenged by aerosol 6 weeks after vaccination with a single dose of vaccine. A 1 μg vaccination dose of CMR or Q-Vax was effective in completely protecting all the mice from the lethal effects of challenge. The calculated dose of CMR needed to protect 50% of mice and guinea pigs was 0.01 and 1.49 μg of vaccine, respectively. The calculated dose of Q-Vax that was needed to protect 50% of mice and guinea pigs was 0.03 and 0.36 μg of vaccine, respectively. These vaccines were extremely potent, considering that they were not living, attenuated, or administered with adjuvants. In these studies, the guinea pig was viewed to be a better model of Q fever, as they more uniformly succumbed to infection. These vaccine efficacy findings were similar to those of Scott et al. [44], who found that a single intraperitoneal dose of 2.5 μg of vaccine protected all A/J mice from sickness and death. Guinea pigs were also completely protected from lethal infection and fever by a killed whole cell vaccine after aerosol challenge [53]. As an alternative to lethal challenge models, vaccine efficacy can also be assessed by determining residual infection in tissues after challenge using real-time quantitative polymerase chain reaction [85].

10.8.2 Nonhuman Primates

Although several studies have demonstrated the efficacy of CMR and WCV in mice [75,86,87], there have been few vaccine efficacy studies in primates. Because challenge studies using human volunteers [83] cannot be done, the use of nonhuman primates represents the most appropriate animal model in which to predict vaccine efficacy in humans. WCV was tested in cynomolgus monkeys by Kishimoto et al. [58]. That study demonstrated that subcutaneous vaccination with 30 μg of Henzerling strain WCV Q fever vaccine was protective, with no signs of clinical illness, after monkeys were challenged by aerosol 6 months after vaccination. However, monkeys vaccinated 12 months before challenge did develop signs of clinical illness, although symptoms were less severe than those found in controls. In that study, vaccinated animals did not develop pneumonia as detected by thoracic radiographic changes, whereas control monkeys developed severe interstitial pneumonia, verified by histopathological examination. Vaccinated monkeys did not exhibit changes in hematology profile or serum chemistry, except for a rise in fibrinogen. However, control monkeys developed neutrophilia and lymphopenia, and increases in alkaline phosphatase, serum glutamic oxalacetic transaminase, total bilirubin, and plasma fibrinogen. Vaccinated monkeys were bacteremic only on the fourth day after challenge, whereas bacteria could be isolated from the blood of control animals for 7 days, beginning 4 days after challenge. Monkeys challenged 12 months after vaccination presented with anorexia and depression. Monkeys developed circulating microagglutinating antibodies to phase I and phase II *C. burnetii* antigens after vaccination which were detectable 4 months later. After challenge, antiphase II and antiphase I antibody titers rose within 1 week and titers were strong 6 weeks after challenge, when the experiment ended.

The cynomolgus animal model was also used to test the efficacy of a licensed (in Australia) cellular Q fever vaccine, Q-Vax, and an investigational vaccine, CMR, in a nonhuman primate nonlethal aerosol challenge model [41]. This study showed that Q-Vax, tested at a single 30 μg dose (the human vaccination dose), and CMR, given as a single 30 or 100 μg dose or two 30 μg doses, were equally efficacious. These results were not surprising in light of a similar study in mice and guinea pigs (described above) where less than 2 μg of Q-Vax or CMR was sufficient to protect 50% of mice and guinea pigs from death after aerosol challenge [40]. This study found signs of illness in monkeys challenged 6 months after vaccination. Although the onset of these signs was similar in all treatment groups, clinical changes in vaccinated monkeys were of less severity and/or of shorter duration than changes seen in control monkeys. Generally, mild radiographic changes in a minority of vaccinated monkeys were observed, whereas radiographic changes developed in most control monkeys, ranging from mild to severe increases in interstitial and bronchial opacity. Significant elevations in alkaline phosphatase, but not in serum glutamic oxalacetic transaminase or bilirubin, were reported. Significant differences in differential blood counts between vaccinated and control animals were found and a drop in hemoglobin and hematocrit in all monkeys during the first 8 days after challenge was seen. Hemoglobin and hematocrit values in the control monkeys also decreased significantly during the second week before recovering 6 weeks after challenge. Except for the groups vaccinated with a single dose of 100 μg CMR or two injections of 30 μg CMR, the majority of vaccinated animals in this study became bacteremic. The duration of bacteremia was longer than the single time point noted in the study by Kishimoto et al. [58]. Bacteremia was noted in all monkeys given placebo, similar to the findings by Kishimoto et al. As expected, the presence of bacteremia in vaccinated monkeys correlated with fever. However, microorganisms were detected in the blood of one-third of infected control animals 14 days after challenge, by which time their mean temperature had returned to normal. Although the presence of fever was not determined in the study by Kishimoto et al. [58], the more recent study found that the duration of significant increases in temperature was decreased by approximately 2 days and maximum temperatures were as much as a full degree lower in the vaccinated groups. Similarly, the proportion of bacteremic monkeys and the duration of bacteremia were less in the vaccinated groups. The presence of clinical symptoms in vaccinated monkeys was not evidence of vaccine failures. Vaccinated monkeys had significantly fewer signs and symptoms of disease and for a shorter period of time than controls.

Single doses of Q fever vaccines stimulate poor antibody responses in human volunteers [88,89]. However, the dependence on cell-mediated immunity for protection against Q fever suggests that vaccines can be protective even with low or undetectable levels of serum antibodies. However, a measurable immune response after vaccination is desirable for eventual FDA vaccine licensure. One rationale for giving a single 100 μg dose or two 30 μg doses of CMR was to evaluate the efficacy of CMR at doses demonstrated to be safe in rodents, yet higher than the typical Q fever WCV vaccine dose (30 μg). In an attempt to generate lasting measurable antibody responses after vaccination, a single 30 μg dose of CMR gave comparable antibody responses to a single 30 μg dose of Q-Vax, whereas a higher (100 μg) CMR dose and a two-dose (primary booster) regimen significantly increased the immunogenicity of the vaccine [41]. Although those vaccination regimens resulted in higher antibody responses than observed in animals given single 30 μg doses of CMR or Q-Vax, the antibody responses were short-lived, with antiphase I and antiphase II antibody levels dropping to baseline by 17 weeks after vaccination.

Aerosol challenge with *C. burnetii* resulted in a classic anamnestic antibody response within 2 weeks in vaccinated monkeys, indicating that the vaccination regimen was sufficient to sensitize the animals [41]. In vaccinated monkeys, antiphase I and antiphase II *C. burnetii* responses rose simultaneously and antibody titers leveled off 2 weeks after challenge. In the control monkeys, the antiphase II response preceded the development of antiphase I antibodies (characteristic of acute Q fever in humans [61]), and within 3 weeks of challenge, the antiphase II response was greater than antibody titers seen in vaccinated monkeys. This likely reflected a moderation of the immune response in vaccinated monkeys as the infection was controlled.

Only vaccination with 100 μg of CMR, but not with Q-Vax or lower doses of CMR, resulted in a significant increase in lymphoproliferative responses over control monkeys. However, the lymphoproliferative responses dropped to baseline by 4 weeks after vaccination. This response was less than reported for human vaccinees, where peripheral blood cells from individuals previously exposed to *C. burnetii* up to 8 years earlier demonstrated marked stimulation with *C. burnetii* antigens *in vitro* [90]. In addition, up to 85% of individuals vaccinated with Q-Vax were found to have converted from a negative proliferative response to a positive one in 6 weeks [88].

10.9 CONCLUSION

Probably more than for any other disease, the use of laboratory animals for studies of Q fever was paramount in beginning to uncover the mysteries surrounding this agent. Animals were used to initially demonstrate that Q fever was caused by an infectious agent that could subsequently be passed to successive animals. Animals were also used to isolate the "Nine Mile agent" from ticks and to show that Q fever was indeed caused by the "Nine Mile agent." The use of laboratory animals was necessary to accomplish those tasks, in part, because *C. burnetii* cannot be cultivated *in vitro*. Since those early days, animal models were developed and used to study the pathogenesis of infection, the efficacy of vaccines and therapeutic agents, to provide a platform to study the immunomodulatory capability of the microorganism, and whether preparations can be safely administered as vaccines.

REFERENCES

1. Davis, G. and H. Cox, A filter-passing infectious agent isolated from ticks. I. Isolation from *Dermacentor andersoni*, reactions in animals, and filtration experiments. *Public Health Rep*, 1938, **53**: 2259–2267.
2. Cox, H.R. and E.J. Bell, The cultivation of *Rickettsia diaporica* in tissue culture and in tissues of developing chick embryos. *Public Health Rep*, 1939, **54**: 2171–2178.
3. Dyer, R.E., A filter-passing infectious agent isolated from ticks. IV. Human infection. *Public Health Rep*, 1938, **53**: 2277–2283.
4. Omsland, A. et al., Host cell-free growth of the Q fever bacterium *Coxiella burnetii*. *Proc Natl Acad Sci USA*, 2009, **106**(11): 4430–4434.
5. Waag, D. et al., Methods for isolation, amplification, and purification of Coxiella burnetii, in *Q Fever: The Biology of Coxiella Burnetii*, J. Williams and H.A. Thompson, editors. Boca Raton: CRC Press, 1991, pp. 73–115.
6. Thompson, H.A., Relationship of the physiology and composition of *Coxiella burnetii* to the Coxiella-host cell interaction, in *Biology of Rickettsial Diseases*, D.H. Walker, editor. Boca Raton: CRC Press, 1988, pp. 51–78.
7. Schramek, S. and H. Mayer, Different sugar compositions of lipopolysaccharides isolated from phase I and pure phase II cells of *Coxiella burnetii*. *Infect Immun*, 1982, **38**(1): 53–57.

8. McCaul, T.F. and J.C. Williams, Developmental cycle of *Coxiella burnetii*: Structure and morphogenesis of vegetative and sporogenic differentiations. *J Bacteriol*, 1981, **147**(3): 1063–1076.
9. Samuel, J.E., Developmental cycle of *Coxiella burnetii*, in *Procaryotic Development*, Y.V. Brun and L.J. Shimkets, editors. Washington, DC: ASM Press, 2000, pp. 427–440.
10. Scott, G.H. and J.C. Williams, Susceptibility of *Coxiella burnetii* to chemical disinfectants. *Ann N Y Acad Sci*, 1990, **590**: 291–296.
11. Stoker, M.G. and P. Fiset, Phase variation of the Nine Mile and other strains of *Rickettsia burneti*. *Can J Microbiol*, 1956, **2**(3): 310–321.
12. Derrick, E.H., Q fever, a new fever entity: Clinical features, diagnosis and laboratory investigation. *Med J Aust*, 1937, **2**: 281.
13. Langley, J.M. et al., Poker players' pneumonia. An urban outbreak of Q fever following exposure to a parturient cat. *N Engl J Med*, 1988, **319**(6): 354–356.
14. Laughlin, T. et al., Q fever: From deer to dog to man. *Lancet*, 1991, **337**(8742): 676–677.
15. Welsh, H.H. et al., Air-borne transmission of Q fever: The role of parturition in the generation of infective aerosols. *Ann. NY Acad. Sci.*, 1958. **70**: 528–540.
16. Welsh, H.H. et al., Q fever in California. IV. Occurrence of *Coxiella burnetii* in the placenta of naturally infected sheep. *Public Health Rep*, 1951, **66**(45): 1473–1477.
17. Lang, G.H., Coxiellosis (Q fever) in animals, in *Q Fever: The Disease*, T. Marrie, editor. Boca Raton: CRC Press, 1990, pp. 23–48.
18. Grant, C.G. et al., Q fever and experimental sheep. From the International Council for Laboratory Animal Science. *Infect Control*, 1985, **6**(3): 122–123.
19. Carcopino, X. et al., Q fever during pregnancy: A cause of poor fetal and maternal outcome. *Ann N Y Acad Sci*, 2009, **1166**: 79–89.
20. Lennette, E.H. and H.H. Welsh, Q fever in California. X. Recovery of *Coxiella burneti* from the air of premises harboring infected goats. *Am J Hyg*, 1951, **54**(1): 44–49.
21. Huebner, R.J. et al., Q fever studies in Southern California. I. Recovery of *Rickettsia burneti* from raw milk. *Public Health Rep*, 1948, **63**: 214–222.
22. Tigertt, W.D., A.S. Benenson, and W.S. Gouchenour, Airborne Q fever. *Bacteriol Rev*, 1961, **25**: 285–293.
23. Marrie, T.J. et al., Route of infection determines the clinical manifestations of acute Q fever. *J Infect Dis*, 1996, **173**(2): 484–487.
24. Fishbein, D.B. and D. Raoult, A cluster of *Coxiella burnetii* infections associated with exposure to vaccinated goats and their unpasteurized dairy products. *Am J Trop Med Hyg*, 1992, **47**(1): 35–40.
25. Marrie, T.J. et al., Exposure to parturient cats: A risk factor for acquisition of Q fever in Maritime Canada. *J Infect Dis*, 1988, **158**(1): 101–108.
26. Derrick, E.H., The course of infection with *Coxiella burneti*. *Med J Aust*, 1973, **1**(21): 1051–1057.
27. Marrie, T., Acute Q fever, in *Q Fever, the Disease*, T. Marrie, editor. Boca Raton: CRC Press, 1990, pp. 125–160.
28. Smith, D.L. et al., The chest x-ray report in Q fever: A report on 69 cases from the 1989 West Midlands outbreak. *Br J Radiol*, 1991, **64**: 1101–1108.
29. Tselentis, Y. et al., Q fever in the Greek Island of Crete: Epidemiologic, clinical, and therapeutic data from 98 cases. *Clin Infect Dis*, 1995, **20**(5): 1311–1316.
30. Smith, D.L. et al., A large Q fever outbreak in the West Midlands: Clinical aspects. *Respir Med*, 1993, **87**(7): 509–516.
31. Heard, S.R., C.J. Ronalds, and R.B. Heath, *Coxiella burnetii* infection in immunocompromised patients. *J Infect*, 1985, **11**(1): 15–18.
32. Raoult, D. et al., Q fever and HIV infection. *AIDS*, 1993, **7**(1): 81–86.
33. Yebra, M. et al., Chronic Q fever hepatitis. *Rev Infect Dis*, 1988, **10**(6): 1229–1230.

34. Koster, F.T., J.C. Williams, and J.S. Goodwin, Cellular immunity in Q fever: Specific lymphocyte unresponsiveness in Q fever endocarditis. *J Infect Dis*, 1985, **152**(6): 1283–1289.
35. Waag, D.M. and J.C. Williams, Immune modulation by *Coxiella burnetii*: Characterization of a phase I immunosuppressive complex differentially expressed among strains. *Immunopharmacol Immunotoxicol*, 1988, **10**(2): 231–260.
36. Read, A.J., S. Erickson, and A.G. Harmsen, The role of CD4+ and CD8+ T cells in clearance of primary pulmonary infection of *Coxiella burnetii. Infect Immun*, 2010, **78**(7): 3019–3026.
37. Peacock, M.G. et al., Serological evaluation of O fever in humans: Enhanced phase I titers of immunoglobulins G and A are diagnostic for Q fever endocarditis. *Infect Immun*, 1983, **41**(3): 1089–1098.
38. Marmion, B.P. et al., Long-term persistence of *Coxiella burnetii* after acute primary Q fever. *QJM*, 2005, **98**(1): 7–20.
39. Marmion, B.P. et al., Q fever: Persistence of antigenic non-viable cell residues of *Coxiella burnetii* in the host—implications for post Q fever infection fatigue syndrome and other chronic sequelae. *QJM*, 2009, **102**(10): 673–684.
40. Waag, D.M., M.J. England, and M.L. Pitt, Comparative efficacy of a *Coxiella burnetii* chloroform:methanol residue (CMR) vaccine and a licensed cellular vaccine (Q-Vax) in rodents challenged by aerosol. *Vaccine*, 1997, **15**(16): 1779–1783.
41. Waag, D.M. et al., Comparative efficacy and immunogenicity of Q fever chloroform: methanol residue (CMR) and phase I cellular (Q-Vax) vaccines in cynomolgus monkeys challenged by aerosol. *Vaccine*, 2002, **20**(19–20): 2623–2634.
42. Ormsbee, R. et al., Limits of rickettsial infectivity. *Infect Immun*, 1978, **19**(1): 239–245.
43. Scott, G.H., G.T. Burger, and R.A. Kishimoto, Experimental *Coxiella burnetii* infection of guinea pigs and mice. *Lab Anim Sci*, 1978, **28**(6): 673–675.
44. Scott, G.H., J.C. Williams, and E.H. Stephenson, Animal models in Q fever: Pathological responses of inbred mice to phase I *Coxiella burnetii. J Gen Microbiol*, 1987, **133**(Pt 3): 691–700.
45. Cerquetti, M.C. et al., Impaired lung defenses against *Staphylococcus aureus* in mice with hereditary deficiency of the fifth component of complement. *Infect Immun*, 1983, **41**(3): 1071–1076.
46. Gervais, F., M. Stevenson, and E. Skamene, Genetic control of resistance to Listeria monocytogenes: Regulation of leukocyte inflammatory responses by the Hc locus. *J Immunol*, 1984, **132**(4): 2078–2083.
47. Boraschi, D. and M.S. Meltzer, Defective tumoricidal capacity of macrophages from A/J mice. II. Comparison of the macrophage cytotoxic defect of A/J mice with that of lipid A-unresponsive C3H/HeJ mice. *J Immunol*, 1979, **122**(4): 1592–1597.
48. Damrow, T.A., J.C. Williams, and D.M. Waag, Suppression of *in vitro* lymphocyte proliferation in C57BL/10 ScN mice vaccinated with phase I *Coxiella burnetii. Infect Immun*, 1985, **47**(1): 149–156.
49. Burnet, F.M. and M. Freeman, Experimenal studies on the virus of Q fever. *Med J Aust*, 1937, **2**: 299.
50. Perrin, T.K. and I.A. Bengtson, The histopathology of experimental Q fever in mice. *Public Health Rep*, 1942, **57**: 790–794.
51. Parker, R.R. and E.A. Steinhaus, American and Australian Q fevers; persistance of the infectious agents in guinea pig tissues after defervescence. *Public Health Rep*, 1943, **8**: 3.
52. La Scola, B., H. Lepidi, and D. Raoult, Pathologic changes during acute Q fever: Influence of the route of infection and inoculum size in infected guinea pigs. *Infect Immun*, 1997, **65**(6): 2443–2447.
53. Russell-Lodrigue, K.E. et al., Clinical and pathologic changes in a guinea pig aerosol challenge model of acute Q fever. *Infect Immun*, 2006, **74**(11): 6085–6091.

54. Sidwell, R.W., B.D. Thorpe, and L.P. Gebhardt, Studies on latent Q fever infections. I. Effects of whole body X-irradiation upon latently infected guinea pigs, white mice and deer mice. *Am J Hyg*, 1964, **79**: 113–124.
55. Sidwell, R.W., B.D. Thorpe, and L.P. Gebhardt, Studies of latent Q fever infections. II. Effects of multiple cortisone injections. *Am J Hyg*, 1964, **79**: 320–327.
56. Russell-Lodrigue, K.E. et al., *Coxiella burnetii* isolates cause genogroup-specific virulence in mouse and guinea pig models of acute Q fever. *Infect Immun*, 2009, **77**(12): 5640–5650.
57. Gonder, J.C. et al., Cynomolgus monkey model for experimental Q fever infection. *J Infect Dis*, 1979, **139**(2): 191–196.
58. Kishimoto, R.A. et al., Evaluation of a killed phase I *Coxiella burnetii* vaccine in cynomolgus monkeys (*Macaca fascicularis*). *Lab Anim Sci*, 1981, **31**(1): 48–51.
59. Waag, D.M. et al., Evaluation of cynomolgus (*Macaca fascicularis*) and rhesus (*Macaca mulatta*) monkeys as experimental models of acute Q fever after aerosol exposure to phase-I *Coxiella burnetii*. *Lab Anim Sci*, 1999, **49**(6): 634–638.
60. Levy, P., D. Raoult, and J.J. Razongles, Q-fever and autoimmunity. *Eur J Epidemiol*, 1989, **5**(4): 447–453.
61. Waag, D. et al., Validation of an enzyme immunoassay for serodiagnosis of acute Q fever. *Eur J Clin Microbiol Infect Dis*, 1995, **14**(5): 421–427.
62. Stoenner, H.G. and D.B. Lackman, The biologic properties of *Coxiella burnetii* isolated from rodents collected in Utah. *Am J Hyg*, 1960, **71**: 45–51.
63. Marrie, T.J. et al., Q fever pneumonia associated with exposure to wild rabbits. *Lancet*, 1986, **1**(8478): 427–429.
64. Quignard, H. et al., La fievre Q chez les petits ruminants. Enquete epidemiologique dans region midi-pyrenees. *Rev Med Vet*, 1982, **133**: 413.
65. Kishimoto, R.A., H. Rozmiarek, and E.W. Larson, Experimental Q fever infection in congenitally athymic nude mice. *Infect Immun*, 1978, **22**(1): 69–71.
66. Atzpodien, E. et al., Valvular endocarditis occurs as a part of a disseminated *Coxiella burnetii* infection in immunocompromised BALB/cJ (H-2d) mice infected with the nine mile isolate of *C. burnetii*. *J Infect Dis*, 1994, **170**(1): 223–226.
67. Andoh, M. et al., SCID mouse model for lethal Q fever. *Infect Immun*, 2003, **71**(8): 4717–4723.
68. Stein, A. et al., Repeated pregnancies in BALB/c mice infected with *Coxiella burnetii* cause disseminated infection, resulting in stillbirth and endocarditis. *J Infect Dis*, 2000, **181**(1): 188–194.
69. La Scola, B. et al., A guinea pig model for Q fever endocarditis. *J Infect Dis*, 1998, **178**(1): 278–281.
70. Meghari, S. et al., Persistent *Coxiella burnetii* infection in mice overexpressing IL-10: An efficient model for chronic Q fever pathogenesis. *PLoS Pathog*, 2008, **4**(2): e23.
71. Luoto, L. and R.J. Huebner, Q fever studies in Southern California. IX. Isolation of Q fever organisms from paturient placentas of naturally infected dairy cows. *Public Health Rep*, 1950, **65**: 541.
72. Ormsbee, R. and B. Marmion, Prevention of *Coxiella burnetii* infection: Vaccines and guidelines for those at risk, in *Q fever. The Disease*, T. Marrie, editor. Boca Raton: CRC Press, 1990, pp. 225–248.
73. Bell, J.F. et al., Recurrent reaction of site of Q fever vaccination in a sensitized person. *Mil Med*, 1964, **129**: 591–595.
74. Benenson, A.S., Q fever vaccine: Efficacy and present status, in *Symposium on Q Fever.*, J.E. Smadel, editor. Walter Reed Army Institute of Medical Science, Publication No. 6. Washington, DC: Government Printing Office, 1959.
75. Williams, J.C. and J.L. Cantrell, Biological and immunological properties of *Coxiella burnetii* vaccines in C57BL/10ScN endotoxin-nonresponder mice. *Infect Immun*, 1982, **35**(3): 1091–1102.

76. Lackman, D. et al., Intradermal sensitivity testing in man with a purified vaccine for Q fever. *Am J Public Health*, 1962, **52**: 87–93.
77. Ackland, J.R., D.A. Worswick, and B.P. Marmion, Vaccine prophylaxis of Q fever. A follow-up study of the efficacy of Q-Vax (CSL) 1985–1990. *Med J Aust*, 1994, **160**(11): 704–708.
78. Elliott, J.J. et al., Comparison of Q fever cellular and chloroform-methanol residue vaccines as skin test antigens in the sensitized guinea pig. *Acta Virol*, 1998, **42**(3): 147–155.
79. Ruble, D.L. et al., A refined guinea pig model for evaluating delayed-type hypersensitivity reactions caused by Q fever vaccines. *Lab Anim Sci*, 1994, **44**(6): 608–612.
80. Wilhelmsen, C.L. and D.M. Waag, Guinea pig abscess/hypersensitivity model for study of adverse vaccination reactions induced by use of Q fever vaccines. *Comp Med*, 2000, **50**(4): 374–378.
81. Freylikhman, O. et al., *Coxiella burnetii* persistence in three generations of mice after application of live attenuated human M-44 vaccine against Q fever. *Ann N Y Acad Sci*, 2003, **990**: 496–499.
82. Smadel, J.E., M.J. Snyder, and F.C. Robins, Vaccination against Q fever. *Am J Hyg*, 1948, **47**: 71–78.
83. Fiset, P., Vaccination against Q fever, in *Vaccines Against Viral and Rickettsial Diseases in Man*, PAHO Science Publication No. 147. Washington, DC: Pan American Health Organization, 1966, pp. 528–531.
84. Marmion, B.P. et al., Vaccine prophylaxis of abattoir-associated Q fever. *Lancet*, 1984, **2**(8417-18): 1411–1414.
85. Zhang, J. et al., Balb/c mouse model and real-time quantitative polymerase chain reaction for evaluation of the immunoprotectivity against Q fever. *Ann N Y Acad Sci*, 2005, **1063**: 171–175.
86. Williams, J.C. et al., Characterization of a phase I *Coxiella burnetii* chloroform–methanol residue vaccine that induces active immunity against Q fever in C57BL/10 ScN mice. *Infect Immun*, 1986, **51**(3): 851–858.
87. Kazar, J. et al., Onset and duration of immunity in guinea pigs and mice induced with different Q fever vaccines. *Acta Virol*, 1986, **30**(6): 499–506.
88. Izzo, A.A., B.P. Marmion, and D.A. Worswick, Markers of cell-mediated immunity after vaccination with an inactivated, whole-cell Q fever vaccine. *J Infect Dis*, 1988, **157**(4): 781–789.
89. Waag, D.M. et al., Low-dose priming before vaccination with the phase I chloroform-methanol residue vaccine against Q fever enhances humoral and cellular immune responses to *Coxiella burnetii. Clin Vaccine Immunol*, 2008, **15**(10): 1505–1512.
90. Jerrells, T.R., L.P. Mallavia, and D.J. Hinrichs, Detection of long-term cellular immunity to *Coxiella burneti* as assayed by lymphocyte transformation. *Infect Immun*, 1975, **11**(2): 280–286.

11 Brucellosis

Bret K. Purcell and Robert Rivard

CONTENTS

11.1 HISTORY/BACKGROUND

Brucellosis is an infectious disease caused by nonmotile, Gram-negative coccobacilli of the genus *Brucella*. Humans acquire this infection through direct or indirect contact with zoonotic infections of wild and domesticated animals [1–4]. However, since the early days of the U.S. biological weapons program, *Brucella* sp. were recognized as possessing the characteristics of an effective weapon. These bacteria are highly pathogenic to both humans and animals, easily grown in the laboratory, and can be disseminated by aerosolization by various methods [5–8].

Brucella sp., in particular *Brucella melitensis*, are highly pathogenic for humans by the aerosol route and since these bacteria also cause zoonotic infections they are an ideal candidate for use as a biological weapon. In fact, the United States began development of *B. suis* as a biological weapon in 1942 and incorporated it into the arsenal of the now-discontinued offensive program [9]. Using classified procedures, the bacteria were prepared and formulated to maintain long-term viability, then placed into munitions and field tested against animal targets in 1944–1945. The United States discontinued its offensive program in 1967 and destroyed all U.S. biological weapon munitions. However, during the development and testing of these weapons, the data generated provided clear evidence that weaponized *Brucella* organisms, if used, pose significant threats to both civilian and military populations [10]. In addition, agricultural assets within the United States represent potential high-yield economic terrorist targets with a brucellosis scenario model estimation of $477.7 million per 100,000 persons exposed [11]. Consequently, when considering the origins of infectious disease outbreaks, the investigators and clinicians should keep in mind that unusual or uncharacteristic epidemics caused by zoonotic pathogens may be the result of an intentional release of a biological weapon [12]. One of the most important diagnostic tools used to help identify the source of these infections is the collection of a thorough history. The careful collection of details of exposure to include contact with animals or animal products, exposure to environments known to harbor potentially infected animals, or the onset of numerous medical casualties with similar disease presentations is critical to the identification of the source of the pathogen. The onset of unusual disease patterns or outbreaks of typically zoonotic infections, particularly if they present as pulmonary infections, should alert the medical staff to the possibility of a biological attack event.

Historical writings suggest the first description of this disease was by Hippocrates over 2000 years ago, where he made the observations of a chronic, relapsing febrile illness closely resembling the clinical disease we know of today as brucellosis [13,14]. British army surgeons in 1751 and again in the 1800s, working on the island of Minorca in the Mediterranean and on the island of Malta, provided important clinical observations of this disease [15,16]. David Bruce [17], for whom the genus *Brucella* was named, isolated the bacterium (*Micrococcus melitensis*) from the spleens of five fatal cases of the human disease in Malta in 1887. "Undulant fever," named by M.L. Hughes [18], described the clinical and pathological features of 844 patients in 1897. A.C. Evans, in 1917, first linked B. Bang's observation of an organism, which he named the "Bacilllus of abortion," causing contagious abortions in cattle to the causative agent of human brucellosis described by Bruce [19,20]. In 1920, *Micrococcus melitensis* was renamed *Brucella melitensis* for the organism's agglutination and cellular morphology profile [21,22].

Brucella bacteria infect a wide variety of animal species to include desert wood rats, dogs, seals, voles, cattle, goats, sheep, and other ruminants causing genital infections, abortion, and fetal death [1,4]. Humans, incidentally, are infected by ingestion of contaminated food products or by coming into contact with infected animals. Several reports indicate that over 500,000 new cases of brucellosis occur annually worldwide to include travel-associated morbidity and significant residual health disability [23–27]. The global distribution of human brucellosis has shifted over the past 10 years from traditional epicenters of Latin America, France, and Israel to central Asia and the near east [3]. The highest incidence of human infections reported in the world is Syria with 1603 cases per million per year followed by Mongolia with 605 cases per million per year and Kyrgyzstan with 362 cases per million per year [28]. Numerous endemic foci of brucellosis have been identified along the U.S./Mexico border which increases the likelihood of cross-border spread into Texas and other areas of the continental United States.

In addition to naturally acquired human brucellosis, laboratory exposure has been shown to be a significant risk to personnel [6,27,29–33]. As biodefense research expands into academic and biotechnology industries in the U.S. laboratory, accidents are anticipated to become more frequent and significant. Strict adherence to biosafety practices, which include the proper use of personal protective equipment, proven engineering controls, good laboratory and microbiology practices and, when possible, vaccinations against particular select agents of interest, will reduce the incidence of laboratory-acquired infections [6,34–36]. However, we must acknowledge that no matter how rigorously our safety and security measures are implemented, there will always be a measure of potential risk of exposure and possible infection of personnel working with select agents. Careful medical monitoring and, ultimately, personal responsibility will be absolutely crucial to all biosafety and biosecurity programs. However, it must be clearly delineated to funding, regulatory, and research entities that these important protective measures should not be so onerous as to impede or stifle critical biodefense research and product development. Balanced, evidence-based decisions and common sense implementation of biosafety and biosecurity practices are essential for a successful program using highly virulent and infectious pathogens.

11.2 MICROORGANISM

11.2.1 Taxonomy

The genus *Brucella* is found within the class α-2 proteobactia, order of Rhizobiales and of the family (III) Brucellaceae [37]. Other alphaproteobacteria that infect animals include *Ehrlichia*, *Rickettsia*, and *Bartonella* ssp., all of which are considered intracellular pathogens and can be transmitted by insect vectors. Brucellae, based upon 16S rRNA gene sequences, also have phylogenetic similarities to *Rhizobium*, *Agrobacterium*, and *Rhodobacter* [38]. Despite minimal DNA variation among the *Brucella* spp. [39], differences in pathogenicity and animal reservoirs have been used to classify and name these bacterial species.

There are at least 10 identified subspecies of *Brucella*, each having a predilection to infect specific animals (Table 11.1). *Brucella melitensis*, *B. suis*, and *B. abortus* compose the list of primary pathogenic strains of human *Brucella* infections. A few

TABLE 11.1
Host Specificity of *Brucella* Species

Brucella Species	Animal Host	Human Pathogenicity
B. melitensis	Sheep, goats	High
B. suis	Swine	High
B. abortus	Cattle, bison	Moderate
B. canis	Dogs	Moderate
B. ovis	Sheep	None
B. neotomae	Desert wood rat	None
B. ceti	Dolphins, whale	None
B. pinnipedialis	Seals	None
B. inopinata	Human breast implant	Rare
B. microti	Common vole, red fox	None

scattered cases of *B. canis* have been documented; however, these infections occur primarily in immunosuppressed or immunocompromised patients [40]. The estimated inhaled infectious dose of *B. canis* is on the order of $>10^6$ organisms and therefore these bacteria are considered to be of low virulence for immunocompetent humans. Occasional infections have been reported for *B. cetacean* and *B. pinnipedias* [41,42]. Recently one case was reported of *B. inopinata* causing infection of a breast implant [43]. Although *Brucella* sp. can be shown experimentally to infect different animal species, many of these infections appear to be self-limited. The animal species-specific nature of *Brucella* strains was demonstrated when cattle were exposed to *B. suis*-infected feral swine and neither the suckling calves receiving milk containing the organism nor calves delivered by infected cows demonstrated infection [44].

11.2.2 Genetics

The genome of *Brucella* spp. are 50%–100% larger than *Bartonella* and *Brucella* spp. can exist in soil for up to 10 weeks by metabolizing plant-based molecules [45–48]. Based upon DNA homology, *Brucella* spp. most likely represent a single species consisting of several clades and biovars separated by small sequence changes composed of single nucleotide polymorphisms [45,49]. Several examples of insertion/deletion events mediated by phage/transposons may account for differences in virulence and host specificities [45]. It has been proposed that all *Brucella* sp. evolved from a common *B. ovis*-like ancestor between 86,000 and 296,000 years ago [50].

11.2.3 Virulence Factors

Analysis of whole-genome-based and fragment lengths of DNA have demonstrated genetic divergence of the Brucellae genus and species [45,48,50,51]. Proteomics, complete genomic sequencing, and multilocus analysis of variable number tandem repeats analyses have identified virulence determinants, pathogenicity islands, and evolutionary relatedness between the *Brucella* ssp. [52–56]. The lack of classic virulence factors (exotoxins or endotoxins), the atypical pathogenicity of the lipopolysaccharide (LPS) of the outer membrane and the inhibition of the host's programmed cell death provide *Brucella* with unusual mechanisms of pathogenesis [23,57]. However, a recent study indicates that urease, type IV secretion system, and LPS O antigen play crucial roles in *B. melitensis* infection by the route of ingestion [58]. Using a mouse model, the investigators constructed mutant strains of *B. melitensis* lacking the *ureABC* genes of *urel* operon, *virB2* or *pmm* genes encoding phosphomannomutase and bacterial cultures of these mutants were introduced intragastically into mice. When compared to wild-type *B. melitensis*, the mutants lacking the *pmm* and *virBe* genes were attenuated in the mesenteric lymph nodes and in the spleen whereas the *ureABC1* mutant was attenuated in the spleen and the liver and not in the mesenteric lymph nodes. This study, using *in vitro* cell culture monolayers, demonstrated that *B. melitensis* can move rapidly through polarized enterocytes if the layer also contained lymphoepithelial cells that are found in the mucosa-associated lymphoid tissue [58].

One prominent feature of brucellosis infections is the lack of significant inflammatory responses by the host to the bacteria. During the infection process, the innate immune response is not activated therefore permitting this "stealth pathogen" to gain relatively unopposed entry into the host and disseminate to a variety of tissues [58–62]. Murine models of *B. abortus* have been found to be devoid of proinflammatory response during infection, and *B. abortus* impairs bovine trophoblastic cell proinflammatory responses [59,60]. These studies also correlate with murine models of *B. melitensis* infection which indicate the organisms directly penetrate the digestive tract mucosa through mucosa-associated lymphoid tissue with little to no intestinal inflammation [58]. These weak local immunological responses by the host to this bacterial invasion and dissemination is thought to be a virulence strategy by the organism similar to *Salmonella enterica* serotype *Typhi* [63,64].

The LPS of Brucellae is quite unusual, both functionally and structurally, from other Gram-negative bacteria [65,66]. The lipid A portion contains 16-carbon fatty acids; however, it lacks the typical Enterobactericeae lipid A 14-carbon myristic acid that may account for the reduced pyrogenicity of *Brucella* LPS when compared to *Escherichia coli* LPS [67]. Host cytokine pathways are not activated by the unusual structure of the *Brucella* LPS, which may explain the lack of intracellular killing by inhibiting cellular defense mechanism activation. The O-polysaccharide portion of LPS from smooth strains of *Brucella* contains a 4,6-dideoxy-4-formamido-alpha-D-mannopyranoside expressed either as a homopolymer of alpha-1,2-linked sugars (A type), or as 3 alpha-1,2 and 2 alpha-1,3-linked sugars (M type). These differences account for taxonomic differences through differences in immunoreactivity of A and M sugar types [68]. Rough mutant vaccine efficacy directed against *B. abortus* and *B. ovis* in the mouse model may be due to a complete LPS core [69].

A comparison of the immune profiles of humans and goats during *B. melitensis* infection has revealed that the bacterial antigens are recognized differently between these two species [70]. This study indicates that fundamental differences in immune responses to this zoonotic disease exist between humans and goats which may have significant implications in host pathogen responses and in attempts to compare pathogenesis using this animal model. Using these differences, serodiagnostic proteins were identified of which 2 were common between humans and goats, 11 were specific for humans, and 16 were specific for goats [70].

11.2.4 Bacterial Identification

The characteristics of a low infectious dose, relatively easy transmission by aerosols created by routine laboratory practices, and a high virulence of this pathogenic organism mandate the use of strict biological safety measures when processing clinical samples to reduce the hazard of working with biological samples in a diagnostics laboratory. Human or animal specimens collected from suspected cases of brucellosis require the use of biological safety level (BSL) 2 laboratory precautions for clinical sample processing. However, the use of BSL-3 laboratory precautions are mandatory when performing manipulation of the bacterial cultures that have been confirmed to be *Brucella* spp. [36,71].

Most Brucellae can be cultivated on trypticase, soy-based, or other enriched media with a typical doubling time of approximately 2 h. These organisms grow very slowly and therefore growth results may not be readily detected for several days to weeks. Lysis centrifugation has been shown to improve recovery of bacteria from both acute and chronic cases of brucellosis [72]. Bacteria can be recovered from agricultural milk and milk products, blood, body fluids and tissues from infected animals or humans, vaginal swabs, fetal membranes, and aborted fetuses. Biovars of *B. abortus* have the additional atmospheric culture requirements of 5%–10% carbon dioxide for growth. Blood cultures collected from *B. melitensis*-infected patients can demonstrate bacterial growth using the BACTEC 9240 blood culture system in approximately 1 week using either the adult Plus Aerobic/F or the pediatric Peds Plus/F bottles [73–77]. Of note, when utilizing BACTEC MYCO/F Lytic media growth rate is reduced [78]. However, blood cultures from infected patients may not always yield positive results [79]. *Brucella melitensis* has a tendency to be more readily cultivatable than *B. abortus*; however, recovery of cultivatable organisms from the blood varies from less than 10% to as high as 90%.

Identification of bacterial isolates from brucellosis infections utilizes both biochemical and serological analyses [80]. *Brucella* ssp. can be cultivated on blood agar plates and do not require the presence of X or V factors for growth; however, different species and biovars may require elevated levels of carbon dioxide for growth. The bacterial plates should be examined on a daily basis for at least 7 days and blood bottles should be blindly subcultured onto agar medium every 7 days for a total of 21 days before these cultures are discarded [81]. The organisms appear as punctuate, nonpigmented, and nonhemolytic colonies in approximately 48 h on sheep blood agar plates. Gram staining of bacterial isolates demonstrates very tiny, poorly staining Gram-negative coccobacilli appearing like “fine sand” [82]. An example of this staining can be seen in Figure 11.1.

Preliminary identification of *Brucella* spp. is performed using Gram, oxidase, urease, and catalase tests in addition to acryflavine agglutination. Biochemical

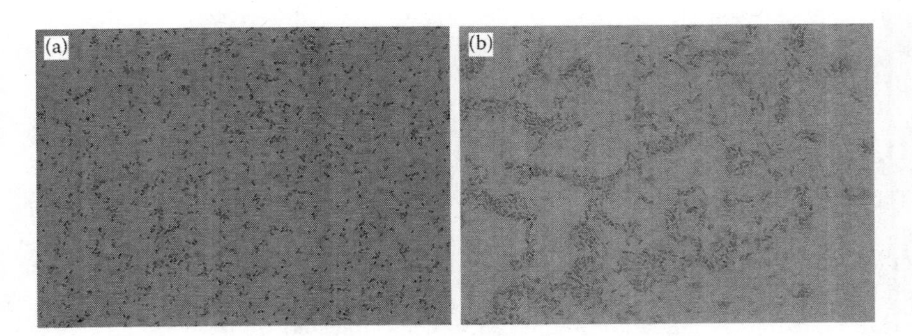

FIGURE 11.1 **(See color insert.)** Gram stains of *Brucella canis* (a) and *Escherichia coli* (b). (Photographs courtesy of Dr. Mark Wolcott and Terry Abshire, Diagnostics Division, U.S. Army Medical Research Institute of Infectious Diseases (USAMRIID), Fort Detrick, Maryland. With permission.)

diagnostics reveal that Brucellae are oxidase and catalase positive, oxidize nitrite to nitrate, and may produce urease [82]. Oxidase results tend to be variable for *B. canis*. *Brucella melitensis*, *B. abortus*, *B. neotomae*, and *B. ovis* demonstrate urease positivity in approximately 15–20 min whereas *B. suis* and *B. canis* are urease positive in less than 5 min. The acriflavine test groups rough and smooth strain Brucellae by auto-agglutination in the presence of acriflavine. *Brucella ovis* and *B. canis* agglutinate whereas *B. abortus*, *B. melitensis*, *B. naeotomae*, and *B. suis* fail to agglutinate in the presence of acriflavine. Presumptive differentiation of *Brucella* species from similar Gram-negative genera and species is outlined in Tables 11.2 and 11.3 [82]. Among the different *Brucella* species and biovars there are differences in their ability to use ribose, lysine, glutamic acid, and orthithine, grow in the presence of thionine and basic fushsin dyes, produce hydrogen sulfide, susceptibility to Tblisi bacteriophage lysis, and be agglutinated by antisera directed against the A and M antigens of the O polysaccharide chain of the specific strain [82]. Phenotypic identification for biotyping *Brucella* spp. includes CO_2 requirement for bacterial growth, H_2S production, growth tolerance on dye media such as basic fuchsin (1:50,000 and 1: 100,000) and thionin (1: 25,000; 1:50,000; 1:100,000), as well as bacterial agglutination using monospecific serum.

Diagnosis of brucellosis has been dependent on the use of a variety of serological tests [80]. These serological assays include the milk ring test, Rose Bengal slide agglutination test, serum agglutination test, *Brucella* microagglutination test, complement fixation test, enzyme-linked immunosorbant assays, fluorescence polarization assay, and the *Brucella* IgM/IgG flow assay [83–86]. These tests are useful to determine whether the clinical symptoms are caused by an acute infection

TABLE 11.2
Presumptive Differentiation of *Brucella* Species from Similar Gram-Negative Genera and Species

Test	*Brucella* spp.	*Acinetobacter* spp.	*Psychobacteria phenylpyruvicusa*[a]
Specimen source	Blood, bone marrow	Various tissues	Various tissues
Morphology by Gram stain	Small coccobaccilli	Broad coccobacilli	Broad coccobacilli
Oxidase	Positive[b]	Negative	Positive
Urea hydrolysis[c]	Positive	Variable	Positive
X or V factor requirements	Negative	Negative	Negative

Source: Adapted from Carvalho Neta AV. et al., 2008, *Infect Immun*, 76:1897–1907.

[a] Formerly *Moraxella phenylpyruvica*.

[b] Oxidase: *B. abortus*, *B. melitensis*, and *B. suis* are all oxidase-positive organisms. *Brucella canis* isolates may be oxidase variable.

[c] Urea hydrolysis: most *Brucella* isolates vigorously hydrolyze urea. Positive values range from greater than or equal to 90% positivity. Negative values range less than or equal to 10% positivity. Variable—range of positivity from 11% to 89%.

TABLE 11.3
Presumptive Differentiation of *Brucella* Species from Similar Gram-Negative Bacteria

Test	*Oligella* spp.	*Haemophilus influenzae*	*Francisella tularensis*
Specimen source	Urinary tract	Various tissues	Various tissues
Morphology by Gram stain	Very small coccobacilli	Small coccobacilli	Very small coccobacilli
Oxidase	Positive	Variable	Negative
Urea hydrolysis[a]	Positive	Variable	Negative
X or V factor requirements	Negative	Positive	Negative[b]

Source: Adapted from Carvalho Neta AV et al., 2008, *Infect Immun*, 76:1897–1907.

[a] Urea hydrolysis: most *Brucella* isolates vigorously hydrolyze urea. Positive values range from greater than or equal to 90% positivity. Negative values range less than or equal to 10% positivity. Variable—range of positivity from 11% to 89%.

[b] Growth enhanced by cysteine.

or from recurrent disease [87]. The case definition for brucellosis, as defined by the Centers for Disease Control and Prevention, may be met if any of the following laboratory criteria are found: (1) bacterial isolation from a clinical specimen, (2) a *Brucella* agglutination titer rise of fourfold or greater between acute and convalescent-phase serum collected more than 2 weeks apart, or (3) immunofluorescence detection of *Brucella* sp. in a clinical specimen [88]. The standard tube agglutination test remains the preferred method worldwide for the confirmation of brucellosis infection [89,90]. This test measures the total quantity of anti-O-polysaccharide antibody and titers of 1:160 or higher are considered diagnostic for an infection. An increase or persistence of IgG antibodies, determined by pretreating the serum with 2-mercaptoethanol to eliminate the presence of IgM antibodies, has been associated with disease relapse or incomplete treatment. Inhibition of agglutination at lower dilutions may occur and therefore test serum should always include dilutions to at least 1:320. Enzyme-linked immunosorbent assays may in the future replace other serological assays and the Coombs' test; however, these assays must first be validated and standardized to meet the rigorous testing requirements for a clinical diagnostic assay [91–93].

Rapid and specific detection of *Brucella* species assays using polymerase chain reaction (PCR) has been developed for use with blood, body fluids, and tissues. The major impediments to these tests are the PCR inhibitory components of the tissues and the presence of relatively few organisms in each clinical specimen, thus requiring elimination of the inhibitors and optimization of DNA extraction methods [94,95]. A comparison between different commercially prepared DNA kits for use with a

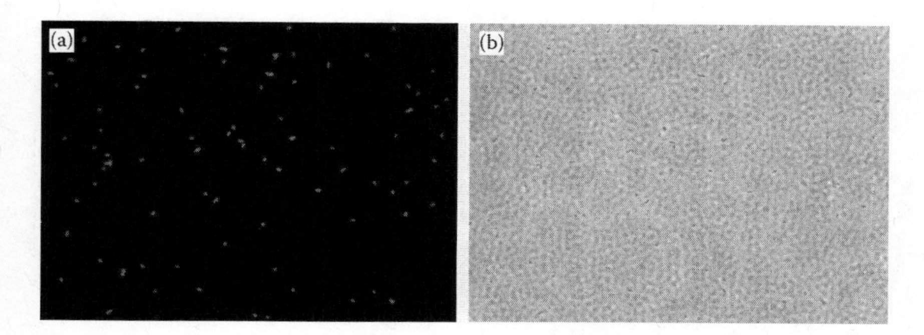

FIGURE 11.2 **(See color insert.)** Direct fluorescent antibody (BRU38) staining of *Brucella suis* found in rhesus monkey lung tissue after aerosol challenge: digital picture displays the stained cells as seen under epifluorescent light source view (a) and then compared to the white light source for view of visible *Brucella* organisms (b). (Photographs courtesy of Dr. M. Wolcott and T. Abshire, USAMRIID, Fort Detrick, Maryland. With permission.)

quantitative real-time PCR assay demonstrated significant differences in DNA yield between kits [96].

An alternative method of detection using a direct fluorescent antibody test is currently under development and may offer a more reliable and rapid method of identifying *Brucella* bacteria in clinical specimens (Figure 11.2). Using this method, clinical samples can be collected and preliminary positive identification may be possible directly from specimens. In Figure 11.2, samples were collected from the lungs of an experimentally infected nonhuman primate and direct fluorescent antibody reagent was used to readily identify *B. suis* bacteria.

11.3 *BRUCELLA* VACCINES

Currently there are no U.S. Food and Drug Administration (FDA)-approved human vaccines for the prevention of brucellosis in the United States. Control of brucellosis spread from animal reservoirs to human populations is managed by disease outbreak identification in domestic agricultural herds with subsequent culling of infected livestock and vaccination of noninfected animal populations [3,97,98]. These measures have significantly reduced domesticated livestock incidences of brucellosis to negligible level with a prevalence of brucellosis of less than one infected animal per 10^6 animals (0.0001%) in livestock herds [99]. In addition, FDA-regulated pasteurization of milk and milk products has resulted in dramatic reductions in cases of food-borne human brucellosis.

Bovine vaccination using the U.S. Department of Agriculture-approved *B. abortus* S19 vaccine had been used effectively to control brucellosis in agricultural animals [100]. In the 1990s, a serological nonreactive vaccine replaced the S19 vaccine that had been found to cause spontaneous abortions in some vaccinated animals and was virulent to humans exposed to the vaccine [101]. In 1996, the *B. abortus* vaccine strain RB 51 was selected to be used in cattle under the State Federal Brucellosis Eradication Program [102,103]. This genetically stable, rough morphology mutant vaccine strain lacks the polysaccharide O-side chains on the bacteria surface and therefore does not stimulate antibody production to antigens used by standard diagnostic tests; however, this vaccine remains pathogenic to humans.

Zinsstag et al. [104] developed and described a livestock-to-human brucellosis transmission model in Mongolia. This model provides an assessment of transmission of brucellosis from livestock-to-humans in the presence and absence of vaccine intervention. The objective of this study was to develop a model for cost-effectiveness analysis of a nation-wide vaccination program for livestock and the impact of this analysis on brucellosis transmission by estimating demographic and transmission parameters between these populations. It was determined that the proportion of protection of a livestock population needed to interrupt human transmission was 0.46 for sheep and 0.66 for cattle using local parameters. The study determined that effective reductions in brucellosis prevalence in cattle could be accomplished using a test and slaughter methodology rather than vaccination scenarios; however, neither method would lead to elimination of disease prevalence in a study population by 10 years [104]. This study does confirm that mass livestock vaccination effectively reduces human brucellosis as outlined in the World Health Organization brucellosis-control project [104,105].

Despite the significant severe adverse reactions in humans associated with *B. abortus* S19 vaccine infections, over 3 million people were vaccinated using the S19 vaccine in the former U.S.S.R. [106]. The *B. melitensis* Rev. 1 vaccine has also been shown to cause brucellosis in humans [107,108]. Although the U.S.S.R. vaccination program did result in a >50% reduction of human brucellosis despite continued presence of disease in domestic agricultural herd populations, approximately 12.5% of vaccinees demonstrated persistent vaccine strain infection. The severe adverse events associated with these vaccines would preclude them from being used under any circumstances within the United States and therefore significant attenuation of virulence factors will need to be made before either of these vaccines is considered for testing in human populations. Other vaccines such as the *B. abortus* strain 19BA, *B. melitensis* 104M, and a phenol-insoluble sodium dodecyl sulfate fraction after infection from *B. abortus* or *B. melitensis* have also been used in China, the former U.S.S.R., and France, but these vaccines were also found to have limited efficacy and significant reactogenicity [103,109,110].

Promotion of replication inhibition and elimination of intracellular survival of *Brucella* are key action components for any successful human vaccine for brucellosis. Studies have revealed that stimulating production of interferon-γ by $\gamma\delta$, $CD8^+$, and $CD4^+$ T cells promotes macrophage bacteriocidal activity directed against intracellular *Brucella* organisms [109]. Infected macrophages are also killed by these activated $CD8^+$ and $\gamma\delta$ T cells. Opsonization by Th1-type antibody isotypes facilitates phagocytosis of *Brucella* bacteria [109] and therefore any new potential brucellosis vaccine must not only stimulate the Th1-type antibody production but also promote a cell-mediated cytotoxic response to this infection [109,111]. Combinations of polyoxidonium or tumor necrosis factor were used as adjuvants with the live *B. abortus* strain 82-PS vaccine to stimulate humoral and cell-mediated responses as detected by *in vitro* T-cell proliferation and phagocytosis assays [112]. Both adjuvants when combined with vaccine afforded 90% protection for guinea pigs challenged with virulent *B. abortus* strain 54-M [112]. In another study, interleukin-15 gene was used as an adjuvant and was combined with a *Brucella* DNA vaccine [113]. The *Brucella* DNA vaccine was composed of three genes encoded *BCSP31*, *SOD*, and ribosomal *L7/L12*. This DNA vaccine was combined with the mouse interleukin-15 gene as separate plasmids that were administered to mice in multiple doses. After vaccination, mice demonstrated enhanced humoral and cell-mediated immunological responses. Immunglobulin IgG1 and IgG2a levels were found to be elevated, with a bias toward IgGa type 1 response over a type 2 IgG1 predominance. This combined DNA vaccine also demonstrated elevated cell-mediated responses in the areas of interferon-γ levels and increased $CD8^+$ T-cell numbers. Challenge experiments were then performed using virulent *B. abortus* strain 2308 bacteria and demonstrated enhanced immunological responses and bacterial clearance in mice exposed to *B. abortus* using bacterial culture of organ tissues and cell-based assays [113].

A recent study used DNA sequencing and comparative genomic analyses to identify specific gene variations between virulent and vaccine strains of *B. abortus* [114]. A total of 24 genes were identified that may be associated with the loss of virulence of the S19 vaccine. Four of these genes were found to encode an outer membrane protein and three proteins associated with erythritol metabolism or uptake [114]. By further characterizing these results, the investigators hope to identify those crucial genes and proteins responsible for attenuated properties of the vaccine strain. Bioinformatic analysis of *Brucella* vaccine strains using the VIOLIN web-based vaccine database and analysis system provides a flexible resource to data and predicts potential *Brucella* vaccine target sites for modification [115].

11.4 HUMAN DISEASE

Brucellosis infection has a diverse and variable course in humans and, like tuberculosis and syphilis, has also been described as one of the "great imitators" [4,8,23,26,90,116]. Therefore, collection of detailed clinical and epidemiological histories, in particular ingestion of contaminated milk products or animal contact, for each patient is crucial for the diagnosis of brucellosis. In the event of a biological attack, the isolation and/or identification of the microorganism will be relied upon for diagnosis. Although several different strains of Brucellae have the potential of infecting humans, the majority of the infections worldwide are caused by two strains, *B. melitensis* and *B. abortus*. The clinical signs and symptoms range from systemic acute febrile illness to an insidious chronic infection. The incubation period after exposure ranges from 3 days to several weeks, and the onset may be abrupt to insidious in nature. Presenting symptoms include fever, fatigue, anorexia, malaise, chills, sweating, joint pain, myalgias, arthralgias, and weight loss. However, individually in some cases, joint pain, lower back pain, the sensation of "burning" feet, or myocardial infarction only were the presenting symptoms [79,117,118].

Neurological and psychological complications occur frequently and include headache, irritability, chorea, meningoencephalitis, transient ischemic attacks, cranial nerve compromise, and depression [26,119]. Patients may experience local pain from focal infections of the joints, bone or genitourinary tract. Pulmonary infections in abattoir workers with brucellosis often result in severe cough and pleuritic chest pain as well as dyspepsia [90,116,120]. Chronic infection often results in weight loss and symptoms persisting from 3 months to a year or more. Hepatomegaly, splenomegaly, or lymphadenopathy may be detected on physical examination of infected patients; however, most patients are found to have normal examinations. Hematological abnormalities such as bone marrow hypoplasia, neutropenia, and pancytopenia are frequently observed particularly in chronic infections [119,121–126]. Dermatological manifestations such as erythematous papular lesions, dermal cysts, Stevens-Johnson syndrome, thrombocytopenic purpura, and erythema nodosum have been seen in infected patients [28,119,123–125].

Bone or joint involvement has been identified in about 30% of patients infected with *B. melitensis* and sacrolititis in 6%–15% of young adults [126–130]. In contrast, monoarticular arthritis involving primarily the knee and hip is the most common form of osteoarticular involvement in children [131]. In adults, large joint arthritis occurs at the same frequency as sacrioliitis; however, unlike septic arthritis caused by pyogenic organisms only mild joint inflammation is found with *B. melitensis* infection, and erythema of the overlying skin of the joint is not common. Exudative synovial fluid aspirated from infected joints demonstrates low numbers of mononuclear cells; consequently, bone destruction is an unusual feature. Although synovial fluid cultures yield positive results in approximately 20% of cases, synovium cultures provided greater bacterial recovery from infected joints. Middle-aged or elderly patients have a higher incidence of spondylitis with accompanying symptoms of lumbar pain, local tenderness, and radiculopathy [132]. Disc space narrowing and epiphysitis of the anterosuperior quadrant of the vertebrae and bridging syndesmophytes are typical radiographic findings; however, bone scans are often negative or weakly positive. Axial skeleton infections are frequent, but paravertebral abscesses or long bone infections are rare [133].

Brucella genitourinary tract infections such as Bartholin's gland abscesses and epididymoorchitis have been documented and, like similar tuberculosis infections, produce "sterile" pyuria on bacterial culture [134–136]. However, with pyleonephritis and cystitis infections *Brucella* can be cultured from the urine. Brucellosis infection during pregnancy can result in placental and fetal infections and increases the risk of spontaneous abortions [137–139].

Pulmonary infections have been described for brucellosis before the discovery of effective antimicrobial therapies. Although chest radiographs usually appear normal, infected patients often complain of dyspnea, cough, and pleuritic chest pain [140]. Focal or diffuse infiltrates, pleural effusions, abscesses, and granulomas may be seen when radiographic changes are noted for cases of brucellosis.

Serological studies reveal frequent elevations in transaminases and mild elevations in alkaline phosphatase and lactate dehydrogenase. Liver abscesses rarely occur; however, hepatitis is often present with biopsies revealing well-formed granulomas or collections of mononuclear cells [116,141]. Bacterial peritonitis has been reported to occur spontaneously in brucellosis-infected patients [142,143]

Central nervous system infections caused by *Brucella* can present with symptoms ranging from a Guillian-Barre syndrome with acute neurobrucellosis involving the dorsal root ganglia and myelitis complicated by subarachnoid hemorrhage to chronic meningoencephalitis [144,145]. Infections of other tissues such as cardiac and skin have been reported. Approximately 80% of deaths from brucellosis are attributed to complications from endocarditis caused by these bacteria [143,146].

11.5 ANIMAL DISEASE

The primary natural hosts for Brucellae human pathogens are bison and elk; however, domesticated animals such as cattle, goats, sheep, and pigs are also known to harbor these infections. The most common source of brucellosis infection in humans is through contact with domesticated animals by the following mechanisms: direct infection during ruminant septic abortions or exposure to the contaminated environment; consumption of contaminated unpasteurized milk or milk products; during the process of slaughtering or butchering of infected animals; or through consumption of contaminated meat products either by handling raw products or inadequate cooking.

Human infection throughout history has been closely associated with the prevalence of disease in domesticated animals within the geographic region. Worldwide, the predominate vehicle of infection is the consumption of contaminated animal products [97]; however, in the United States the lack of infected domesticated animals and the use of strict food regulations in the commercial collection and preparation of animal products limit U.S. exposures to mostly occupational (such as slaughterhouse) workers, veterinarians, cattlemen, and shepherds. The transmission of *B. abortus* and *B. melitensis* in other countries is attributed to consumption of unpasteurized yoghurts, soft cheeses, and ice cream [147]. Other incidences of infection have risen from environmental and animal exposures. Respiratory infections have been observed in persons camping in the deserts of Kuwait during the lambing season [148]. Placenta and blood during the time of infection-induced abortion may contain up to 10^{10} *Brucella* organisms per gram of tissue [149]. Consequently, the extremely high concentrations of organisms found in infected animal fluids during parturition can be aerosolized resulting in incidental human infection and the continued propagation of the disease in other uninfected animals.

In the United States, fewer than 200 cases of infection occur per year (0.04 cases per 100,000 population) but in the Middle East, particularly countries bordering the Mediterranean Sea, Jordan, Kuwait, and Iran, the incidence of infection is higher [150–152]. In the Middle East and Mongolia, consumption of camel milk and undercooked traditional delicacies such as liver are likely sources of infections in these regions [153]. *Brucella melitensis* in cattle has emerged as a serious problem in Saudi Arabia, Kuwait, Israel, and southern Europe because the *B. abortus* vaccine strain does not provide adequate protection. In Columbia and Brazil, *B. suis* biovar 1 infections in cattle rather than pigs has been found to be an important source of infection [154].

11.6 ANIMAL MODELS

11.6.1 Rodent and Avian Models

A broad range of animal models for brucellosis has been developed spanning from murine models of infection to nonhuman primate models of inhalational disease. Recent studies have utilized several routes of infection, including intranasal, oral, intraperitoneal, and aerosolized, in murine models of virulent and nonvirulent *Brucella* strains [155–158]. Murine models are often preferred for initial research studies due to: (1) the overall lower costs associated with the use of mice rather than larger animals; (2) institutional animal use committee limitations requiring the use of the evolutionarily lowest species of animal for initial studies; (3) the ethical acceptability of using mice; and (4) historical research evidence of infectivity by different strains of *Brucella.*

In 2001, researchers described the histopathological features of mice inoculated intranasally with *B. melintensis* [159]. Two subsequent papers describe the use of aerosolized delivery of *Brucella* bacteria in murine models of pulmonary infection. Inhalational dissemination poses the greatest risk in terms of overall infectivity as well as morbidity and mortality during bioterrorism events; consequently, research studies using this route of delivery provide the most appropriate and reliable measures for therapeutics and vaccine testing of potential biological threats. BALB/C mice were aerosol challenged with virulent strains of either *B. abortus* or *B. melintensis* to test the efficacy of potential live attenuated vaccine strains used to vaccinate test animals [160]. In 2007, Olsen et al. [155] characterized an aerosolized challenge murine model using *B. abortus* (S2308) and *B. melitensis* (S16M) strains with dosages ranging from 10^3 to 10^{10} colony-forming units (cfu). The optimal inhalational dose for both strains was determined to be 10^9 cfu and subsequent challenges of mice vaccinated with RB51 (SRB51) demonstrated significant protection afforded by this vaccine by reduced lung, liver, and spleen colonization [155]. Further murine model development and refinement by Smither et al. [158] in 2009 demonstrated that retained doses of at least 10^3 cfu of *B. melitensis* were sufficient to produce infection in BALB/c mice and that this model was successfully used to evaluate the live attenuated animal vaccine *B. melitensis* Rev. 1.

Izadjoo et al. [157] describe the development of a male BALB/c murine model to evaluate genitourinary tract pathogenicity of potential vaccines. The recently identified pathogen, *Brucella microti*, has been characterized and evaluated using both *in vitro* intramacrophagic replication studies and murine models of infection [2]. Other murine models such as Swiss albino mice [161], C57BL/6 and 129/Sv [162,163], IRF-1 [164], as well as Sprague-Dawley rats [165] have been successfully used in studies ranging from vaccine challenge experiments and quorum-sensing regulators of virulence factors to pathogenic temporal analysis modeling and vertical transmission studies. Testing of antimicrobial therapeutics has been successfully conducted using murine and rat models of infection [166,167]. Guinea pigs [168–170] and chicken embryos [171] have also been used in brucellosis modeling and infection experiments.

11.6.2 Caprine and Sheep Animal Models

Caprine (goat) models have been used extensively in veterinarian medicine as surrogates to bovine and ovine brucellosis. This model has been used to evaluate infection and disease progression as well as efficacy testing of vaccines. Goats experimentally infected by intravenous or uterine arteries were found to have placentitis with *B. abortus* found in erythrophagocytic trophoblasts of the placent-

Bio[A]type	MLVA clusters/ clade [10]	Species	Subspecies	Representative strains	Human pathology[B]	Animal models[C]				

A	A1	*F. tularensis*	*Tularensis*	SCHU S4 (North America)			X	X	X	X
	A2	*F. tularensis*	*Tularensis*	ATCC 6223 (North America)						
	A2	*F. tularensis*	*Tularensis*	(North America)				X	X	
	NA	*F. tularensis*	*Mediaasiatic*	GIEM 543 (Central Asia)		X				
	NA	*F. tularensis*	*Novicida*	ATCC 15452 (North America)		X		X		X
B	B1	*F. tularensis*	*Holartica*	Eurasia		X		X		
	B2	*F. tularensis*	*Holartica*	Scandinavia/ North America		X		X		
	B3	*F. tularensis*	*Holartica* LVS	Eurasia	NA	X	X	X	X	X
	B3	*F. tularensis*	*Holartica*	GIEM 503 Eurasia/ North America		X		X		
	B4	*F. tularensis*	*Holartica*	North America/ Sweden		X		X		
	B5	*F. tularensis*	*Holartica*	Japan		X		X	X	
NA	NA	*F. philomiragia*	NA	Eurasia						

2011.015R

A. Biotype as originally determined from virulence in the rabbit and ferment glycerol [149].
B. Color of figure indicates relative virulence for humans. Red being the most harmful. Green being the least harmful; hospital bed indicates virulence for immune-compromised only.
C. Denotes frequently used experimental animal models including the mouse, rat, nonhuman primate and the hamster; the hamster symbol is representative of several small rodents that include hamsters, voles, guinea pigs, and Sprague Dawley or Fischer 344 rats.

FIGURE 9.1 Common animal models for F. *tularensis.*

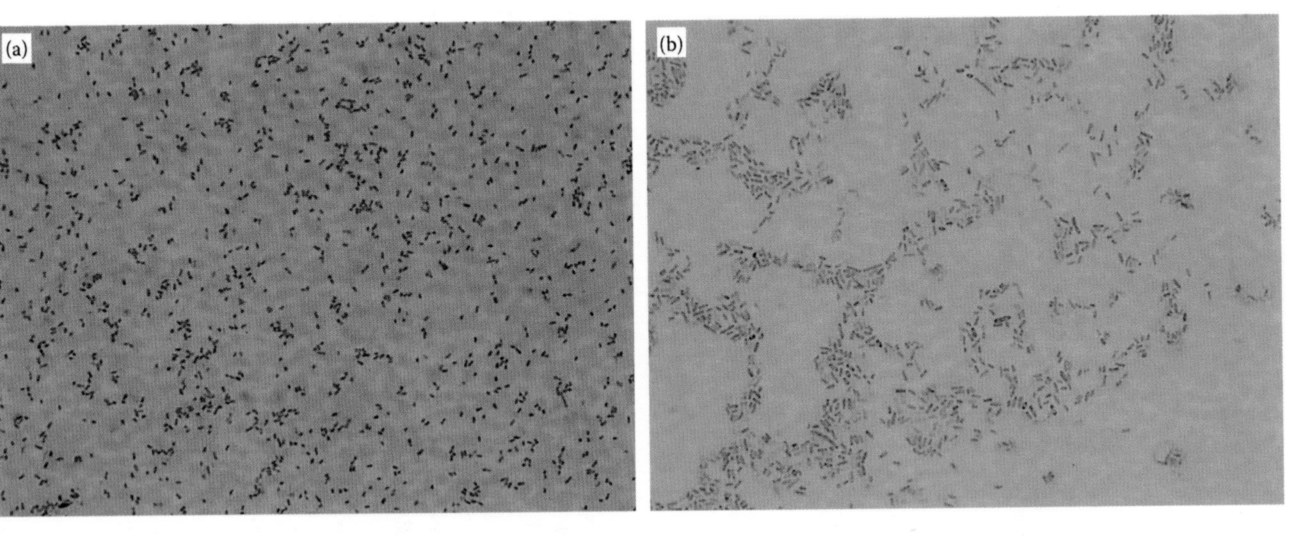

FIGURE 11.1 Gram stains of *Brucella canis* (a) and *Escherichia coli* (b). (Photographs courtesy of Dr. Mark Wolcott and Terry Abshire, Diagnostics Division, U.S. Army Medical Research Institute of Infectious Diseases (USAMRIID), Fort Detrick, Maryland. With permission.)

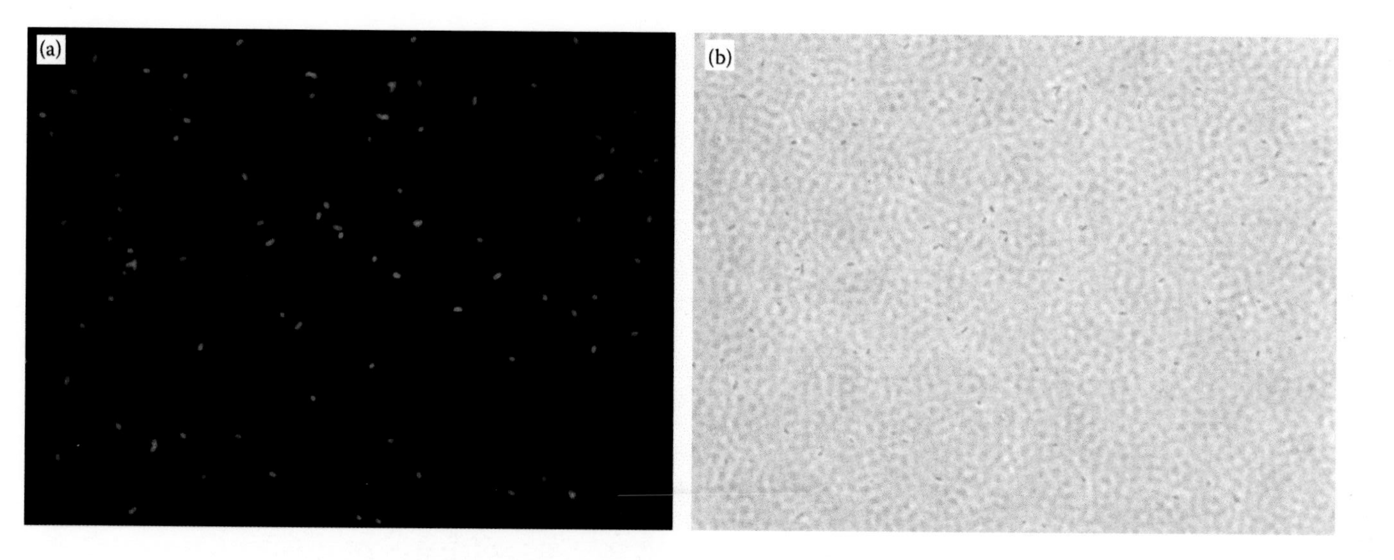

FIGURE 11.2 Direct fluorescent antibody (BRU38) staining of *Brucella suis* found in rhesus monkey lung tissue after aerosol challenge: digital picture displays the stained cells as seen under epifluorescent light source view (a) and then compared to the white light source for view of visible *Brucella* organisms (b). (Photographs courtesy of Dr. M. Wolcott and T. Abshire, USAMRIID, Fort Detrick, Maryland. With permission.)

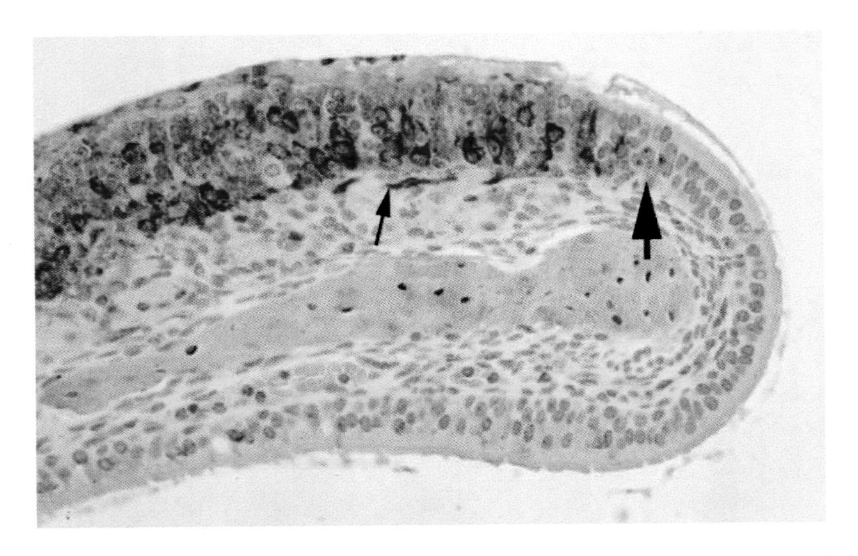

FIGURE 12.1 Immunohistochemical stain of the nasal turbinate of a mouse infected with Venezuelan equine encephalitis virus by aerosol. The red stain indicates abundant viral infection of the olfactory epithelium that lines this part of the turbinate (the block arrow marks the border with respiratory epithelium), and the respiratory lining of the remainder of the turbinate is uniformly uninfected. Note that the virus is also present in the beginning of an olfactory nerve (thin arrow).

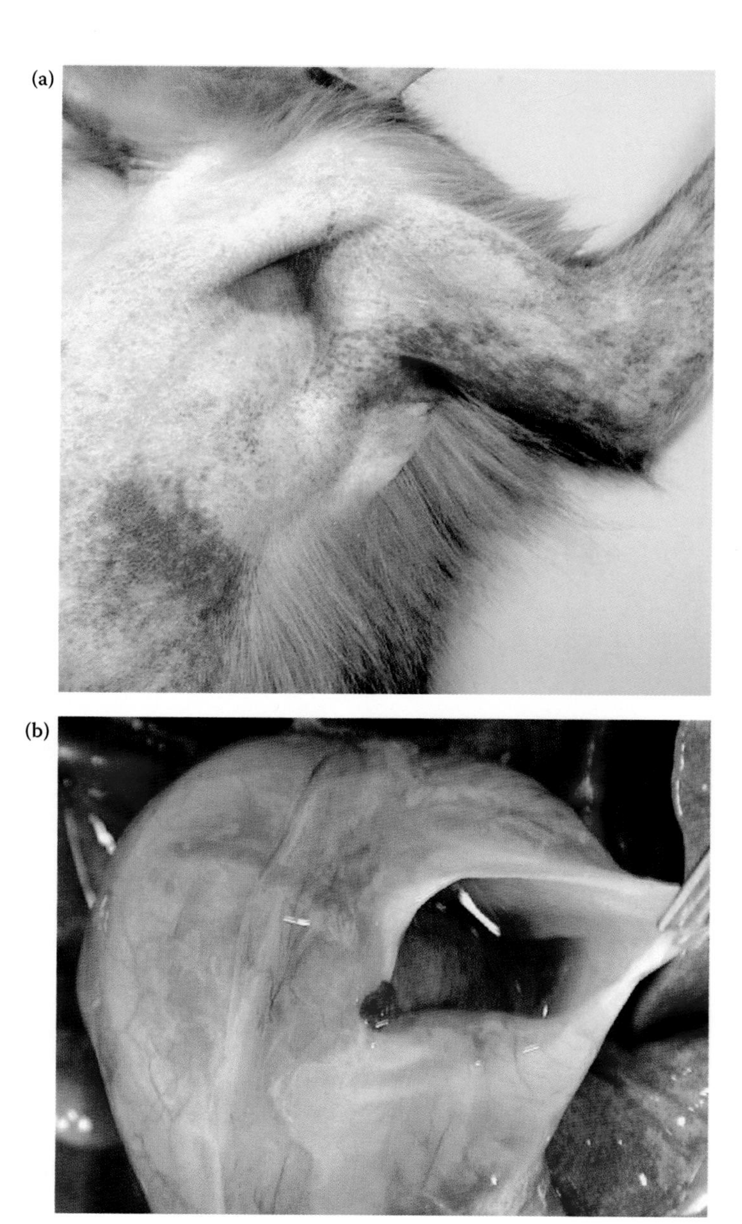

FIGURE 14.1 Representative gross necropsy lesions from nonhuman primates experimentally infected with hemorrhagic fever viruses. (a) Typical petechial rash of the left arm and chest of a rhesus macaque 11 days after infection with the Marburg virus (Musoke strain). (b) Accumulation of fluid in the pericardial cavity of a cynomolgus monkey 13 days after infection with the Lassa virus (Josiah strain).

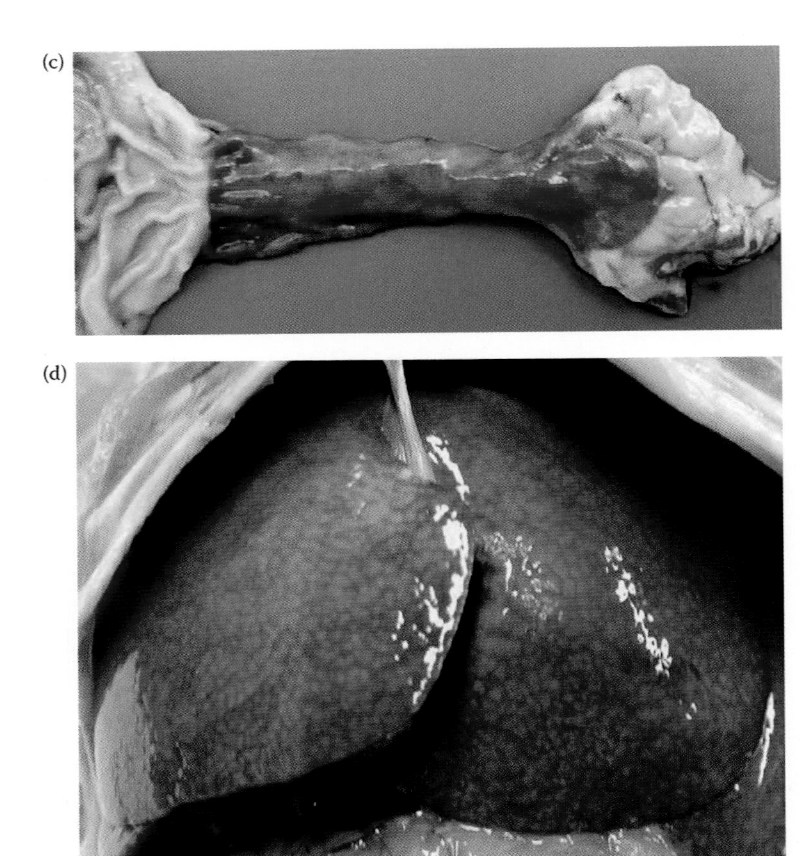

FIGURE 14.1 Continued. (c) Marked congestion of the duodenum at the gastroduodenal junction of a cynomolgus monkey 5 days after infection with the Zaire ebolavirus. (d) Reticulation and discoloration of the liver 11 days after infection with the Lassa virus (Josiah strain).

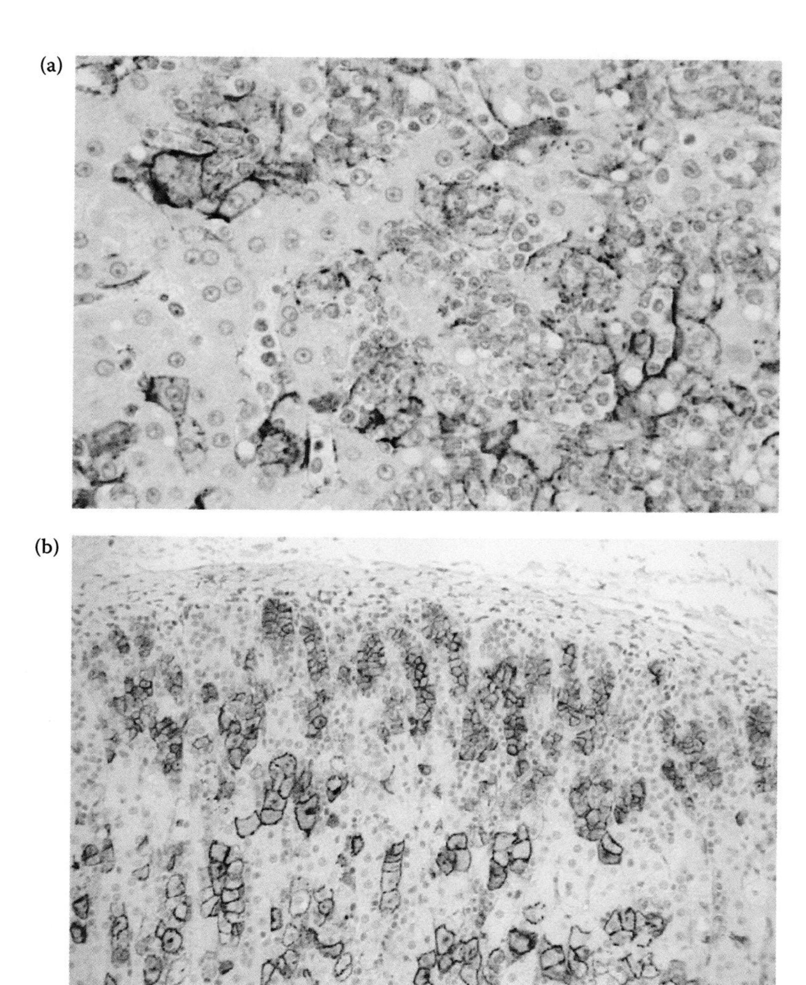

FIGURE 14.2 Immunohistochemical staining patterns and histopathology of nonhuman primates experimentally infected with hemorrhagic fever viruses. (a) Prominent immunostaining of cells within the zona glomerulosa and zona fasciculata of the adrenal gland of a cynomolgus monkey 15 days after infection with the Lassa virus (Josiah strain) (immunoperoxidase staining; original magnification, 20×). (b) Immunopositive hepatocytes peripheral to and within an inflammatory foci of lymphocytes, macrophages, and fewer neutrophils in the liver of a cynomolgus monkey 15 days after infection with the Lassa virus (Josiah strain) (immunoperoxidase staining; original magnification, 40×).

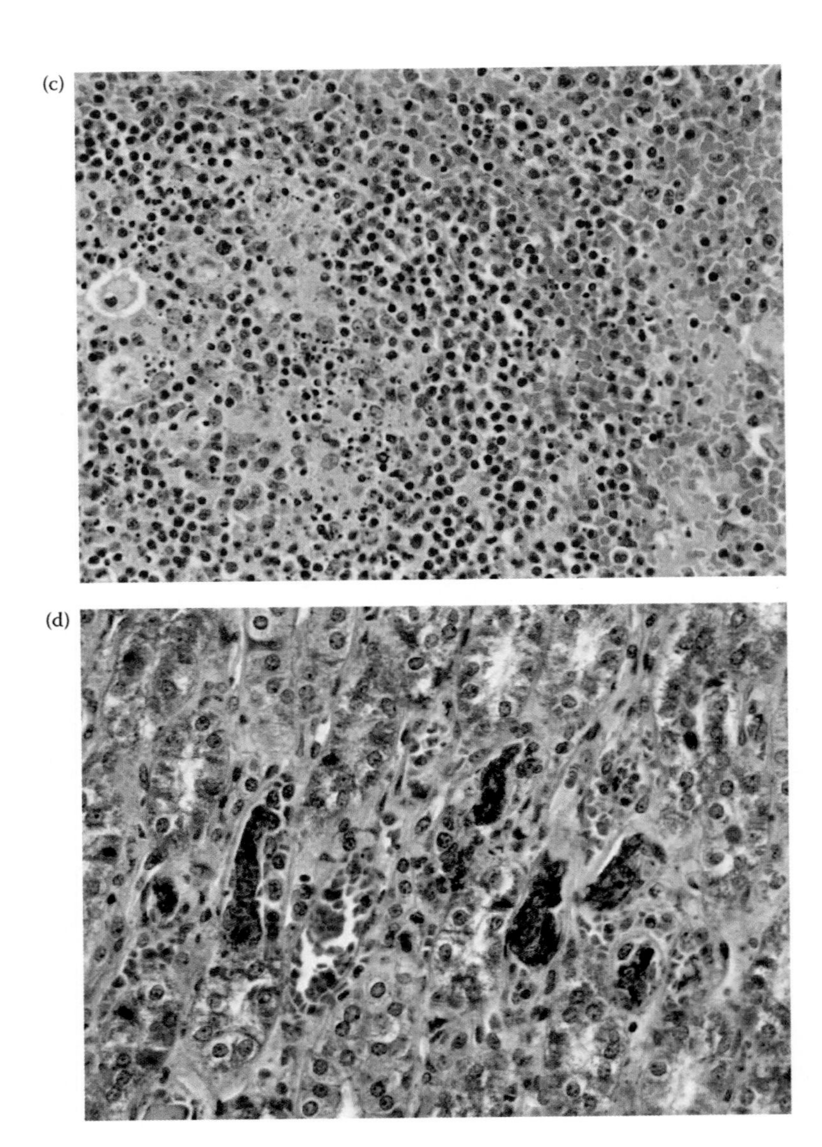

FIGURE 14.2 Continued. (c) Phosphotungstic acid hematoxylin stain of the kidney from cynomolgus monkey showing an abundance of polymerized fibrin in medullary vessels 12 days after infection with the Ivory Coast ebolavirus (Original magnification, 20×). (d) Necrosis and apoptosis of lymphocytes with concomitant lymphoid depletion and hemorrhage in the spleen of rhesus monkey 7 days after infection with the Marburg virus (Musoke strain) (hematoxylin and eosin stain; original magnification, 40×).

ome [172]. This study suggests that replication of the *B. abortus* in trophoblasts precedes placentome and fetal infection and that the infected trophoblasts are the source of the bacteria infecting other tissue sites. A subsequent study using pregnant goats exposed to *B. abortus* demonstrated spontaneous abortions and multiple tissue infections to include milk and uterine fluids [173].

Vaccine efficacy studies using *B. melitensis* deletion mutants have been successfully conducted using a pregnant caprine model [174], and challenge experiments have also been conducted using licensed vaccines and pregnant ewes [175]. In the caprine model, the two unmarked deletion mutant vaccine candidates, BMΔ*asp24* and BMΔ*virB2*, did not cause abortion or colonized fetal tissues when administered to pregnant angora goats and only BMΔ*asp24* was isolated from maternal tissues. Challenge experiments using wild-type *B. melitensis* demonstrated that BMΔ*asp24* provided more protection than BMΔ*virB2* and both attenuated strains provided protection from abortion and infection in the majority of pregnant goats [174]. In the pregnant ewe model, the cattle vaccine RB51 did not afford protection of sheep to abortion caused by challenge experiments using *B. melitensis* [175].

11.6.3 Nonhuman Primate Models

In the 1920s, researchers successfully infected nonhuman primate models with brucellosis [176,177]. These initial studies evaluated the pathogenesis of various *Brucella* species in experimentally infected monkeys. Stumptail macaque monkeys (*Macaca arctodies*) were challenged with *B. canis* by the oral, conjunctival, and intravenous routes and found to harbor bacteria in the peripheral blood up to 7 weeks after inoculation and some tissues were found to be infected up to 5 weeks after challenge [178]. Infected monkeys were found to develop high levels of neutralizing antibody, and focal granulomatous lesions similar to those seen in human infection were seen in the lymphoid tissue, spleen, and liver of some animals.

Cynomolgus philipinensis monkeys were used in a series of vaccination experiments using a variety of purified antigens from *B. melintensis* in combination with inoculations using live-attenuated *B. melitensis* Rev. 1 vaccine [179–181]. Chen and Elberg [179] demonstrated that nonhuman primates vaccinated either by the oral or by the cutaneous route produced high levels of anti-*Brucella* antibodies by both complement fixation and passive hemagglutination assays; however, of the six cynomolgus monkeys, three orally and three cutaneously vaccinated, only those monkey vaccinated by skin puncture using a bifurcated needle and viable Rev. 1 bacteria demonstrated complete protection against subcutaneous challenge with 1600 cfu of virulent *B. melintensis*. However, despite the added protection from using a live vaccine, a vaccine strain-associated bacteremia occurred 14 days after cutaneous vaccination for two of the three monkeys [179]. A subsequent study using a combined method of antigen vaccination with fraction 1 in adjuvant 65 before administration of the live vaccine Rev. 1 resulted in significant protection of postvaccination bacteremia [180]. In addition, this study again re-enforced the previous findings that protection afforded by Rev. 1 vaccination was directly related to development of protective antibodies stimulated by the live vaccine [180]. In 1976, Chen and Elberg [181] again demonstrated that the use of a combined vaccine method which included a soluble fraction 1 from *B. melitensis* and the live vaccine Rev. 1 could provide protection against subcutaneous challenge with virulent *B. melitensis* strain 6015 in six of six monkeys. The six monkeys tested were divided into two groups: Group 1 received soluble fraction 1 and a second vaccination using Rev. 1 vaccine 4 weeks later and Group 2 received both vaccines simultaneously. Although the number of test subjects was small in each group, both cohorts demonstrated significant *B. melitensis* antibody titers and were protected after challenge with virulent *B. melitensis.*

Rhesus macaque monkey models have also been developed utilizing aerosol exposure of virulent *Brucella* strains to better understand the pathogenesis and disease progression from an inhalational route of exposure. These models more closely replicate the method of bioweapon dispersion that would result in the highest morbidity and mortality among human and animal populations exposed to these agents. Mense et al. [182] described the pathological changes found in rhesus macaques exposed to 10^2–10^5 cfu of *B. melitensis* strain 16M with dose-related hepatic and splenic inflammation. Histopathological analysis revealed inflammation in the liver, spleen, kidneys, testes, and epididymides of challenged monkeys and these findings were consistent with human brucellosis pathology. An interesting finding from this study indicated that higher inhalational doses of organisms resulted in a more rapid and profound inflammatory response with clearing of the bacteria whereas lower inhaled doses of *Brucella* yielded bacteria recoverable from a variety of tissue sites 9 weeks after challenge [182].

A rhesus macaque aerosol challenge model has also been used to evaluate *B. suis* pathogenesis by the inhalational route and current as well as possible future diagnostic algorithms used to detect *Brucella* in cases of human brucellosis [183]. The averaged inhaled dose of *B. suis* used in this study was 5.6×10^8 cfu. As in human cases, temperature (monitored by telemetry in the nonhuman primate model), blood chemistries, or complete blood count values provide no consistent changes that could be attributed specifically to brucellosis infection. Gross pathology and histopathological studies were also considered nonspecific and could not be shown to be due to *Brucella* early infection; however, nonspecific parenchymal necrosis and lymphocytic infiltration of the rhesus liver was consistent with human brucellosis findings. Routine *Brucella* culture of exposed nonhuman primates using swabs, bronchoalveolar lavage fluid, and animal tissues did recover viable *B. suis* bacteria. Real-time PCR tests of noninvasive diagnostic samples such as pharyngeal and nasal swabs and bronchoalveolar lavage fluid up to 7 days after exposure and tissues obtained during necropsy were successful in detecting the presence of *Brucella* DNA [183]. The authors propose using PCR as an ideal screening method for outbreak investigations.

11.7 SUMMARY

Brucellosis still remains a significant agricultural and human disease risk worldwide. Adding to these risks is the potential of intentional dissemination of virulent *Brucella* strains into human or livestock populations by state-sponsored biological weapon programs or terrorist groups. In our current turbulent world of societal, political, and economic upheavals, major outbreaks of this disease, whether they are directed against human populations or the agricultural centers, may impose insurmountable stressors on the local society with far-reaching ramifications on overall geopolitical global stability. Development of effective therapeutic prophylaxis and treatment measures to not only naturally occurring brucellosis but intentionally modified, multidrug-resistant strains will be a key to mitigating these potentially disastrous outcomes.

DEDICATION

This chapter is dedicated to those military service members who have served, are currently serving, or have lost their lives to protect the People of the United States, our life, liberty, and Constitution.

REFERENCES

1. Morris JG, Southwick FS. 2010. *Brucella*, voles, and emerging pathogens. *J Infect Dis.* 202(1):1–2.
2. Jimenez de Bagues MP, Ouahrani-Bettache S, Quintana JF, Mitjana O, Hanna N, Bessoles S, Sanchez F. et al., 2010. The new species *Brucella microti* replicates in macrophages and causes death in murine models of infection. *J Infect Dis.* 202(1):3–10.
3. Pappas G, Papadimitriou P, Akritidis N, Christou L, Tsiano EV. 2006. The new global map of human brucellosis. *Lancet Infect Dis*, 6:91–99.
4. Godfroid J, Cloekaert A, Liautard J-P, Kohler S, Fretin D, Walravens K, Garin-Bastuji B, Letesson J-J. 2005. From the discovery of the Malta fever's agent to the discovery of a marine mammal reservoir, brucellosis has continuously been a re-emerging zoonosis. *Vet Res.* 36:313–326.
5. Pappas G, Panagopoulou P, Christou L, Akritidis N. 2006. *Brucella* as a biological weapon. *Cell Mol Life Sci.* 63:2229–2236.
6. Yagupsky P, Baron EJ. Laboratory exposures to Brucellae and implications for bioterrorism. *Emerg Infect Dis.* 11(8):1180–1185.
7. Christopher GW, Agan MB, Cieslak TJ, Olson PE. History of U.S. military contributions to the study of bacterial zoonoses. *Mil Med.* 170(4 Suppl):39–48.
8. Maloney GE. 2009. *CBRNE—Brucellosis.* http://emedicine.medscape.com/article/830118.
9. Martin JW, Christopher GW, Eitzen EM. 2007. History of biological weapons: From poisoned darts to intentional epidemics. In: Dembek ZF, ed., *Medical Aspects of Biological Warfare.* Washington, DC: Department of Defense, Office of The Surgeon General, US Army, Borden Institute, pp. 1–38.
10. Department of the Army. 1977. *US Army Activity in the US Biological Warfare Programs, Vols 1 and 2.* Washington, DC: Department of the Army Headquarters, unclassified.
11. Kaufmann AF, Meltzer MI, Schmid GP. 1997. The economic impact of a bioterrorist attack: Are prevention and post-attack intervention programs justifiable. *Emerg Infect Dis.* 3(2):83–94.
12. Kortepeter MG, Parker GW. 1997. Potential biological weapon threats. *Emerg Infect Dis.* 5(4):523–527.
13. Wright SG. 2000. Brucellosis. In: G. Thomas Strickland, ed., *Hunter's Tropical Medicine and Emerging Infectious Disease*, 8th ed. Philadelphia: WB Sanders, p. 416.

14. Capasso L. 2002. Bacteria in two-millenia-old cheese, and related epizoonoses in Roman populations. *J Infect.* 45(2):122–127.
15. Evans AC. 1950. Comments on the early history of human brucellosis. In: Larson CH, Soule MH, eds. *Brucellosis*. Baltimore, MD: Waverly Press, pp. 1–8.
16. Cleghorn G. 1751. *Observations of the Epidemical Diseases of Minorca* (*From the Years 1744 to 1749*). London, England: D. Wilson; cited in: Evans AC. 1950. Comments on the early history of human brucellosis. In: Larson CH, Soule MH, eds. *Brucellosis.* Baltimore, MD: Waverly Press, pp. 1–8.
17. Bruce D. 1887. Note on the discovery of a micro-organism in Malta fever. *Practitioner (London)*, 39:161–170; cited in: Evans AC. 1950. Comments on the early history of human brucellosis. In: Larson CH, Soule MH, eds. *Brucellosis*. Baltimore, MD: Waverly Press, pp. 1–8.
18. Hughes ML. 1897. *Mediterranean, Malta or Undulant Fever.* London, England: Macmillan and Co; cited in: Evans AC. 1950. Comments on the early history of human brucellosis. In: Larson CH, Soule MH, eds. *Brucellosis.* Baltimore, MD: Waverly Press, pp. 1–8.
19. Evans AC. 1950. Comments on the early history of human brucellosis. In: Larson CH, Soule MH, eds. *Brucellosis*. Baltimore, MD: Waverly Press, 1–8.
20. Bang B. 1897. Die Aetiologie Des Seuchenhaften ("Infectiosen") Verwerfens. *Z Thiemed* (Jena). 1:241–278.
21. Madkour MM. 1989. Historical aspects of brucellosis. In: Madkour MM, ed. *Brucellosis.* Boston: Butterworths, pp. 1–10.
22. Elser PH, Hagius SD, Davis DS, DelVecchio VG, Enright FM. 2002. Characterization of the caprine model for ruminant brucellosis. *Vet Microbiol.* 90:425–431.
23. Pappas G, Akritidis N, Bosilkovski M, Tsianos E. 2005. Brucellosis. *N Engl J Med.* 352:2325–2336.
24. Memish ZA, Balkhy HH. 2004. Brucellosis and international travel. *J Travel Med.* 11:49–55.
25. Solera J, Lozano E, Martinez-Alfaro E, Espinosa A, Castillejos ML, Abad L. 1999. Brucellar spondylitis: Review of 35 cases and literature survey. *Clin Infect Dis.* 29:1440–1449.
26. Ghaffarpour M, Khoshroo A, Harirchian MH, Sikaroodi H, Pourmahmoodian H, Jafari S, Hejazi SS. 2007. Clinical, epidemiological, laboratory and imaging aspects of brucellosis with and without neurological involvement. *Acta Medica Iranica.* 45(1):63–68.
27. Wunschel M, Olszowski AM, Weissgerber P, Wulker N, Kluba T. 2011. Chronic brucellosis: A rare cause of septic loosening of arthroplasties with high risk of laboratory-acquired infections. *Z Orthop Unfall.* 149(1):33–36; doi: 10.1055/s-0030-1249851, PMID: 20419627.
28. Franco MP, Mulder M, Gilman RH, Smits HL. 2007. Human brucellosis. *Lancet Infect Dis.* 7:775–786.
29. Singh K. 2009. Laboratory-acquired infections. *CID.* 49:142–147.
30. Olle-Goig JE, Canela-Soler J. 1987. An outbreak of *Brucella melitensis* infection by airborne transmission among laboratory workers. *Am J Publ Health.* 77:335–338.
31. Memish ZA, Mah MW. 2001. Brucellosis in laboratory workers at a Saudi Arabian hospital. *AJIC.* 29(1):48–52.
32. Staszkiewicz J, Lewis CM, Colville J, Zervos M, Band J. 1991. Outbreak of *Brucella melitensis* among microbiology laboratory workers in a community hospital. *J Clin Microbiol.* 29(2):287–290.
33. Fiori PL, Mastrandrea S, Rappelli P, Cappuccinelli P. 2000. *Brucella abortus* infection acquired in microbiology laboratories. *J Clin Microbiol.* 38(5):2005–2006.
34. Hawley RJ, Eitzen EM. 2001. Biological weapons—A primer for microbiologists. *Ann Rev Microbiol.* 55:235–253.

35. Rusnak JM, Kortepeter MG, Hawley RJ, Anderson AO, Boudreau E, Eitzen E. 2004. Risk of occupationally acquired illness from biological threat agents in unvaccinated laboratory workers. *Biosecur Bioterror.* 2:281–293.
36. U.S. Department of Health and Human Services, Centers for Disease Control and Prevention, and National Institutes of Health. 2007. *Biosafety in Microbiological and Biomedical Laboratories (BMBL)*, 5th ed. Washington, DC: U.S. Government Printing Office. http://www.cdc.gov/biosafety/publications/bmbl5/index.htm
37. Bergey DH, Hold JG. 1994. *Bergey's Manual of Determinative Bacteriology*, 9th ed. Baltimore, MD: Williams and Wilkins.
38. Moreno E, Stackebrandt E, Dorsch M, Wolters J, Busch M, Mayer H. 1990. *Brucella abortus* 16S rRNA and lipid A reveal a phylogenetic relationship with members of the alpha-2 subdivision of the class Proteobacteria. *J Bacteriol.* 172:3569–3576.
39. Verger JM, Grimont F, Grimont PA, Grayon M. 1985. *Brucella*, a monospecific gensu as shown by deoxyribonucleic acid hybridization. *Int J Syst Bacteriol.* 35:292–295.
40. Lucero NE, Corazza R, Almuzara MN, Reynes E, Escobar GI, Boeri E, Ayala SM. 2010. Human *Brucella canis* outbreak linked to infection in dogs. *Epidemiol Infect.* 138(2):280–285.
41. Foster G, Osterman BS, Godroid J, Jacques I, Cloeckaert A. 2007. *Brucella ceti* sp. *nov.* and *Brucella pinnipedialis* sp. *nov.* for Brucella strains with cetaceans and seals as their preferred hosts. *Internat J Systematic Evol Microbiol.* 57:2688–2693.
42. Center for Food Security and Public Health. 2009. *Brucellosis in Marine Animals.* www.cfsph.iastate.edu/Factsheets/pdfs/brucellosis_marine.pdf
43. Tiller RV, Gee JE, Lonsway DR, Gribble S, Bell SC, Jennison AV, Bates J, Coulter C, Hoffmaster AR, De BK. 2010. Identification of an unusual *Brucella* strain (BO2) from a lung biopsy in a 52 year-old patient with chronic destructive pneumonia. *BMC Microbiol.* 10:23–34; PMID: 20105296.
44. Ewalt DR, Payeur JB, Rhyan JC, Geer PL. 1997. *Brucella suis* biovar 1 in naturally infected cattle: A bacteriological, serological, and histological study. *J Vet Diagn Invest.* 9(4):417–420.
45. Ficht T. 2010. *Brucella* taxonomy and evolution. *Future Microbiol.* 5(6):859–866.
46. Paulsen IT, Seshadri R, Nelson KE, Eisen JA, Heidelberg JF, Read TD, Dodson RJ. et al., 2002. The *Brucella suis* genome reveals fundamental similarities between animal and plant pathogens and symbionts. *Proc Natl Acad Sci.* 99(20):13148–13153.
47. Moreno E, Moriyon I. 2002. *Brucella melitensis*: A nasty bug with hidden credentials for virulence. *Proc Natl Acad Sci.* 99(1):1–3.
48. Grimont F, Verger JM, Cornelis P, Limet J, Lefèvre M, Grayon M, Régnault B, Van Broeck J, Grimont PA. 1992. Molecular typing of *Brucella* with cloned DNA probes. *Res Microbiol.* 143(1):55–65.
49. Moreno E, Moriyon I. 2001. Genus *Brucella*. In: Dworkin M, Falkow S, Rosenberg E, Schleifer KH, Stackebrandt I. eds. *The Prokaryotes: An Evolving Electronic Resource for the Microbiological Community*. New York: Springer.
50. Foster JT, Beckstrom-Sternberg SM, Pearson T, Beckstrom-Sternberg JS, Chain PS, Roberto FF, Hnath J, Brettin T, Keim P. 2009. Whole-genome-based phylogeny and divergence of the genus *Brucella*. *J Bacteriol.* 191(8):2864–2870.
51. Maquart M, Le Fleche P, Foster G, Tryland M, Ramisse F, Djonne B, Al Dahouk S. et al., 2009. MLVA-16 typing of 295 marine *Brucella* isolates from different animal and geographic origins identifies 7 major groups within *Brucella ceti* and *Brucella pinnipedialis*. *BMC Microbiol.* 9:145–149.
52. Mujer CV, Wagner MA, Eschenbrenner M, Horn T, Kraycer JA, Redkar R, Hagius S, Elzer P, DelVecchio VG. 2002. Global analysis of *Brucella melitensis* proteomes. *Ann NY Acad Sci.* 969:97–101.

53. DelVecchio VG, Kapatral V, Redkar RJ, Patra G, Mujer C, Los T, Ivanova N. et al., 2002. The genomic sequence of the facultative intracellular pathogen *Brucella melitensis. Proc Natl Acad Sci USA.* 99:443–448.
54. Halling SM, Peterson-Burch BD, Bricker BJ, Zuerner RL, Qing Z, Li L-L, Kapur V, Alt DP, Olsen SC. 2005. Completion of the genome sequences of *Brucella abortus* and comparison to the highly similar genomes of *B. melitensis* and *B. suis. J Bacteriol.* 187:2715–2726.
55. Bricker BJ, Ewalt DR, Halling SM. 2003. *Brucella* "HOOF-Prints": Strain typing by multi-locus analysis of variable number tandem repeats (VNTRs). *BMC Microbiol.* 3:15.
56. Bricker BJ, Ewalt DR. 2005. Evaluation of the HOOF-Print assay for typing *Brucella abortus* strains isolated from cattle in the United States: Results with four performance criteria. *BMC Microbiol.* 5:37.
57. Gross A, Terraza A, Ouahrani-Bettache S, Liautard JP, Dornand J. 2000. *In vitro Brucella suis* infection prevents programmed cell death of human monocytic cells. *Infect Immun.* 68:342–351.
58. Paixao TA, Roux CM, den Hartigh AB, Sankaran-Walters S, Dandekar S, Santos RL, Tsolis RM. 2009. Establishment of systemic *Brucella melitensis* infection through the digestive tract requires urease, the type IV secretion system and lipopolysaccharide O antigen. *Infect Immun.* 77(10):4197–4208.
59. Barquerp-Calvo E, Chaves-Olarte E, Weiss DS, Guzman-Verri C, Chacon-Diaz C, Rucavado A, Moriyon I, Moreno E. 2007. *Brucella abortus* uses a stealthy strategy to avoid activation of the innate immune system. *PLoS One.* 2:e631.
60. Carvalho Neta AV, Stynen AP, Paixao TA, Miranda KL, Silva FL, Roux CM, Tsolis RM. et al., 2008. Modulation of the bovine trophoblastic innate immune response by *Brucella abortus. Infect Immun.* 76:1897–1907.
61. Cirl C, Wieser A, Yandav M, Duerr S, Schubert S, Fischer H, Stappert D. et al., 2008. Subversion of Toll-like receptor signaling by a unique family of bacterial Toll/interleukin-1 receptor domain-containing proteins. *Nat Med.* 14:399–406.
62. Salcedo SP, Marchesini MI, Lelouard H, Fugier E, Jolly G, Balor S, Muller A. et al., 2008. `*Brucella* control of dendritic cell maturation is dependent on the TIR-containing protein Btpl1. *PLoS Pathog.* 4:e21.
63. Tsolis RM, Young GM, Solnick JV, Baumler AJ. 2008. From bench to bedside: Stealth of enteroinvasive pathogens. *Nat. Rev. Microbiol.* 6:883–892.
64. Winter SE, Raffatellu EM, Wilson RP, Russmann WH, Baumler AJ. 2008. The Salmonella enteric serotype Typhi regulator TviA reduces interleukin-8 production in intestinal epithelial cells by repressing flagellin secretion. *Cell Microbiol.* 10:247–261.
65. Bundle DR, Cherwonogrodzky JW, Caroff M, Perry MB. 1987. The lipopolysaccharides of *Brucella abortus* and B melitensis. *Ann Inst Pasteur Microbiol.* 138(1):92–98.
66. Moreno E, Borowiak D, Mayer H. 1987. *Brucella* lipopolysaccharides and polysaccharides. *Ann Inst Pasteur Microbiol.* 138(1):102–105.
67. Goldstein J, Hoffman T, Frasch C, Lizzio EF, Beining RR, Hochstein D, Lee YL, Angus RD, Golding B. 1992. Lipopolysaccharide (LPS) from *Brucella abortus* is less toxic than that from Escherichia coli, suggesting the possible use of *B. abortus* or LPS from *B. abortus* as a carrier in vaccines. *Infect Immun.* 60(4):1385–1389.
68. Cherwonogrodzky JW, Perry MB, Bundle DR. 1987. Identification of the A and M antigens of *Brucella* as the O-polysaccharides of smooth lipopolysaccharides. *Can J Microbiol.* 33(11):979–981.
69. Alton GG, Jones LM, Pietz DE. 1975. Laboratory techniques in brucellosis. *Monogr Ser World Health Organ.* 55:1–163.

70. Liang L, Leng D, Burk C, Nakajima-Sasaki R, Kayala MA, Atluri VL, Pablo J. et al., 2010. Large-scale immune profiling of infected humans and goats reveals differential recognition of *Brucella melitensis* antigens. *PLoS Negl Trop Dis.* 4(5):1–10.
71. U.S. Department of Health and Human Services, Centers for Disease Control and Prevention, and National Institutes of Health. 2007. Appendix A—Primary containment for biohazards: Selection, installation and use of biological safety cabinets. In: *Biosafety in Microbiological and Biomedical Laboratories* (*BMBL*), 5th ed. Washington, DC: U.S. Government Printing Office. http://www.cdc.gov/biosafety/publications/bmbl5/BMBL5_appendixA.pdf
72. Mantur BG, Mangalgi SS. 2004. Evaluation of conventional Castaneda and lysis centrifugation blood culture techniques for diagnosis of human brucellosis. *J Clin Microbiol.* 42:4327–4328.
73. Ruiz J, Lorente I, Perez J, Simarro E, Martinez-Campos L. 1997. Diagnosis of brucellosis by using blood cultures. *J Clin Microbiol.* 35:2417–2418.
74. Yagupsky P, Peled N, Press J, Abu-Rashid M, Abramson, O. 1997. Rapid detection of *Brucella melitensis* from blood cultures by a commercial system. *Eur J Clin Microbiol Infect Dis.* 16:605–607.
75. Yagupsky P, Peled N, Press J, Abramson O, Abu-Rashid M. 1997. Comparison of BACTEC 9240 Peds Plus medium and Isolator 1.5 microbial tube for detection of *Brucella melitensis* from blood cultures. *J Clin Microbiol.* 35:1337–1384.
76. Yagupsky P. 1999. Detection of Brucellae in blood cultures. *J Clin Microbiol.* 37:3437–3442.
77. Ozturk R, Mert A, Kocak F, Koksal F, Tabak F, Bilir M, Aktuglu Y. 2002. The diagnosis of brucellosis by use of BACTEC 9240 blood culture system. *Diagn Microbiol Infect Dis.* 44:133–135.
78. Yagupsky P. 2004. Use of BACTEC MYCO/F LYTIC medium for detection of *Brucella melitensis* bacteremia. *J Clin Microbiol.* 42:2207–2208.
79. Mantur BG, Biradar MS, Bidri RC, Mulimani MS, Veerappa K, Kariholu P, Patil SB, Mangalgi SS. 2006. Protean clinical manifestations and diagnostic challenges of human brucellosis in adults: 16 years' experience in an endemic area. *J Med Microbiol.* 55(Pt 7):897–903.
80. Godroid J, Nielsen K, Saegerman C. 2010. Diagnosis of brucellosis in livestock and wildlife. *Croat Med J.* 51:296–305.
81. ASM (American Society for Microbiology) *Sentinel Laboratory Guidelines for Suspected Agents Of Bioterrorism.* Brucella *species*, revised October 15, 2004; accessed May 14, 2006 from www.asm.org/asm/images/pdf/brucella101504.pdf
82. Weyant RS, Popovic T, Bragg SL. 2001. Basic laboratory protocols for the presumptive identification of *Brucella* species. Washington, DC: Centers for Disease Control and Prevention (CDC) https://lrn.hr.state.or.us/home/bt/docs/brucella/levelaprocedures.pdf
83. World Organisation for Animal Health (OIE). 2004. *Manual of Diagnostic Tests and Vaccines for Terrestrial Animals*, 5th ed. Paris: OIE, 2004. http://www.oie.int/international-standard-setting/terrestrial-manual/access-online/
84. World Intellectual Property Organization Differential diagnostic assay for brucellosis. www.freepatentsonline/5190860.html
85. Nielsen K. 2002. Diagnosis of brucellosis by serology. *Vet Microbiol.* 90:447–459.
86. *Brucella* serology. FAO BruNet Publication. www.vircell/fileadmin/bibiografia/brucellacapt/evaluation_of_in-house_and_commercial_immunoassays.pdf
87. Versalovic J. *Manual of Clinical Microbiology*, 7th Ed., Chapter 44, Volume 1. Herndon, VA: American Society for Microbiology Press, May, 2011.
88. Centers for Disease Control and Prevention. 1997. Case definitions for infectious conditions under public health surveillance. *MMWR Recomm Rep.* 46(RR10): 1–55.

89. Young EJ. 1991. Serologic diagnosis of human brucellosis: Analysis of 214 cases by agglutination tests and review of the literature. *Rev Infect Dis.* 13:359–372.
90. Mantur BG, Amarnath SK, Shinde RS. 2007. Review of clinical and laboratory features of human brucellosis. *Indian J Med Microbiol.* 25(3):188–202.
91. Al-Shamahy HA, Wright SG. 1998. Enzyme-linked immunosorbent assay for *Brucella* antigen detection in human sera. *J Med Microbiol.* 47:169–172.
92. Al Dahouk S, Tomaso H, Nockler K, Neubauer H, Frangoulidis D. 2003. Laboratory-based diagnosis of brucellosis—A review of the literature. Part I: Techniques for direct detection and identification of *Brucella* spp. *Clin Lab.* 49:487–505.
93. Al Dahouk S, Tomaso H, Nockler K, Neubauer H, Frangoulidis D. 2003. Laboratory-based diagnosis of brucellosis—A review of the literature. Part II: Serological tests for brucellosis. *Clin Lab.* 49:577–589.
94. Dahouk SA, Tomaso H, Nockler K, Neubauer, H. 2004. The detection of *Brucella* spp. Using PCR-ELISA and real-time PCR assays. *Clin Lab.* 50:387–394.
95. Matero P, Hemmila H, Tomaso H, Piiparinen H, Rantakokko-Jalava K, Nuotio L, Nikkari S. 2011. Rapid field detection assays for *Bacillus anthracis*, *Brucella* spp., *Francisella tularensis* and *Yersinia pestis. Clin Microbiol Infect.* 17(1):34–43; PMID: 2032255.
96. Tomaso H, Kattar M, Eickhoff M, Wernery U, Al Dahouk S, Straube E, Neubauer H, Scholz HC. 2010. Comparison of commercial DNA preparation kits for detection of Brucellae in tissue using quantitative real-time PCR. *BMC Infect Dis.* 10:100–104.
97. Alausa OK, Corbel MJ, Elberg SS, Gargani G, Gubina EA, Shi-Lang L et al. 1986. Joint FAO/WHO expert committee on brucellosis: Sixth report. *World Health Organization Tech Rep Ser.* 740:1–132.
98. Corbel MJ. 1997. Brucellosis: An overview. *Emerg Infect. Dis.* 3(2):213–221.
99. U.S. Department of Agriculture, Animal and Plant Health Inspection Service, Veterinary Services. 2010. *National Brucellosis Surveillance Strategy*, December, pp. 1–8. http://www.aphis.usda.gov/animal_health/animal_diseases/brucellosis/downloads/natl_bruc_surv_strategy.pdf
100. Ragan VE. 2002. The Animal and Plant Health Inspection Service (APHIS) brucellosis eradication program in the United States. *Vet Microbiol.* 90(1–4):11–18.
101. Ficht TA, Kahl-McDonagh MM, Arenas-Gamboa AM, Rice-Ficht AC. 2009. Brucellosis: The case for live, attenuated vaccines. *Vaccine.* 27(Suppl 4):D40–D43.
102. U.S. Department of Agriculture. *Brucella abortus Strain RB51 Vaccine Licensed for Use in Cattle*, APHIS document. http://www.aphis.usda.gov/animal_health/animal_dis_spec/cattle/downloads/rb51_vaccine.pdf
103. Moriyon I, Grillo MJ, Monreal D, Gonzalez D, Marin C, Lopez-Goni I, Mainar-Jaime RC, Moreno E, Blasco JM. 2004. Rough vaccines in animal brucellosis: Structural and genetic basis and present status. *Vet Res.* 35:1–38.
104. Zinsstag J, Roth F, Orkhon D, Chimed-Ochir G, Nansalmaa M, Kolar J, Vounatsou P. 2005. A model of animal–human brucellosis transmission in Mongolia. *Prev Vet Med.* 69:77–95.
105. Kolar J. 1977. *Brucella* vaccine production in Mongolia. World Health Organization, Assignment Report on WHO Project MOG BLG 001, SEA/Vaccine/89, 40pp.
106. Vershilova PA. 1961. The use of live vaccine for vaccination of human beings against brucellosis in USSR. *Bull World Health Organ.* 24:85–89.
107. Spink WW, Hall JW, Finstad J, Mallet E. 1962. Immunization with viable *Brucella* organisms. Results of a safety test in humans. *Bull World Health Organ.* 26:409–419.
108. Blasco JM, Diaz R. 1993. *Brucella melitensis* Rev-1 vaccine as a cause of human brucellosis. *Lancet.* 342(8874):805.
109. Perkins SD, Smither SJ, Atkins HS. 2010. Towards a *Brucella* vaccine for humans. *FEMS Microbiol Rev.* 34:379–394.

110. Hadjichristodoulou C, Voulgaris P, Toulieres L, Babalis T, Manetas G, Goutziana G, Kastritis I, Tselentis I. 1994. Tolerance of human brucellosis vaccine and the intradermal reaction test for brucellosis. *Eur J Clin Microbiol.* 13:129–134.
111. Fugier E, Pappas G, Gorvel JP. 2007. Virulence factors in brucellosis: Implications for aetiopathogenesis and treatment. *Expert Rev Mol Med.* 9:1–10.
112. Denisov AA, Korobovtseva YS, Karpova OM, Tretyakova AV, Mikhina LV, Ivanov AV, Salmakov KM, Borovick RV. 2010. Immunopotentiation of live brucellosis vaccine by adjuvants. *Vaccine.* 28S:F17–F22.
113. Hu X-D, Chen S-T, Li J-Y, Yu D-H, Yi-Zhang, Cai H. 2010. An IL-15 adjuvant enhances the efficacy of a combined DNA vaccine against *Brucella* by increasing the CD8+ cytotoxic T cell response. *Vaccine.* 28:2408–2415.
114. Crasta OR, Folkerts O, Fei Z, Mane SP, Evans C, Marino-Catt S, Bricker B, Yu GX, Du L, Sobral BW. 2008. Genomic sequence of *Brucella abortus* vaccine strain S19 compared to virulent strains yield candidate virulence genes. *PLoS One.* 3(5):e2193.
115. Yongqun H, Zuoshuang X. 2010. Bioinformatics analysis of *Brucella* vaccines and vaccine targets using VIOLIN. *Immunome Res.* 6(Suppl 1):S5, 1–12.
116. Young EJ. 1983. Human brucellosis. *Rev Infect Dis.* 5(5):821–842.
117. Mantur BG, Akki AS, Mangalgi SS, Patil SV, Gobbur RH, Peerapur BV. 2004. Childhood brucellosis—A microbiological, epidemiological and clinical study. *J Trop Pediatr.* 50:153–157.
118. Bingol A, Togay-Isiday C. 2006. Neurobrucellosis an exceptional cause of transient ischemic attacks. *Eur J Neurol.* 13:544–548.
119. Yetkin MA, Bulut C, Erdinc FS, Oral B, Tulek N. 2006. Evaluation of the clinical presentations in neurobrucellosis. *Int. J Infect Dis.* 10:446–452.
120. Karimi A, Alborzi A, Rasooli M, Kadivar MR, Nateghian AR. 2003. Prevalence of antibody to *Brucella* species in butchers, slaughterers and others. *East Mediterr Health J.* 9:178–184.
121. Crosby E, Llosa L, Miro QM, Carrillo C, Gotuzzo E. 1984. Hematologic changes in brucellosis. *J Infect Dis.* 150:419–424.
122. Hatipoglu CA, Yetkin A, Ertem GT, Tulek N. 2004. Unusual clinical presentations of brucellosis. *Scand J Infect Dis.* 36:694–697.
123. Yildirmak Y, Palanduz A, Telhan L, Arapoglu M, Kayaalp N. 2003. Bone marrow hypoplasia during *Brucella* infection. *J Pediatr Hematol Oncol.* 25:63–64.
124. Gurkan E, Baslamisli F, Guvene B, Bozkurt B, Unsal C. 2003. Immune thrombocytopenic purpura associated with *Brucella* and *Toxoplasma* infections. *Am J Hematol.* 74:52–54.
125. Makokopakis E, Christias E, Kofteridis D. 2003. Acute brucellosis presenting with erythema nodosum. *Eur J Epidemiol.* 18:913–915.
126. Gotuzzo E, Alarcón GS, Bocanegra TS, Carrillo C, Guerra JC, Rolando I, Espinoza LR. 1982. Articular involvement in human brucellosis: A retrospective analysis of 304 cases. *Semin Arthritis Rheum.* 12:245–255.
127. Alarcon GS, Bocanegra TS, Gotuzzo E, Espinoza LR. 1987. The arthritis of brucellosis: A perspective one hundred years after Bruce's discovery. *J Rheumatol.* 14:1083–1085.
128. Mousa AR, Muhtaseb SA, Almudallal DS, Khodeir SM, Marafie AA. 1987. Osteoarticular complications of brucellosis: A study of 169 cases. *Rev Infect Dis.* 9:531–543.
129. Aydin M, Fuat YA, Savas L, Reyhan M, Pourbagher A, Turunc TY, Ziya Demiroglu Y, Yologlu NA, Aktas A. Scintigraphic findings in osteoarticular brucellosis. *Nucl Med Commun.* 6:639–647.
130. Cascio A, Iaria C. Campenni A, Blandino A, Baldari S. 2004. Use of sulesomab in the diagnosis of brucellar spondylitis. *Clin Microbiol. Infect.* 10:1020–1022.

131. Tsolia M, Drakonaki S, Messaritaki A, Farmakakis T, Kostaki M, Tsapra H, Karpathios T. 2002. Clinical features, complications and treatment outcome of childhood brucellosis in central Greece. *J Infect.* 44:257–262.
132. Howard CB, Alkrinawi S, Gadalia A, Mozes M. 1993. Bone infection resembling phalangeal microgeodic syndrome in children: A case report. *J Hand Surg (Br).* 18:491–493.
133. Rotes-Querol J. 1957. Osteo-articular sites of brucellosis. *Ann Rheum Dis.* 16:63–68.
134. Ibrahim AIA, Shetty SD, Saad M, Bilal NE. 1988. Genito-urinary complications of brucellosis. *Br J Urol.* 61:294–298.
135. Kelalis PP, Greene LF, Weed LA. 1962. Brucellosis of the urogenital tract: A mimic of tuberculosis. *J Urol.* 88:347–353.
136. Peled N, David Y, Yagupsky P. 2004. Bartholin's gland abscess caused by *Brucella melitensis. J Clin Microbiol.* 42:917–918.
137. Lubani MM, Dudin KI, Sharda DC, Abu Sinna NM, Al-Shab T, Al-Refe'ai AA, Labani SM, Nasrallah A. 1988. Neonatal brucellosis. *Eur J Pediatr.* 147:520–522.
138. Giannacopoulos I, Eliopoulou MI, Ziambaras T, Papanastasiou DA. 2002. Transplacentally transmitted congenital brucellosis due to *Brucella abortus. J Infect.* 45:209–210.
139. Khan MY, Mah MW, Memish ZA. 2001. Brucellosis in pregnant women. *Clin Infect Dis.* 32:1172–1177.
140. Buchanan TM, Faber LC, Feldman RA. 1974. Brucellosis in the United States, 1960–1972: An abattoir-associated disease, I: Clinical features and therapy. *Medicine (Baltimore).* 53:403–413.
141. Solera J, Martinez-Alfaro E, Espinosa A. 1997. Recognition and optimum treatment of brucellosis. *Drugs.* 53:245–256.
142. Refik MM, Isik AT, Doruk H, Cömert B. 2003. *Brucella*: A rare causative agent of spontaneous bacterial peritonitis. *Indian J Gastroenterol.* 22:190.
143. Peery TM, Belter LF. 1960. Brucellosis and heart disease, II: Fatal brucellosis. *Am J Pathol.* 36:673–697.
144. Namiduru M, Karaoglan I, Yilmaz M. 2003. Guillain-Barre syndrome associated with acute neurobrucellosis. *Int J Clin Pract.* 57:919–920.
145. Goktepe AS, Alaca R, Mohur H, Coskun U. 2003. Neurobrucellosis and a demonstration of its involvement in spinal roots via magnetic resonance imaging. *Spinal Cord.* 41:574–576.
146. Reguera JM, Alarcon A, Miralles F, Pachon J, Juarez C, Colmenero JD. 2003. Brucellosis endocarditis: Clinical, diagnostic, and therapeutic approach. *Eur J Clin Microbiol Infect Dis.* 22:647–650.
147. Eckman MR. 1975. Brucellosis linked to Mexican cheese. *J Am Med Assoc.* 232:634–637.
148. Mousa AR, Elhag KM, Khogali M, Marafie AA. 1988. The nature of human brucellosis in Kuwait: Study of 379 cases. *Rev Infect Dis.* 10(1):211–217.
149. Anderson TD, Cheville NF, Meador VP. 1986. Pathogenesis of placentitis in the goat inoculated with *Brucella abortus*, II: Ultrastructural studies. *Vet Pathol.* 23(3):227–239.
150. Dajani YF, Masoud AA, Barakat HF. Epidemiology and diagnosis of human brucellosis in Jordan. 1989. *J Trop Med Hyg.* 92(3):209–214.
151. Mousa AM, Elhag KM, Khogali M, Sugathan TN. 1987. Brucellosis in Kuwait: A clinico-epidemiological study. *Trans R Soc Trop Med Hyg.* 81(6):1020–1021.
152. Roushan, MRH, Mohrez M, Gangi SMS, Amiri MJS, Hajiahmadi M. 2004. Epidemiological features and clinical manifestations in 469 adult patients with brucellosis in Babol, Northern Iran. *Epidemiol Infect.* 132:1109–1114.
153. Malik GM. 1997. A clinical study of brucellosis in adults in the Asir region of southern Saudi Arabia. *Am J Trop Med.* 56(4):375–377.

154. Lopez MA. 1989. Brucellosis in Latin America. In: Young EJ, Corbel MJ, eds. *Brucellosis: Clinical and Laboratory Aspects*. Boca Raton: CRC Press, pp. 151–161.
155. Olsen SC, Waters WR, Stoffregen WS. 2007. An aerosolized *Brucella* spp. challenge model for laboratory animals. *Zoonoses Public Health.* 54(8):281–285.
156. Mense MG, Van De Verg LL, Bhattacharjee AK, Garrett JL, Hart AJ, Linder LE, Hadfield TL, Hoover DL. 2001. Bacteriologic and histologic features in mice after intranasal inoculation of *Brucella melitensis. Am J Vet Res.* 62(3):398–405.
157. Izadjoo MJ, Mense MG, Bhattacharjee AK, Hadfield TL, Crawford RM, Hoover DL. 2008. A study on the use of male animal models for developing a live vaccine for brucellosis. *Transbound Emerg Dis.* 55(3–4):145–151.
158. Smither SJ, Perkins SD, Davies C, Stagg AJ, Nelson M, Atkins HS. 2009. Development and characterization of mouse models of infection with aerosolized *Brucella melitensis* and *Brucella suis. Clin Vacc Immunol.* 16(5):779–783.
159. Greenfield RA, Drevets DA, Machado LJ, Voskuhl GW, Cornea P, Bronze MS. 2002. Bacterial pathogens as biological weapons and agents of bioterrorism. *Am J Med Sci.* 323:299–315.
160. Kahl-McDonagh MM, Arenas-Gamboa AM, Ficht TA. 2007. Aerosol infection of BALB/c mice with *Brucella melitensis* and *Brucella abortus* and protective efficacy against aerosol challenge. *Infect Immun.* 75:4923–4932.
161. Kaushik P, Sing DK, Kumar SV, Tiwari AK, Shukla G, Dayal S, Chaudhuri P. 2010. Protection of mice against *Brucella abortus* 544 challenge by vaccination with recombinant OMP28 adjuvated with CpG oligonucleotides. *Vet Res Commun.* 34:119–132.
162. Trant CGM, Lacerda TLS, Carvalho NB, Azervedo V, Rosinha GMS, Salcedo SP, Gorvel J-P, Oliveira SC. 2010. The *Brucella abortus* phosphoglyerate kinase mutant is highly attenuated and induces protection superior to that of vaccine strain 19 in immunocompromised and immunocompetent mice. *Infect Immun.* 78(5):2283–2291.
163. Rajashekara G, Glover DA, Krepps M, Splitter GA. 2005. Temporal analysis of pathogenic events in virulent and avirulent *Brucella melitensis* infections. *Cellular Microbiol.* 7(10):1459–1473.
164. Rambow-Larsen AA, Rajashekara G, Petersen E, Splitter G. 2008. Putative quorum-sensing regulator BlxR of *Brucella melitensis* regulates virulence factors including the type IV secretion system and flagella. *J Bacteriol.* 190(9):3274–3282.
165. Baek BK, Lee BO, Hur J, Rahman MS, Lee SI, Kakoma I. 2005. Evaluation of the Sprague-Dawley rat as a model for vertical transmission of *Brucella abortus. Can J Vet Research.* 69:305–308.
166. Atkins HS, Spencer S, Brew SD, Jenner DC, Sefton AM, MacMillan AP, Brooks TJ, Simpson AJ. 2010. Evaluation of azithromycin, tovafloxacin and grepafloxacin as prophylaxis against experimental murine *Brucella melitensis* infection. *Int J Antimicrob Agents.* 36(1):66–68.
167. Yumuk Z, Dundar V. 2005. The effect of long-term ethanol feeding on efficacy of doxycycline plus refampicin in the treatment of experimental brucellosis caused by *Brucella melitensis* in rats. *J Chemother.* 17(5):509–513.
168. Popow J, Jezyna C. 1969. Morphological studies of guinea pig lungs in experimental brucellosis caused by different *Brucella* strains. *Arch Hyg Bakteriol.* 153(2):157–161.
169. Lu SL. 1984. A study of delayed hypersensitivity and desensitization in guinea pigs after *Brucella* infection. *Dev Biol Stand.* 56:317–321.
170. Van de Verg LL, Hartman AB, Bhattacharjee AK, Tall BD, Yuan L, Sasala K, Hadfield TL, Zollinger WD, Hoover DL, Warren RL. 1996. Outer membrane protein of *Neisseria meningitis* as a mucosal adjuvant for lipopolysaccharide of *Brucella melitensis* in mouse and guinea pig intranasal immunization models. *Infect Immun.* 64(12):5263–5268.

171. Detilleux PG, Cheville NF, Deyoe BL. 1988. Pathogenesis of *Brucella abortus* in chicken embryos. *Vet Pathol.* 25:138–146.
172. Anderson TD, Meador VP, Cheville NF. 1986. Pathogenesis of placentitis in the goat inoculated with *Brucella abortus*. I. Gross and histopathologic lesions. *Vet Pathol.* 23(3):219–226.
173. Meador VP, Deyoe BL. 1986. Experimentally induced *Brucella abortus* infection in pregnant goats. *Am J Vet Res.* 47(11):2337–2342.
174. Kahl-McDonagh MM, Elzer PH, Hagius SD, Walker JV, Perry WL, Seabury CM, den Hartigh AB. et al., 2006. Evaluation of novel *Brucella melitensis* unmarked deletion mutants for safety and efficacy in the goat model of brucellosis. *Vaccine.* 24(24):5169–5177.
175. el Idrissi AH, Benkirane A, el Maadoudi M, Bouslikhane M, Berranda J, Zerouali A. 2001. Comparison of the efficacy of *Brucella abortus* strain RB51 and *Brucella melitensis* Rev. 1 live vaccines against experimental infection with *Brucella melitensis* in pregnant ewes. *Rev Sci Tech.* 20(3):741–747.
176. Fleischner EC, Vecki M, Shaw EB, Meyer KF. 1921. The pathogenicity of *B. abortus* and *B. melitensis* for monkeys. *J Infect Dis.* 29:662–668.
177. Huddleson F, Hallman ET. 1929. The pathogenicity of species of the genus *Brucella* for monkeys. *J Infect Dis.* 45:293–303.
178. Percy DH, Egwu IN, Jonas AM. 1972. Experimental *Brucella canis* infection in the monkey (*Macaca arctoides*). *Can J Comp Med.* 36:221–225.
179. Chen TH, Elberg SS. 1970. Immunization against *Brucella* infections: Immune response of mice, guinea pigs, and *Cynomolgus philipinensis* to live and killed *Brucella melitensis* strain Rev. I administered by various methods. *J Infect Dis.* 122(6):70–82.
180. Chen TH, Elberg SS. 1973. Immunization against *Brucella* infections. Priming of *Cynomolgus philipinensis* with purified antigen of *Brucella melitensis* prior to injection of Rev. 1 vaccine. *J Comp Pathol.* 83:357–367.
181. Chen TH, Elberg SS. 1976. Priming of *Macaca cynomolgus philippinensis* with purified antigen of *Brucella melitensis* before injection of Rev. I vaccine. *J Infect Dis.* 134(3):294–296.
182. Mense MG, Borschel RH, Wilhelmsen CL, Pitt ML, Hoover DL. 2004. Pathogenic changes associated with brucellosis experimentally induced by aerosol exposure in rhesus macaques (*Macaca mulatta*). *Am J Vet Res.* 65(5):644–652.
183. Yingst SL, Huzella LM, Chuvala L, Wolcott M. 2010. A rhesus macaque (*Macaca mulatta*) model of aerosol-exposure brucellosis (*Brucella suis*): Pathology and diagnostic implications. *J Med Microbiol.* 59:724–730.

12 Alphaviruses

William D. Pratt, Donald L. Fine, Mary Kate Hart, Shannon S. Martin, and Douglas S. Reed

CONTENTS

12.1 BACKGROUND

In the Americas during the 1930s, three distinct viruses were recovered from dying horses or burros that were exhibiting signs of encephalitis. Later found to be the cause of temporally associated encephalitis in humans, these viruses, eastern equine encephalitis virus (EEEV), Venezuelan equine encephalitis virus (VEEV), and western equine encephalitis virus (WEEV), were named after the diseases they cause (EEE, VEE, and WEE, respectively). The viruses are closely related, single-stranded, positive-sense RNA viruses and have been assigned as members of the *Alphavirus* genus of the family *Togaviridae*. The natural cycles of infection for EEEV, VEEV, and WEEV are similar—transmission cycles are maintained between mosquitoes and their vertebrate hosts, with occasional outbreaks into the surrounding equine and human populations. There are differences seen in this pattern with the IAB and IC varieties of VEEV, where equines act as amplifying hosts and the outbreaks become epizootic in character [1]. In humans, there are also differences seen among the viruses with respect to the medical consequences from infection of the central nervous system (CNS). EEEV causes the most severe encephalitis, with case fatality rates ranging up to 30–40%; WEEV causes similarly severe encephalitis, but the case fatality rates are lower (5–10%); and VEEV rarely causes severe encephalitis except in young children. The disease more frequently seen in human adults infected with VEEV is an incapacitating febrile disease of rapid onset and slow recovery.

VEEV, WEEV, and EEEV are all infectious by the aerosol route. Before the widespread use of vaccines for laboratory workers, VEEV was a frequent cause of laboratory infections, with most of these known or thought to be caused by aerosol exposure [2]. Laboratory infections with EEEV and WEEV have not been as numerous; however, two incidences out of seven cases of laboratory-acquired WEEV infection resulted in fatalities. These "new-world" alphaviruses are endemic in the Americas, with VEEV isolated from Central and South America, WEEV from the western two-thirds of North America, and EEEV on the eastern seaboard of North America and South America. Because these viruses can be infectious by aerosol, can be grown to high viral titers in cell culture or in embryonated eggs, and are relatively stable, they are considered to be potential biological warfare agents. The National Institute of Allergy and Infectious Diseases Biological Defense Research Agenda [3] categorizes VEEV, WEEV, and EEEV as category B priority pathogens, and they are considered select agents (except for WEEV) in the Centers for Disease Control and Prevention's Select Agent Program. For an extensive review of the medical aspects of these viruses in the context of biological warfare or bioterrorism, the reader is invited to consult the most recent review in *Medical Aspects of Biological Warfare* [4].

With the heightened interest in biodefense, there is a need to understand the nature of the biological threat from these viruses and to develop medical countermeasures against the threat. Licensure of medical countermeasures against the aerosol threats posed by VEEV, WEEV, and EEEV will be possible under the Food and Drug Administration guidelines outlined in the "Animal Rule" [5]. This will require well-defined animal models that are relevant to the human disease. This chapter reviews the information that is currently available about the animal models and pathogenesis of VEEV, WEEV, and EEEV in these models and hopes to stimulate efforts in areas of research in which there are major gaps in our knowledge and understanding.

12.2 STRUCTURE AND BIOLOGY OF ALPHAVIRUSES

Alphaviruses are relatively simple, structured, enveloped viruses with an icosahedral symmetry. The viral genome of each consists of a single positive-sense strand of genomic RNA, which is enclosed in a nucleocapsid composed of 240 copies of a single species of capsid protein. The nucleocapsid is surrounded by a lipid bilayer from the host cell membrane and is closely associated with the 240 pairs of virus-encoded glycoproteins, E1 and E2, which traverse the lipid bilayer [6]. In the viral particle, the two glycoproteins are heterodimers that are associated into 80 trimeric spikes. As the most exterior portion of the viral particle, the glycoprotein spikes are the primary determinants for cell tropism and virulence [7] and serve as the target for neutralizing antibody [8]. Viral infections are initiated when the glycoproteins bind to receptors on the host cell surface, with subsequent viral entry into the cell through receptor-mediated endocytosis. On acidification of the endosome, it is thought that the glycoprotein heterodimers undergo conformational changes that expose a putative fusion domain that is in a highly conserved region of the E1 glycoprotein, which allows fusion between the lipid bilayers of the virus and cell, and disassembly of the virion [9]. A proposed alternative model is that infection occurs at the cell membrane without membrane fusion or the disassembly of the viral protein shell through a pore-like structure that forms on changes in pH [10]. In both models, the genomic RNA is released from the nucleocapsid into the cytoplasm for viral replication.

The alphavirus genome contains the genes for four nonstructural proteins (nsP1–nsP4) and three structural proteins (capsid, E1, and E2). The genomic RNA serves as mRNA for the translation of the nonstructural proteins, but not for the structural proteins. The nonstructural proteins function, in part, as the viral polymerase to transcribe a negative-sense strand of RNA from the genomic RNA. The negative-stranded RNA, in turn, serves as the template for generating new genomic RNA and for generating 26S subgenomic, positive-sense RNA, which serves as the mRNA for the structural proteins. The polypeptide precursor for the structural proteins translated from the 26S RNA contains four regions from the N to C terminus: C (capsid), PE2 (E3–E2), 6K, and E1. The first region translated, C, contains protease activity, which acts to cleave the capsid protein from the polypeptide as it is being formed. Capsid protein specifically binds to a single copy of new genomic RNA and induces nucleocapsid formation. The next region of the polypeptide, the E3 portion of PE2, is translocated into the lumen of the rough endoplasmic reticulum and is followed by the remainder of PE2. It appears that signal sequences in the 6 K polypeptide are responsible for membrane translocation of the E1 region of the polypeptide into the lumen of the RER, where E1 is cleaved to form heterodimers with PE2. PE2 and E1 move together through the Golgi apparatus, where they are glycosylated. Typically, PE2 undergoes additional cleavage by a host cell furin protease after the heterodimers leave the Golgi to form the mature E2–E1 heterodimer and the free, unassociated E3 peptide. The formation of viral spikes, the association of nucleocapsid with glycoproteins, and the assembly of viral particles all occur on the cell surface of mammalian cells [11] and can also occur on intracellular membranes in insect cells [12].

12.3 ANTIGENIC RELATIONSHIPS

Alphaviruses were originally grouped into seven antigenic complexes on the basis of serological cross-reactivity. The EEEV, VEEV, and WEEV antigenic complexes formed three of these complexes, and the VEEV complex was further subdivided into subtype (I, II, III, IV, V, and VI) and variety (IAB, IC, ID, IE, IF, IIIA, IIIB, IIIC, IIID). Phylogenetic studies generated from E1 amino acid sequences have refined the VEEV complex into seven different species: VEEV containing the IAB, IC, ID, and IE varieties; Mosso das Pedras virus (IF); Everglades virus (II); Pixuna virus (IV); Cabassou virus (V); and Rio Negro virus (AG80-663; VI) [13]. The WEEV complex includes WEEV and several viruses of little pathogenic consequence to humans: Aura virus, Fort Morgan virus, Highlands J virus, Sindbis virus, and Whataroa virus. It is worth noting that Sindbis virus is the prototypic alphavirus and has been extensively studied in mice. The EEEV complex only includes EEEV but is divided into North and South American antigenic varieties on the basis of hemagglutination inhibition tests, and it is further subdivided into four major lineages on the basis of phylogenetic analysis [14].

12.4 ANIMAL MODELS AND PATHOGENESIS

Protection from alphaviruses appears to be primarily, if not exclusively, mediated by antibodies. Studies conducted to identify a significant role for cytotoxic T cells in alphavirus infections have demonstrated lytic activity under some circumstances, but transfer of cells generated mixed results, sometimes reducing viral burden but not providing protection from death [15–17]. In contrast, the passive transfer of antibodies has protected against peripheral challenge with alphaviruses [18] and, in some cases, against aerosol challenge [19]. The development of vaccines is preferred over the development of antibody-based therapeutics for the alphaviruses because studies in animals indicated that passive transfer of antibodies fails to protect the animal if given after onset of clinical illness [20] or after an intracranial (i.c.) infection [21].

12.5 VENEZUELAN EQUINE ENCEPHALITIS VIRUS

Mice, rats, hamsters, guinea pigs, rabbits, macaques, horses, and burros have all been used to study key features of VEE [22–25]. VEE in humans is typically manifested by flu-like symptoms [26–29], and the CNS is affected only in a minority of naturally occurring cases, usually in the young. In fatal human cases of VEE, systemic pathological changes include widespread congestion, edema and hemorrhage, lymphoid necrosis, hepatocellular degeneration, and interstitial pneumonia [30]. CNS changes include meningoencephalitis and myelitis, with infiltrates of lymphocytes, mononuclear cells, and neutrophils. Multiple regions of the brain are typically affected.

12.5.1 Mouse Model of VEEV Infection

The mouse has been the most extensively used animal model of VEE, and mice are similarly affected by the epizootic and enzootic strains of VEEV. Mice exhibit a biphasic illness and develop both extraneural and CNS infection before dying within 1 week after infection by the subcutaneous (s.c.), intraperitoneal (i.p.), aerosol, intranasal (i.n.), or i.c. route [31–33]. Early signs of disease can include a rough coat appearance, lethargy, a hunched appearance, decreased peer interaction, and weight loss within 4–5 days, which can rapidly progress to signs of ataxia, paresis, and/or paralysis. The median lethal dose (LD_{50}) of a IA strain of VEEV (Trinidad donkey [TrD]) in mice ranges from less than 1 plaque-forming unit (pfu) to approximately 30 pfu, depending on the route of infection [32]. Probit analyses in BALB/c mice with sex, route of infection, and dose as covariates showed no significant differences in LD_{50} between the sexes and between the routes (W.D. Pratt and S.L. Norris, unpublished data, 2006). LD_{50} values for male and female BALB/c mice were 6.7 and 1.8 pfu, respectively, for the s.c. route of virus infection, and were 9.0 and 7.4 pfu, respectively, for the aerosol route of infection. The mean survival times for these groups of mice were not significantly different and ranged from 7.6 to 8.8 days. Using enzootic strains of VEEV, probit analyses indicated LD_{50} values for aerosol challenge to be approximately 30 pfu for a IE strain of VEEV (68U201) and between 440 and 1200 pfu for a IIIA strain of VEEV (Mucambo), depending on the mouse strain tested (M.K. Hart and R.F. Tammariello, unpublished data, 2006).

Potential vaccine candidates are typically evaluated for safety in the mouse model. Initial screenings for determining and comparing the safety of live attenuated vaccine candidates or the degree of attenuation of virus from infectious DNA clones with individual mutations were done in 8- to 10-week-old C57BL/6 mice using the s.c. route of virus inoculation [34]. Further evaluation of safety has used different routes of virus inoculation, younger mice, and different strains of mice. Hart et al. noticed a higher degree of attenuation in the recombinant VEEV, V3526, as compared to the live attenuated vaccine, TC-83, using the aerosol route of virus infection in 4- to 8-week-old C3H/HeN mice [35], and this was further demonstrated using the i.c. route of infection in 6- to 8-week-old C3H/HeN mice [32]. In determining the safety of their recombinant Sindbis/VEE virus vaccine candidates and to compare their attenuation to their VEEV parents, Paessler et al. [36,37] used the s.c. and i.c. routes of virus administration in 6-day-old NIH Swiss mice. They further evaluated the safety of one of their live chimeric alphavirus vaccine candidates, SIN/ZPC, in 11- to 18-week-old mice with selective immunodeficiencies (B-cell deficient, and $\alpha\beta$ T-cell receptor (TCR), $\gamma\delta$ TCR, and interferon-γ receptor knockout) on a C57BL/6 background [38]. For safety evaluation, newborn mice have been found to be too sensitive to be used to screen live attenuated VEE viruses (W.D. Pratt, unpublished data), but they can be useful in evaluating persistence of infectious virus [38] or residual infectivity after virus inactivation [39,40]. Safety tests of live attenuated viruses in mouse models are usually incorporated into the evaluation of their immunogenicity and efficacy as vaccine candidates [34,37].

The mouse model is also used as a primary screening for immunogenicity and efficacy testing, and down-selection of VEE vaccine candidates [34,37,39–42]. Typically, 6- to 8-week-old female mice from one of various genetic backgrounds (BALB/c, C57BL/6, NIH Swiss, or C3H/HeN) are used. Vaccines are usually administered by the s.c. route, but in some cases, for example, DNA-based vaccines may be administered by intradermal (i.d.) injection [41] or gene gun [42]. The route of infection for the challenge virus can also vary from routes such as s.c. or i.p., which are less stringent, to aerosol, i.n., or i.c., which are considered more stringent based on the pathogenesis (see Section 12.5.2).

Initial studies comparing three genetically different inbred mouse strains (BALB/c, C57Bl/6, and C3H/HeN) observed that all three strains had high-titer antibody responses to live attenuated vaccine, TC-83, or formalin-inactivated vaccine, C-84 (M.K. Hart and R.F. Tammariello, unpublished data, 2005) [31]. The mouse strains varied in the dominant isotypes of their responses, with BALB/c and C3H/HeN mice tending to make mostly IgG2a and IgG1 antibodies, whereas the C57Bl/6 mice had IgG2b titers that were higher than their IgG2a and IgG1 titers. Several protective epitopes have been identified by using monoclonal antibodies; most of these are present on the E2 glycoprotein between amino acids 180 and 220, although some have also been described on the E1 glycoprotein [18,43].

Vaccination with either TC-83 or C-84 elicited protective immunity from a TrD challenge in all three mouse strains when the challenge virus was administered s.c., but it proved to be more difficult to protect C3H/HeN mice from an aerosol challenge [31]. This could not be predicted by examining the serological responses by isotype, but the antibody responses in secretions indicated an association between antiviral IgA titers and mucosal protection [31,35]. Improved mucosal protection was achieved in C3H/HeN mice given a new vaccine candidate, V3526, by aerosol. Potential live attenuated vaccines are now routinely tested for safety by i.n. administration after the observation that the TC-83 vaccine was lethal to a majority of C3H/HeN mice when administered by this route [44].

12.5.2 Pathogenesis in the Mouse Model

Many aspects of the pathogenesis of the IAB strain of VEEV (TrD) have been studied in mice, including those relating to the course of natural infection after vector transmission and those relating to aerosol infection. In addition, mice have been used to study the immunological and apparent immunopathological responses to VEEV infection.

Footpad inoculation has been used to model the infection by mosquito vectors, and dendritic cells in the dermis were the first cell type to be infected by this route [45]. Infected dendritic cells transport VEEV to the draining lymph node, where initial replication occurs. Virus can be detected in the lymph node by 4 hours postinfection (PI), and viral titers of 10^6–10^7 pfu per gram were present in the draining lymph nodes by 6 h PI [46,47]. By 12–24 h PI, VEEV was already present in the blood and in other lymphoid tissues such as the spleen, nondraining lymph nodes, and thymus [47]. Virus seeding of nonlymphoid tissues, including heart, lung, liver, pancreas, kidney, adrenal gland, and salivary gland, also occurs at this time. Clearance of virus from the blood and peripheral tissues occurs relatively rapidly, being essentially completed by about 3–4 days PI. However, invasion of the brain takes place before peripheral clearance. Virus can be detected in the brains of mice infected by the footpad or s.c. route by 48 h PI [47–49].

Mice have proven especially useful in characterizing a number of virulence factors in VEEV. For instance, the E2 glycoprotein of VEEV appears to play an important role in specifically targeting dendritic cells. Even a single-point mutation in E2 can prevent infection of dendritic cells *in vivo* and actually make the virus avirulent in mice [45]. Infection of macrophages may also be important in the early pathogenesis of VEEV. Early infection of mononuclear phagocytes by virulent VEEV occurs in lymph nodes [48], and the virulence of different VEEV strains correlates with their ability to replicate rapidly in macrophages [7]. Additional mutations in E2 or in the E1 glycoprotein have also been shown to affect other steps in the spread of VEEV throughout the body, including its entry into the brain.

VEEV invasion of the brain is a critical event in the pathogenesis of VEEV and has been the subject of several studies [7,44,48–51]. Virus in the blood of mice can seed the perivascular areas in the connective tissue underlying the olfactory neuroepithelium within 18 h of peripheral infection [49], and it can gain access to the brain via the olfactory nerves that originate in and transit this region. This appears to be the main route for VEEV into the mouse brain. Entry via the blood–brain barrier does not appear to be an important means of CNS invasion by VEEV in mice.

Because of the susceptibility of humans to aerosol infection with VEEV, this route of exposure has been another focus of animal studies. Mice infected with VEEV by aerosol administration exhibit early massive infection of the olfactory mucosae compared to natural infection. The bipolar or olfactory sensory neurons in the olfactory neuroepithelium are in direct contact with the environment [52], and these cells are a strong and specific target of aerosolized virulent VEEV (Figure 12.1) [25,44,48,49,53]. These cells, via their axonal processes that form the olfactory nerves, provide direct access to the olfactory bulbs of the brain. By both routes of inoculation, then, VEEV can invade the brain by the olfactory system. Compared to peripheral inoculation of VEEV, though, invasion of the brain occurs much faster by the aerosol route.

Within the brain, neurons are the major target of virulent VEEV in mice [25,44,48,49,53]. From the olfactory bulbs, VEEV spreads first to structures in the brain that receive efferent connections from the olfactory bulbs, and then to remaining regions of the brain in a generally rostral to caudal fashion. Infection then proceeds into the spinal cord, where anterior horn neurons are the major target. The patterns of virus distribution through the brain appear identical after infection by

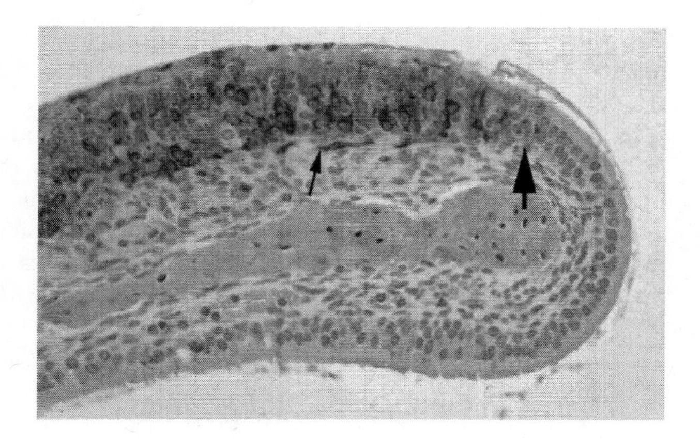

FIGURE 12.1 **(See color insert.)** Immunohistochemical stain of the nasal turbinate of a mouse infected with Venezuelan equine encephalitis virus by aerosol. The red stain indicates abundant viral infection of the olfactory epithelium that lines this part of the turbinate (the block arrow marks the border with respiratory epithelium), and the respiratory lining of the remainder of the turbinate is uniformly uninfected. Note that the virus is also present in the beginning of an olfactory nerve (thin arrow).

either the peripheral or the aerosol route. In addition to neurons, cells in the brains of mice with morphological features of glial cells also appear to be infected, although not to the same extent as neurons.

The mechanism of neuronal cell death in alphavirus infections of the brain has been the subject of extensive investigation. It is believed that the increased susceptibility to VEE and other alphavirus encephalitides among children could be explained by the increased susceptibility of immature neurons to undergoing apoptosis after infection [54]. A mouse model with the prototype alphavirus, Sindbis virus, was used to investigate the roles of neuronal infection and apoptosis in the pathogenesis of alphavirus encephalitis. Some strains of Sindbis virus cause fatal encephalitis in newborn mice, but weanling mice are resistant [55]. Older mice, however, are susceptible to a more neurovirulent strain [55,56]. The greater susceptibility of newborn mice to certain strains correlated with widespread apoptosis of neurons in the CNS. Infected neurons of older mice did not exhibit significant apoptosis with these less neurovirulent strains; however, they did after infection with the more virulent Sindbis virus strain [56,57]. Experimental mouse models of VEEV infection have also studied the role of apoptosis in neuronal destruction. Separate studies reported that apoptosis was the mechanism of neuronal cell death in mice infected with VEEV [55,58], although death by necrosis also appeared to occur. Inflammation of the brain is a significant component of VEEV infection in the CNS. Mice, like other animals and humans with VEE, exhibit meningoencephalitis characterized by cellular infiltrates in the meninges, perivascular spaces, and neuropil [23,25,30,44]. These infiltrates are composed predominantly of lymphocytes, but they are usually mixed with some histiocytic cells and neutrophils. Neutrophils are more prominent early in infection. The possibility that immunopathological mechanisms might contribute to the pathogenesis of VEEV was suggested years ago, when it was shown that treatment with antithymocyte serum prolonged the survival time of infected mice [59]. Further support for the notion that VEE involves immune-mediated disease was provided by a study that showed that SCID mice with VEE lack the cellular inflammatory changes in the brain that immunocompetent mice have, and that they survived about 3 days longer than immunocompetent mice [60].

Experiments have also implicated astrocytes for having a role in the neurodegeneration seen in VEE [53,61]. One study reported that degeneration of neurons occurred in portions of the brain associated with astrogliosis in which viral infection was not apparent [53]. This study also reported that the brains of VEEV-infected mice expressed increased levels of apoptosis-signaling molecules. In a separate study, VEEV infection of primary astrocytes resulted in increased expression of the proinflammatory mediators, tumor necrosis factor α and inducible nitric oxide synthase [61]. These experiments indicate that astrocytes may contribute to the damage to neurons that occurs in VEE.

12.5.3 Use of Telemetry in the Mouse Model

Having a small animal model that is predictive of human responses to experimental vaccines as well as a sensitive indicator of vaccine efficacy is highly desirable. In contrast to nonhuman primates (NHPs), the most appropriate model for predicting how humans will respond to VEEV vaccine candidates [62] is the mouse. Mice are an inexpensive model and are not associated with the ethical concerns linked to studies with NHPs. Development of telemetric implants small enough for use in mice has provided a system that increases the sensitivity of detection of murine responses to vaccination and challenge. Recent studies using inbred and outbred mouse strains implanted with telemetry devices have demonstrated this increased sensitivity of physiological response detection. In one study, the onset of clinical symptoms associated with VEEV TrD infection was identified up to 2 days sooner than identification of symptoms via cage-side observations [63]. Furthermore, the telemeters were capable of detecting changes in diurnal temperature patterns that are similar to the biphasic disease that develops in mice after VEEV TrD infection [47]. Such changes cannot be detected by cage-side observations.

A common analysis used to evaluate fever responses to a vaccine or challenge material is group mean fever hours [34,64,65]. An analysis of group mean data collected from two strains of mice vaccinated with an experimental VEEV vaccine, V3526, showed no significant difference between groups of V3526-inoculated mice, suggesting V3526 did not induce a fever response. Examination of individual temperature profiles resulted in the same conclusion for BALB/c mice (Figure 12.2a and b). However, examination of individual animal temperature profiles after vaccination of CF-1 mice, the outbred strain, suggested otherwise. Forty percent of CF-1 mice demonstrated a loss of diurnal temperature rhythm for 6–9 days after vaccination. The loss of diurnal rhythm was not associated with an increase in temperature, but rather the range of diurnal temperatures was reduced from approximately 2°C to approximately 1°C (Figure 12.2c and d). Increases in activity after vaccination were also observed in CF-1 mice 4–5 days after vaccination; changes in activity levels were not observed in BALB/c mice. Regarding the frequency of these response, note that the percentage of outbred mice with changes in temperature and activity are similar to the percentage of humans, NHPs, and equines that exhibit febrile responses after vaccination with V3526 [63,65]. Approximately 30% of humans, 50% of NHPs, and 20% of equines developed a fever of short duration after V3526 vaccination. From this study it was concluded that use of telemetry in outbred mouse strains, specifically CF-1 mice, is sufficiently sensitive to detect subtle changes in temperature and activity that may be reflective of the febrile response induced by vaccination in large animal models and humans. These studies further demonstrated that analysis of individual animal data is essential for detecting and predicting responses to vaccination as analysis of group means did not reveal physiological changes seen in individual mice that were in fact predictive of outcomes in larger species and humans.

As the CF-1 mouse model showed increased sensitivity to V3526 which was only detectable with implanted telemetry devices, the utility of CF-1 mice to evaluate vaccine efficacy was evaluated. BALB/c mice were also evaluated for comparison. CF-1 mice implanted with telemeters were vaccinated with V3526 and challenged 28 days later with 1×10^4 pfu VEEV TrD by the s.c. route. Cage-side observations after challenge indicated all mice remained healthy during the 28-day observation period. However, the temperature and activity data collected by the telemeters revealed 40% of mice experienced altered diurnal temperature rhythms similar to those observed after vaccination (Figure 12.3a). Activity data in these mice also deviated from the normal pattern for several days after challenge. In comparison, BALB/c mice

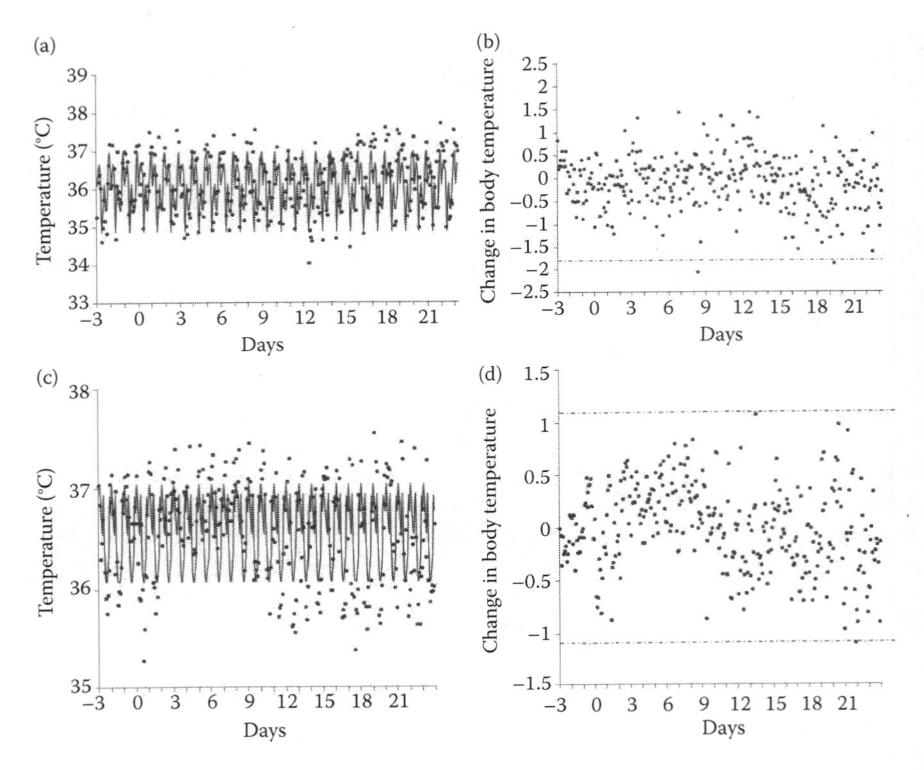

FIGURE 12.2 Body temperature was monitored via implanted telemeters during a baseline period (day −7 to −1) and during the 28-day post-V3526 vaccination period. Only day −3 through day 24 are shown to allow better resolution of the data. Body temperature (a) and relative change in body temperature (b) for a representative BALB/c mouse vaccinated subcutaneously with 1×10^4 pfu V3526. Body temperature (c) and relative change in body temperature (d) for a representative CF-1 mouse vaccinated subcutaneously with 1×10^4 pfu V3526. The averaged hourly body temperatures for each mouse (a and c, solid dots) during days −3 to −1 were used to establish an autoregressive integrated moving average model. This 24-h model was extrapolated forward in time to forecast body temperature values for the remaining study days (day 0 to day 28, a and c, solid line). Changes in body temperature represent the difference between the hourly body temperature data and the forecast values. The solid dots (b and d) represent the difference between the hourly body temperature data and the forecast values. Dashed lines represent 3 SD of the mean change in body temperature.

evaluated in this study had no temperature or activity patterns that differed from baseline (Figure 12.3b). These studies further demonstrated the utility of CF-1 mice implanted with telemeters as a highly sensitive mouse model for VEEV.

12.5.4 Other Small-Animal Models of VEEV Infection

The pathological effects of VEEV infection have been examined in numerous small laboratory animals other than the mouse. These include hamsters, rats, guinea pigs,

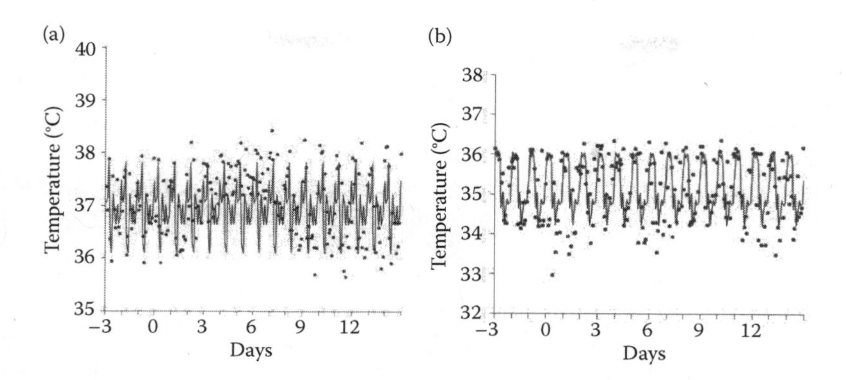

FIGURE 12.3 Body temperature was monitored via telemetric implants during the baseline period (day −35 to day −29), after vaccination with V3526 (day −28 to day −1) and during the Venezuelan equine encephalitis virus TrD post-challenge period (day 0 to day 28). Only day −3 through day 14 are shown to allow better resolution of the data. Body temperature for a BALB/c mouse (a) and a CF-1 mouse (b) vaccinated s.c. with 1×10^4 pfu V3526 and s.c. challenged with 1×10^4 pfu TrD 28 days later are shown. The average hourly body temperatures for each mouse (solid dots) during the baseline period were used to establish an autoregressive integrated moving average model. This 24-h model was extrapolated forward in time to forecast body temperature values for the vaccination and challenge periods (solid line).

and rabbits. Comparative early studies indicated that VEEV infection by injection or aerosol can result in generalized necrosis of both myeloid and lymphoid tissues, and that the virus can exhibit lymphomyelotropism and neurotropism in different animals [24]. Hamsters, guinea pigs, and rabbits present with severe lymphomyelotropic effects, and the infection is lethal before they exhibit signs of neurological disease [23,66–69]. In contrast, mice and rats develop a fatal neurological disease.

Hamsters are profoundly susceptible to the targeting of lymphoid and myeloid tissues by VEEV, which results in extensive immune system damage and subsequent bacterial overgrowth and endotoxic shock [23,70]. Hamsters die before the encephalitic phase develops, unless they are treated with antibiotics (to prevent bacterial proliferation), in which case death is caused by encephalitis [70]. Ultrastructure studies indicated that virtually all cells in the lymphoreticular tissues examined (thymus, lymph nodes, spleen, bone marrow, and ileum) are affected by the time the hamsters become moribund [66].

Adult hamsters, because of their great sensitivity to VEEV, were used recently to test attenuation of a panel of novel VEE vaccine candidate strains of mutant viruses [34], and previously to evaluate the virulence of VEEV subtypes. In the latter test, a histopathology comparison was performed in hamsters using IAB, IE, II, and III strains of virulent VEEV and two nonlethal viruses, TC-83 (VEEV IAB) vaccine and IV strain of VEEV [68]. Similar findings to those previously described for the IAB strain of VEEV were observed using the IE strain of VEEV, but this virus also infected brain tissues. Infection with the II strain of VEEV did not induce significant changes in the spleen or Peyer's patches, although the bone marrow and brain lesions were similar to those caused by the IA strain of VEEV. The virulent III strain of

VEEV caused minimal or no damage to the spleen, bone marrow, and Peyer's patches, and death was attributed to the observed brain lesions. Lesions were not identified in animals infected with TC-83 or the IV strain of VEEV, except for a transient lymphopenia followed by hyperplasia. Fluorescent antibody testing confirmed the presence of viral antigen in affected tissues and also detected antigen in the acinar cells of the pancreas, as also reported in mice [71] and guinea pigs [72]. Subsequent studies indicated viral infection of islet cells, as well as acinar cells, resulting from viremia and the spread of virus from the spleen [73].

Nine virulent subtype I strains of VEEV were tested for lethality in adult white rats [20]. IAB and IC strains of VEEV caused death in rats, with the IC strain, V-198, being the only one that exhibited 100% lethality at a dose of 10^6 pfu. Viral replication was observed in the thymus, spleen, and brain, but not in bone marrow or livers of infected rats. Similar to mice, rats become moribund approximately 1 week after infection and exhibit signs of neurological disease.

Unvaccinated guinea pigs infected with the TrD strain of VEEV by aerosol, s.c., or i.p. usually die within 2–6 days, although some animals survive for as long as 12 days [24]. Rabbits generally die by 4 days after infection. Both rabbits and guinea pigs had widespread necrosis of lymphoid and myeloid tissues [24]. Their spleens were slightly enlarged but had few white blood cells present, and significant drops in white blood cells were evident in the blood by 2 days PI. The sinusoids of the spleen were empty, the pulp cords were swollen with red blood cells, and the Malpighian follicles were necrotic. In contrast to mice, encephalomyelitis was rarely observed in guinea pigs. In a very early study, a marked difference was observed among several species, with VEEV inducing lymphomyelopoietic changes in rabbits and guinea pigs, neurotropic changes in NHPs, and both types of changes in mice, regardless of the route of administration [24]. The guinea pig has served well as a model for preliminary pathogenesis studies of VEEV because they, like horses, have a more variable susceptibility to the different epizootic or enzootic subtypes [74]. Powers et al. [75] and Greene et al. [76] used this variability in the guinea pig model to assess the role that nonenvelope and envelope genes contribute to virulence.

12.5.5 NHP Models of VEEV Infection

NHPs are commonly used as models of human disease caused by infectious agents, and it was appreciated in early studies with VEEV that the disease in NHPs was very similar to what had been reported for humans [24,25,77]. The prevalent species used for VEEV studies has been either the rhesus (*Macaca mulatta*) or the cynomolgus (*Macaca fascicularis*) macaque, with the latter species being predominantly used in most recent studies [34,62,78–81]. Both species of macaques develop fever, viremia, and lymphopenia within 1–2 days after a parenteral infection with an epizootic strain of VEEV [62,77,81]. Fever persists for up to 5–6 days, and clinical signs of encephalitis do not begin to appear until late in the course of the disease, if they appear at all. Encephalitic signs include loss of balance, slight tremors, and severe, prolonged hypothermia. In comparison, the febrile response after aerosol exposure to a virulent infectious clone of TrD (V3000) appears to occur a little later (36–72 h) after exposure [62]. However, fever duration and severity were greater in aerosol-exposed macaques compared to that in parenterally exposed macaques when given equivalent challenge doses of V3000 (~10^8 pfu). Other clinical signs were essentially the same between the two routes of exposure, including leukopenia and viremia. Cerebrospinal fluid from animals exposed by either route was positive for IgG specific for VEEV, and the levels of antibody correlated with the severity of the febrile response. A similar response including fever, viremia, and lymphopenia was reported for rhesus macaques infected with an epizootic IC strain of VEEV [77]. By either route, VEEV exposure is only rarely lethal in healthy macaques, similar to what is reported for humans.

Because VEEV is rarely lethal, the use of s.c. implanted devices that transmit physiological data by radiotelemetry has greatly facilitated research studies in VEE using NHPs [34,62,79]. These systems continuously monitor and record data, such as body temperature, from exposed animals, while greatly improving the safety of personnel and limiting animal handling. The temperature data collected by radiotelemetry can be modeled mathematically to account for diurnal variation that is normally seen in NHPs and can provide a model by which significant deviations from normal can be identified, and quantitatively and qualitatively measured (Figure 12.4).

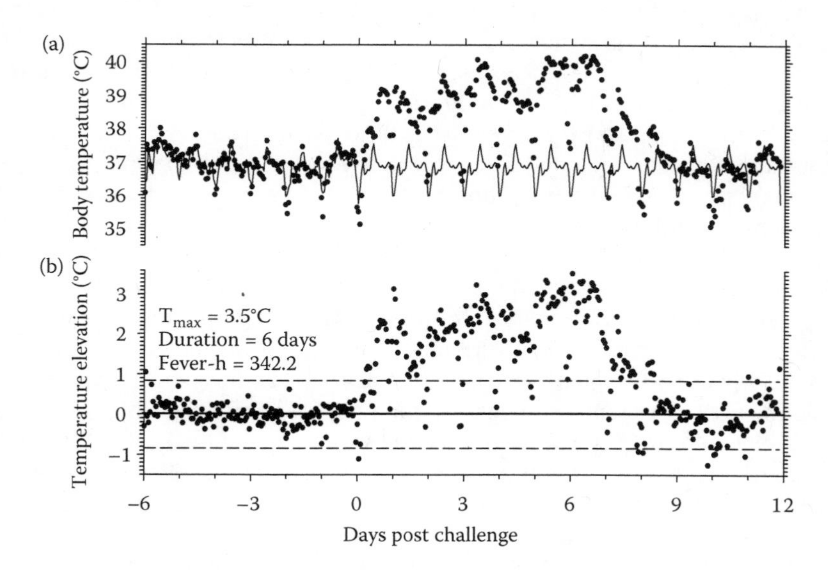

FIGURE 12.4 Body temperatures were monitored by radiotelemetry during a baseline period (day −10 to day 0) and during the 21 days after challenge. Body temperature (a) and analysis (b) for an unvaccinated monkey challenged by aerosol with 4.0 × 10^8 pfu of V3000 (Adapted from Pratt, W. D. et al., *Vaccine*, 2003, **21**(25–26): 3854–3862; Pratt, W.D. et al., *Vaccine*, 2009, **27**(49): 6814–6823.). The average hourly body temperatures for the monkey (a, solid dots) were subjected to a baseline training period (day −10 to day −3) to fit an autoregressive integrated moving average model. The 24-h training period model was extrapolated forward in time to forecast body temperature values for the remaining day −2 to day 12 time (a, solid line). The residual temperatures (b, solid dots) represent the difference between the hourly body temperature data and forecast values. Residual temperature data above 3 SD (dashed lines) were used to compute fever duration (number of days with over 18 h of significant temperature elevation) and fever hours (sum of the significant temperature elevations) over the challenge period.

There are few reports in the literature detailing the response of macaques to exposure with enzootic strains of VEEV. In 1968, Verlinde [82] reported on cynomolgus macaques infected with Mucambo virus. After s.c. infection only one of two macaques developed a febrile response that did not begin until day 5 and was mild, lasting only 2 days. Virus was not detectable in the blood of either animal, and neither animal showed any signs of encephalitis. Subsequent pathological examination showed no evidence of virus in the CNS. In contrast, after i.c. injection of the virus, fever onset was seen after 2–3 days and persisted for 6 days. Both animals became viremic and developed a slight paralysis of the feet and hands. The animals recovered from the infection by day 12 and were subsequently killed; pathological examination found multiple lesions in the CNS with areas of neuronal damage and perivascular infiltrations of mononuclear cells, particularly in the thalamus. Monath et al. [77] reported that rhesus macaques injected s.c. with viruses of either a ID or IE strain of VEEV became viremic and weakly leukopenic but did not mount a febrile response. Taken together, these results suggested that enzootic strains do not cause significant disease in macaques.

More recent studies using radiotelemetry devices in macaques have compared some of the epizootic and enzootic strains of VEEV, and these strains are highlighted in Figure 12.5. Cynomolgus macaques injected s.c. with a dose of 10^6 pfu of a virulent infectious clone of 68U201 (IE1009) rapidly developed a fever profile similar to that seen with epizootic VEEV (V3000) given s.c. at a high dose [62], but more rapid and shorter than that seen with an equivalent dose of V3000 (Figure 12.5a) [34,64]. Macaques exposed by high aerosol doses (~10^8 pfu) of either 68U201 virus or V3000 showed similar clinical signs of infection [34,64,79]. The febrile response after aerosol exposure to either of these viruses was biphasic and persisted for 4–5 days (Figure 12.5a). Lymphopenia and viremia were similarly severe and occurred during the febrile period (Figure 12.5b and c). A different range of responses was seen after exposure to a IIIA strain of VEEV, in which macaques inoculated s.c. with a virulent infectious clone of Mucambo virus (3A3500) developed only a slight fever that was not significantly different from temperatures of mock-infected controls (Figure 12.5a) [79]. Onset of fever was delayed in macaques exposed by the aerosol route to Mucambo virus; however, the fever was severe and prolonged (Figure 12.5a) [79]. Lymphopenia after Mucambo virus challenge was moderate, regardless of route of infection (Figure 12.5b), and viremia was low compared to levels seen with V3000 and 68U201 (Figure 12.5c). Macaques exposed to aerosols of Mucambo developed clinical signs indicative of encephalitis including loss of balance, tremors, and prolonged severe hypothermia [79].

Pathological examination of macaques exposed to virulent epizootic strains has shown clear evidence of encephalitis. Gleiser et al. [25] reported that necropsy of rhesus macaques parenterally injected with TrD revealed lesions throughout the brain. The most intense lesions were particularly concentrated in the thalamus, and the principal lesions observed were lymphocytic perivascular cuffing and glial nodules. Danes et al. [83,84] carefully examined VEEV infection in rhesus macaques challenged i.n., which is similar to the aerosol route of infection. In these companion studies, macaques were divided into three groups: group 1 received a laminectomy of the lamina cribrosa of the ethmoidal bone with a replacement bone graft and was

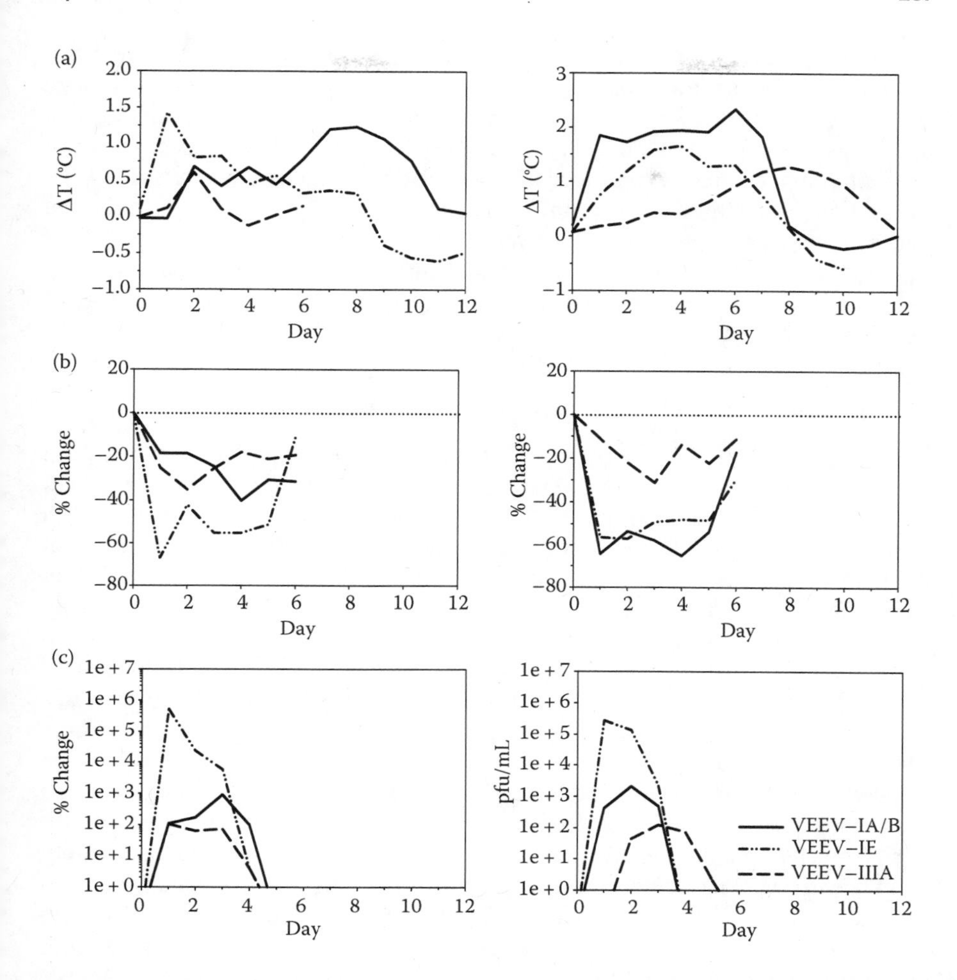

FIGURE 12.5 Cynomolgus macaques develop fever (a), lymphopenia (b), and viremia (c) in response to s.c. or aerosol infection with epizootic or enzootic strains of Venezuelan equine encephalitis virus (VEEV) (Adapted from Pratt, W.D. et al., *Vaccine*, 2003, **21**(25–26): 3854–3862; Reed, D.S. et al., *J. Infect Dis*, 2004, 189(6): 1013–1017.). Groups of macaques with telemetry implants for recording body temperature were either inoculated s.c. with approximately 10^6 pfu of virus or aerosol exposed to approximately 10^8 pfu of virus. Virulent infectious clones of VEEV (V3000, clone of the IA/B strain, TrD; IE1009, clone of the IE strain, 68U201; 3A3500, clone of the IIIA strain, Mucambo virus) were used for s.c. inoculations, and V3000, 68U201, and Mucambo virus were used for aerosol exposures.

challenged i.n. 8 weeks after the operation with virulent VEEV (most probably TrD); group 2 received no operation and was challenged i.n. with virulent VEEV; and group 3 was challenged with virulent VEEV through a tracheostomy. In group 2, VEEV reached the olfactory bulb and tract of the macaques faster and replicated to higher levels than in the macaques with disrupted olfactory nerves (group 1). Although the authors felt that VEEV had a predilection for the olfactory system that was similar to that seen in mice [48,49], they did not feel that spread to the brain through the olfactory tract was any more significant than the virus spread through the bloodstream during viremia. They based this determination on findings of similar histological signs and similar levels and presence of virus in various areas of the brain (hippocampus, gyrus, occipital lobe, medulla) between the macaques from groups 1 and 3 and those from group 2. The authors also felt that their results showed that the brain as a whole could not be regarded as a target organ for VEEV. This is supported by the different levels of virulence and lethality caused by encephalitis and seen between the macaque and the mouse, where studies support the view that the mouse brain is a target organ [44,48,49].

12.5.6 Application of the Animal Models of VEEV Infection

From the literature, it is clear that mice and macaques represent the two most important animal models of VEE. The advantages of using mice include their small size and relatively low cost, the availability of well-defined genetic strains with immunological or other features particularly suited to certain studies, the wide availability of mouse reagents, and the significant database that already exists for VEE in mice. Considering the similarities that exist between mice and humans infected with VEEV, these advantages make the mouse a highly useful animal model of VEE. However, it is not without its disadvantages. Whereas fatal encephalitis caused by virulent VEEV is relatively rare in humans, mice are exquisitely sensitive to neuroinvasion by VEEV, with 100% of mice succumbing from encephalitis when virus is administered by most routes of infection. At least some of this sensitivity may be because mice are much more "olfactory" animals than humans or macaques—a point relevant given that VEEV may invade the brain via the olfactory system. To whatever extent this is true and pertinent to VEE, there remains some uncertainty about how relevant particular features of the disease in mice are to humans.

Not surprisingly, then, macaques have recently been used more extensively in VEE studies. The use of macaques does have the disadvantage in terms of expense and limited availability, and there is some concern regarding the doses of VEEV required to cause disease in macaques. However, macaques are phylogenetically closer to humans than the other models. There is a greater similarity in the clinical and pathological manifestations of VEE between humans and macaques than between humans and mice. In particular, macaques are less likely than mice to develop fatal encephalitis, mirroring what has been reported in humans. The ability to continuously monitor physiological changes in macaques exposed to VEEV using radiotelemetry is another significant advantage that the mouse model has yet to provide. The main advantage of the hamster model, as discussed, is the animals' usefulness in testing the degree of attenuation of various VEEV strains to levels much more sensitive than what is seen with mice.

A good example of how the advantages of all three animal models of VEE can be best used is provided by studies with V3526, the genetically engineered VEEV that is currently under development for use as a human vaccine. V3526 was conceived on paper through rational design [85], was selected in and studied extensively in mice [32,34,44,85,86], was tested for attenuation in hamsters [34], and has been shown to be safe and efficacious in macaques [34,64]. The development and testing of V3526 has been the driving force behind much of the recent animal work on VEEV, and the V3526 paradigm had become our standard for the development and evaluation of additional vaccines against EEEV and WEEV. Recently added to this paradigm has been the use of telemetry in the mouse model to further examine safety.

12.6 EASTERN EQUINE ENCEPHALITIS VIRUS

In comparison to VEEV, most animal studies with EEEV were performed before the advent of many of the modern experimental techniques now in use, and they have mainly focused on the peripheral route of infection. Because of this, there has been a recent significant focus on animal modeling and pathogenesis studies that are focused on the aerosol or i.n. route of infection and are more relevant to the development of biodefense and bioterrorism countermeasures [87–91]. Insights into the human disease caused by EEEV have been gained by studying macaques, mice, hamsters, guinea pigs, and other species of animals. These studies underscore important features of the human disease, which in its more severe manifestations includes fever, seizures, coma, and meningoencephalitis characterized by neuronal damage and vasculitis affecting mainly the basal nuclei and thalamus but also the cortex, hippocampus, and brainstem [92–94].

12.6.1 Mouse Model of EEEV Infection

EEEV infection by the parenteral route in young mice presents similarly to that seen with VEE in mice—a biphasic course of disease manifested by initial virus replication in peripheral tissues followed by viremia, CNS invasion, and encephalitis, with death occurring in 4–6 days [95,96]. In mice infected with EEEV by an aerosol or i.c. route the course of disease appeared to be more rapid, and the mice typically died between days 2 and 4 after infection [90]. In studies comparing the two diseases in mice, it was noted that VEEV-infected mice exhibited signs of disease, such as piloerection and weight loss, significantly earlier than mice infected with EEEV [95]. Whereas the disease in VEEV-infected mice progressed with ataxia, paresis, and/or paralysis to death, the EEEV-infected mice appeared healthy during the first days after infection, and only exhibited signs of disease 12–24 h before death. Interestingly, the neurological signs of disease observed in the later stages of EEEV infection were seizures with no signs of paralysis. Another feature of EEEV infection in mice that contrasts VEEV infection is the finding that EEEV replication in lymphoid tissues is severely restricted and this probably contributes to the lack of early clinical signs of infection.

Probit analysis of the survival data indicated that the LD_{50} of a North American strain of EEEV, NJ1959, delivered to 8- to 9-week-old BALB/c mice by either 1 or >6 μm aerosol particles was 6.2×10^2 and 2.9×10^3 pfu, respectively [90]. Similar analysis of a South American strain of EEEV, ArgM, delivered by 1 μm aerosol particles indicated that the LD_{50} was 4.6×10^2 pfu, but the LD_{50} for this strain at a particle size >6 μm could not be determined (>1.4×10^4 pfu) because mortality was not ≥50% in any of the dose groups in this study. Analysis of another North American strain of EEEV, FL91-4679, delivered to 6- to 8-week-old mice by aerosol indicated that the LD_{50} was approximately 500 pfu in two mouse strains (M.K. Hart and R.F. Tammariello, unpublished data, 2005), and challenge studies to evaluate potential vaccine candidates using this strain typically use a dose of 10^5 pfu by aerosol. A comparative LD_{50} study by other routes of infection, using an EEEV identified only as the parent strain, indicated that the LD_{50} was less than 1 pfu by the i.c. route, approximately 400 pfu by the i.p. route, and approximately 1250 pfu when administered s.c. to 8–12 g weanling Swiss mice [21].

In mice, EEEV causes encephalitis but generally fails to induce the vasculitis that is a component of the human disease [97,98]. Typical of alphaviruses, newborn mice are more susceptible to EEEV, with resistance to peripheral infection developing by approximately 4–8 weeks of age [97–101]. This has been attributed to an inability of the virus to pass through blood vessels in older mice; however, a very recent study indicates that this resistance may be the result of a marked reduction in the osteoblast population that occurs with the maturation of the skeletal system [96]. In this study, metaphyseal osteoblasts were found to be an early site for intense viral replication in 5-week-old C57BL/6 mice, and the authors proposed this as a primary site for viral amplification that would lead to viremia and CNS infection. Fewer osteoblasts are found in older mice, which are more resistant to peripheral inoculation with EEEV. However, older mice are susceptible to EEEV delivered i.n., i.c., or by aerosol [90,97–99,102].

Using immunohistochemistry and *in situ* hybridization to track the spread of EEEV in mice, Vogel et al. [96] made several interesting observations that contrast with what has been seen in VEEV infections. In this study, 5-week-old female mice were inoculated with EEEV (strain FL91-4679) in the footpad and followed in a time-course study for 4 days. As expected, virus replicated near the site of inoculation before hematogenous spread to select extraneural tissues and to the CNS. Whereas VEEV-infected mice exhibit extraneural virus amplification in the draining lymph nodes, the primary site for extraneural amplification of EEEV appears to be within osteoblasts in active growth areas of bones [96]. Interestingly, although viral antigen was found in the draining popliteal lymph node associated primarily with dendritic cells, weak staining by *in situ* hybridization indicated that the virus was not actively replicating at this site. Another difference from VEEV pathogenesis seen by these authors was in the site of initial infection of the CNS. In VEEV infections, the virus first infects the olfactory neuroepithelium before spreading to the brain through the olfactory tract and limbic structures [49]. In EEEV infections, the pattern appears to be more rapid, random, and multifocal, indicating a hematogenous route of CNS infection [96]. Neurons in many parts of the brain are infected in susceptible mice, and in particular, the neurons of the cerebral cortex [96,98]. Despite the numerous neurons infected, the neuropathological changes in EEE-infected mice were relatively minimal, lacking both vasculitis and perivascular cuffing [98].

12.6.2 Hamster Models of EEEV Infection

Virulent EEEV also causes lethal disease in Syrian hamsters, with infection of multiple peripheral organs as well as the brain [21,103]. The LD_{50} of a "parent" strain of EEEV was determined to be approximately 1 pfu when injected i.p. [21]. More recently, Paessler et al. [104] evaluated hamsters as a potential animal model for EEEV because they exhibit a vascular disease that more closely resembles human disease. The North American strain, 79-2138, isolated from mosquitoes in Massachusetts in 1979, was inoculated s.c. into 6- to 8-week-old female golden hamsters. Two to three days after infection with 10^3 pfu, hamsters exhibited vomiting, lethargy, and anorexia and were observed to be pressing their heads against the cage walls. Stupor, coma, and respiratory signs were observed over the course of days 4 and 5 after infection, with deaths occurring between days 5 and 6. A serial pathogenesis study conducted by this group indicated the presence of virus in the serum by day 1, which persisted for more than 3 days, with viremia peaking on day 2. Examination of the brain and visceral organs indicated the presence of virus in the brain, lungs, liver, kidneys, spleen, and muscle within 1 day of infection. Viral infection of the heart was observed by 2 days after infection. Viral titers in the visceral organs peaked between days 2 and 3, and clearance of the virus correlated with the production of neutralizing antibodies. However, virus was not cleared from the brain, in which titers continued to increase until death. Vasculitis and microhemorrhages were important features in the brains of infected golden hamsters, mimicking the histological findings in human EEE [92]. Vasculitis and associated hemorrhage were observed as early as 24 h after infection in the basal nuclei and brainstem, preceding the appearance of viral antigen in the brain, as well as apparent neuronal damage, neuronophagia, microgliosis, and infiltration of the brain parenchyma by macrophages and neutrophils. Other parts of the brain were affected later in the infection. Vasculitis also affected many extraneural tissues.

12.6.3 Other Small-Animal Models of EEEV Infection

EEEV causes disease with encephalitis when administered i.c., s.c., i.d., intramuscularly, or by aerosol in guinea pigs [90,101,105,106]. In a recent study in which guinea pigs were infected with either the NJ1959 (North American) or ArgM (South American) strain of EEEV by the aerosol route of infection, animals developed clinical signs, such as decreased activity and dorsal tremors, within 18–72 h [90]. Over the course of the next 3–4 days, these signs rapidly progressed to circling, recumbency, and coma with the mean time to death between 5 and 6 days. In follow-on serial sampling studies, virus was detected in the lungs, blood, and liver by 24 h PI. By day 3, the virus was at much lower titer levels in these tissues or was not detected. In contrast, virus titers in the brain rapidly increased over the first 4 days of infection. Concomitant pathological data showed virus in the olfactory neuroepithelium and olfactory bulbs by 1–2 days PI with transneuronal spread to all regions of the brain by day 4. Neurons appeared to be the primary target for virus infection and the main microscopic lesions in the brain were neuronal necrosis, vasculitis, and inflammation of the meninges and neuropil.

Rabbits show lesions similar to guinea pigs, though with less severe neutrophilic infiltration and apparently greater microgliosis [106]. Male Fisher-Dunning rats (250–275 g) were infected s.c. with 10^6 pfu of EEEV strain Arth 167, which did not produce a lethal disease and was not further evaluated [20].

12.6.4 NHP Models of EEEV Infection

Much of the early work with EEEV in NHPs was done in conjunction with studies with WEEV. Hurst [107] demonstrated that EEEV or WEEV injected i.c. into rhesus macaques was invariably fatal, whereas parenteral inoculation resulted in either inapparent infection or fatal encephalitis. Febrile responses were good predictors of subsequent clinical signs of encephalitis, which included tremors, rigidity of the neck, general weakness or loss of muscle control, and greater than normal salivation. Although animals were viremic early after infection, by the time the febrile response had waned and neurological signs had begun, the virus could no longer be isolated from the blood [108].

In 1939, Wyckoff and Tesar [109] published the findings from rhesus macaques inoculated with EEEV or WEEV via a variety of routes. Similar to Hurst [107], the authors found that i.c. inoculation was invariably fatal, whereas challenge by other routes was highly variable. No disease was seen in macaques infected s.c., i.v., or intraocularly. For EEEV, fever was seen 3 days after i.n. inoculation, and death occurred 2–3 days after fever began. As with the prior reports, neurological signs were not seen until after the fever had peaked. Neurological signs reported include shivering, convulsions, lethargy, paralysis, and coma. None of the macaques that developed neurological signs recovered from the disease. In contrast to what was reported by Hurst [107], however, several macaques did develop a fever but did not progress to show signs of encephalitis, and they subsequently recovered. In more recent studies, cynomolgus macaques exposed by aerosol to EEEV developed a fever and were moribund within 48–72 h of fever onset [109]. Clinical signs included development of neurological signs as the fever waned, as well as a prominent leukocytosis.

A recent natural history study reported the development of an aerosol model of EEEV infection based on the cynomolgus macaque with neurological signs consistent with human cases [89]. This study defined a 10^7 pfu dose of the FL91-4679 strain of EEEV as a suitable aerosol challenge dose for vaccine efficacy studies. Prominent clinical signs of disease were fever, a dramatic granulocytosis, and an increase in alkaline phosphatase. Within 5–9 days after infection, macaques had developed tremors, become comatose, and were promptly euthanized. Viremia during the course of infection in these animals was either transient or undetectable. Gross necropsy results were unremarkable and the pathology reports from this study focused on key features in the brains of these macaques [91]. EEEV antigen was widespread, but was especially present in neurons of the corpus striatum, pons, frontal cortex, mesencephalon, thalamus, medulla oblongata regions of the brain. Specific histological features in these regions included widespread neuronal necrosis, perivascular cuffing, and inflammatory cell infiltration. Vasculitis was prominently featured in the cerebral blood vessels, as was areas of edema and hemorrhage. Measurement of viral titers found the presence of EEEV in all tissues that were examined; however, significantly higher concentrations were found in the CNS and in particular the diencephalon region of the brain where viral titers peaked at 7.5×10^8 pfu/g (Figure 12.6).

As an alternative to traditional NHP species, Adams et al. [87] identified the common marmoset (*Callithrix jacchus*) as a useful model of human EEE. In this study, marmosets received 10^6 pfu of the FL93-393 North American strain of EEEV intranasally [87]. By day 1–2 PI marmosets became anorexic, and by the time of

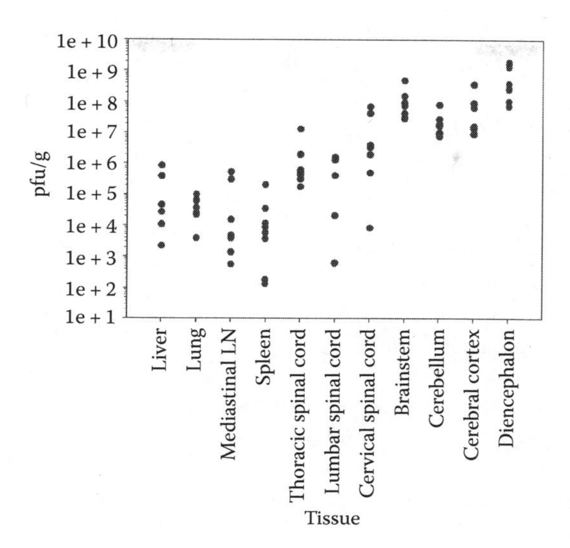

FIGURE 12.6 High virus titers in the brain and central nervous system (CNS) of cynomolgus macaques that succumbed to eastern equine encephalitis virus (EEEV) infection. Tissue samples were taken at necropsy and homogenized in phosphate-buffered saline, diluted and assessed for the presence of virus by plaque assay using Vero cells. The graph shows virus titers (pfu per gram of tissue) for peripheral tissues as well as different regions of the CNS and brain of 10 naive macaques that succumbed to EEEV infection after aerosol exposure.

death or euthanasia (day 4–5) they had developed fever, leukocytosis, and dramatic weight loss. Tremors and coma were not reported but findings such as inactivity, somnolence, excessive or absent eyelid movement, and depressed posture appeared to be leading neurological signs. There were no gross pathological lesions noted upon necropsy, but histopathological findings showed moderate meningoencephalitis with perivascular hemorrhage in the cerebral cortex. Although, EEEV antigen was detected in the brain, liver, and muscle of the marmosets at the time of death, none of the marmosets develop detectable viremia during the course of disease.

Another recent study described using owl monkeys (*Aotus nancymaae*) given 10^4 pfu of the FL93-393 strain of EEEV by either the i.n. or s.c. route of infection [88]. Although none of the animals displayed clinical signs of disease, those inoculated s.c. did develop viremia within 24 h PI which lasted on average for 3.3 days. On day 6, two animals from each group were euthanatized for histological examination, but tissues showed no significant gross or histological findings. Taken together, this study suggests the potential use of this species of New World monkey as an NHP model, especially if higher challenge doses of EEEV induce similar clinical and pathological signs to those seen in other NHP species.

12.6.5 Comparison of EEE Models

The recent report that golden hamsters infected with EEEV uniformly exhibit neurotropism, encephalopathy, and fatal outcome with pathological features more similar to the human disease than other animals seems to make the hamster a valuable animal model, especially for pathogenesis studies [104]. In particular, the early onset of vascular damage in the brains of hamsters before direct viral damage of neurons makes it potentially useful for exploring the basis of similar findings in humans. The uniform susceptibility of hamsters to peripheral infection also makes this species a good candidate for vaccine protection studies. The small size and relative economy of using hamsters are benefits surpassed only by using mice, which develop resistance to peripheral EEEV infection at a young age and do not manifest the vasculopathy of hamsters and humans. The pathological consequences of EEEV in NHPs have been insufficiently studied to comment on how appropriate this model may be for neuropathogenesis studies. As with VEEV, the dose required to reproducibly cause disease in macaques should be considered with caution regarding the utility of this model. Nonetheless, mice and macaques, despite their resistance to peripheral infection, will likely continue to be used for EEE studies involving intracerebral or aerosol infection. Additional studies with novel mouse strains could identify an animal model more useful for EEE research than the older studies have indicated. Other animal models would also benefit from additional investigation.

12.7 WESTERN EQUINE ENCEPHALITIS VIRUS

As with EEEV, recent animal studies using WEEV are limited. Studies with WEEV, the "virus of equine encephalomyelitis from California," in the 1930s tested the susceptibility of rabbits, guinea pigs, mice, rats, and NHPs, detecting WEEV in the brains of each of those animal models.

12.7.1 Mouse Models of WEEV Infection

Since the first publication of this chapter, there have been a number of studies done to develop mouse models of WEE that are applicable to biodefense research [110–112]. Logue et al. [110] compared virulence variations among six North American strains (McMillan, BFS-2005, Imperial 181, 85-452NM, 71V1658, Montana-64) in the Arbovirus Reference Collection at the Centers for Disease Control and Prevention, Fort Collins, Colorado, USA, in 6- to 8-week-old CD1 mice using the s.c. route of infection. Out of these strains, the McMillan strain was found to be the most virulent—an s.c. inoculation of 10^3 pfu was found to be consistently lethal, as were other routes of infection (aerosol—10^5 pfu, i.n.—10^5 pfu, and i.c.—10^3 pfu). The mean survival time from these different routes of infection ranged from 1.9 days in mice infected by the i.c. route to 4.0–4.6 days in mice infected by the s.c., aerosol, or i.n. routes. In another comparative study using 7- to 20-week-old female BALB/c mice, Nagata et al. [111] assessed the virulence of eight strains of WEEV given at 1.5×10^3 pfu by the i.n. route. The survival times from these viruses ranged from 4–6 days for the McMillan, Fleming, and California strains, to 7–8 days for the CBA-87 strain, to 7–12 days for the B11, 71V-1658, Mn520, and Mn548 strains. A separate group of investigators compared the McMillan strain to two other North American strains, CO92 and TBT235, in 9-week-old female NIH Swiss mice, and found similar outcomes; the McMillan strain was virulent by either the i.n. or i.p.

route of infection (mean survival time of 3.2 and 4.6 days, respectively), but the other viruses were virulent by only the i.n. route of infection [113]. In our own studies, probit analysis with a South American strain of WEEV, CBA-87, indicated an LD_{50} of 40 pfu in the BALB/c and C57BL/6 mouse strains when the virus was administered by aerosol (M.K. Hart and R.F. Tammariello, unpublished data, 2005). Early signs of disease can include a rough coat, a hunched appearance, lethargy, decreased peer interaction, and weight loss, which rapidly progress to signs of ataxia, paresis, and/or paralysis. Viremia inconsistently appears in the mouse [114] and does not appear to correlate with mortality [110]. In the CD1 mouse model developed by Logue et al. [110] (McMillan strain of WEEV, 6- to 8-week-old mice, infected s.c. or by aerosol), moderate levels of virus (10^3–10^4 pfu/g) were detected in the brains of sampled animals by 24 h, and by 72–96 h high levels of virus (10^5–10^8 pfu/g) in the brains were achieved [110]. Histological lesions in the brain consisted of laminar or multifocal areas of neuronal necrosis and edema, which were randomly distributed but were most prominent in the forebrain areas. Other lesions included edema in perivascular spaces, and marginated lymphocytes, reactive endothelium, and apoptotic nuclei in the small blood vessels of the brain.

Younger mice can have a different target-organ pathology profile [115]. In a study comparing 1- to 2-day-old suckling mice to 3-week-old Swiss mice infected s.c. with the B-629 strain of WEEV found that the suckling mice developed a rapidly progressive disease and died within 48 h. The pathological changes exhibited in these mice were largely limited to the skeletal muscle, bone marrow, cartilage, and other mesodermal tissues, but did not involve the brain. Conversely, the 3-week-old mice developed the typical lesions consisting of widespread meningoencephalitis with necrosis and perivascular lymphoid cuffing, consistent with other adult mouse models [110]. Intramuscular injection of 15-day-old mice with WEEV produced encephalitic disease, without evidence of spread via local nerves, in 80–90% of the examined mice [101]. The remaining 10–20% of the mice had signs of flaccid paralysis with evidence of progression by local nerves, but not by diffuse hematoencephalic spread. WEEV appeared to traverse some blood vessels in the mice at this age to gain access to the CNS. The virus appeared to be unable to traverse these vessels in 3-week-old mice but did seem to progress via the nerves in the inoculated muscle, with 70–80% of mice exhibiting flaccid paralysis, 10–20% developing encephalitis, and a few mice with no signs of CNS illness [101,105]. After 1 month of age, the local nerves of mice were generally resistant to WEEV as well.

12.7.2 Hamster Models of WEEV Infection

Golden Syrian hamsters are susceptible to lethal infection from WEEV strain, B-11, by the i.p. route, and have been used in potency assays for WEE vaccine [116] and in cross-protection studies with VEEV and EEEV [117]. In a study with an unidentified strain of WEEV, hamsters infected by the i.c., i.p., i.d., i.n., or respiratory routes exhibited lethal disease characterized by incoordination, shivering, tachypnea, conjunctivitis, and collapse [118]. Survival times ranged from 4–5 days (i.c., i.n., respiratory infections) to 6 days (i.p., i.d. infections). Regardless of the route of infection, virus was detected in the brain by 24 h PI, and peaked at titers of 10^8–10^9 pfu/g of brain tissue by 72h. Brain lesions, such as neuronal necrosis, perivascular cuffing, hemorrhage, astrocytosis, and microgliosis, were found in most major areas of the brain. A recent study developed a hamster model using 100–110 g female Syrian golden hamsters with the VR-70 strain of WEEV given by the i.p. or i.n. route of infection [119]. Average survival times for i.n. infection was 3.0–3.4 days and for i.p. infection was 3.0–5.0 days and was dose dependent with lower doses tending to lengthen survival time. The virus was isolated from the serum, spleen, liver, and brain, with serum titers (10^4–10^5 pfu/mL) peaking by day 2 PI and titers from brain tissue (10^9 pfu/g) peaking by day 4 PI.

12.7.3 Other Small-Animal Models of WEEV Infection

The guinea pig (300 g) became the most widely used animal model for WEE in early studies. Young guinea pigs were susceptible to WEEV infection by the i.c., i.n., i.p., s.c., and i.v. routes [105,120]. A difference in susceptibility of WEEV infection seen between the oral feeding by stomach tube and the intranasal distillation led to the suggestion that WEEV infected the nasal mucosa [120]. Age appears to play a role in susceptibility, as older guinea pigs (350 g) develop resistance to peripheral infection with WEEV [105,120]. The disease course in guinea pigs has an early febrile period in which virus is detected in the serum, which later progresses to a period of prostration, during which the fever declines and virus is no longer detected in the serum [120]. The typical disease course has a normal temperature for the first day of disease, a slight increase on the second or third day, and fever peaking on the third or fourth day. Temperature then drops as the animal develops other symptoms and becomes prostrate, dying between days 4 and 6 [120]. Throughout the early stage of disease, the virus is detected in the spleen, liver, kidneys, and salivary glands but not in the saliva, urine, or fecal material of infected guinea pigs. In the late stage of the disease, the virus persists in the salivary glands and appears in the adrenals [120]. As described previously in this chapter, histopathological findings in guinea pigs infected with WEEV by various routes are reported to be qualitatively similar to those in guinea pigs infected with EEEV [106]. Male Fisher-Dunning rats (250–275 g) were infected s.c. with 10^6 pfu of WEEV 72V4768, which did not produce a lethal disease and was not further evaluated [20].

12.7.4 NHP Models of WEEV Infection

Similar to EEEV, there are few recent data on WEEV infection of macaques. Most of the existing data are from studies done in the 1930s, looking at the disease caused by injection of WEEV by a variety of routes in rhesus macaques. In 1932, Howitt [121] described the result of i.n. and i.c. inoculation of WEEV. By either route, the macaques became viremic. Although the i.n. inoculated macaque never demonstrated any other signs of disease, the intracerebrally injected macaque developed a fever and was prostrate by day 7 after inoculation. In his study of EEEV and WEEV, Hurst [107] demonstrated that WEEV injected i.c. into rhesus macaques was invariably fatal, whereas parenteral inoculation resulted in either inapparent infection or fatal encephalitis. Hurst [107] also observed that, although the onset of fever and general disease course was more protracted with WEEV, the outcome was the same as inoculation with EEEV when fever appeared. The nine macaques that survived parenteral infection with WEEV were subsequently challenged by i.c. inoculation, and only three survived, indicating that the parenteral inoculation did not induce adaptive immune responses. As was mentioned earlier in this chapter, Wyckoff and Tesar [109] published the findings of inoculating rhesus macaques with WEEV by a variety of routes. Similar to their findings with EEEV and with Hurst's findings, i.c. or i.n. inoculation was fatal, whereas challenge by other routes usually resulted in few to no signs of disease. Onset of fever and disease course were more protracted with WEEV than was seen with EEEV. Fever was usually associated with neurological signs, including shivering, convulsions, lethargy, paralysis, and coma, and was more protracted with death between days 9 and 11. Macaques that developed signs indicating encephalitis did not recover.

In work done more recently, both rhesus and cynomolgus macaques were found to be susceptible to aerosolized WEEV (Figure 12.7a and b). Radiotelemetry indicated that fever onset does not begin until 4–5 days after infection [120]. An increase in heart rate was seen after exposure that roughly corresponded with the onset of fever. In fatal cases, the fever peaked and then declined; however, the heart rate remained high until animals were moribund. In agreement with the prior studies, neurological signs began as the fever peaked and gradually worsened as the fever waned. In contrast to the prior studies done by Hurst [107] or Wyckoff and Tesar [109], not all macaques that developed fever and signs indicative of neurological involvement succumbed to infection. Surprisingly, no virus was isolated from the blood at any point after exposure in either species of macaque. Viral isolation and histological examination of macaques that succumbed to infection found evidence of viral infection only in tissues of the CNS [120]. Unlike what was reported in macaques exposed to VEEV, lymphopenia was not seen after exposure to WEEV. More commonly seen was a general increase in leukocytes after exposure, particularly monocytes and segmented neutrophils. The severity of the increase in granulocytes corresponded with outcome, similar to what was seen with EEEV in cynomolgus macaques [89]. Analysis of serum samples found that elevations in serum glucose levels correlated with the severity of both the neurological signs and the febrile response.

12.8 SUMMARY

Viruses causing equine encephalitis consist of three antigenically and phylogenetically distinct but related viruses within the genus *Alphavirus*. These viruses, VEEV, EEEV, and WEEV, are cycled through nature by mosquito vectors and cause periodic epizootics in equines and occasional severe and lethal infections in humans. These viruses are also highly infectious as aerosols and are considered biological threat agents. In this review of animal studies with these viruses, it should be clear that there are major gaps in our knowledge and understanding of the animal models that are pertinent to the aerosol nature of the biological threat from EEEV and WEEV, in particular. It is our hope to stimulate interest in these areas of research and to give insight into how to proceed by fully describing the applicable efforts made for VEEV.

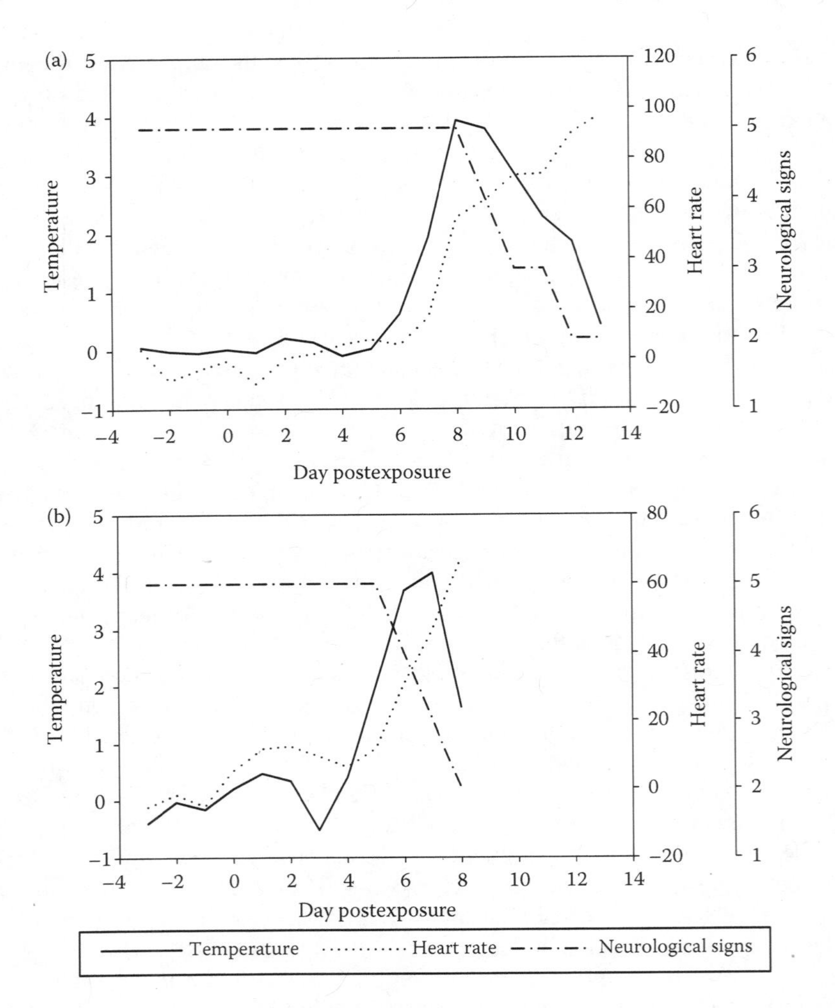

FIGURE 12.7 Cynomolgus (a) and rhesus (b) macaques develop fever and encephalitis after aerosol exposure to western equine encephalitis virus (WEEV) (Adapted from Reed, D.S. et al., J. Infect Dis, 2005. **192**(7): 1173–1182.). Macaques with telemetry implants for recording body temperature and heart rate were aerosol exposed to WEEV. Temperature and heart rate values shown are the average daily residual values obtained by subtracting predicted from actual values. Neurological signs were assessed daily by the principal investigator, technicians, and animal caretakers and were scored as follows: 5 = normal, 4 = depression, 3 = occasional tremors/seizures, 2 = frequent tremors/seizures, and 1 = comatose.

REFERENCES

1. Weaver, S.C. et al., Venezuelan equine encephalitis. *Annu Rev Entomol,* **49**: 141–174.
2. Chosewood, L.C. and D.E. Wilson, eds. *Biosafety in Microbiological and Biomedical Laboratories*, 5th ed. Washington, DC: U.S Government Printing Office, 2007.

3. National Institute of Allergy and Infectious Diseases (NIAID). *NIAID Biodefense Research Agenda for Category B and C Priority Pathogens*. U.S. Department of Health and Human Services, National Institutes of Health, National Institute of Allergy and Infectious Diseases, NIH Publication No. 03-5315, January 2003.
4. Steele, K. et al., Alphavirus encephalitides, in *Medical Aspects of Biological Warfare*, Z.F. Dembeck, ed. Washington, DC: U.S. Government Printing Office, 2007, pp. 241–270.
5. U.S. Food and Drug Administration (FDA), New drug and biological drug products; Evidence needed to demonstrate effectiveness of new drugs when human efficacy studies are not ethical or feasible, Final Rule (Animal Rule): 67 FR 37988 [May 31, 2002]. *Fed Regist*, 2002, **67**(105): 37988–37998.
6. Zhang, W. et al., Placement of the structural proteins in Sindbis virus. *J Virol*, 2002, **76**(22): 11645–11658.
7. Grieder, F.B. and H.T. Nguyen, Virulent and attenuated mutant Venezuelan equine encephalitis virus show marked differences in replication in infection in murine macrophages. *Microb Pathog*, 1996, **21**(2): 85–95.
8. Roehrig, J.T., J.W. Day, and R.M. Kinney, Antigenic analysis of the surface glycoproteins of a Venezuelan equine encephalomyelitis virus (TC-83) using monoclonal antibodies. *Virology*, 1982, **118**(2): 269–278.
9. White, J., M. Kielian, and A. Helenius, Membrane fusion proteins of enveloped animal viruses. *Q Rev Biophys*, 1983, **16**(2): 151–195.
10. Paredes, A.M. et al., Conformational changes in Sindbis virions resulting from exposure to low pH and interactions with cells suggest that cell penetration may occur at the cell surface in the absence of membrane fusion. *Virology*, 2004, **324**(2): 373–386.
11. Simons, K. and H. Garoff, The budding mechanisms of enveloped animal viruses. *J Gen Virol*, 1980, **50**(1): 1–21.
12. Gliedman, J. B., J.F. Smith, and D.T. Brown, Morphogenesis of Sindbis virus in cultured Aedes albopictus cells. *J Virol*, 1975, **16**(4): 913–926.
13. Weaver, S.C. et al., Family *Togaviridae*, in *Virus Taxonomy: Eighth Report of the International Committee on Taxonomy of Viruses*, C.M. Fauquet et al., eds. San Diego, CA: Elsevier Academic Press, 2005, pp. 645–653.
14. Arrigo, N.C., A.P. Adams, and S.C. Weaver, Evolutionary patterns of eastern equine encephalitis virus in North versus South America suggest ecological differences and taxonomic revision. *J Virol*, 2010, **83**(2): 1014–1025.
15. Mullbacher, A. and R.V. Blanden, H-2-linked control of cytotoxic T-cell responsiveness to alphavirus infection. Presence of H-2Dk during differentiation and stimulation converts stem cells of low responder genotype to T cells of responder phenotype. *J Exp Med*, 1979, **149**(3): 786–790.
16. Griffin, D.E. and R.T. Johnson, Role of the immune response in recovery from Sindbis virus encephalitis in mice. *J Immunol*, 1977, **118**(3): 1070–1075.
17. Jones, L.D. et al., Cytotoxic T-cell activity is not detectable in Venezuelan equine encephalitis virus-infected mice. *Virus Res*, 2003, **91**(2): 255–259.
18. Mathews, J.H. and J.T. Roehrig, Determination of the protective epitopes on the glycoproteins of Venezuelan equine encephalomyelitis virus by passive transfer of monoclonal antibodies. *J Immunol*, 1982, **129**(6): 2763–2767.
19. Phillpotts, R.J., L.D. Jones, and S.C. Howard, Monoclonal antibody protects mice against infection and disease when given either before or up to 24 h after airborne challenge with virulent Venezuelan equine encephalitis virus. *Vaccine*, 2002, **20**(11–12): 1497–1504.
20. Jahrling, P.B., A. DePaoli, and M.C. Powanda, Pathogenesis of a Venezuelan encephalitis virus strain lethal for adult white rats. *J Med Virol*, 1978, **2**(2): 109–116.

21. Brown, A. and J.E. Officer, An attenuated variant of eastern encephalitis virus: Biological properties and protection induced in mice. *Arch Virol*, 1975, **47**(2): 123–138.
22. Kissing, R.E. et al., Venezuelan equine encephalomyelitis virus in horses. *Am J Hyg*, 1956, **63**: 274–287.
23. Jackson, A.C., S.K. SenGupta, and J.F. Smith, Pathogenesis of Venezuelan equine encephalitis virus infection in mice and hamsters. *Vet Pathol*, 1991, **28**(5): 410–418.
24. Victor, J., D.G. Smith, and A.D. Pollack, The comparative pathology of Venezuelan equine encephalomyelitis. *J Infect Dis*, 1956, **98**: 55–66.
25. Gleiser, C.A. et al., The comparative pathology of experimental Venezuelan equine encephalomyelitis infection in different animal hosts. *J Infect Dis*, 1962, **110**: 80–97.
26. Rivas, F. et al., Epidemic Venezuelan equine encephalitis in La Guajira, Colombia, 1995. *J Infect Dis*, 1997, **175**(4): 828–832.
27. Weaver, S.C. et al., Re-emergence of epidemic Venezuelan equine encephalomyelitis in South America. *VEE Study Group. Lancet*, 1996, **348**(9025): 436–440.
28. Watts, D.M. et al., Venezuelan equine encephalitis febrile cases among humans in the Peruvian Amazon river region. *Am J Trop Med Hyg*, 1998, **58**(1): 35–40.
29. Bowen, G.S. et al., Clinical aspects of human Venezuelan equine encephalitis in Texas. *Bull Pan Am Health Organ*, 1976, **10**(1): 46–57.
30. de la Monte, S. et al., The systemic pathology of Venezuelan equine encephalitis virus infection in humans. *Am J Trop Med Hyg*, 1985, **34**(1): 194–202.
31. Hart, M.K. et al., Venezuelan equine encephalitis virus vaccines induce mucosal IgA responses and protection from airborne infection in BALB/c, but not C3H/HeN mice. *Vaccine*, 1997, **15**(4): 363–369.
32. Ludwig, G.V. et al., Comparative neurovirulence of attenuated and non-attenuated strains of Venezuelan equine encephalitis virus in mice. *Am J Trop Med Hyg*, 2001, **64**(1–2): 49–55.
33. Phillpotts, R.J. and A.J. Wright, TC-83 vaccine protects against airborne or subcutaneous challenge with heterologous mouse-virulent strains of Venezuelan equine encephalitis virus. *Vaccine*, 1999, **17**(7–8): 982–988.
34. Pratt, W.D. et al., Genetically engineered, live attenuated vaccines for Venezuelan equine encephalitis: Testing in animal models. *Vaccine*, 2003, **21**(25–26): 3854–3862.
35. Hart, M.K. et al., Improved mucosal protection against Venezuelan equine encephalitis virus is induced by the molecularly defined, live-attenuated V3526 vaccine candidate. *Vaccine*, 2000, **18**(26): 3067–3075.
36. Paessler, S. et al., Recombinant sindbis/Venezuelan equine encephalitis virus is highly attenuated and immunogenic. *J Virol*, 2003, **77**(17): 9278–9286.
37. Paessler, S. et al., Replication and clearance of Venezuelan equine encephalitis virus from the brains of animals vaccinated with chimeric SIN/VEE viruses. *J Virol*, 2006, **80**(6): 2784–2796.
38. Paessler, S. et al., Alpha-beta T cells provide protection against lethal encephalitis in the murine model of VEEV infection. *Virology*, 2007, **367**(2): 307–323.
39. Martin, S.S. et al., Evaluation of formalin inactivated V3526 virus with adjuvant as a next generation vaccine candidate for Venezuelan equine encephalitis virus. *Vaccine*, 2010, **28**(18): 3143–3151.
40. Martin, S.S. et al., Comparison of the immunological responses and efficacy of gamma-irradiated V3526 vaccine formulations against subcutaneous and aerosol challenge with Venezuelan equine encephalitis virus subtype IAB. *Vaccine*, 2010, **28**(4): 1031–1040.
41. Dupuy, L.C. et al., Directed molecular evolution improves the immunogenicity and protective efficacy of a Venezuelan equine encephalitis virus DNA vaccine. *Vaccine*, 2009, **27**(31): 4152–4160.

42. Perkins, S.D., L.M. O'Brien, and R.J. Phillpotts, Boosting with an adenovirus-based vaccine improves protective efficacy against Venezuelan equine encephalitis virus following DNA vaccination. *Vaccine*, 2006, **24**(17): 3440–3445.
43. Schmaljohn, A.L. et al., Non-neutralizing monoclonal antibodies can prevent lethal alphavirus encephalitis. *Nature*, 1982, **297**(5861): 70–72.
44. Steele, K.E. et al., Comparative neurovirulence and tissue tropism of wild-type and attenuated strains of Venezuelan equine encephalitis virus administered by aerosol in C3H/HeN and BALB/c mice. *Vet Pathol*, 1998, **35**(5): 386–397.
45. MacDonald, G.H. and R.E. Johnston, Role of dendritic cell targeting in Venezuelan equine encephalitis virus pathogenesis. *J Virol*, 2000, **74**(2): 914–922.
46. Davis, N.L. et al., A molecular genetic approach to the study of Venezuelan equine encephalitis virus pathogenesis. *Arch Virol Suppl*, 1994, **9**: 99–109.
47. Grieder, F.B. et al., Specific restrictions in the progression of Venezuelan equine encephalitis virus-induced disease resulting from single amino acid changes in the glycoproteins. *Virology*, 1995, **206**(2): 994–1006.
48. Vogel, P. et al., Venezuelan equine encephalitis in BALB/c mice: kinetic analysis of central nervous system infection following aerosol or subcutaneous inoculation. *Arch Pathol Lab Med*, 1996, **120**(2): 164–172.
49. Charles, P.C. et al., Mechanism of neuroinvasion of Venezuelan equine encephalitis virus in the mouse. *Virology*, 1995, **208**(2): 662–671.
50. Ryzhikov, A.B. et al., Spread of Venezuelan equine encephalitis virus in mice olfactory tract. *Arch Virol*, 1995, **140**(12): 2243–2254.
51. Ryzhikov, A.B. et al., Venezuelan equine encephalitis virus propagation in the olfactory tract of normal and immunized mice. *Biomed Sci*, 1991, **2**(6): 607–614.
52. Morrison, E.E. and R.M. Costanzo, Morphology of the human olfactory epithelium. *J Comp Neurol*, 1990, **297**(1): 1–13.
53. Schoneboom, B.A. et al., Inflammation is a component of neurodegeneration in response to Venezuelan equine encephalitis virus infection in mice. *J Neuroimmunol*, 2000, **109**(2): 132–146.
54. Griffin, D.E., The Gordon Wilson lecture: Unique interactions between viruses, neurons and the immune system. *Trans Am Clin Climatol Assoc*, 1995, **107**: 89–98.
55. Griffin, D.E. et al., Age-dependent susceptibility to fatal encephalitis: Alphavirus infection of neurons. *Arch Virol Suppl*, 1994, **9**: 31–39.
56. Lewis, J. et al., Alphavirus-induced apoptosis in mouse brains correlates with neurovirulence. *J Virol*, 1996, **70**(3): 1828–1835.
57. Ubol, S. et al., Neurovirulent strains of alphavirus induce apoptosis in BCL-2-expressing cells: role of a single amino acid change in the E2 glycoprotein. *Proc Natl Acad Sci USA*, 1994, **91**(11): 5202–5206.
58. Jackson, A.C. and J.P. Rossiter, Apoptotic cell death is an important cause of neuronal injury in experimental Venezuelan equine encephalitis virus infection of mice. *Acta Neuropathol (Berl)*, 1997, **93**(4): 349–353.
59. Woodman, D.R., A.T. McManus, and G.A. Eddy, Extension of the mean time to death of mice with a lethal infection of Venezuelan equine encephalomyelitis virus by antithymocyte serum treatment. *Infect Immun*, 1975, **12**(5): 1006–1011.
60. Charles, P.C. et al., Immunopathogenesis and immune modulation of Venezuelan equine encephalitis virus-induced disease in the mouse. *Virology*, 2001, **284**(2): 190–202.
61. Schoneboom, B.A. et al., Astrocytes as targets for Venezuelan equine encephalitis virus infection. *J Neurovirol*, 1999, **5**(4): 342–354.
62. Pratt, W.D. et al., Use of telemetry to assess vaccine-induced protection against parenteral and aerosol infections of Venezuelan equine encephalitis virus in non-human primates. *Vaccine*, 1998, **16**(9–10): 1056–1064.

63. Martin, S.S. et al., Telemetric analysis to detect febrile responses in mice following vaccination with a live-attenuated virus vaccine. *Vaccine*, 2009, **27**(49): 6814–6823.
64. Reed, D.S. et al., Genetically engineered, live, attenuated vaccines protect nonhuman primates against aerosol challenge with a virulent IE strain of Venezuelan equine encephalitis virus. *Vaccine*, 2005, **23**(24): 3139–3147.
65. Fine, D.L. et al., Venezuelan equine encephalitis virus vaccine candidate (V3526) safety, immunogenicity and efficacy in horses. *Vaccine*, 2007, **25**(10): 1868–1876.
66. Walker, D.H. et al., Lymphoreticular and myeloid pathogenesis of Venezuelan equine encephalitis in hamsters. *Am J Pathol*, 1976, **84**(2): 351–370.
67. Austin, F.J. and W.F. Scherer, Studies of viral virulence. I. Growth and histopathology of virulent and attenuated strains of Venezuelan encephalitis virus in hamsters. *Am J Pathol*, 1971, **62**(2): 195–210.
68. Jahrling, P.B. and F. Scherer, Histopathology and distribution of viral antigens in hamsters infected with virulent and benign Venezuelan encephalitis viruses. *Am J Pathol*, 1973, **72**(1): 25–38.
69. Dill, G.S., Jr., C.E. Pederson, Jr., and J.L. Stookey, A comparison of the tissue lesions produced in adult hamsters by two strains of avirulent Venezuelan equine encephalomyelitis virus. *Am J Pathol*, 1973, **72**(1): 13–24.
70. Gorelkin, L. and P.B. Jahrling, Virus-initiated septic shock. Acute death of Venezuelan encephalitis virus-infected hamsters. *Lab Invest*, 1975, **32**(1): 78–85.
71. Kundin, W.D., C. Liu, and P. Rodina, Pathogenesis of Venezuelan equine encephalomyelitis virus. I. Infection in suckling mice. *J Immunol*, 1966, **96**(1): 39–48.
72. Hruskova, J. et al., Subcutaneous and inhalation infection of guinea pigs with Venezuelan equine encephalomyelitis virus. *Acta Virol*, 1969, **13**(5): 415–421.
73. Gorelkin, L. and P.B. Jahrling, Pancreatic involvement by Venezuelan equine encephalomyelitis virus in the hamster. *Am J Pathol*, 1974, **75**(2): 349–362.
74. Scherer, W.F. and J. Chin, Responses of guinea pigs to infections with strains of Venezuelan encephalitis virus, and correlations with equine virulence. *Am J Trop Med Hyg*, 1977, **26**(2): 307–312.
75. Powers, A.M. et al., The use of chimeric Venezuelan equine encephalitis viruses as an approach for the molecular identification of natural virulence determinants. *J Virol*, 2000, **74**(9): 4258–4263.
76. Greene, I.P. et al., Venezuelan equine encephalitis virus in the guinea pig model: Evidence for epizootic virulence determinants outside the E2 envelope glycoprotein gene. *Am J Trop Med Hyg*, 2005, **72**(3): 330–338.
77. Monath, T.P. et al., Experimental studies of rhesus monkeys infected with epizootic and enzootic subtypes of Venezuelan equine encephalitis virus. *J Infect Dis*, 1974, **129**(2): 194–200.
78. Monath, T.P. et al., Recombinant vaccinia—Venezuelan equine encephalomyelitis (VEE) vaccine protects nonhuman primates against parenteral and intranasal challenge with virulent VEE virus. *Vaccine Res*, 1992, **1**: 55–68.
79. Reed, D.S. et al., Aerosol infection of cynomolgus macaques with enzootic strains of venezuelan equine encephalitis viruses. *J Infect Dis*, 2004, **189**(6): 1013–1017.
80. Hammamieh, R. et al., Blood genomic profiles of exposures to Venezuelan equine encephalitis in cynomolgus macaques (*Macaca fascicularis*). *Virol J*, 2007, **4**: 82.
81. Koterski, J. et al., Gene expression profiling of nonhuman primates exposed to aerosolized Venezuelan equine encephalitis virus. *FEMS Immunol Med Microbiol*, 2007, **51**(3): 462–472.
82. Verlinde, J.D., Susceptibility of cynomolgus monkeys to experimental infection with arboviruses of group A (Mayaro and Mucambo), group C (Oriboca and Restan) and an unidentified arbovirus (Kwatta) originating from Surinam. *Trop Geogr Med*, 1968, **20**(4): 385–390.

83. Danes, L. et al., The role of the olfactory route on infection of the respiratory tract with Venezuelan equine encephalomyelitis virus in normal and operated *Macaca* rhesus monkeys. I. Results of virological examination. *Acta Virol*, 1973, **17**(1): 50–56.
84. Danes, L. et al., Penetration of Venezuelan equine encephalomyelitis virus into the brain of guinea pigs and rabbits after intranasal infection. *Acta Virol*, 1973, **17**(2): 138–146.
85. Davis, N.L. et al., Attenuated mutants of Venezuelan equine encephalitis virus containing lethal mutations in the PE2 cleavage signal combined with a second-site suppressor mutation in E1. *Virology*, 1995, **212**(1): 102–110.
86. Hart, M.K. et al., Onset and duration of protective immunity to IA/IB and IE strains of Venezuelan equine encephalitis virus in vaccinated mice. *Vaccine*, 2001, **20**(3–4): 616–622.
87. Adams, A.P. et al., Common marmosets (*Callithrix jacchus*) as a nonhuman primate model to assess the virulence of eastern equine encephalitis virus strains. *J Virol*, 2008, **82**(18): 9035–9042.
88. Espinosa, B.J. et al., Susceptibility of the *Aotus nancymaae* owl monkey to eastern equine encephalitis. *Vaccine*, 2009, **27**(11): 1729–1734.
89. Reed, D.S. et al., Severe encephalitis in cynomolgus macaques exposed to aerosolized eastern equine encephalitis virus. *J Infect Dis*, 2007, **196**(3): 441–450.
90. Roy, C.J. et al., Pathogenesis of aerosolized eastern equine encephalitis virus infection in guinea pigs. *Virol J*, 2009, **6**: 170.
91. Steele, K.E. and N.A. Twenhafel, Review paper: Pathology of animal models of alphavirus encephalitis. *Vet Pathol*, 2010, **47**(5): 790–805.
92. Deresiewicz, R.L. et al., Clinical and neuroradiographic manifestations of eastern equine encephalitis. *N Engl J Med*, 1997, **336**(26): 1867–1874.
93. Farber, S.e.a., Encephalitis in infants and children caused by the virus of the eastern variety of equine encephalitis. *J Am Med Assoc*, 1940, **114**: 1725–1731.
94. Bastian, F.O. et al., Eastern equine encephalomyelitis. Histopathologic and ultrastructural changes with isolation of the virus in a human case. *Am J Clin Pathol*, 1975, **64**(1): 10–13.
95. Gardner, C.L. et al., Eastern and Venezuelan equine encephalitis viruses differ in their ability to infect dendritic cells and macrophages: Impact of altered cell tropism on pathogenesis. *J Virol*, 2008, **82**(21): 10634–10646.
96. Vogel, P. et al., Early events in the pathogenesis of eastern equine encephalitis virus in mice. *Am J Pathol*, 2005, **166**(1): 159–171.
97. Murphy, F.A. and S.G. Whitfield, Eastern equine encephalitis virus infection: Electron microscopic studies of mouse central nervous system. *Exp Mol Pathol*, 1970, **13**(2): 131–146.
98. Liu, C. et al., A comparative study of the pathogenesis of western equine and eastern equine encephalomyelitis viral infections in mice by intracerebral and subcutaneous inoculations. *J Infect Dis*, 1970, **122**(1): 53–63.
99. Morgan, I., Influence of age on susceptibility and on immune response of mice to eastern equine encephalomyelitis virus. *J Exp Med*, 1941, **74**: 115–132.
100. Olitsky, P.K. and C.G. Harford, Intraperitoneal and intracerebral routes in serum protection tests with the virus of equine encephalomyelitis. *J Exp Med*, 1938, **68**: 173–189.
101. Sabin, A.B. and P.K. Olitsky, Variations in pathways by which equine encephalomyelitic viruses invade the CNS of mice and guinea pigs. *Pro Soc Exp Biol Med*, 1938, **38**: 595–597.
102. Sidwell, R.W. and D.F. Smee, Viruses of the *Bunya* and *Togaviridae* families: potential as bioterrorism agents and means of control. *Antiviral Res*, 2003, **57**(1–2): 101–111.
103. Dremov, D.P. et al., Attenuated variants of eastern equine encephalomyelitis virus: Pathomorphological, immunofluorescence and virological studies of infection in Syrian hamsters. *Acta Virol*, 1978, **22**(2): 139–145.

104. Paessler, S. et al., The hamster as an animal model for eastern equine encephalitis—and its use in studies of virus entrance into the brain. *J Infect Dis*, 2004, **189**(11): 2072–2076.
105. Sabin, A.B. and P.K. Olitsky, Age of host and capacity of equine encephalomyelitic viruses to invade the CNS. *Pro Soc Exp Biol Med*, 1938, **38**: 597–599.
106. Hurst, E.W., The histology of equine encephalomyelitis. *J Exp Med*, 1934, **59**: 529–543.
107. Hurst, E.W., Infection of the rhesus monkey (*Macaca mulatta*) and the guinea-pig with the virus of equine encephalomyelitis. *J Path Bact*, 1936, **42**: 271–302.
108. Wu, D. et al., The relationship of Sindbis virus assembly and the viral protein 6K with intermediate filaments (in Chinese). *Wei Sheng Wu Xue Bao*, 1990, **30**(6): 417–421.
109. Wyckoff, R.W.G. and W.C. Tesar, Equine encephalomyelitis in monkeys. *J Immunol*, 1939, **37**: 329–343.
110. Logue, C.H. et al., Virulence variation among isolates of western equine encephalitis virus in an outbred mouse model. *J Gen Virol*, 2009, **90**(Pt 8): 1848–1858.
111. Nagata, L.P. et al., Infectivity variation and genetic diversity among strains of western equine encephalitis virus. *J Gen Virol*, 2006, **87**(Pt 8): 2353–2361.
112. Wu, J.Q. et al., Complete protection of mice against a lethal dose challenge of western equine encephalitis virus after immunization with an adenovirus-vectored vaccine. *Vaccine*, 2007, **25**(22): 4368–4375.
113. Atasheva, S. et al., Chimeric alphavirus vaccine candidates protect mice from intranasal challenge with western equine encephalitis virus. *Vaccine*, 2009, **27**(32): 4309–4319.
114. Forrester, N.L. et al., Western equine encephalitis submergence: Lack of evidence for a decline in virus virulence. *Virology*, 2008, **380**(2): 170–172.
115. Aguilar, M.J., Pathological changes in brain and other target organs of infant and weanling mice after infection with non-neuroadapted western equine encephalitis virus. *Infect Immun*, 1970, **2**(5): 533–542.
116. Cole, F.E., Jr. and R.W. McKinney, Use of hamsters of potency assay of eastern and western equine encephalitis vaccines. *Appl Microbiol*, 1969, **17**(6): 927–928.
117. Cole, F.E., Jr. and R.W. McKinney, Cross-protection in hamsters immunized with group A arbovirus vaccines. *Infect Immun*, 1971, **4**(1): 37–43.
118. Zlotnik, I. et al., The pathogenesis of western equine encephalitis virus (W.E.E.) in adult hamsters with special reference to the long and short term effects on the C.N.S. of the attenuated clone 15 variant. *Br J Exp Pathol*, 1972, **53**(1): 59–77.
119. Julander, J.G. et al., Effect of exogenous interferon and an interferon inducer on western equine encephalitis virus disease in a hamster model. *Virology*, 2007, **360**(2): 454–460.
120. Reed, D.S. et al., Aerosol exposure to western equine encephalitis virus causes fever and encephalitis in cynomolgus macaques. *J Infect Dis*, 2005. **192**(7): 1173–1182.
121. Howitt, B.F., Equine encephalomyelitis. *J Infect Dis*, 1932, **51**: 493–510.

13 Orthopoxviruses

Peter B. Jahrling and Victoria Wahl-Jensen

CONTENTS

13.1 BACKGROUND

Despite the eradication of smallpox, variola virus (VARV) remains a public health concern, because of the possibility that clandestine stocks of VARV may be in the hands of bioterrorists [1]. The impact of a VARV attack in the human population now would be even more catastrophic than it was during the last century: vaccination programs were abandoned worldwide around 1976, the prevalence of immunosuppressed populations has grown, and mobility, including intercontinental air travel, has accelerated the pace of viral spread worldwide. It is for these reasons that considerable investment is being made in the development of improved medical countermeasures (MCM) against smallpox, including new vaccines and antiviral drugs [2,3].

It is generally recognized that the development and licensure of such countermeasures will depend on animal models for the demonstration of protective efficacy. These animal models should be faithful to the human disease, and ideally would utilize the actual etiological agent (VARV), and not a surrogate. Although a significant proportion of such developmental work can be accomplished using surrogate orthopoxvirus (OPV) in rodents and primates, increased confidence in countermeasures against VARV can be obtained only by efficacy testing in nonhuman primate models using VARV.

The fact that VARV naturally infects only humans complicates the development of animal models for smallpox using VARV. VARV can infect a variety of laboratory animals experimentally, with the exception of recent studies using monkeys [4], VARV infection does not result in lethal systemic disease [5]. Further refinement of the VARV primate model is desirable, including the natural aerogenic route of exposure [6].

This chapter focuses on animal models for OPV disease that promise to have utility in countermeasure development for smallpox. Nonhuman primate models using VARV or monkeypox virus (MPXV) are most relevant to this objective, and to providing insight into the pathophysiology of smallpox in humans; yet nonhuman primate studies are expensive, and use of VARV requires the most stringent levels of biosafety (biosafety level 4, BSL-4) and biosecurity. As such, it is restricted to the two relevant World Health Organization (WHO) Collaborating Centres, the Centers for Disease Control and Prevention (CDC) in Atlanta, USA, and the State Research Centre of Virology and Biology (SRC VB VECTOR) in the Russian Federation. MPXV, though less restricted, still requires biosafety level 3 (BSL-3) biocontainment and is a select agent [7], implying that there are restrictions on its use. As a result of these restrictions, the use of small animal models for OPV disease, using ectromelia virus (ECTV), cowpox virus (CPXV), rabbitpox virus (RPXV), and vaccinia virus (VACV), has a place in efforts to understand and develop countermeasures for the human pathogens [8].

Animal models for OPV disease must address the critical balance between direct viral interaction with host target cells and the protective immune response. Poxviruses, more than most other viral pathogens, express a variety of immunomodulatory proteins and apoptosis inhibitors, which can tip the balance toward virulence. In some cases, these virus–host interactions may be species-specific and may not be reliably generalized. A detailed description of the immunobiology of OPV infections is beyond the scope of this chapter.

13.2 ECTROMELIA AND VACCINIA VIRUS ANIMAL MODELS

Much of what is believed about VARV pathogenesis is inferred from studies with ECTV in mice [9]. ECTV is a natural pathogen of mice, and, following its initial discovery in the 1930s, Fenner [10] used the model to elucidate the concept of primary and secondary viremia, which parallels exanthamous disease in humans. In the 1970s, ECTV/mouse models were used to demonstrate the role of T cells and macrophages in cell-mediated immunity and recovery from acute disease. In the 1980s, ECTV infections of inbred mouse strains were used to identify genetic determinants of resistance and susceptibility [9,11]. More recently, the availability of various knockout strains of inbred mice has facilitated the investigation of virus–host relationships.

Susceptibility of mice to ECTV is genetically determined. C57BL/6 mice are relatively resistant; the median lethal dose (LD_{50}) is $>10^6$ plaque-forming units (PFU) via footpad inoculation. In contrast, for A/J mice the LD_{50} is <0.01 PFU [11]. Genetic resistance relates in part to the granule exocytosis pathway of effector T cells [12]. BALB/c and DBA/2 strain mice are, like A/J mice, highly susceptible via both dermal and aerosol routes. The genetics of resistance/susceptibility are complex and can also vary with the ECTV strain.

ECTV infections in susceptible mice are initiated by dermal abrasions. The virus replicates locally, and then migrates to internal organs via the afferent lymphatics and draining lymph nodes and the bloodstream (primary viremia). The virus replicates in major organs, especially the liver and spleen, resulting in secondary viremia within 4–5 days. Depending on the mouse strain, replication in the skin may lead to exanthema as early as 6 days after exposure. In A/J mice, death occurs before exanthema, as a consequence of severe liver necrosis. Following aerosol exposure, there is a severe primary pneumonia [13,14]. The ECTV A strain mouse model has been used recently to evaluate various analogs of the antiviral drug cidofovir against lethal infection [13,14]. In this study, the octadecyloxyethyl derivative of cidofovir (CDV), administered orally, protected 100% of mice challenged via aerosol with 2.3×10^4 PFU, and completely blocked viral replication in the spleen and liver. Under these same conditions, unmodified cidofovir was without effect. Studies using a synthesized glycerol ester of CDV, 1-O-octadecyl-2-O-benzyl-sn-glycero-3-CDV (ODBG-CDV) showed that, while ODBG-CDV was equally effective as hexadecyloxypropyl-cidofovir (HDP-CDV, also known as CMX001) in preventing death in a lethal mouse model of ectromelia, this modified version of CDV yielded an altered transport of the compound. Specifically, higher concentrations of ODBG-CDV were observed in the lung, an important target organ for early replication of virus. The altered transport of drug also resulted in lower levels of ODBG-CDV in the liver suggesting that it may reduce first pass removal of drug by the liver following oral drug administration [13].

VACV also infects mice; the outcome depends on murine genetics, VACV strains, doses, and routes of exposure. Infection of C57BL/6 mice by VACV strain Western Reserve (WR) via the intranasal route in doses $>10^4$ PFU is lethal [15]. This model was utilized to demonstrate that the vaccinia virus gene *E3L* (which provides interferon, IFN, resistance *in vitro*) is required for pathogenesis in the intact animal. BALB/c mice are somewhat more resistant, although head-to-head comparisons have not been reported, and lethality is dose-dependent: in one published titration, 10^7 PFU of VACV strain WR was 100% lethal, whereas 10^4 PFU killed 20% of mice [16]. BALB/c mice were also used to rank VACV strains for virulence; the New York City Board of Health strain was more virulent than the WR strain, with LD_{50}s via the intranasal routes of $10^{4.0}$ and $10^{4.8}$ PFU, respectively. Neither strain was lethal via tail scarification, subcutaneous, or oral routes [17]. VACV derived from the Wyeth vaccine was less virulent via intranasal exposure ($LD_{50} > 10^7$ PFU).

SKH-1 hairless mice have been used to establish dermal infections using VACV. The severity of systemic infection can be quantified by counting skin lesions, and this model has been used to demonstrate the efficacy of 5% cidofovir, applied topically, in reducing both skin lesions and viral burdens in the lungs, kidneys, and spleen [18]. Intranasal infection of BALB/c mice with VACV strain WR leads to pneumonia, weight loss, and death. Cidofovir administration (100 mg/kg, intraperitoneally) initiated 1 day after intranasal exposure protected all treated mice; in contrast, placebo controls all died within 8 days of exposure. Cidofovir markedly improved lung consolidation scores and reduced viral burdens in the liver, spleen, and brain; peak titers were 30- to 1000-fold lower than in placebo controls [19].

13.3 COWPOX VIRUS ANIMAL MODELS

Cowpox virus (CPXV) is a zoonotic pathogen that infects a wide range of host species. Rodents are believed to be the reservoir host of CPXV [20]; however, the virus can infect domestic animals, pet rats, various captive zoo animals, and humans [21–25].

Studies analogous to those with ECTV and VACV have been performed using CPXV strain Brighton Red, which is lethal for BALB/c mice under defined conditions. Disease patterns and lethality following aerosol or intranasal exposure vary with age/weight of the mice: 100% of 10–12 g mice infected intranasally with 2×10^6 PFU (100 LD_{50}) succumbed with a mean time to death of 8 days, whereas only 50% of 7-week-old mice succumbed when given the same dose [26]. Lethally infected mice died with bilateral viral pneumonitis and viral burdens of $>10^9$ PFU/g in the lungs. This model has been used to test prophylactic and therapeutic efficacy of various treatment regimens for protection against systemic disease: mice treated with a single dose of cidofovir administered intraperitoneally (100 mg/kg) were uniformly protected against an intranasal challenge ($2–5 \times 10^6$ PFU) when the drug was given 4 days prior to exposure, and as late as 4 days after exposure. Five days or more after exposure, cidofovir was less effective [26]. However, 50% of animals were protected when cidofovir was administered 6 days following (aerosol) exposure, after animals already showed signs of disease [26]. In contrast, vaccinia immune globulin was totally ineffective in reducing mortality and IFN-α B/D (5×10^7 U/kg) was effective prior to exposure and 1 day after exposure, but not later. Infection by tail scarification was protective when initiated 8 days prior to challenge, but its efficacy diminished as the prechallenge interval was reduced; vaccination was ineffective when initiated 2 days after challenge. This observation, which conflicts with epidemiological data suggesting vaccine efficacy up to 4 days after exposure in humans, may reflect the higher challenge dose in the animal model. It also illustrates the danger in extrapolating from rodent models to humans.

Although important insight into virulence and protective immune responses can be obtained from these murine models, virus–host interactions must be assessed individually and cannot be generalized [27]. For example, the requirement for IFN-γ after infection with ECTV versus VACV is very different. In ECTV-infected mice, transfer of immune splenocytes from IFN-γ knockout (k/o) mice is highly effective in reducing the titer of virus in the liver and spleen; but in an analogous experiment, VACV-immune splenocytes are ineffective [28]. Thus, despite the apparent similarity of these two model OPV infections, recovery involves diverse and somewhat unpredictable host immune responses. Cytolytic T-cell functions can be beneficial, detrimental, or neutral [28], and this balance will be unique to each virus–host system. This type of difficulty is an important consideration in the study of specialized pathogens like OPVs when they are outside their natural hosts.

In addition to murine models of CPXV infection, recent advances have been made using marmosets (*Callithrix jacchus*). Marmosets that were infected intravenously or intranasally developed an acute and systemic disease and died within 1–3 days after onset of clinical signs [29]. The infectious dose (ID) 50 for calpox in marmosets is 8.3×10^2 PFU for the intranasal route; this is approximately 10,000-fold lower than MPXV and VARV doses needed for macaque models. A dose of 8.3×10^3 PFU was uniformly lethal. In contrast to macaque models of MPXV and VARV, the marmoset model using calpox produced relatively few skin lesions. Marmosets provide several advantages over macaque models including small size of animals, easy handling, high reproduction rate in captivity, and relatively inexpensive purchase price and keeping. A major disadvantage of marmosets is the lack of specific reagents.

13.4 RABBITPOX VIRUS ANIMAL MODELS

Rabbitpox is a disease caused by infection with rabbitpox virus (RPXV). It is important to distinguish between RPXV, which is a VACV strain classified in the genus *Orthopoxvirus*, and myxoma virus, which is in a distinct genus, *Lepovipoxvirus*. Although the myxoma virus produces lethal disease in New Zealand White rabbits with many similarities to that caused by RPXV, the myxoma virus is more distantly related to human pathogens, and is therefore less relevant to human smallpox than the animal models for OPVs discussed elsewhere in this chapter.

It has been shown that rabbits exposed to RPXV via the aerosol route develop a disease syndrome similar to humans with smallpox [30–32]. In these studies, the Utrecht strain of RPXV was shown to be somewhat more virulent than the Rockefeller Institute strain. The Utrecht strain produced a lethal infection in New Zealand White rabbits, with death occurring 7–12 days after exposure; higher doses resulted in a more fulminant disease course, but data suggested that little more than a single RPXV particle was sufficient to cause infection. Rabbits typically remained healthy for a 4- to 6-day incubation period, followed by fever, weakness, rapid weight loss, and profuse, purulent discharges from the eyes and nose. A bright erythema appeared on the lips and tongue, coinciding with a generalized skin rash. The number of lesions varied from a few to a confluence; in some cases, death occurred before rash developed. The lesions started as red papules, converting to a pseudopustule with caseous contents. Death usually occurred before true scabs could form, and was presaged by a rapid fall in body temperature. High RPXV burdens were detected in all visceral tissues, peaking between days 5 and 8, at titers of 10^8 PFU/g in the lungs and 10^7 PFU/g in the spleen and adrenal gland. In some instances, early deaths in rabbits were correlated with a blood coagulation defect [33], analogous to the hemorrhagic form of human smallpox [34]. There is some evidence that infected rabbits become contagious only in the late stages of disease, despite the presence of virus in nasopharyngeal fluids earlier, as described for human smallpox.

RPXV infection of the rabbit has parallels to human smallpox. In more recent studies, intradermal inoculation of rabbits has resulted in a similar disease pattern [35]. Whereas a viral dose of 1×10^2 PFU given intradermally results in systemic infection, a higher dose (5×10^3 PFU) is required for lethality. Initially, the intradermal injection site becomes swollen, leading to necrosis by 5 days after infection. Fever begins by day 3, followed by increased respiration rate by day 4, and secondary lesions, including eye and nasal discharges, by day 7, accompanied by weight loss. Proximal to death, respiration rate decreases and heart rate increases, and the animal falls into respiratory distress by day 7 or 8.

Sophisticated analysis of these immunological events is hindered by the lack of reagents and inbred rabbit strains, a situation that leads to preference for model studies using mice. However, evaluation of virulence genes by genetic manipulation of the RPXV genome and testing in rabbits can now be approached [36], and it is also plausible that the RPXV rabbit model could be developed further and adopted as a stepping stone for prioritizing testing of candidate therapeutics and vaccines in the available primate models for human smallpox. Indeed, New Zealand White rabbits have recently been used to study efficacy of CMX001 in both a prophylactic and a therapeutic setting against RPXV [37,38]. Additionally, because RPXV can be transmitted via naturally generated aerosols between rabbits, these efficacy studies of CMX001 administered after signs of disease development during animal-to-animal transmission were possible and demonstrated the utility of CMX001 as an effective therapeutic in symptomatic rabbitpox disease.

13.5 MONKEYPOX VIRUS ANIMAL MODELS

MPXV is a significant human pathogen that produces many of the signs and symptoms of smallpox, although its potential for transmission from person to person is less than that of smallpox [39,40]. There is evidence that MPXV strains of West African origin are less virulent than those that arise sporadically in Central Africa, specifically in the Democratic Republic of Congo [41–43]. The name "monkeypox" may be a misnomer, as the virus is maintained in nature by association with rodent reservoirs, including squirrels [44,45]. In 2003, MPXV was inadvertently imported into the United States in a shipment of rodents originating in Ghana that included an infected giant Gambian rat [46,47]. The rat infected a number of prairie dogs held in the same facility, and a chain of transmission ensued that involved hundreds of prairie dogs and spread to more than 75 human cases in 11 states. Prairie dogs involved in the U.S. MPXV outbreak were shown to have pulmonary consolidation, enlarged lymph nodes, and multifocal plaques in the gastrointestinal wall [48]. This outbreak of monkeypox rekindled interest in MPXV, not only as a surrogate for smallpox but as a disease entity in its own right.

Experimental infection of ground squirrels with the U.S. strain of MPXV was reported to kill all squirrels exposed intraperitoneally to $10^{5.1}$ PFU, or intranasally to $10^{6.1}$ PFU, within 6–9 days [49]. Systemic infections with high viral burdens were reported; major histological findings included centrilobular necrosis of the liver, splenic necrosis, and interstitial inflammation in the lungs. It is possible that MPXV infection of squirrels might be developed into a useful animal model for testing countermeasures for monkeypox and smallpox. To date, this model has been used to characterize the clinical and pathological features of experimental infection with the Central and West African clades of MPXV, further confirming the enhanced virulence of the Central African clade [50]. Recent advancements in the development of prairie dog models for human monkeypox have been reported [51,52], as have alternative rodent models, including African dormice (*Graphiurus* sp.) [53].

MPXV infection of primates has been accomplished via the aerosol [54,55], intramuscular [56], intratracheal, intrabronchial, and intravenous routes of exposure [57–63]. Most of the earlier reported studies used cynomolgus macaques, either *Macaca iris* or *Macaca fascicularis* [64], although rhesus monkeys (*Macaca mulatta*) may also be suitable [65]. Aerosol exposures are most appropriate for modelling primary exposures following a biological warfare attack. Natural transmission of MPXV (and VARV) probably occurs by a combination of aerosol, fomites, and mucosal exposures. Aerosol exposures require BSL-4 biocontainment in a class III cabinet, and are somewhat less readily controlled in comparison with intravenous exposures.

Experimental MPXV infection of cynomolgus monkeys by the aerosol route (calculated inhaled dose of 30,000 PFU) resulted in 5 of 6 monkeys dying (on days

9, 10, 10, 11, and 12, respectively; mean time to death of 10.4 days), with significant fevers (>102.5°F), mild enanthema, coughs, and leukocytosis with absolute and relative monocytosis [66]. The virus was isolated from buffy coat cells of febrile animals, and, at necropsy, high titers of the virus (>10^6 PFU/g) were isolated from the lungs and spleen [54]. Histopathological examinations attributed death to severe fibrinonecrotic bronchopneumonia; immunohistochemistry indicated abundant MPXV antigen in samples of affected airway epithelium and surrounding interstitium. The clinical parameters measured in monkeys exposed to aerosolized MPXV occur in a sequence similar to humans, but are accelerated in comparison to humans [67]. More recently, a novel respiratory model of MPXV was developed that utilizes a bronchoscope in conjunction with a liquid MicroSprayer aerosolizer and a high-pressure syringe [68]. Whereas this method of virus inoculation delivers a larger particle size than traditional collision nebulizers, it is highly reproducible and produces a systemic form of disease reminiscent of ordinary smallpox. In comparison to intravenous exposure studies, lower doses of virus were able to produce systemic disease. Three doses of MPXV Zaire '79 were used: 3.42×10^6, 8.37×10^6, and 3.53×10^7 PFU. Interestingly, terminal lesions were strikingly similar between the groups, although kinetics to disease signs were more rapid in the higher dose groups. Several animals in the study also developed hemorrhagic disease; however, this was not dependent exclusively on the dose of virus administered. The microsprayer delivery also has the added benefits being portable and less expensive, and does not require extensive training to operate. The obvious disadvantage, however, is that this route of inoculation completely bypasses the upper airway.

Intravenous exposure of cynomolgus macaques to MPXV also resulted in uniform systemic infection; disease severity was related to dose [69]. Cynomolgus monkeys infected by the intravenous route with 1×10^7 PFU of MPXV (Zaire '79 strain, CDC V79-I-005) develop a low-grade fever beginning on day 3. Pox lesions first appear on days 4 and 5, with death first occurring on day 8 and a mean time to death of 12 days, which was 4–8 days after onset of the rash. This is shorter than the 10–14 days seen with human monkeypox. Mortality occurred in 11 of 12 (92%) infected monkeys, compared with 10% mortality in the human disease. Pox lesions were found in all animals, and were graded as "grave" on the WHO scoring system (>250 lesions). Hands, feet, mouth, and soft pallet were fully involved. All monkeys followed this pattern of progressing through the typical stages of lesion development, with those that lived long enough ultimately developing scabs. Weight loss was also seen in all animals. Laboratory findings were largely unremarkable, except for a terminal rise in blood urea nitrogen and creatinine.

At necropsy, animals had significant organ involvement, both from the viewpoint of gross pathological lesions and virus replication. Virus replication in the lungs, liver, and spleen was greater than 10^8 PFU, and blood had 10^5 PFU of virus, which was cell-associated. Plasma was free of infectious virus. Virus titers in the kidneys were slightly higher than titers in blood, suggesting that the kidneys are a site of viral replication; however, significant virus burdens (above the contained blood) were not detected in the brain.

To determine the effect of infectious dose on disease progression, lower doses of 10^5 and 10^6 PFU per animal were evaluated. Mortality was not seen at lower doses, but all animals became sick and developed lesions. The number of lesions, based on the WHO scoring system, was dose-dependent, ranging from mild at 10^5 PFU, through moderate at 10^6 PFU, to severe at 10^7 PFU. Infected animals showed significant increases in white blood cells, but this was not dose-dependent. There was a drop in platelets that was dose-dependent, reaching a low on days 2–8, but then returning to normal ranges. Pulmonary function was not significantly impaired at the lower virus doses. All animals developed low-grade fevers (<104°F) by days 3 and 4. Poxvirus lesions were first seen between days 4 and 5, continued to increase in magnitude until days 10–12, and then resolved over the next 2 weeks in surviving animals. Animals infected with lethal doses of 10^7 PFU or greater had lesion counts exceeding 1500 lesions. Viral loads in blood, measured as genomes per milliliter of whole blood by quantitative polymerase chain reaction, could be detected at 24 h after infection, and increased to $>10^7$ genomes/mL prior to death. Surviving animals, either those given lower infectious doses or those treated successfully with antiviral chemotherapy or vaccination, had viral loads that never exceeded 10^6 genomes/mL. Albumin decreased in a dose-dependent manner, with monkeys infected with 10^7 PFU falling to levels of 1.5 g/dL, while total serum protein remained within normal limits.

The intravenous MPXV challenge model was used to test the efficacy of a candidate vaccine, the highly attenuated modified vaccinia Ankara (MVA), in comparison and in combination with the licensed Dryvax vaccine [70]. Monkeys were vaccinated in week 0 with MVA or Dryvax, and, in week 8, the MVA-immunized monkeys were boosted with either MVA or Dryvax. After challenge with MPXV in week 16, the placebo controls developed more than 500 pox lesions and became gravely ill; 2 of 6 died. In contrast, none of the monkeys receiving Dryvax or MVA/Dryvax developed illness; monkeys in the MVA/MVA group remained healthy but developed an average of 16 lesions. None of the vaccinated monkeys developed significant viremia as detected by quantitative polymerase chain reaction [71], in contrast with placebo controls, which developed virus titers $>10^8$ genomes/mL in blood. In the course of immunizing these monkeys, it was observed that MVA elicited higher enzyme-linked immunoadsorbent assay titers within 10 days of immunization than Dryvax recipients. To determine whether the immune response to MVA was sufficient to be protective this early, monkeys were immunized with a single dose of MVA or Dryvax, and challenged on day 10. In contrast with controls, which developed >500 lesions each and became gravely ill, none of the MVA or Dryvax recipients became ill; isolated lesions (3–6 per animal) appeared in both groups. MVA and Dryvax both limited viral replication to titers lower than the artificial viremia created by intravenous infection with MPXV, whereas challenge virus replication titers exceeded 10^8 genomes/mL. Analogous vaccine efficacy studies for Dryvax in comparison with a cell culture-derived VACV against an aerosol challenge demonstrated solid protection [72]. Rhesus monkeys were used in a similar intravenous challenge model to evaluate a DNA vaccine strategy with a combination of four genes (*L1R*, *A27L*, *A33R*, and *B5R*); results were promising [65]. Recently, these genes have been expressed in an alphavirus replicon, with similar promising results [73].

There has been some reluctance to accept the intravenous challenge model, on the grounds that the challenge should be via the "natural route." The counter argument to this concern is that protection against an overwhelming intravenous dose is a very stringent criterion, and protection against intravenous challenge may predict efficacy against peripheral challenge routes. However, because of these concerns, and because intravenous challenge sets the bar too high for antiviral drug evaluations, alternative exposure models, including intratracheal routes, are being explored. As the dose–response curve is very steep, intravenous administration of the virus is an advantage in calibrating the inoculum dose. Aerogenic or mucosal routes of exposure would require larger numbers of animals.

Despite these limitations, the intravenous MPXV model was used to demonstrate the efficacy of a number of candidate antiviral drugs, including cidofovir [74] and ST-246 [60]; these studies are outside the scope of this chapter.

13.6 VARIOLA VIRUS ANIMAL MODELS

The development of an animal model in which VARV produces a disease similar to human smallpox is necessary to provide the most convincing demonstration of protective efficacy of vaccines and antiviral drugs for smallpox [6,75,76]. Because of the species-specificity of VARV, it was not surprising that attempts to infect and produce disease with VARV in rodents and rabbits were unsuccessful [77]. Indeed, even in primates, early experiments with VARV resulted in mild, but self-limited infections. Cynomolgus macaques, exposed to aerosols containing 2×10^8 pock-forming units, developed a rash after a 6-day incubation period; virus replicated in the lungs and secondary sites of replication were established in lymph nodes before viremia occurred [64]. In the same study [31], 109 rhesus monkeys were exposed; all developed fever by day 5 and rash between days 7 and 11, but only 2 died. Bonnet macaques (*Macaca radiata*) were also resistant to disease following infection [78]; 0 of 14 died. However, the same authors demonstrated that cortisone treatment rendered monkeys susceptible; 14 of 16 died, as did 1 untreated but pregnant monkey. In human populations, pregnant women suffered the highest mortality following smallpox infections [79].

The historical record thus suggested that there were no known models suitable for modeling the pathogenesis of VARV in humans [80]. However, infection of macaques was known to produce skin lesions and evidence of systemic infection, and a primate model was used to license MVA in Germany in the 1960s [81]. It was therefore reasonable to test alternate VARV strains in higher doses by a variety of routes to seek a model for lethal smallpox. Aerosol exposures of cynomolgus monkeys to either the Yamada or Lee VARV strains ($10^{8.5}$ PFU) resulted in infection but no serious disease [3]; however, subsequent studies, in which monkeys were exposed to either Harper or India 7124 VARV strains intravenously, resulted in acute lethality [4]. Doses lower than 10^9 PFU resulted in decreased lethality, and quantifiable parameters of disease severity diminished with declining dose.

In monkeys dying after VARV infection, the end-stage lesions resembled terminal human smallpox. Our understanding of the pathophysiology of human smallpox is imprecise, as the disease was eradicated before the development of modern tools of virology and immunology, but insight derived from the nonhuman primate models may inspire reinvestigation of archived specimens using modern techniques such as immunohistochemistry and cDNA microarrays, which were used in the primate model reports [4,82]. A recent review of all pathology reports published in English in the last 200 years [34] suggested that, in general, otherwise healthy patients who died of smallpox usually succumbed to renal failure, shock secondary to volume depletion, and difficulty with oxygenation and ventilation as a result of viral pneumonia and airway compromise, respectively. Degeneration of hepatocytes might have caused a degree of compromise, but liver failure was not usually the proximal cause of death.

End-stage lesions in monkeys inoculated with VARV closely resembled human pathology [4]. Experimental infection permitted evaluation of multiple parameters at intermediate time points prior to death. Monkeys inoculated intravenously had a demonstrable artificial viremia immediately after inoculation. Following an eclipse phase of several days, virus in the blood was associated only with monocytic cells. Animals that died had profound leukocytosis, thrombocytopenia, and elevated serum creatinine levels. High viral burdens in target tissues were associated with organ dysfunction and multisystem failure. Distribution of viral antigens by immunohistochemistry correlated with the presence of replicating viral particles demonstrated by electron microscopy, and with pathology in the lymphoid tissues, skin, oral mucosa, gastrointestinal tract, reproductive system, and liver. Histological evidence of hemorrhagic diathesis was corroborated by elevations in D-dimers. Apoptosis of T cells in lymphoid tissue was documented—a probable consequence of viral replication in macrophages and the resultant cytokine storm. "Toxemia," described by clinicians as the terminal event in human smallpox, is probably a consequence of an over-stimulation of the innate immune response, including interleukin-6 and IFN-γ, as much as it is a result of direct viral damage to target tissues.

More recently, a serial sampling study was performed on cynomolgus macaques exposed to doses of VARV strain Harper calibrated to induce ordinary or hemorrhagic disease [83]. Several key differences were noted between these models. In the ordinary smallpox model, lymphoid and myeloid hyperplasias were consistently found whereas lymphocytolysis and hematopoietic necrosis developed in hemorrhagic smallpox. Viral antigen accumulation, as assessed immunohistochemically, was mild and transient in the ordinary smallpox model. In contrast, in the hemorrhagic model antigen distribution was widespread and included tissues and cells not involved in the ordinary model. Hemorrhagic smallpox developed only in the presence of secondary bacterial infections—an observation also commonly noted in historical reports of human smallpox.

The availability of cDNA microarrays to study human gene expression patterns permitted analysis of peripheral blood samples from the monkeys [84]. VARV elicited striking and temporally coordinated patterns of gene expression (features that represent an IFN response), cell proliferation, and immunoglobulin expression, correlated with viral dose and modulation of the host immune response. Surprisingly, there was a virtual absence of a tumor necrosis factor-α/nuclear factor-κB response, suggesting that VARV gene products may ablate this response. Although the interaction of VARV with the human immune system can only be approximated in the monkey models, the extrapolation from nonhuman primates to humans is less tenuous than that from rodents to humans. Whether MPXV in monkeys is more faithful to human smallpox than VARV in monkeys is a focus of intense investigation, and both nonhuman primate models may provide insight into the development of diagnostic, prophylactic, and therapeutic strategies.

13.7 CONCLUSION: CHALLENGES TO COMPLIANCE WITH THE ANIMAL RULE

This chapter has focused on animal models for human OPV disease and their utility for demonstrating efficacy of candidate medical countermeasures in compliance with the U.S. Food and Drug Administration (FDA) Animal Efficacy Rule. Guidance provided by the FDA in the form of "essential elements to address" sets an impossible standard; it is unlikely that any one model will ever satisfy all requirements of the Rule simultaneously. Different models will be required to address specific indications.

Rodent and rabbit models have several advantages over primates in that they are less expensive, less heterogeneous, and often better characterized immunologically. Less well-characterized rodents such as squirrels and prairie dogs are less useful for efficacy testing of MCM but do have value in matching virus species with their natural hosts. Greater statistical power can be achieved with rodents compared with primates, and use of surrogate OPVs that are not select agents and do not require BSL-3 or BSL-4 biocontainment can be handled more readily than MPXV and VARV. However, at some point, compliance with the Rule demands that efficacy be demonstrated with the authentic pathogen in an animal model that recapitulates the human disease.

The necessary leap of faith from animals to humans is facilitated by demonstration of MCM efficacy in primates. But recapitulation of the human disease is impossible if full compliance with other "essential elements to address" is demanded. VARV is uniquely adapted to humans and host restriction can be overcome only by huge exposure doses by the intravenous route. Lower doses by alternative routes result in infection but no overt disease. However, the FDA considers the primary study endpoint to be enhancement of survival or prevention of major morbidity. Surrogates including reductions in viremia or lesion counts are considered helpful but are not accepted as primary endpoints. Another challenge is relating the pathophysiology of the model disease course to humans, especially when death associated with human smallpox was described only vaguely as "toxemia." The model infections are far better characterized than the human condition. In time it may be possible to correlate human MPXV from outbreaks in the Democratic Republic of Congo with the nonhuman primate model but it will require yet another leap of faith to extrapolate to human smallpox.

Further refinement of the nonhuman primate models will continue now that a thorough evaluation of data obtained from previous experiments has been accomplished. Refinements include alternative routes of exposure such as aerosol, intrabronchial, intratracheal, and droplet exposures. Special attention will be paid to elucidation of biomarker patterns that could be used in a clinical setting as triggers for early intervention, thus increasing the likelihood of treatment success. Additional refinements include use of telemetry and medical imaging such as magnetic resonance, positron emission tomography, single photon emission computed tomography, and computed tomography (CT) [85]. Sequential sampling of individual animals over time mitigates variability by permitting each animal to serve as its own control and facilitates achievement of statistical power using a reduced number of animals.

Hopefully, refinement of the animal models and the tools used to evaluate them will be accompanied by an evolution of the Animal Rule. One optimistic development is the Medical Countermeasure Initiative recently announced by the FDA which includes a pillar for regulatory science for MCM development and evaluation [86]. This signals an awareness of the need to work in partnership with the research community to establish sufficient confidence in animal models to make important regulatory decisions that will enhance biopreparedness for the nation.

REFERENCES

1. Henderson, D.A. et al., Smallpox as a biological weapon: Medical and public health management. Working Group on Civilian Biodefense. *JAMA: The Journal of the American Medical Association*, 1999, **281**(22): 2127–2137.
2. Russell, P.K., Project BioShield: What it is, why it is needed, and its accomplishments so far. *Clin Infect Dis*, 2007, **45**(Suppl 1): S68–S72.
3. LeDuc, J.W. and P.B. Jahrling, Strengthening national preparedness for smallpox: An update. *Emerging Infectious Diseases*, 2001, **7**(1): 155–157.
4. Jahrling, P.B. et al., Exploring the potential of variola virus infection of cynomolgus macaques as a model for human smallpox. *Proceedings of the National Academy of Sciences of the United States of America*, 2004, **101**(42): 15196–15200.
5. Fenner, F. et al., Smallpox and its eradication. *World Health Organization*, **1988**(2165): 296–309.
6. U.S. Department of Health and Human Services, *Guidance for Industry Animal Models—Essential Elements to Address Efficacy Under the Animal Rule*, Draft Guidance. Rockville, MD: U.S. Food and Drug Administration. 2009.
7. Centers for Disease Control and Prevention (CDC) and Department of Health and Human Services (HHS), Possession, use, and transfer of select agents and toxins. Final rule. *Federal Register*, 2008, **73**(201): 61363–61366.
8. Chapman, J.L. et al., Animal models of orthopoxvirus infection. *Veterinary Pathology*, 2010, **47**(5): 852–870.
9. Buller, M. and G. Palumbo, Poxvirus pathogenesis. *Microbiological Reviews*, 1991, **55(1)**(1970): 80–122.
10. Fenner, F., The clinical features and pathogenesis of mousepox (infectious ectromelia of mice). *Journal of Pathology and Bacteriology*, 1948, **60**: 529–552.
11. Buller, R.M., The BALB/c mouse as a model to study orthopoxviruses. *Current Topics in Microbiology and Immunology*, 1985, **122**: 148–153.
12. Mullbacher, A. et al., Granzymes are the essential downstream effector molecules for the control of primary virus infections by cytolytic leukocytes. *Proceedings of the National Academy of Sciences of the United States of America*, 1999, **96**(24): 13950–13955.
13. Hostetler, K.Y. et al., Oral 1-O-octadecyl-2-O-benzyl-sn-glycero-3-cidofovir targets the lung and is effective against a lethal respiratory challenge with ectromelia virus in mice. *Antiviral Research*, 2007, **73**(3): 212–218.
14. Buller, R.M. et al., Efficacy of oral active ether lipid analogs of cidofovir in a lethal mousepox model. *Virology*, 2004, **318**(2): 474–481.
15. Brandt, T.A. and B.L. Jacobs, Both carboxy- and amino-terminal domains of the vaccinia virus interferon resistance gene, E3L, are required for pathogenesis in a mouse model. *Journal of Virology*, 2001, **75**(2): 850–856.
16. Alcami, A. and G.L. Smith, A soluble receptor for interleukin-1 beta encoded by vaccinia virus: A novel mechanism of virus modulation of the host response to infection. *Cell*, 1992, **71**(1): 153–167.

17. Lee, M.S. et al., Molecular attenuation of vaccinia virus: Mutant generation and animal characterization. *Journal of Virology*, 1992, **66**(5): 2617–2630.
18. Quenelle, D.C., D.J. Collins, and E.R. Kern, Cutaneous infections of mice with vaccinia or cowpox viruses and efficacy of cidofovir. *Antiviral Research*, 2004, **63**(1): 33–40.
19. Smee, D.F., K.W. Bailey, and R.W. Sidwell, Treatment of lethal vaccinia virus respiratory infections in mice with cidofovir. *Antiviral Chemistry & Chemotherapy*, 2001, **12**(1): 71–76.
20. Chantrey, J. et al., Cowpox: Reservoir hosts and geographic range. *Epidemiology and Infection*, 1999, **122**(3): 455–460.
21. Kurth, A. et al., Cowpox virus outbreak in banded mongooses (*Mungos mungo*) and jaguarundis (*Herpailurus yagouaroundi*) with a time-delayed infection to humans. *PLoS One*, 2009, **4**(9): e6883.
22. Baxby, D. and M. Bennett, Poxvirus zoonoses. *Journal of Medical Microbiology*, 1997, **46**(1): 17–20, 28–33.
23. Herder, V. et al., Poxvirus infection in a cat with presumptive human transmission. *Veterinary Dermatology*, 2011, **22**(2): 220–224.
24. Ninove, L. et al., Cowpox virus transmission from pet rats to humans, France. *Emerging Infectious Diseases*, 2009, **15**(5): 781–784.
25. Campe, H. et al., Cowpox virus transmission from pet rats to humans, Germany. *Emerging Infectious Diseases*, 2009, **15**(5): 777–780.
26. Bray, M. et al., Cidofovir protects mice against lethal aerosol or intranasal cowpox virus challenge. *The Journal of Infectious Diseases*, 2000, **181**(1): 10–19.
27. Mullbacher, A. et al., Can we really learn from model pathogens? *Trends in Immunology*, 2004, **25**(10): 524–528.
28. Mullbacher, A. and R.V. Blanden, T-cell-mediated control of poxvirus infection in mice. *Progress in Molecular and Subcellular Biology*, 2004, **36**: 39–55.
29. Kramski, M. et al., A novel highly reproducible and lethal nonhuman primate model for orthopox virus infection. *PLoS One*, 2010, **5**(4): e10412.
30. Lancaster, M.C. et al., Experimental respiratory infection with poxviruses. II. Pathological studies. *British Journal of Experimental Pathology*, 1966, **47**(5): 466–471.
31. Westwood, J.C. et al., Experimental respiratory infection with poxviruses. I. Clinical virological and epidemiological studies. *British Journal of Experimental Pathology*, 1966, **47**(5): 453–465.
32. Nalca, A. et al., Evaluation of orally delivered ST-246 as postexposure prophylactic and antiviral therapeutic in an aerosolized rabbitpox rabbit model. *Antiviral Research*, 2008, **79**(2): 121–127.
33. Boulter, E.A., H.B. Maber, and E.T. Bowen, Studies on the physiological disturbances occurring in experimental rabbit pox: An approach to rational therapy. *British Journal of Experimental Pathology*, 1961, **42**: 433–444.
34. Martin, D.B., The cause of death in smallpox: An examination of the pathology record. *Military Medicine*, 2002, **167**(7): 546–551.
35. Adams, M.M., A.D. Rice, and R.W. Moyer, Rabbitpox virus and vaccinia virus infection of rabbits as a model for human smallpox. *Journal of Virology*, 2007, **81**(20): 11084–11095.
36. McFadden, G., Poxvirus tropism. *Nature Reviews Microbiology*, 2005, **3**(3): 201–213.
37. Rice, A.D. et al., Efficacy of CMX001 as a prophylactic and presymptomatic antiviral agent in New Zealand white rabbits infected with rabbitpox virus, a model for orthopoxvirus infections of humans. *Viruses*, 2011, **3**(2): 63–82.
38. Rice, A.D. et al., Efficacy of CMX001 as a post exposure antiviral in New Zealand white rabbits infected with rabbitpox virus, a model for orthopoxvirus infections of humans. *Viruses*, 2011, **3**(1): 47–62.
39. Fine, P.E. et al., The transmission potential of monkeypox virus in human populations. *International Journal of Epidemiology*, 1988, **17**(3): 643–650.

40. Jezek, Z. et al., Human monkeypox: Secondary attack rates. *Bulletin of the World Health Organization*, 1988, **66**(4): 465–470.
41. Hutson, C.L. et al., Dosage comparison of Congo Basin and West African strains of monkeypox virus using a prairie dog animal model of systemic orthopoxvirus disease. *Virology*, 2010, **402**(1): 72–82.
42. Chen, N. et al., Virulence differences between monkeypox virus isolates from West Africa and the Congo basin. *Virology*, 2005, **340**(1): 46–63.
43. Likos, A.M. et al., A tale of two clades: Monkeypox viruses. *The Journal of General Virology*, 2005, **86**(Pt 10): 2661–2672.
44. Khodakevich, L. et al., The role of squirrels in sustaining monkeypox virus transmission. *Tropical and Geographical Medicine*, 1987, **39**(2): 115–122.
45. Khodakevich, L., Z. Jezek, and K. Kinzanzka, Isolation of monkeypox virus from wild squirrel infected in nature. *Lancet*, 1986, **1**(8472): 98–99.
46. Ligon, B.L., Monkeypox: A review of the history and emergence in the Western hemisphere. *Seminars in Pediatric Infectious Diseases*, 2004, **15**(4): 280–287.
47. Perkins, S., Monkeypox in the United States. *Contemporary Topics in Laboratory Animal Science/American Association for Laboratory Animal Science*, 2003, **42**(5): 70, 72.
48. Langohr, I.M. et al., Extensive lesions of monkeypox in a prairie dog (*Cynomys* sp). *Veterinary Pathology*, 2004, **41**(6): 702–707.
49. Tesh, R.B. et al., Experimental infection of ground squirrels (*Spermophilus tridecemlineatus*) with monkeypox virus. *Emerging Infectious Diseases*, 2004, **10**(9): 1563–1567.
50. Sbrana, E. et al., Comparative pathology of North American and central African strains of monkeypox virus in a ground squirrel model of the disease. *The American Journal of Tropical Medicine and Hygiene*, 2007, **76**(1): 155–164.
51. Hutson, C.L. et al., A prairie dog animal model of systemic orthopoxvirus disease using West African and Congo Basin strains of monkeypox virus. *Journal of General Virology*, 2009, **90**(Pt 2): 323–333.
52. Knight, J., Prairie-dog model offers hope of tackling monkeypox virus. *Nature*, 2003, **423**(6941): 674.
53. Schultz, D.A. et al., Experimental infection of an African dormouse (*Graphiurus kelleni*) with monkeypox virus. *Virology*, 2009, **383**(1): 86–92.
54. Zaucha, G.M. et al., The pathology of experimental aerosolized monkeypox virus infection in cynomolgus monkeys (*Macaca fascicularis*). *Laboratory Investigation*, 2001, **81**(12): 1581–1600.
55. Nalca, A. et al., Experimental infection of cynomolgus macaques (*Macaca fascicularis*) with aerosolized monkeypox virus. *PLoS One*, 2010, **5**(9): e12880.
56. Wenner, H.A. et al., Studies on the pathogenesis of monkey pox. II. Dose-response and virus dispersion. *Archiv fur die gesamte Virusforschung*, 1969, **27**(2): 166–178.
57. Stittelaar, K.J. et al., Modified vaccinia virus Ankara protects macaques against respiratory challenge with monkeypox virus. *Journal of Virology*, 2005, **79**(12): 7845–7851.
58. Johnson, R.F. et al., Comparative analysis of monkeypox virus infection of cynomolgus macaques by the intravenous or intrabronchial inoculation route. *Journal of Virology*, 2011, **85**(5): 2112–2125.
59. Brown, J.N. et al., Characterization of macaque pulmonary fluid proteome during monkeypox infection: Dynamics of host response. *Molecular & Cellular Proteomics*, 2010, **9**(12): 2760–2771.
60. Jordan, R. et al., ST-246 antiviral efficacy in a nonhuman primate monkeypox model: Determination of the minimal effective dose and human dose justification. *Antimicrobial Agents and Chemotherapy*, 2009, **53**(5): 1817–1822.
61. Edghill-Smith, Y. et al., Smallpox vaccine-induced antibodies are necessary and sufficient for protection against monkeypox virus. *Nature Medicine*, 2005. **11**(7): 740–747.

62. Huggins, J. et al., Nonhuman primates are protected from smallpox virus or monkeypox virus challenges by the antiviral drug ST-246. *Antimicrobial Agents and Chemotherapy*, 2009, **53**(6): 2620–2625.
63. Buchman, G.W. et al., A protein-based smallpox vaccine protects non-human primates from a lethal monkeypox virus challenge. *Vaccine*, 2010, **28**(40): 6627–6636.
64. Hahon, N., Smallpox and related poxvirus infections in the simian host. *Bacteriological Reviews*, 1961, **25**: 459–476.
65. Hooper, J.W. et al., Smallpox DNA vaccine protects nonhuman primates against lethal monkeypox. *Journal of Virology*, 2004, **78**(9): 4433–4443.
66. Jahrling, P.B., G.M. Zaucha, and J.W. Huggins, Countermeasures to the reemergence of smallpox virus as an agent of bioterrorism, in *Emerging Infections*, W.M. Scheld, W.A. Craig, and J.M. Hughes, editors. Washington, DC: ASM Press, 2000, pp. 187–200.
67. Breman, J.G. and D.A. Henderson, Diagnosis and management of smallpox. *The New England Journal of Medicine*, 2002, **346**(17): 1300–1308.
68. Goff, A.J. et al., A novel respiratory model of infection with monkeypox virus in cynomolgus macaques. *Journal of Virology*, 2011, **85**(10): 4898–4909.
69. Stittelaar, K.J. et al., Antiviral treatment is more effective than smallpox vaccination upon lethal monkeypox virus infection. *Nature*, 2006, **439**(7077): 745–748.
70. Earl, P.L. et al., Immunogenicity of a highly attenuated MVA smallpox vaccine and protection against monkeypox. *Nature*, 2004, **428**: 182–185.
71. Kulesh, D.A. et al., Smallpox and pan-orthopox virus detection by real-time 3-minor groove binder TaqMan assays on the roche LightCycler and the Cepheid smart Cycler platforms. *Journal of Clinical Microbiology*, 2004, **42**(2): 601–609.
72. Jahrling, P.B., Medical countermeasures against the re-emergence of smallpox virus, in *Biological Threats and Terrorism: Assessing the Science and Response Capabilities*, S.L. Knobler and L.A. Pray, editors. Washington, DC: National Academy Press, 2002, pp. 50–53.
73. Hooper, J.W. et al., Molecular smallpox vaccine delivered by alphavirus replicons elicits protective immunity in mice and non-human primates. *Vaccine*, 2009, **28**(2): 494–511.
74. Wei, H. et al., Coadministration of cidofovir and smallpox vaccine reduced vaccination side effects but interfered with vaccine-elicited immune responses and immunity to monkeypox. *Journal of Virology*, 2009, **83**(2): 1115–1125.
75. U.S. Food and Drug Administration (FDA) and Department of Health and Human Services (HHS), New drug and biological drug products; evidence needed to demonstrate effectiveness of new drugs when human efficacy studies are not ethical or feasible. Final rule. *Federal Register*, 2002, **67**(105): 37988–37998.
76. U.S. Food and Drug Administration (FDA) and Department of Health and Human Services (HHS), Investigational new drug safety reporting requirements for human drug and biological products and safety reporting requirements for bioavailability and bioequivalence studies in humans. Final rule. *Federal Register*, 2010, **75**(188): 59935–59963.
77. Marennikova, S.S., Field and experimental studies of poxvirus infections in rodents. *Bulletin of the World Health Organization*, 1979, **57**(3): 461–464.
78. Rao, A.R. et al., Experimental variola in monkeys. I. Studies on disease enhancing property of cortisone in smallpox. A preliminary report. *Indian Journal of Medical Research*, 1968, **56**(12): 1855–1865.
79. Rao, A.R. et al., Pregnancy and smallpox. *Journal of the Indian Medical Association*, 1963, **40**: 353–363.
80. Committee on the Assessment of Future Scientific Needs for Variola Virus, I.o.M., Scientific needs for live variola virus, in *Assessment of Future Scientific Needs for Live Variola Virus*. Washington, DC: National Academy Press, 1999, pp. 81–85.

81. Hochstein-Mintzel, V. et al., [An attenuated strain of vaccinia virus (MVA). Successful intramuscular immunization against vaccinia and variola (author's translation)]. Zentralblatt fur Bakteriologie, Parasitenkunde, Infektionskrankheiten und Hygiene. *Erste Abteilung Originale. Reihe A: Medizinische Mikrobiologie und Parasitologie*, 1975, **230**(3): 283–297.
82. Rubins, K.H. et al., The temporal program of peripheral blood gene expression in the response of nonhuman primates to Ebola hemorrhagic fever. *Genome Biology*, 2007, **8**(8): R174.
83. Wahl-Jensen, V. et al., Progression of pathogenic events in cynomolgus macaques infected with variola virus. *Plos One*, 2011, **6**(10): e24832.
84. Rubins, K.H. et al., The host response to smallpox: Analysis of the gene expression program in peripheral blood cells in a nonhuman primate model. *Proceedings of the National Academy of Sciences of the United States of America*, 2004, **101**(42): 15190–15195.
85. Li, K. et al., Potential applications of conventional and molecular imaging to biodefense research. *Clin Infect Dis*, 2005, **40**(10): 1471–1480.
86. U.S. Department of Health and Human Services and U.S. Food and Drug Administration, *Advancing Regulatory Science for Public Health*. Rockville, MD: Government Printing Office, 2010.

14 Animal Models for Viral Hemorrhagic Fevers

Kelly L. Warfield and Thomas W. Geisbert

CONTENTS

14.1 INTRODUCTION

Historically, the term viral hemorrhagic fever (VHF) refers to a clinical illness or syndrome characterized by high fever and a bleeding diathesis caused by a virus in one of four virus families [1]. The four virus families that cause VHF are Arenaviridae, Bunyaviridae, Flaviviridae, and Filoviridae. Currently, 12 specific viruses cause VHF, but the number is likely to expand as new viruses emerge. While all of the VHFs are caused by small RNA viruses with lipid envelopes [1], the viruses are biologically, geographically, and ecologically diverse. Most of the VHFs are zoonoses (transmissible from animal to man). The ecology and host reservoir of the viruses that cause VHF, except for the Filoviridae, are well-defined. Transmission to humans may occur from contact with the infected reservoir, a bite from an infected arthropod, aerosols generated from infected rodent excreta, or direct contact with infected patients or animal carcasses [2]. With the exception of the flaviviruses and Rift

Valley fever (RVF), which are not considered transmissible from person to person, infected humans can spread VHF infection to close contacts [1,2].

The signs, symptoms, clinical course, mortality, and pathogenesis of the hemorrhagic fever (HF) viruses vary among the virus families, the viruses within a specific virus family, and the virus species or strains of a particular virus. Hemorrhage may not necessarily occur in each individual case of VHF, and, if seen, may be a late event in the course of the disease. However, hemorrhage and circulatory shock is seen as a clinical manifestation amongst the patients in most VHF outbreaks. Other symptoms such as fever, headache, generalized myalgia, prostration conjunctivitis, rash, lymphadenopathy, pharyngitis, and edema are common to most VHF outbreaks.

Importantly, most of the VHFs are serious public health threats, and are classified as biosafety level 3 or level 4 (BSL-3 or BSL-4) agents by the *Biosafety in Microbiological and Biomedical Laboratories* manual of the Centers for Disease Control and Prevention [3]. Without the application of specific diagnostic testing, VHF syndromes can be difficult to differentiate from one another. This is a particularly problematic issue as several VHF agents are classified as category A pathogens and have known capability for use as bioweapons. Some VHF agents have a history of state-sponsored weaponization, to include Marburg, Ebola, Junin, Machupo, yellow fever virus, and others [1,4–7]. Unfortunately, there are currently no available or approved vaccines or effective therapeutics for most of the VHF agents [1]. Other features that characterize the VHFs as serious bioweapon threats include high morbidity and mortality rates, the potential for person-to-person transmission, low infective dose and highly-infectious-by-aerosol dissemination, and the feasibility of large-scale production [1]. The amount of fear by the public and sensationalism by the media that are generated during outbreaks also contribute to the consideration of these viruses as bioweapons.

This chapter will briefly discuss the basic biology of the Arenaviridae, Bunyaviridae, Filoviridae, and Flaviridae. As the diversity of the causative viruses for VHF is so great, it is beyond the scope of this chapter to adequately discuss each of the viruses that are causative of HF. Therefore, we will give an overview of each virus family that has members which cause VHF and then review in greater detail a small number of virus members from each of the Arenaviridae, Bunyaviridae, and Flaviviridae families. Both the Marburg virus (MARV) and the Ebola virus (EBOV), the only two members of the Filoviridae family, will be discussed thoroughly. The authors have chosen to give particular attention to the animal models for studying filovirus infections for several reasons. In our opinion, the Filoviridae represent the most serious and lethal bioweapon threat of the VHFs. Filoviruses fulfill all of the criteria for an effective VHF bioweapon listed above, as well as several unique characteristics that make them particularly lethal. Because of their unique pathogenesis, filoviruses have the highest mortality of all the HF viruses. Natural EBOV outbreaks are occurring with increasing frequency due to increased contact between man and infected nonhuman primates. The virtually endemic status for EBOV in Africa may afford greater accessibility to persons wishing to acquire the virus for nefarious purposes. The natural host and ecology for both MARV and EBOV remain unknown, although new findings are highly suggestive that fruit bats are at least one of the reservoirs [8,9]. Additionally, filoviruses have an extremely rapid rate of replication in an infected host, high infectivity, demonstrated ease of transmissibility through various modes of transmission, and relative environmental stability [10,11]. These combined factors make them extremely dangerous from both public health and bioweapon threat perspectives.

14.2 FILOVIRIDAE

EBOV and MARV, members of the family Filoviridae, cause an acute and rapidly progressive HF with mortality rates up to 90% [10,12]. These viruses are fast-acting, with death often occurring within 7–10 days after infection; however, the incubation period is considered to be 2–21 days [1,13]. Filoviruses are transmitted through contact with a reservoir such as fruit bats, bodily fluids, or tissues of humans or infected animals such as nonhuman primates or swine [14–19]. Historically, nosocomial transmission often occurs through reuse of incorrectly sterilized needles and syringes, emergency surgical interventions for undiagnosed bleeding when there has been failure to make a correct diagnosis, or while nursing an infected patient through contact with blood, vomit, other infected secretions, or infected tissues [10]. Additionally, filoviruses have also been documented to be transmissible by aerosol [20–22]. Another disconcerting property of the filoviruses is that they can be fairly stable, even when treated under harsh environmental conditions, and can survive in dried human blood for several days [11,22,23].

14.2.1 History

The first recognized filovirus outbreak took place in Marburg, Germany, during 1967, caused by the import of MARV-infected monkeys from Uganda [24,25]. While only a few natural outbreaks of MARV have been recognized, nearly 20 confirmed outbreaks of EBOV have occurred [26]. However, MARV should not be underestimated; a serious outbreak occurred during 1998–2000 in the Durba region of Africa, associated with a gold mining operation, and a second outbreak in 2005 in the Uige Province in northern Angola, West Africa, where the mortality rates were 80%–90% [27–29]. It is possible that the limited number of MARV outbreaks reflect the lack of ecological disturbances that have so recently characterized the increased numbers of EBOV outbreaks. The first recognized outbreak of EBOV, identified in 1976, was named after a small river in the Yambuku region of Zaire. Two simultaneous outbreaks occurred during 1976 in northern Zaire (currently the Democratic Republic of Congo) and southeast Sudan. As EBOV is transmitted through bodily fluids, transmission often occurs while nursing an infected patient, contact with blood, vomit, or other bodily fluid. This was a major cause of spread of disease among the 1976 outbreaks [10]. Lasting from June to November, the Sudan outbreak caused 151 deaths from 284 suspected and confirmed cases of EBOV with a mortality rate of 53% [10]. A second outbreak was observed from August to October 1976, in the Yambuku region of northeast Zaire, caused by a more lethal virus species (mortality rate of 88%; 280 deaths from 318 probable/confirmed cases) [10]. Transmission of disease during the Zaire outbreak was often associated with providing care for infected patients or receiving injections at the local hospital, suggesting that the use of contaminated needles contributed to the spread of disease. The institution of barrier nursing slowed the incidence of transmission in the two outbreaks. Later cases were identified and isolated much more rapidly than earlier ones, thus decreasing the risk of transmission to care-giving family members. Two slightly different viruses were found in these outbreaks and were named based on the location where they were isolated: Zaire virus (ZEBOV) and Sudan virus (SEBOV). It is now becoming clear, however, that there have been multiple separate, but virtually continuous, geographic and temporally spaced outbreaks in both human and great apes [30,31]. New proposed taxonomy defines two genera including *Marburgvirus* and *Ebolavirus* having six species including *Marburg marburgvirus, Bundibugyo ebolavirus (BEBOV), Sudan ebolavirus (SEBOV), Taï Forest ebolavirus, Zaire ebolavirus,* and a tentative third genus named *Lloviu cuevavirus* [32].

14.2.2 Biochemical Properties

The filoviruses are so named for their filamentous appearance. The shape of these viruses is highly variable from straight rods to an "S" shape and the so-called Shepherd's crook structures. The negative-sense RNA genome of the filoviruses encodes seven structural proteins and one nonstructural protein [33–35]. The transmembrane glycoprotein (GP) is located on the exterior of the virion and infected cells. It is thought to function in receptor binding and membrane fusion with the host cells, thus allowing viral entry [33]. Of the four structural proteins (VP24, VP30, VP35, and VP40), two are known to associate with genomic RNA. Together with the RNA-dependent RNA polymerase protein (L) and the nucleoprotein (NP), VP30 and VP35 form a ribonucleoprotein complex with the genomic RNA [33,36]. NP, VP35, and L are essential for replication and encapsidation of the viral genome; VP30 appears to be required for efficient transcription [33,37]. Additionally, VP35 can function as an interferon (IFN) antagonist, effectively blocking the IFN response mounted by the host [33,38,39]. The remaining structural proteins (VP24 and VP40) are membrane associated; they are thought to be on the interior of the membrane [33,36]. VP40 has been shown to associate with cellular membranes and drives viral assembly and budding; the MARV VP40 also appears to interfere with IFN responses in infected cells [33,40–46]. The role of VP24 is mostly unknown, but it may function as a minor matrix protein and also as an IFN antagonist [47–49]. The nonstructural protein, secreted GP (sGP), is found in the blood of viremic patients and animals [50]. Interestingly, MARV lacks sGP, despite its close relationship with EBOV [50,51]. There is some evidence that sGP contributes to immune evasion by interfering with a signal transduction pathway required for neutrophil activation after binding neutrophils. This claim is controversial, as it is uncertain whether sGP actually binds the neutrophils [52–54].

14.2.3 Diagnostics

In any infectious disease, whether a natural outbreak or bioterrorism event, rapid recognition and accurate diagnosis is the key to epidemic containment and control. Once the disease agent is identified, its natural behavior, pathogenesis, and effective countermeasures become known. The “first responder” for Ebola HF will most likely be a healthcare professional. The professional must be observant, take a careful, thorough history, perform a thorough medical examination, take appropriate precautionary measures, and request or apply the appropriate diagnostic criteria and tests. In the case of filoviruses, for which there is no effective vaccine or therapeutic treatment, quarantine remains the one effective public health protective measure in the public health arsenal. Maintenance of appropriate quarantine and barrier nursing techniques are absolutely essential. It is more prudent to sacrifice a small degree of absolute diagnostic accuracy and proceed with barrier nursing and quarantine procedures in a suspect case of filovirus infection (which is appropriate for all VHFs and many other contagious diseases), while waiting for confirmatory assays, than to wait for absolute confirmatory diagnostics and delay implementation of control strategies [13].

A variety of accurate diagnostic tests are available for EBOV; some are simple, some are complex, and which ones are used is often situational, such as the geographic location where the disease is suspected. Virus isolation, the “gold standard” of diagnostics, may not be practical, or even possible, for every case; it may only be necessary to establish the primary or index cases. Once the case definition is established, expediency that allows a small degree of error may be more acceptable than prolonged or complex tests. In the diagnosis of the filoviruses, false positives are more acceptable than false negatives. The various diagnostic tests for EBOV and MARV have included IgG and IgM antibody tests, immunofluorescent antibody tests, antigen-capture enzyme-linked immunosorbent assay (ELISA) along with Western Blot confirmation, radioimmunoprecipitation assays (RIPAs), reverse transcriptase polymerase chain reaction (RT-PCR), and a fluorogenic 5′-nuclease assay [55]. A combination of multiple tests is recommended, as all provide complementary information useful in epidemiological studies, and include genetic information as well, which becomes important in these studies [56]. Antigen capture ELISA is quick, easy, robust, and adaptable to large numbers of samples. Virus isolation takes 1–2 weeks in a BSL-4 facility and a “cold-chain,” to preserve the sample, must be maintained when shipping to an approved reference laboratory. Virus isolation is still the gold standard for diagnostics, and is recommended, at a minimum, for the initial and selected cases (particularly those that differ from the developed case definition) in the outbreak to establish a sample library for future epidemiological reference and studies. ELISA and RT-PCR are the most useful initial rapid diagnostic techniques for antigen detection in the acute clinical setting [1].

14.2.4 Human Disease

Relatively little is known about filovirus HF pathogenesis in humans, as it must be manipulated under BSL-4 conditions. Additionally, the limited number of known outbreaks throughout history has restricted the opportunity to study pathogenesis in infected humans. Filoviruses cause a HF disease that is highly lethal with case mortality rates up to 90%. Most of the clinical information on Ebola HF is derived from SEBOV and ZEBOV cases, which are the two most virulent human pathogenic species. In contrast, the Reston ebolavirus (REBOV) has not caused any known disease in humans and for the Cote d'Ivoire ebolavirus (CIEBOV) only a single nonlethal case has been reported [57]. Outbreaks with MARV have been less frequent; however, the latest outbreaks in Durba and Angola showed very high mortality rates [27–29]. The incubation period ranges from 2 to 21 days after which infected individuals present with symptoms such as fever, chills, headache, myalgia, and anorexia [58,59]. Additional symptoms such as abdominal pain, sore throat, nausea, vomiting, cough, arthralgia, diarrhea, and pharyngeal and conjunctival injections may follow after this initial period [59]. At the peak of the illness, patients become dehydrated, apathetic, and disoriented, and may develop a hemorrhagic manifestations including a characteristic, nonpuritic, maculopapular centripetal rash associated with varying degrees of erythema and desquamation, as well as gastrointestinal bleeding [58,59]. Fatal cases develop clinical signs early during infection and death commonly occurs between days 6 and 16 [58,59]. The following laboratory parameters can be associated with the disease: leucopoenia, thrombocytopoenia, lymphopenia, elevated serum transaminase levels [aspartate aminotransferase (AST) > alanine aminotransferase (ALT)], hyperproteinaemia, and proteinuria [59–61]. Prothrombin and partial thromboplastin times are prolonged and fibrin split products are detectable. Late in the disease course, secondary bacterial infection may lead to elevated white blood cell counts. There is an increased risk of abortion for pregnant women [61].

In both humans as well as animal models, the virus infects antigen-presenting cells, primarily macrophages and dendritic cells (DCs), in the early course of disease; infection of these cells is currently considered the primary event in the pathogenesis of filovirus infection [62–66]. Endothelial cells are infected, although the exact kinetics and role of their infection in filovirus pathogenesis is disputed [62,67–69]. Many cellular receptors have been proposed to mediate EBOV and/or MARV binding and entry to the cell, including the asialoglycoprotein receptor on hepatocytes, human macrophage C-type lectin specific for galactose and *N*-acetylgalactosamine, DC-specific ICAM-3-grabbing nonintegrin (DC-SIGN), the related DC-SIGNR, and the α-folate receptor [70–75]. However, the exact nature of and requirement for these cellular receptors is unknown and it is possible that the heavy glycosylation of GP alone can bind and mediate entry through cell-surface lectins [76]. This is supported by the fact that several C-type lectin molecules, including the human macrophage C-type lectin specific for galactose and *N*-acetylgalactosamine and DC-SIGN are believed to bind EBOV GP and mediate entry into macrophages, DC, and endothelial cells [71,74]. Most significantly, EBOV is able to evade the host immune response while replicating in the critical antigen-presenting cells. Filoviruses are capable of infecting and replicating to very high levels in human and nonhuman primate DCs. Filovirus infection impairs the normal DC responses to infection and the DCs fail to become activated or mature, thus prohibiting normal innate and adaptive immune responses [64,65,69]. Additionally, the activation of macrophages may be impaired; however, the ability of monocytes and macrophages to respond to filovirus infection is controversial [62,65,67,77]. Abbarent cytokine and chemokine production are predictors of fatal infection [78,79].

Whereas clinical signs such as fever, petechiae and hemorrhage are well-known, internal symptoms are relatively unexplored. However, EBOV survivors have been documented to have an early and short-lived rise in various serum cytokines, suggesting that initiation of effective innate immune responses is key to overcoming EBOV infection [80–83]. Infected monocytes and macrophages may be the major mediators of this inflammatory response, whereby cytokine and chemokine secretion increases the permeability of endothelial layer and causes induction of shock [62,67]. The inflammatory response is quickly followed by a T cell response and an increase in the markers of the activation of cytotoxic T cells [80,82]. Additionally, an early and increasing EBOV-specific IgG and transient IgM antibody response is seen in EBOV survivors; the EBOV-specific IgG is detectable up to 2 years after infection [80,82,84]. Patients who develop rapid immune responses seem to clear circulating EBOV antigen rapidly, indicating that rapid induction of the appropriate immune responses in humans can result in survival from filovirus infection [82]. In contrast, victims of EBOV fail to mount a substantial cellular or humoral immune response. While T cells are activated and IFN-γ is secreted early in infection, these responses quickly disappear and are not observed at the time of death; the disappearance of T cells in the periphery is presumably due to apoptosis by a yet unidentified mechanism [21,82,85]. Even with the clinical observations, little is known from humans as to which organs and cell types are early or late targets of EBOV infection.

14.2.5 Animal Models

Several animal species have been modeled for use in vaccine, pathogenesis, and therapeutic studies for filovirus infections. Various species are sensitive to natural isolates of filoviruses. This includes the suckling mouse (which can be used to isolate the virus), certain immunodeficient mouse strains (including SCID and IFN-deficient mice), hamster, and various species of nonhuman primates, including the African green monkey, rhesus and cynomolgus macaques, baboon, chimpanzee, and gorilla. Other animal species including guinea pigs, immunocompetent mice, and adult hamsters have required multiple passages to adapt the virus in order to establish lethal infection models [86–94]. The selection of appropriate animal models for the study of Ebola and Marburg HF pathogenesis has been difficult for a number of reasons, not the least of which is the lack of adequate, well-preserved, and well-characterized human clinical and autopsy data to use for modeling. Despite nearly 1500 fatal cases of Ebola HF and 450 cases of Marburg HF, only a limited number of tissues from fewer than 30 cases of EBOV and 10 cases of MARV infection have been examined, respectively.

14.2.5.1 Nonhuman Primates

EBOV and MARV infections cause complex disease characterized by pathophysiological shock, multiple organ dysfunction, disruption of the coagulation system, disseminated intravascular coagulation (DIC), and profound acute immune suppression. There are also probably endocrine disruptions caused by involvement of the hypothalamic–pituitary–adrenal axis, likely involving disrupted adrenal gland dysfunction, given the extent of virus-infected cells within the adrenal gland [95]. The only species that reliably reproduces all of the complex interactions of the clinical, histopathological, and pathophysiological aspects of the disease in humans is the nonhuman primate. This is not to say that the other models are not valuable in identifying individual pathways of pathogenic and host response mechanisms; in many cases, the simpler models are easier to work with in defining basic mechanistic concepts. However, the nonhuman primate is considered the final reliable predictor of vaccine and/or therapeutic efficacy, or of definitive complex pathogenesis pathways. There are differences in susceptibilities and clinical manifestations to EBOV and MARV between nonhuman primate species. Nutritional status, age differences among study animals, and concurrent diseases, particularly bacterial and parasitic infections, impacted early studies before purpose-bred monkeys became the norm for studies.

EBOV (Zaire virus) has been modeled in African green monkeys (*Chlorocebus aethiops*, formerly *Cercopithecus aethiops*), cynomolgus macaques (*Macaca fascicularis*), rhesus macaques (*Macaca mulatta*), and hamadryad baboons (*Papio hamadryas*) [26,96–98]. MARV has been modeled in the above monkey species as well as in squirrel monkeys (*Saimiri scireus*) [98]. The pathological data is generally consistent among nonhuman primates, with some minor variations. For instance, African green monkeys fail to present with the macular cutaneous rash following either EBOV or MARV infection, despite the fact that a rash is a characteristic feature of disease in the other established monkey models (Figure 14.1) as well as in human disease [85,95,99–102]. Also, baboons show hemorrhage, rather than fibrin deposition, as a manifestation of DIC [103,104].

Limited data are present on EBOV infection with other viral species and strains. For SEBOV, the infection appears to be uniformly lethal in cynomolgus macaques; however, the course of SEBOV is more prolonged than that of EBOV (Zaire) [97,105–110]. Although REBOV does not appear to cause lethal disease in humans, the infection of macaques causes severe morbidity and mortality, although African green monkeys appear to be resistant [110–113]. CIEBOV (aka Taï Forest ebolavirus) appears to be less virulent in both humans and macaques, as no human deaths are attributed to this virus species and it does not cause uniform lethality in cynomolgus macaques [107,114]. Pathogenesis of the other EBOV species, including Cote d'Ivoire (Figure 14.2), Reston, and Bundibugyo, appears to be similar to the more studied Zaire virus, except that the disease course is extended and mortality is not always uniform. Cynomolgus or rhesus macaques have become the preferred species for filovirus vaccine and therapeutic work, and are used almost exclusively in the more recent filoviral studies. The challenge doses [10–1000 plaque-forming units (pfu)] and routes (intramuscular or subcutaneous) in these models are designed to mirror an accidental laboratory exposure. For EBOV, a challenge dose of 1000 pfu proves lethal within 5–7 days after challenge in cynomolgus macaques and 7–10 days after challenge in rhesus macaques [26,96,97]. Due to the potential use of the filoviruses as a bioweapon, more recent focus has examined the ability of aerosol exposure to cause morbidity and mortality in nonhuman primates (macaques and African green monkeys) [21,108,115–118].

Extensive analysis of EBOV pathogenesis in Zaire virus-infected cynomolgus macaques was carried out using serial sampling studies [63,66,69]. Clinical symptoms of disease including fever, macular cutaneous rashes, anorexia, mild dehydration, diarrhea, depression, and bleeding generally manifest between 3 and

4 days following infection [66,69]. Over the course of disease, the total white blood cell counts and prevalence of granulocytes in the leukocyte population increase with a concomitant loss of monocytes and NK and T cells [66,69,119]. As would be expected of a VHF, platelet counts decreased over the course of EBOV infection, as did levels of hemoglobin, hematocrit, and erythrocytes [66,69]. Other hemorrhagic manifestations, including bleeding from the gastrointestinal (Figure 14.1) and urogenital tracts, petechia, and hemorrhage from injection sites and mucous membranes, may develop during the peak of the illness (usually 5–10 days after onset of symptoms).

Nonhuman primates are susceptible to human isolates of MARV directly from blood or organ homogenates without passage [120–122]. Experimentally, the incubation period for MARV in monkeys is 2–6 days with death typically occurring between 8 and 11 days following infection, but varies according to the amount, route, and viral isolate used as the inoculum. The initial sign of MARV infection is fever, which may begin as early as 2 days following infection, and other clinical signs including anorexia, rash, huddling, weight loss, dehydration, diarrhea, prostration, failure to respond to

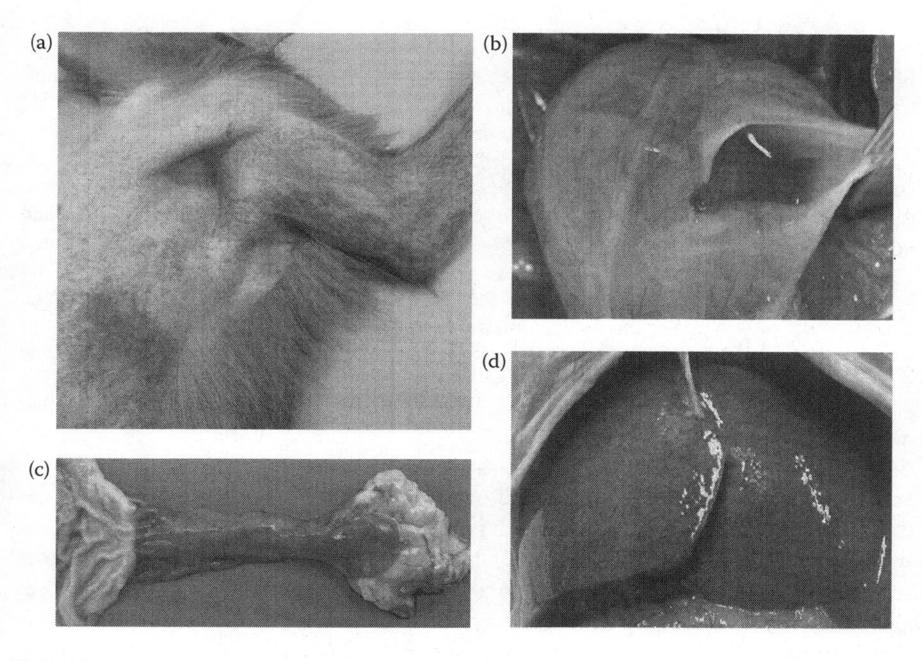

FIGURE 14.1 **(See color insert.)** Representative gross necropsy lesions from nonhuman primates experimentally infected with hemorrhagic fever viruses. (a) Typical petechial rash of the left arm and chest of a rhesus macaque 11 days after infection with the Marburg virus (Musoke strain). (b) Accumulation of fluid in the pericardial cavity of a cynomolgus monkey 13 days after infection with the Lassa virus (Josiah strain). (c) Marked congestion of the duodenum at the gastroduodenal junction of a cynomolgus monkey 5 days after infection with the Zaire ebolavirus. (d) Reticulation and discoloration of the liver 11 days after infection with the Lassa virus (Josiah strain).

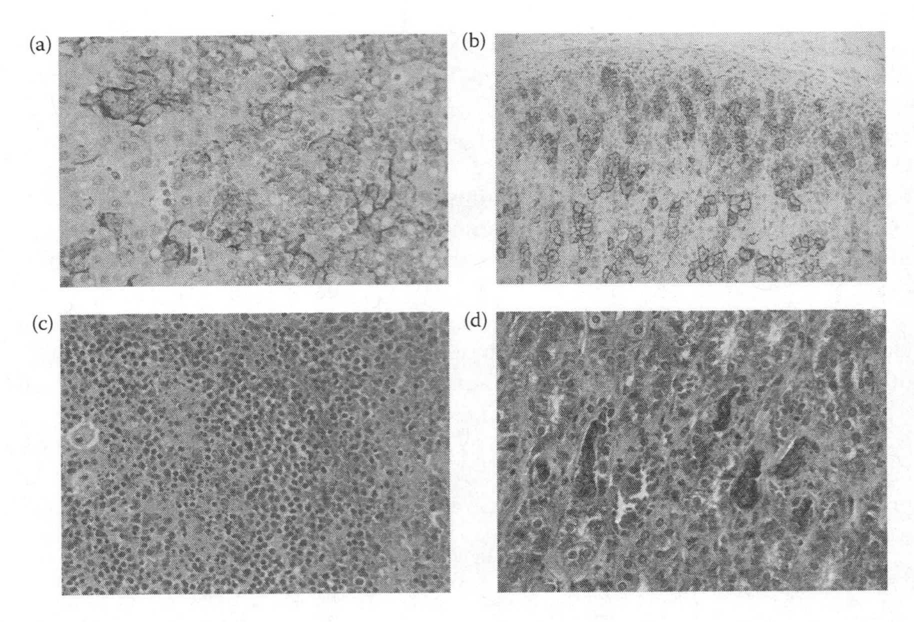

FIGURE 14.2 (**See color insert.**) Immunohistochemical staining patterns and histopathology of nonhuman primates experimentally infected with hemorrhagic fever viruses. (a) Prominent immunostaining of cells within the zona glomerulosa and zona fasciculata of the adrenal gland of a cynomolgus monkey 15 days after infection with the Lassa virus (Josiah strain) (immunoperoxidase staining; original magnification, 20×). (b) Immunopositive hepatocytes peripheral to and within an inflammatory foci of lymphocytes, macrophages, and fewer neutrophils in the liver of a cynomolgus monkey 15 days after infection with the Lassa virus (Josiah strain) (immunoperoxidase staining; original magnification, 40×). (c) Phosphotungstic acid hematoxylin stain of the kidney from cynomolgus monkey showing an abundance of polymerized fibrin in medullary vessels 12 days after infection with the Ivory Coast ebolavirus (original magnification, 20×). (d) Necrosis and apoptosis of lymphocytes with concomitant lymphoid depletion and hemorrhage in the spleen of rhesus monkey 7 days after infection with the Marburg virus (Musoke strain) (hematoxylin and eosin stain; original magnification, 40×).

stimulation, and bleeding from body orifices develop later. In macaques, the maculopapular rash (Figure 14.1), which is centered on hair follicles, usually develops between 4 and 6 days following MARV infection. It is seen principally on the flexor surfaces of the arms and thighs, and to a lesser extent on the thorax, abdomen, face, and neck. The course of MARV in nonhuman primates is rapidly and almost invariably fatal once clinical signs appear. MARV is present in the blood, saliva, and urine. Early hematological changes include lymphocytosis (Figure 14.2) resulting in profound lymphopenia, variable neutrophilia, and profound thrombocytopenia beginning around day 5 or 6 of the infection. Eosinopenia has been observed in African green monkeys, but has not yet been described in other nonhuman primate species [120].

The principal gross necropsy lesions in monkeys following MARV infection are similar to those seen following EBOV infection. Variably present, but more often than not, are splenomegaly, enlarged fatty liver, enlarged mesenteric lymph nodes, consolidated hemorrhagic areas in the lungs, and vascular congestion. There is petechial and/or ecchymotic hemorrhage in the organs and vessels of the entire gastrointestinal tract, reproductive tract, adrenal gland, pancreas, liver, spleen, brain, and heart. Hemorrhagic effusion in the pericardium, pleural, and abdominal cavity is variably but often present, as is hemorrhage from body orifices. The primary microscopic lesions of MARV in nonhuman primates occur in the liver, spleen, and mononuclear phagocytic system, which include fixed and circulating macrophages. Lesions include depletion and apoptosis of lymphoid tissue and multifocal necrosis of liver tissue with the presence of apoptotic bodies, swelling of Kupffer cells, hypertrophy of the macrophage and DCs in lymphoid tissues, perivascular hemorrhages in the brain, and sometimes also pulmonary hemorrhage [95,97,120,121,123–125]. Foci of MARV-infected cells lesions are observed in pancreatic islets of cynomolgus monkeys [126]. Depletion to complete destruction of the white pulp elements of the spleen are observed, along with red pulp disruption by masses of necrotic debris. The damage within the spleen and lymph nodes is due to follicular apoptosis and is noted in both MARV- and EBOV-infected nonhuman primates [95,97,120,127].

One of the benefits of the nonhuman primate model is the ability of the monkeys to succumb to the natural human isolates of EBOV and MARV. The pathogenesis of EBOV and MARV infections in nonhuman primates closely mirrors that seen in human disease [97,128,129]. Additionally, clinical symptoms such as fever, rash, bleeding, anorexia, dehydration, and recumbency have been observed in both monkeys and humans [66]. However, this model also has its inconsistencies, as compared to human disease. The nearly uniform mortality observed in cynomolgus and rhesus macaques following filovirus infections does not accurately model that observed in natural human outbreaks (53%–88%). As another example, cynomolgus monkeys rapidly and uniformly succumb to REBOV, which is apparently avirulent in humans [113,130].

Although the majority of filovirus studies have been performed in cynomolgus and rhesus macaques, the baboon may be a model to consider for further future development. The baboon species is not a 100% lethal model [131–133]. The baboon survivability rate from EBOV infection is probably more consistent with human survival percentages than with other nonhuman primate models. For this reason, a case may be made that baboons make a better model for the testing of candidate lead therapies for filoviruses [131–133]. However, although the 100% lethal monkey model is an extremely stringent one, it is desirable from a statistical and numerical point of view in order to test vaccine and therapeutic efficacy with the smallest numbers of nonhuman primates possible. In future studies, however, it may be desirable to challenge nonhuman primates with sublethal doses, performing basic studies, so that prognostic clinical indicators of survivability can be explored. The use of rhesus rather than cynomolgus macaques for vaccine studies should also be considered for the following reasons [90,95,96,134]. Rhesus monkeys have been the standard for the pharmaceutical industry, especially in the case of the human immunodeficiency virus. Many of the immunological markers for rhesus macaques are well defined and commercially available; rhesus macquaes can be typed across more than 21 major histocompatibility complex class I alleles, unlike cynomolgus macaques. Nonhuman primate models appear to have complex, primate-specific, lethal, pathophysiological manifestations, consistent with the human disease characteristics that the current rodent models do not possess. Therefore, the nonhuman primate models are currently the model of choice for final vaccine, therapeutic, and pathogenesis studies of filoviruses.

14.2.5.2 Mouse

Suckling mice, as well as adult knockout mice lacking IFN responses or those with severe combined immunodeficiency (SCID), are sensitive to wild-type filoviruses [135,136]. EBOV was adapted, through serial passage, to effect disease and death in mice [135–138]. Bray et al. [136,139] visually plaque-picked EBOV and passed it through mouse spleens, reselecting visually identical plaques in order to adapt EBOV to immunocompetent BALB/c mice. Sequence analysis shows only five amino acid changes in the entire viral genome relative to the precursor virus and an isolate from the 1976 outbreak; these changes occur in NP, VP35, VP24, and L [129]. Mouse adaptation of MARV was elusive until the use of a different approach where initial passage of the virus population occurred in SCID mice until the time to death after inoculation was similar to that observed in nonhuman primates [140] and then the virus was passaged in immunocompetent BALB/c mice until lethality was observed [141]. Serial passage of MARV strains "Ci67" and "Ravn" in liver homogenates of infected SCID mice was performed for 15 passages and for 10 passages, respectively, prior to the transfer of viruses in the liver homogenates into immunocompetent mice [140–142]. The lethal mouse MARV Ci67 viral variant contained 11–13 nucleotides differences between progenitor and the lethal immunocompetent mouse viral variant [142]; in contrast, the adaptation of MARV Ravn appeared to require extensive biased hypermutagenesis that generated approximately five times the number of nucleotide differences from the progenitory virus with a majority of Thy to Cyt transitions [141,142]. There are currently no mouse models for the remaining filoviruses [143].

When injected into mice intraperitoneally, the mouse-adapted ZEBOV and MARV are lethal at low doses ($LD_{50} < 50$ pfu), but not following mucosal (e.g., aerosol or intranasal), intramuscular, or subcutaneous inoculation unlike nonhuman primates [136,139,141]. The reason for this is unknown, but undoubtedly is related to the viral pathogenesis, more specifically due to EBOV and MARV tropism and induction of innate immunity [144,145]. Interestingly, lymphotoxin-α knockout mice, which do not have lymph nodes, but have normal innate and adaptive immune responses, apparently develop lethal illnesses after subcutaneous infections of mouse-adapted EBOV, suggesting multiple factors in the immune system beyond the simplistic model of innate versus adaptive may be instrumental in filovirus protection [135]. While the subcutaneous route does not induce disease, it is capable of protecting mice from subsequent intraperitoneal challenge with mouse-adapted ZEBOV [135,136,138].

The disease course in the murine model is similar to that observed in nonhuman primates and guinea pigs. An inflammatory cytokine response is observed early in the course of infection, and the cellular disease targets are conserved across the models; lymphopenia and neutrophilia are observed in all models [136,139–141,146]. The largest difference of the mouse models is the lack of fibrin deposits and DIC in the mouse after infection [96,129,146]. However, the prevalence of the fibrin deposits and

DIC in humans and even in nonhuman primates is disputed, as the strain of virus or the monkey species evaluated appears to influence their production [129]. Additionally, mice fail to develop the rash characteristic of filovirus infection in other models [96,141]. However, the cellular hallmarks of disease in mice are very similar to the pathogenesis observed in the nonhuman primate and guinea pig models, making the murine model an effective means of evaluation of the course of disease, efficacy of vaccines, and mechanisms of potential therapeutic drugs [129,136,139,141,146–149]. In fact, the evaluation of most candidate EBOV vaccines began in mice before moving forward to testing in nonhuman primates [129,146,147,150–154]. Additionally, the mouse models have been used extensively to evaluate potential drug targets [140,149,155–157]. Importantly, the lethality of mouse-adapted EBOV has been tested in two other animal models. The mouse-adapted isolate has slightly decreased virulence in monkeys compared to the wild-type ZEBOV isolate from which it is derived, as only one of three cynomolgus macaques succumbed to virus infection, although the two remaining animals showed signs of severe disease prior to recovery [137]. Interestingly, guinea pigs succumb to infection with the mouse-adapted ZEBOV, but not the wild-type ZEBOV, which required eight passages before complete lethality in ZEBOV-infected guinea pigs was observed [129].

An effective mouse model for study of human diseases is quite desirable as mice are easily manipulated, require very little in way of housing, and are readily available for large-scale experiments. One distinct advantage of any murine model is the availability of genetically engineered "knockout" mice, which can easily help determine mechanisms of immunity; for instance, the use of knockout mice has helped determine which immune components are needed for protective immune responses to EBOV infection [136,139,154]. Performing thorough pathogenesis and immunology studies in mice with significant numbers per time point is less taxing, more cost-effective, and potentially safer than similar studies in nonhuman primates. The mouse model has been used as a means to test and refine candidate vaccines and therapeutics prior to testing in nonhuman primates. Given all the advantages of the mouse model, especially their availability and the wide array of reagents with which their immune response can be evaluated, it is particularly useful for preliminary diagnostics and evaluation of potentially protective immune responses. Caution should be used during studies with the mouse model as confirmation in guinea pigs and nonhuman primates is nearly always necessary due to differences in the animal models and to known human pathology. As the mouse does not experience DIC, which often occurs in monkeys and humans, or bystander apoptosis, which appears to play a critical role in the pathogenesis of the human disease, this creates some limitations in its use as a model, and may be partially responsible for its inability to accurately predict survival of EBOV infections of primates following vaccination or therapeutic regimens.

14.2.5.3 Hamster

Suckling hamsters are susceptible to naturally occurring isolates of EBOV and MARV. Little work has been done on developing a hamster model for EBOV. In contrast, early MARV work was done with the hamster model, where nine serial passages were required to establish the desired lethality in the adult hamster model. Organ changes were consistent with those in other model systems, including liver, spleen, and lung damage [123,158]. However, the character of the MARV infection changed dramatically upon high passage in adult hamsters, when the tenth passage of MARV caused severe brain lesions in suckling hamsters inoculated intracranially or intraperitoneally. The increased neurovirulence was probably caused by repeated selection for an unknown enhanced virulence factor present in MARV, but this very severe encephalitis lesion has not been reported in any other model system. Severe central nervous system lesions also occurred when adult hamsters were inoculated intracranially with high passage MARV, but only perivascular hemorrhages occurred in the central nervous system of adult hamsters when they were inoculated intraperitoneally with this same adapted strain [123,158]. The neurotropism identified in the suckling hamster during attempts to establish the original MARV models is troubling and should be characterized to determine whether it is a species-specific adaptation associated with mutation of VP24, as has been suggested by work in the EBOV guinea pig model [87]. Any developed vaccine will likely need to undergo neurovirulence testing in nonhuman primates prior to approval, especially since encephalitis was documented in three of the original seven fatal Marburg HF cases during the original 1967 MARV outbreak [121].

The hamster model is among the least developed of animal models for EBOV or MARV. Furthermore, the immunological reagents that are available for characterizing pathogenic and immunological events following filovirus infections in hamsters are limited. Hamsters are difficult to work with under BSL-4 conditions due to their temperament, and they are prone to biting. Due to the ease of use of other rodent models, the use of hamsters in filovirus studies has not been widespread. With the limited dexterity imposed by multiple layers of protective gloves and the potential safety hazard involved using filoviruses, there is little reason at this time to recommend the hamster over other safer, better characterized, and more conventional models for routine studies of EBOV or MARV. However, the hamster model may be useful to study MARV-induced effects on the central nervous system, as central nervous system lesions and symptoms do occur in humans and the hamster is the only model in which the development of these lesions has been consistently observed.

14.2.5.4 Guinea Pig

The guinea pig has also been frequently used as a model in filovirus infections. Both EBOV and MARV strains have been adapted to cause both disease and lethality in inbred and outbred guinea pigs, by passaging the human isolate of interest through guinea pigs [86–88,90,92,159,160]. Initial infection of the guinea pig with EBOV and MARV results in a febrile, nonfatal illness. Challenge of guinea pigs results in the infection of mononuclear phagocytes, as well as damage to the liver, spleen, adrenal glands, and kidneys [86,88,92,103,104,161,162]. High viremias, fever, lethargy, and anorexia precede death, which usually occurs 8–14 days after infection. Hemorrhage does consistently occur [86,136,139], although signs of bleeding from bodily orifices and mucosal surfaces are observed more often in outbred Hartley guinea pigs than in inbred strain 13 guinea pigs (K.L. Warfield and T.W. Geisbert, unpublished observations).

The adaptation to lethality of the different EBOV and MARV strains has taken a range of passages from two up to nine serial passages through guinea pigs [86–88,90–92,120,123,159,160]. The histological lesions in guinea pigs are similar to those observed in nonhuman primates, although slightly more granulomatous, especially in the liver [88,92,104,161,162]. This suggests that, unlike the nonhuman primate, the guinea pig macrophage/monocyte responses to filovirus infections remain, at least partially, functionally intact. As the guinea pig monocytic response still has the capability of responding to tissue necrosis, the inflammatory response of guinea pigs to filovirus infections is clearly different from that of nonhuman primates. However, the sequence of infection and tissue predilection is the same [86,88,92].

Elegant work by Ryabchikova and colleagues [88,92,103,104,162] has shown that repeated passage of filoviruses may increase the virulence in both guinea pigs and nonhuman primates and cause disease in rodents that is more similar to humans than lower passages. Characterization of serial passage and adaptation of EBOV to the guinea pig has demonstrated that, by continuous selective passage, this inflammatory response can be overcome by performing more passages, and that important mutations occur in VP24 in the fifth to ninth passages which decrease the inflammatory response of the guinea pig and allow the virus to multiply unchecked. VP24 is proposed to be significant for the process of MARV adaptation to guinea pigs, and for species adaptation of the filoviruses [87]. A single paper describes minimal encephalitic glial nodule formation in 5 of 15 experimentally infected guinea pigs [163]. No other reports could be located that associate this lesion with MARV in the guinea pig in the more recent literature. It must also be kept in mind that due to the danger of the organism and the increased danger during tissue collection, the brain is often not examined in routine studies.

The principal shortcomings in using guinea pigs as a model are a considerable lack of available characterized and defined immunological reagents, variable fibrin deposition, and DIC compared to the coagulopathy that is often observed in humans, and the failure, thus far, to predict the efficacy of antifilovirus therapeutics or vaccines in nonhuman primates. As with the mouse model, the ease with which guinea pigs can be manipulated, stored, and obtained makes them attractive subjects in vaccine and therapeutic studies. Proof of concept and efficacy studies are often evaluated in rodents before moving to nonhuman primates, which are more expensive, harder to obtain, and much more difficult and hazardous to house and manipulate. Certainly, despite some dissimilarities of this model to human and nonhuman primate disease, it will continue to be useful in screening novel treatments for filoviruses, especially when optimizing dose, schedule, and composition of lead therapeutics and vaccines prior to final testing in nonhuman primates.

14.2.5.5 Other Animal Models

Species considered resistant to EBOV, which have been used for the production of antibodies for immunodiagnostics or immune or hyperimmune sera to treat filovirus-infected individuals, include: horse, goat, sheep, rabbit, and the immunocompetent mouse [20,144,156,164–169].

14.2.6 Summary

To date, various species of the nonhuman primates have most faithfully reproduced the coagulopathy, hemorrhagic, and pathophysiological shock syndrome demonstrated in filovirus-infected humans. The immunological phenomena, which are clearly a key event in the pathogenesis of filovirus infection, have still not been well defined or characterized in the nonhuman primate. Most of that data has come from other models, primarily the mouse. The filoviruses must be adapted to affect disease and morbidity in mice and guinea pigs. Utilization of these animal models, as well as *in vitro* experiments, has and will continue to furnish a great deal of insight regarding the cellular targets and pathogenesis of these dangerous viruses.

14.3 ARENAVIRIDAE

The family Arenaviridae consists of 22 recognized viruses and 9 newly discovered putative species that are divided into the Old and New World arenaviruses [170,171]. Most of the arenaviruses have rodent reservoirs, with the exception of the Tacaribe virus that has only been isolated from the Artibeus fruit-eating bat [172]. At least seven arenaviruses are associated with VHF in humans. Lassa, Lujo, Junin, Guanarito, and Sabia viruses are known to cause HF in western Africa, southern Africa, Argentina, Venezuela, and Brazil, respectively, whereas Machupo and Chapare viruses have caused HF in Bolivia. Arenaviruses have enveloped, spherical virions that encapsidate the viral genome, and consist of two single-stranded, negative-sense RNA molecules encoding the five viral proteins: GP1, GP2, NP, Z, and L [173]. In nature, arenaviruses are classically transmitted to humans by inhalation of aerosols present in rodent urine and feces, by ingestion of food contaminated with rodent excreta, or by direct contact of rodent excreta with abraded skin and mucous membranes. Person-to-person transmission of arenaviruses happens infrequently but can occur by direct contact with infectious blood and bodily fluids, and has been reported for both Lassa fever and Bolivian HF [174–177]. As a family, the Arenaviridae have several noteworthy characteristics. Arenaviruses that cause disease in humans have the capacity to induce persistent infection in their natural hosts with chronic viremia and viruria; the epidemiological implications are obvious. Viral replication is not associated with extensive cell damage or cytopathic effect either *in vitro* or *in vivo*.

Of the Old World arenaviruses that cause disease in humans, the Lassa virus is the most significant pathogen. Lassa infection is endemic in West African countries and it has been estimated that there are 300,000–500,000 cases of Lassa infection annually with approximately 5000 deaths [178]. The Lassa virus is maintained and spread by *Mastomys* species, more commonly known as the multimammate rat [178]. The symptoms of Lassa infection include fever, headache, malaise, myalgia, retrosternal pain, cough, and gastrointestinal symptoms. About 70% of patients experience pharyngitis, and over half of these have exudates in the tonsillar fauces [179]. A minority of cases present with classic symptoms of bleeding, neck/facial swelling, and shock. Most patients have serum chemistries consistent with acute hepatitis, although icterus is rare. Clinically, the degree of AST elevation is correlated with the outcome of human disease. The disease is particularly severe in pregnant women and their offspring. Deafness is a common sequela, occurring in ~30% of convalescent patients [179,180]. Meningoencephalitis and pericarditis have also been reported as have pleural and pericardial effusions [179]. In contrast to Argentine HF and Bolivian HF (caused by Junin and Machupo viruses, respectively), patients with Lassa fever generally have much higher viremia of longer duration. Additionally, glycoprotein-neutralizing antibodies require 2–6 months to form, and usually never achieve significant levels [179]. The fatality rates of hospitalized cases are between 15% and 20% [178,180,181]. There are no vaccines or preventative modalities for Lassa fever. The antiviral drug ribavirin was shown to reduce mortality from Lassa fever in high-risk patients [182] and, along with supportive care, is the recommended treatment for Lassa virus infection [183]. Laboratory diagnosis of Lassa infection is based on isolation of virus, antigen detection by ELISA, or detection of viral RNA by RT-PCR from patient sera. Alternately, the presence of or increase in virus-specific IgM or IgG by ELISA or indirect fluorescent antibody can be used as a laboratory diagnostic [184]. ELISA and RT-PCR are the most useful diagnostic techniques for antigen detection in the acute clinical setting, especially when the causative agent is unknown.

Several animal models have been developed for studying Lassa fever including nonhuman primates, mice, and guinea pigs. At least five species of nonhuman primates have been shown to be susceptible to parenteral or aerosol Lassa virus challenge including African green monkeys (*Chlorocebus aethiops*) [185], hamadryad baboons (*Papio hamadryas*) [186], cynomolgus macaques (*Macaca fascicularis*) [185,187,188], rhesus macaques (*Macaca mulatta*) [185,189–192], and common marmosets (*Callithrix jacchus*) [193]. Since these models have been established, the rhesus and cynomolgus macaques, which both succumb to Lassa virus challenge, have been the most widely used to study the pathogenesis of and immunity to Lassa virus infections. There appear to be differences in pathogenicity of Lassa in nonhuman primates associated with both the strain of virus, dose of virus, and species of host. For example, whereas the Josiah strain of Lassa virus causes a uniformly lethal disease in cynomolgus monkeys, it does not cause a uniformly lethal disease in rhesus macaques with around 60% of them succumbing to a comparable Lassa challenge [185]. Moreover, whereas the Josiah strain of Lassa virus causes uniform lethality in cynomolgus macaques, the AV strain of Lassa virus does not cause a uniformly lethal disease in cynomolgus monkeys [192]. A challenge dose appears to play an important role in disease in macaques regardless of Lassa strain, as several studies have shown that lower challenge doses can produce more severe and more lethal infections than higher challenge doses [185,192]. Pathological studies of Lassa virus-infected monkeys have shown many similarities to infected humans. Virus-induced lesions (Figure 14.1) were reported in the brain, lungs, liver, spleen, kidneys, and lymphatic tissues, but the extent of histopathological lesions was not severe enough to explain death on the basis of cytopathic effect in Lassa virus-infected monkeys (Figure 14.2) [189–191,194]. Subsequent studies showed that hematological and immunological alterations were associated with Lassa virus disease (Figure 14.2), and these were likely initiated by dysfunction of platelets, T lymphocytes, and endothelial cells that lead to Lassa virus-induced shock and death [188,192]. Squirrel monkeys (*Saimiri scirreus*) [185,195] and Capuchin monkeys (*Cebus* species) [185] have also been evaluated as models of Lassa fever; however, disease does not appear to be as severe as with other nonhuman primate species and ranges from minimal disease to low mortality [185].

Immunocompetent laboratory mice are resistant to the Lassa virus. Several Russian publications have reported lethal infection of CBA/calac [196] and C3H/Sn [197] mice by intracerebral challenge with the Lassa (Josiah strain) virus; however, this does not mimic a natural route of infection and produces a disease that has little resemblance to human Lassa fever. Recent studies have shown that mice expressing humanized instead of murine major histocompatibility complex class I failed to control Lassa infection and developed severe Lassa fever [198]. The rodent model most studied for the Lassa virus is the guinea pig. Unlike the case with filoviruses, the Lassa virus can produce a lethal disease in guinea pigs without adaptation by serial passage [199,200]. The pathogenicity of the Lassa virus in guinea pigs is dependent on both the strain of virus and the strain of guinea pig. The Josiah strain of Lassa virus produces a uniformly lethal infection in strain 13 guinea pigs but only produces a lethal infection in about 30% of outbred Hartley guinea pigs [185]. Several isolates of the Lassa virus that were recovered from severely ill humans and produced lethal infection in cynomolgus macaques failed to produce lethal disease in guinea pigs, thus confirming the importance of virus and host strains [201]. Moreover, pathological studies showed that histological lesions in Lassa virus-infected guinea pigs were relatively mild and lacked the lung or liver damage noted in nonhuman primates [199,200]. Although Lassa infection in guinea pigs may not faithfully reproduce clinical disease and pathology seen in humans and nonhuman primates, guinea pigs have proven useful for screening candidate therapeutics and vaccines [202–204].

Infection with New World arenaviruses (Junin, Machupo, Guanarito, Sabia, and Chapare) that cause HF in humans begins with an insidious onset. Clinical signs of disease include fever, headache, malaise, myalgia, nausea, and vomiting. Hemorrhagic and neurological symptoms are frequently observed in severe cases and include petechial rashes, bleeding from the nose, gums, and gastrointestinal tract, tremor and lethargy. Lymphopenia and thrombocytopenia are also prominent clinical findings. Approximately one third of untreated cases will develop more severe hemorrhagic and/or neurological symptoms including diffuse ecchymoses, bleeding from mucous membranes and venipuncture sites, convulsions, and coma [205,206]. Mortality rates range from 5% to 30%. A live attenuated Junin vaccine was developed and is currently in use in Argentine [206,207]. Ribavirin in conjunction with supportive care is approved for use in treating HF caused by the New World arenaviruses [183] and was successfully used to treat a laboratory-acquired Sabia virus infection [208].

The pattern of human disease where Junin or Machupo infection of some patients is dominated by hemorrhagic signs whereas other cases are dominated by neurological manifestations recapitulates well in nonhuman primates. Rhesus monkeys are highly susceptible to the Junin virus [185,209–212] and infection of rhesus macaques with low-passage isolates of Junin viral strains results in distinctive hemorrhagic or neurological disease that correlates with clinical illness patterns present in the humans from whom the viral strains were obtained [209]. Infection of African green monkeys [213,214], cynomologus monkeys [215], or rhesus monkeys [215–220] with the Machupo virus typically results in a lethal infection. Approximately 20% of macaques infected with the Machupo virus survive the initial viremic phase of illness and are aviremic; however, between 3 and 5 weeks after infection, these animals develop neurological signs with severe intention tremors, ataxia, and coma and most of these animals succumb [215–220]. For the Guanarito virus, a single study evaluated rhesus monkeys as a potential model of human disease [221]. Although the animals showed clinical signs of illness consistent with human disease all animals survived challenge.

Marmosets have also been developed as models for Argentine HF and show clinical symptoms and pathology consistent with human infection including hemorrhages, thrombocytopenia, hepatocellular necrosis, and severe lymphoid depletion [222–224]. Infection of marmosets with the Machupo virus also produces a uniformly lethal infection that resembles human disease, although hemorrhagic manifestations were not reported [217]. Infection of other New World monkeys including capuchin monkeys (*Cebus* species), howler monkeys (*Alouetta caraya*), and squirrel monkeys (*Saimiri sciureus*) with Junin and Machupo viruses have only produced mild or inapparent clinical disease [185].

A number of rodent models have been developed for studying the South American HF viruses, Junin, Machupo, and Guanarito. Whereas immunocompetent mice are resistant to challenge with these viruses, guinea pigs are mostly susceptible. For the Junin virus, pathogenicity is dependent on virus strain but is independent of host background with inbred strains 2 and 13 and outbred strain Hartley guinea pigs being equally susceptible to various strains of Junin [225–228]. Some strains of Junin such as Espindola and Romero produce uniformly lethal infection in guinea pigs whereas others appear to be attenuated. As seen with Junin virus infection of either inbred strain 13 or outbred Hartley guinea pigs, challenge of animals with the Guanarito virus produces a uniformly lethal infection without adaptation [229]. In contrast to Junin and Guanarito viruses, adaptation of the Machupo virus by serial passage is required to produce a uniformly lethal infection in guinea pigs [185].

14.4 BUNYAVIRIDAE

The members of the family Bunyaviridae are encoded by a negative-sensed, trisegmented RNA genome within enveloped virions. All of the bunyaviruses are arthropod-borne [230], except the hantaviruses that are transmitted by chronically infected rodent carriers [231]. The bunyaviruses cause a plethora of diseases, most of which are not HFs. However, the Bunyaviridae contains several viruses that can cause HF, and are considered to be bioterrorism threats, including the phlebovirus RVF, the nairovirus Crimean Congo HF, and the hantaviruses.

One of the most significant threats in the Bunyaviridae family of viruses is the RVF virus (RVFV). RVFV was developed as a bioweapon by the U.S. offensive biological weapons program prior to its termination in 1969 [7]. It is classified as a category A, or high priority, pathogen by the National Institute of Allergy and Infectious Diseases (http://www.niaid.nih.gov/dmid/biodefense/bandc_priority.htm) and a category B, or moderate priority, pathogen by the Centers for Disease Control and Prevention (http://www.bt.cdc.gov/agent/agentlist.asp). It is transmitted by mosquitoes and potentially other biting insects, direct contact from infected blood or tissues (especially those associated with abortions), and aerosols [232,233]. Ingestion of contaminated raw animal milk has also been implicated epidemiologically [234]. Despite high levels of viremia and isolation of low titers of virus from throat washings, there are no reported cases of person-to-person transmission of RVFV [234]. Outbreaks of RVFV occur primarily in Africa and the Middle East, and the virus is named after the epizootic infections that occurred on farms in the Rift Valleys of Kenya [235]. The infection usually affects domestic animals such as cattle, sheep, goats, buffalo, and camels, but can also infect a broad number of species including rhesus monkeys, cats, squirrels, rats, and mice [236]. Several genera of mosquitoes in the United States have the capacity to act as vectors of RVFV [237,238]. The incubation period for RVF is 2–7 days. Symptoms include fever, retro-orbital pain, photophobia, jaundice, weakness, back pain, dizziness, and weight loss. Less than 1% of patients develop HF or encephalitis, and retinitis occurs in approximately 10%. The case fatality rate is ~1% of the humans infected [239]. Diagnosis of RVF is similar to that of other VHFs, and is accomplished by virus isolation, RT-PCR for viral genomes, or detection of RVFV-specific IgM or IgG. The treatment for RVF is fairly nonspecific and includes supportive care and ribavirin [184]. Prevention of RVF includes broad measures such as mosquito control and use of personal protective equipment for veterinary and laboratory personnel. Immunization against RVFV has proven effective in some situations with both attenuated and inactivated vaccines [240–244]. However, no licensed RVFV vaccine is currently available in the United States [1].

Several animal models are available for study of RVFV. Ruminants (e.g., sheep, goats, and cattle), nonhuman primates, and rodents including hamsters, mice, rats, guinea pigs, and gerbils have also been established as animals for studies of RVFV [245]. As RVFV has a broad host range, little to no passage in laboratory animals is required for adaptation. The virulence of RVFV in each animal model is divergent based on the viral isolate, inoculation dose, route of inoculation, and age of the animal, as would be predicted based on natural outbreaks [243,246–253].

Findlay [236] first described a hamster model for RVF in 1932, where hamsters were shown to be highly susceptible to infection (50% infectious dose <1 pfu) and developed a rapidly fatal disease similar to that seen in RVFV-infected ruminants and humans [254]. Hallmarks of the infection include fulminant fatal disease characterized by extreme lethargy, anorexia, high levels of viremia and viral titers in tissues, especially the liver, severe histopathological changes in hepatocytes and lymphocytes of the spleen, and gastrointestinal bleeding [236,254,255].

The characteristics of RVFV infection in laboratory mice were originally described in a series of publications in 1956 [256–261]. Infection of mice results in high levels of viremia and infection of a wide variety of tissues and cell types [262]. A major hallmark of the disease course in mice is severe hepatitis, most likely due to destruction of infected hepatocytes by apoptosis [262]. Mortality of a majority of the infected animals occurs within 3–6 days with the remaining mice developing neuroinvasion and panencephalitis despite clearance of the virus from the blood and liver [262]. This mouse model appears to mimic dominant pathological results of severe human RVF disease and, thus, has been widely used for exploring pathogenesis and immune responses to RVFV, as well as therapeutics and preventative measures [263–272].

Infection of laboratory rats has also been used to investigate viral and host genetic determinants of virulence as well as to investigate RVFV vaccines [246,273–277]. Early studies showed that Lewis and MAXX rats were partially resistant, in contrast to Wistar-Furth rats that were highly susceptible to RVF disease [273–275]. The Wistar-Furth rats died of hepatitis or encephalitis within 3–5 days of RVFV infection and this susceptibility was linked to a single Mendelian dominant gene [275]. However, more recent studies found that Lewis, but not Wistar-Furth, rats (although from a different source) are susceptible to RVFV infection, and these studies, in contrast to previous studies, indicated that IFN resistance was not important for RVFV susceptibility in rats [274,276,278,279].

Rhesus monkeys are considered to be the most comparable model to human infections because, similar to human infections, rhesus monkeys injected intravenously with a virulent RVFV strain develop fever, viremia, decreased leukocyte counts, hepatitis, and show elevated liver function tests [243,248,250–252]. Approximately 20% of the RVFV-infected monkeys succumb to disease characterized by an HF syndrome [243,247,248,250–252,280,281]. Epidemiological studies have shown that monkeys are naturally infected with HF viruses including RVFV, and laboratory studies have often used rhesus macaques to study the biology and interventions for RVFV [282].

14.5 FLAVIVIRIDAE

Viruses that cause West Nile and yellow fever, dengue HF, Omsk HF, and Kyasanur Forest disease are among those in the Flaviviridae family. The flaviviruses have isometric, enveloped virions that encapsidate a positive-sensed, single-stranded RNA genome. All of the viruses in this family are arthropod-borne (either by mosquitoes or ticks). The geographic distribution depends upon the virus strain. Flaviviruses are found in most parts of the world, including Asia, Africa, Americas, India, and the Pacific. The yellow fever virus was pursued for weaponization by Japan and the United States in the first half of the twentieth century [1,4,6,7]. However, there is an effective, licensed vaccine for the yellow fever virus and this virus is not considered among the highest biowarfare threat agents.

The dengue virus is the only pathogen considered a category A priority for biodefense threats by the National Institute of Allergy and Infectious Diseases (http://www.niaid.nih.gov/dmid/biodefense/bandc_priority.htm). In contrast, the dengue virus is excluded as a probable serious VHF biowarfare threat by the American Medical Association Consensus group because it is not transmissible as a small particle aerosol and primary dengue only rarely causes VHF [1]. Dengue viral infections have a broad range of clinical outcomes from absolutely no symptoms to undifferentiated fever, dengue fever, dengue HF (DHF), or dengue shock syndrome (DSS). The dengue virus is endemic in at least 112 countries and causes 100 million cases of dengue fever and half a million cases of DHF each year [283,284]. The virus is transmitted by the *Aedes aegypti* mosquito and replicates primarily in mononuclear phagocytic cells of the infected human [285,286]. Similar to other VHFs, the diagnostic tools for dengue include detection of virus-specific IgM or viral genomes by PCR [286,287]. Other diagnostics to allow for early and rapid diagnosis, as well as clinical prognosis, are being developed [286–288]. There are four serotypes (DEN 1–4) of the dengue virus, which are classified and diagnosed according to biological, antigenic, serological, and immunological criteria [289].

The nature of dengue virus infections is complicated. Whereas a primary infection may be asymptomatic or have a wide range of manifestations from mild acute febrile illness to classical dengue fever, immunity to the same serotype appears to be life-long; however, infection with a second heterotypic serotype can lead to enhanced and severe disease (DHF/DSS) [287,290]. The development of DHF and DSS is immune-mediated, where antibodies from a prior exposure cause enhanced viral replication (antibody-dependent enhancement) in early stages of a secondary heterotypic infection by mediating increased virus uptake and replication due to affinity for the antigen–antibody complexes for Fc receptors on B lymphocytes and professional antigen-presenting cells [290–292]. Therefore, the development of a dengue virus vaccine requires long-lasting protection against all four serotypes for prevention of vaccine-induced DHF or DSS. Several live attenuated vaccines are being developed and will have to prove efficacy against disease, safety during vaccination, and no increase in severe disease following dengue exposure [287,293].

A number of rodent and monkey animal models have been used to study dengue virus infections and pathogenesis. Hamsters have been used for testing and development of dengue virus vaccines, but do not support high levels of replication nor development of disease similar to dengue infection in humans [294–296]. From very early to more recent studies, it has been well documented that the dengue virus replicates in immunocompromised mice and replication in immunocompetent mice requires adaptation of the virus through serial passage. Infection models include intracerebral inoculations of suckling mice, systemic infections of IFN-α/β and -γ receptor deficient mice, and SCID mice engrafted with human cells [297–302]. An important advance in recent years showed a DEN-2 mouse model where intravenous injection of immunocompetent mice induced neurological and hematological changes, more similar (as compared to previous models) to those seen in humans [303]. More recently, two new mouse models have been established that appear to be much more relevant to the pathogenesis of human disease. In the first study, the first animal model of antibody-enhanced severe DENV infection was developed using AG129 mice challenged with both mouse-adapted and clinical isolates of DENV, and passively transferred low doses of either serotype-specific or cross-reactive antibodies enhanced disease *in vivo* [304]. The second study demonstrates an HF model using a unique nonadapted DEN-2 clinical isolate (D2Y98P) in AG129 mice [305]. Both models appear to imitate important characteristics of human disease pathology including viremia, organ damage, disease kinetics, vascular leakage, increased circulating chemokine and cytokine levels, thrombocytopenia, and disseminated infection of tissue phagocytes [304,305].

The dengue virus was previously shown to infect cynomolgus and rhesus macaques as well as chimpanzees [306–311]. Dengue virus infections of nonhuman primates caused by a subcutaneous route of injection, which is intended to be similar to that of a mosquito bite, result in productive infections with transient viremia observed and subsequent detectable antibody and T cell responses in convalescent animals [291,306,307]. These models have been commonly used to test both treatments and preventatives for dengue infection [312]. The establishment of an animal model that recapitulates human dengue fever has been a major obstacle in understanding the pathogenesis of dengue fever. A recent study by Onlamoon and colleagues [313] demonstrated classic dengue fever induced in rhesus macaques by intravenous injection of a high challenge dose ($1 \times 10e7$ pfu). In all animals, a classic dengue hemorrhage correlating with peak viremia developed 3–5 days after infection. Only mild changes in blood chemistry values were noted, whereas changes in immune cells were more apparent including modest thrombocytopenia, notable neutropenia concomitant with slight decrease of hemoglobin and hematocrit, significant elevation of D-dimers, and transient depletion of T and B lymphocytes with a bimodal pattern for platelet–monocyte and platelet–neutrophil aggregates [313]. Additional studies will be required to expand the understanding of the pathogenesis in this model and correlate with human disease; however, the model seems to most closely mimic human dengue fever and, thus, should further testing of new therapeutics and vaccines to combat dengue virus infection and disease.

14.6 CONCLUSIONS AND FUTURE DIRECTIONS

Many animal models are available for studying the disease processes, as well as treatment and prevention modalities, for the HFs caused by members of the Arenaviridae, Bunyaviridae, Flaviviridae, and Filoviridae families. Research to develop, define, and refine animal models for the VHFs is critically important from both a public health and biodefense perspective. From a public health/natural outbreak perspective, it is not likely that vast increases in the knowledge base of the pathophysiology in humans will occur, especially for the arenaviruses, bunyaviruses, and filoviruses. The understanding of these infections is, and will continue to be, complicated because of the small numbers of patients infected during outbreaks, the remoteness of the regions where outbreaks occur, and the difficulties in accessing quality healthcare and obtaining samples from infected individuals. Both the rodent and nonhuman primate models are necessary and invaluable tools for understanding the biology of these dangerous viruses, especially due to the general lack of knowledge of the biology and human responses following viral infection. It is especially important to continue to develop and refine rodent models so they will accurately predict the outcome of therapeutic and vaccine trials, and all but final testing of products in nonhuman primates can be avoided. Currently, there are difficulties in comparing animal models used by different researchers at various institutions around the world due to the use of multiple and varied virus strains, doses of virus administered, routes of inoculation, and sources of experimental animals. Efforts should be made by researchers of the HF viruses to standardize and share reagents and combine research efforts to achieve these goals.

REFERENCES

1. Borio, L., Inglesby, T., Peters, C. J., Schmaljohn, A. L., Hughes, J. M., Jahrling, P. B., Ksiazek, T. et al., Hemorrhagic fever viruses as biological weapons: Medical and public health management, *JAMA* 287(18), 2391–2405, 2002.
2. LeDuc, J. W., Epidemiology of hemorrhagic fever viruses, *Rev Infect Dis* 11(Suppl 4), S730–S735, 1989.
3. U.S. Department of Health and Human Services, Centers for Disease Control and Prevention, and National Institutes of Health, *Biosafety in Microbiological and Biomedical Laboratories*, 4 ed., Government Printing Office, Washington, DC, 1999.
4. Williams, P. and Wallace, D., *Unit 731, Japan's Secret Biological Warfare in World War II*, Free Press, New York, 1989, pp. 38–40.
5. Alibek, K. and Handelman, S., *Biohazard*, Random House, New York, 1999.
6. Miller, J., Engelberg, S., and Broad, W. J., *Germs: Biological Weapons and America's Secret War*, Simon & Schuster, Waterville, ME, 2001.

7. Center for Nonproliferation Studies, Chemical and biological weapons: Possession and programs past and present. http://cns.miis.edu/research/cbw/possess.htm, 2000.
8. Leroy, E. M., Kumulungui, B., Pourrut, X., Rouquet, P., Hassanin, A., Yaba, P., Delicat, A., Paweska, J. T., Gonzalez, J. P., and Swanepoel, R., Fruit bats as reservoirs of Ebola virus, *Nature* 438(7068), 575–576, 2005.
9. Towner, J. S., Pourrut, X., Albarino, C. G., Nkogue, C. N., Bird, B. H., Grard, G., Ksiazek, T. G., Gonzalez, J. P., Nichol, S. T., and Leroy, E. M., Marburg virus infection detected in a common African bat, *PLoS One* 2(1), e764, 2007.
10. Feldmann, H. and Klenk, H. D., Marburg and Ebola viruses, *Adv Virus Res* 47, 1–52, 1996.
11. Geisbert, T. W., Marty, A. M., and Jahrling, P. B., Viral hemorrhagic fevers, in *Physician's Guide to Terrorist Attack*, M. J. Roy, ed., Humana Press, Inc., Totowa, NJ, 2003.
12. Feldmann, H., Klenk, H. D., and Sanchez, A., Molecular biology and evolution of filoviruses, *Arch Virol Suppl* 7, 81–100, 1993.
13. Peters, C. J. and Khan, A. S., Filovirus diseases, *Curr Top Microbiol Immunol* 235, 85–95, 1999.
14. Centers for Disease Control and Prevention (CDC), Update: Filovirus infection associated with contact with nonhuman primates or their tissues, *MMWR Morb Mortal Wkly Rep* 39(24), 404–405, 1990.
15. Brown, D. W., Threat to humans from virus infections of non-human primates, *Rev Med Virol* 7(4), 239–246, 1997.
16. Centers for Disease Control and Prevention (CDC), Update: Filovirus infections among persons with occupational exposure to nonhuman primates, *MMWR Morb Mortal Wkly Rep* 39(16), 266–267; 273, 1990.
17. Mwanatambwe, M., Yamada, N., Arai, S., Shimizu-Suganuma, M., Shichinohe, K., and Asano, G., Ebola hemorrhagic fever (EHF): Mechanism of transmission and pathogenicity, *J Nippon Med Sch* 68(5), 370–375, 2001.
18. Pinzon, J. E., Wilson, J. M., Tucker, C. J., Arthur, R., Jahrling, P. B., and Formenty, P., Trigger events: Enviroclimatic coupling of Ebola hemorrhagic fever outbreaks, *Am J Trop Med Hyg* 71(5), 664–674, 2004.
19. Barrette, R. W., Metwally, S. A., Rowland, J. M., Xu, L., Zaki, S. R., Nichol, S. T., Rollin, P. E. et al., Discovery of swine as a host for the Reston ebolavirus, *Science* 325(5937), 204–206, 2009.
20. Jaax, N., Jahrling, P., Geisbert, T., Geisbert, J., Steele, K., McKee, K., Nagley, D., Johnson, E., Jaax, G., and Peters, C., Transmission of Ebola virus (Zaire strain) to uninfected control monkeys in a biocontainment laboratory, *Lancet* 346(8991–8992), 1669–1671, 1995.
21. Johnson, E., Jaax, N., White, J., and Jahrling, P., Lethal experimental infections of rhesus monkeys by aerosolized Ebola virus, *Int J Exp Pathol* 76(4), 227–236, 1995.
22. Belanov, E. F., Muntianov, V. P., Kriuk, V. D., Sokolov, A. V., Bornotov, N. I., Plankov, O. V., and Sergeev, A. N., Survival of Marburg virus infectivity on contaminated surfaces and in aerosols, *Vopr Virusol* 41, 32–34, 1996.
23. Frolov, V. G. and Gusev, Iu.M., Stability of Marburg virus to lyophilization process and subsequent storage at different temperatures [in Russian], *Vopr Virusol* 41, 275–277, 1996.
24. Martini, G. A. and Siegert, R., eds., *Marburg Virus Disease*, Springer-Verlag, Berlin, 1971.
25. Smith, D. H., Johnson, B. K., Isaacson, M., Swanapoel, R., Johnson, K. M., Killey, M., Bagshawe, A., Siongok, T., and Keruga, W. K., Marburg-virus disease in Kenya, *Lancet* 1(8276), 816–820, 1982.
26. Geisbert, T. W. and Hensley, L. E., Ebola virus: New insights into disease aetiopathology and possible therapeutic interventions, *Expert Rev Mol Med* 6(20), 1–24, 2004.
27. Zeller, H., Lessons from the Marburg virus epidemic in Durba, Democratic Republic of the Congo (1998–2000) [in French], *Med Trop (Mars)* 60(2 Suppl), 23–26, 2000.

28. Colebunders, R., Sleurs, H., Pirard, P., Borchert, M., Libande, M., Mustin, J. P., Tshomba, A. et al., Organisation of health care during an outbreak of Marburg haemorrhagic fever in the Democratic Republic of Congo, 1999, *J Infect* 48(4), 347–353, 2004.
29. Towner, J. S., Khristova, M. L., Sealy, T. K., Vincent, M. J., Erickson, B. R., Bawiec, D. A., Hartman, A. L. et al., Marburgvirus genomics and association with a large hemorrhagic fever outbreak in Angola, *J Virol* 80(13), 6497–6516, 2006.
30. Walsh, P. D., Abernethy, K. A., Bermejo, M., Beyers, R., De Wachter, P., Akou, M. E., Huijbregts, B. et al., Catastrophic ape decline in western equatorial Africa, *Nature* 422(6932), 611–614, 2003.
31. Whitfield, J., Ape populations decimated by hunting and Ebola virus, *Nature* 422(6932), 551, 2003.
32. Kuhn, J. H., Becker, S., Ebihara, H., Geisbert, T. W., Johnson, K. M., Kawaoka, Y., Lipkin, W. I. et al., Proposal for a revised taxonomy of the family Filoviridae: Classification, names of taxa and viruses, and virus abbreviations, *Arch Virol* 155(12), 2083–2103.
33. Wilson, J. A., Bray, M., Bakken, R., and Hart, M. K., Vaccine potential of Ebola virus VP24, VP30, VP35, and VP40 proteins, *Virology* 286(2), 384–390, 2001.
34. Sanchez, A., Kiley, M. P., Holloway, B. P., and Auperin, D. D., Sequence analysis of the Ebola virus genome: Organization, genetic elements, and comparison with the genome of Marburg virus, *Virus Res* 29(3), 215–240, 1993.
35. Volchkov, V. E., Becker, S., Volchkova, V. A., Ternovoj, V. A., Kotov, A. N., Netesov, S. V., and Klenk, H. D., GP mRNA of Ebola virus is edited by the Ebola virus polymerase and by T7 and vaccinia virus polymerases, *Virology* 214(2), 421–430, 1995.
36. Elliott, L. H., Kiley, M. P., and McCormick, J. B., Descriptive analysis of Ebola virus proteins, *Virology* 147(1), 169–176, 1985.
37. Muhlberger, E., Weik, M., Volchkov, V. E., Klenk, H. D., and Becker, S., Comparison of the transcription and replication strategies of Marburg virus and Ebola virus by using artificial replication systems, *J Virol* 73(3), 1999.
38. Basler, C. F., Wang, X., Muhlberger, E., Volchkov, V., Paragas, J., Klenk, H. D., Garcia-Sastre, A., and Palese, P., The Ebola virus VP35 protein functions as a type I IFN antagonist, *Proc Natl Acad Sci U S A* 97(22), 12289–12294, 2000.
39. Basler, C. F., Mikulasova, A., Martinez-Sobrido, L., Paragas, J., Muhlberger, E., Bray, M., Klenk, H. D., Palese, P., and Garcia-Sastre, A., The Ebola virus VP35 protein inhibits activation of interferon regulatory factor 3, *J Virol* 77(14), 7945–7956, 2003.
40. Ruigrok, R. W., Schoehn, G., Dessen, A., Forest, E., Volchkov, V., Dolnik, O., Klenk, H. D., and Weissenhorn, W., Structural characterization and membrane binding properties of the matrix protein VP40 of Ebola virus, *J Mol Biol* 300(1), 103–112, 2000.
41. Dessen, A., Volchkov, V., Dolnik, O., Klenk, H. D., and Weissenhorn, W., Crystal structure of the matrix protein VP40 from Ebola virus, *Embo J* 19(16), 4228–4236, 2000.
42. Swenson, D. L., Warfield, K. L., Kuehl, K., Larsen, T., Hevey, M. C., Schmaljohn, A., Bavari, S., and Aman, M. J., Generation of Marburg virus-like particles by co-expression of glycoprotein and matrix protein, *FEMS Immunol Med Microbiol* 40(1), 27–31, 2004.
43. Kolesnikova, L., Bamberg, S., Berghofer, B., and Becker, S., The matrix protein of Marburg virus is transported to the plasma membrane along cellular membranes: Exploiting the retrograde late endosomal pathway, *J Virol* 78(5), 2382–2393, 2004.
44. Kolesnikova, L., Bugany, H., Klenk, H. D., and Becker, S., VP40, the matrix protein of Marburg virus, is associated with membranes of the late endosomal compartment, *J Virol* 76(4), 1825–1838, 2002.
45. Bavari, S., Bosio, C. M., Wiegand, E., Ruthel, G., Will, A. B., Geisbert, T. W., Hevey, M., Schmaljohn, C., Schmaljohn, A., and Aman, M. J., Lipid raft microdomains: A gateway for compartmentalized trafficking of Ebola and Marburg viruses, *J Exp Med* 195(5), 593–602, 2002.

46. Valmas, C., Grosch, M. N., Schumann, M., Olejnik, J., Martinez, O., Best, S. M., Krahling, V., Basler, C. F., and Muhlberger, E., Marburg virus evades interferon responses by a mechanism distinct from Ebola virus, *PLoS Pathog* 6(1), e1000721, 2010.
47. Han, Z., Boshra, H., Sunyer, J. O., Zwiers, S. H., Paragas, J., and Harty, R. N., Biochemical and functional characterization of the Ebola virus VP24 protein: Implications for a role in virus assembly and budding, *J Virol* 77(3), 1793–1800, 2003.
48. Licata, J. M., Johnson, R. F., Han, Z., and Harty, R. N., Contribution of Ebola virus glycoprotein, nucleoprotein, and VP24 to budding of VP40 virus-like particles, *J Virol* 78(14), 7344–7351, 2004.
49. Reid, S. P., Leung, L. W., Hartman, A. L., Martinez, O., Shaw, M. L., Carbonnelle, C., Volchkov, V. E., Nichol, S. T., and Basler, C. F., Ebola virus VP24 binds karyopherin alpha1 and blocks STAT1 nuclear accumulation, *J Virol* 80(11), 5156–5167, 2006.
50. Sanchez, A., Ksiazek, T. G., Rollin, P. E., Miranda, M. E., Trappier, S. G., Khan, A. S., Peters, C. J., and Nichol, S. T., Detection and molecular characterization of Ebola viruses causing disease in human and nonhuman primates, *J Infect Dis* 179(Suppl 1), S164–S169, 1999.
51. Rollin, P. E., Williams, R. J., Bressler, D. S., Pearson, S., Cottingham, M., Pucak, G., Sanchez, A. et al., Ebola (subtype Reston) virus among quarantined nonhuman primates recently imported from the Philippines to the United States., *J Infect Dis* 179(Suppl 1), S108–S114, 1999.
52. Maruyama, T., Buchmeier, M. J., Parren, P. W. H. I., and Burton, D. R., Ebola virus, neutrophils, and antibody specificity, *Science* 282, 845, 1998.
53. Sui, J. and Marasco, W. A., Evidence against Ebola virus sGP binding to human neutrophils by a specific receptor, *Virology* 303(1), 9–14, 2002.
54. Kindzelskii, A. L., Yang, Z., Nabel, G. J., Todd, R. F., 3rd, and Petty, H. R., Ebola virus secretory glycoprotein (sGP) diminishes Fc gamma RIIIB-to-CR3 proximity on neutrophils, *J Immunol* 164(2), 953–958, 2000.
55. Henchal, E. A., Teska, J. D., Ludwig, G. V., Shoemaker, D. R., and Ezell, J. W., Current laboratory methods for biological threat agent identification, *Clin Lab Med* 21, 7.1–7.13, 2001.
56. Rollin, P. E. and Ksiazek, T. G., Ebola haemorrhagic fever, *Trans R Soc Trop Med Hyg* 92(1), 1–2, 1998.
57. Le Guenno, B., Formenty, P., and Boesch, C., Ebola virus outbreaks in the Ivory Coast and Liberia, 1994–1995, *Curr Top Microbiol Immunol* 235, 77–84, 1999.
58. Hartman, A. L., Towner, J. S., and Nichol, S. T., Ebola and Marburg hemorrhagic fever, *Clin Lab Med* 30(1), 161–177, 2010.
59. Roddy, P., Thomas, S. L., Jeffs, B., Nascimento Folo, P., Pablo Palma, P., Moco Henrique, B., Villa, L. et al., Factors associated with Marburg hemorrhagic fever: Analysis of patient data from Uige, Angola, *J Infect Dis* 201(12), 1909–1918.
60. MacNeil, A., Farnon, E. C., Wamala, J., Okware, S., Cannon, D. L., Reed, Z., Towner, J. S. et al., Proportion of deaths and clinical features in Bundibugyo Ebola virus infection, Uganda, *Emerg Infect Dis* 16(12), 1969–1972, 2010.
61. Casillas, A. M., Nyamathi, A. M., Sosa, A., Wilder, C. L., and Sands, H., A current review of Ebola virus: Pathogenesis, clinical presentation, and diagnostic assessment, *Biol Res Nurs* 4(4), 268–275, 2003.
62. Stroher, U., West, E., Bugany, H., Klenk, H. D., Schnittler, H. J., and Feldmann, H., Infection and activation of monocytes by Marburg and Ebola viruses, *J Virol* 75(22), 11025–11033, 2001.
63. Geisbert, T. W., Young, H. A., Jahrling, P. B., Davis, K. J., Kagan, E., and Hensley, L. E., Mechanisms underlying coagulation abnormalities in Ebola hemorrhagic fever: Overexpression of tissue factor in primate monocytes/macrophages is a key event, *J Infect Dis* 188(11), 1618–1629, 2003.

64. Mahanty, S., Hutchinson, K., Agarwal, S., McRae, M., Rollin, P. E., and Pulendran, B., Cutting edge: Impairment of dendritic cells and adaptive immunity by Ebola and Lassa viruses, *J Immunol* 170(6), 2797–2801, 2003.
65. Bosio, C. M., Aman, M. J., Grogan, C., Hogan, R., Ruthel, G., Negley, D., Mohamadzadeh, M., Bavari, S., and Schmaljohn, A., Ebola and Marburg viruses replicate in monocyte-derived dendritic cells without inducing the production of cytokines and full maturation, *J Infect Dis* 188(11), 1630–1638, 2003.
66. Geisbert, T. W., Hensley, L. E., Larsen, T., Young, H. A., Reed, D. S., Geisbert, J. B., Scott, D. P., Kagan, E., Jahrling, P. B., and Davis, K. J., Pathogenesis of Ebola hemorrhagic fever in cynomolgus macaques: Evidence that dendritic cells are early and sustained targets of infection, *Am J Pathol* 163(6), 2347–2370, 2003.
67. Gupta, M., Mahanty, S., Ahmed, R., and Rollin, P., Monocyte derived human macrophages and peripheral blood mononuclear cells infected with Ebola virus secrete MIP-1 alpha and TNF-alpha and inhibit Poly-IC induced IFN-alpha *in vitro*, *Virology* 284(20), 20–25, 2001.
68. Schnittler, H. J. and Feldmann, H., Viral hemorrhagic fever—A vascular disease? *Thromb Haemost* 89(6), 967–972, 2003.
69. Geisbert, T. W., Young, H. A., Jahrling, P. B., Davis, K. J., Larsen, T., Kagan, E., and Hensley, L. E., Pathogenesis of Ebola hemorrhagic fever in primate models: Evidence that hemorrhage is not a direct effect of virus-induced cytolysis of endothelial cells, *Am J Pathol* 163(6), 2371–2382, 2003.
70. Chan, S. Y., Empig, C. J., Welte, F. J., Speck, R. F., Schmaljohn, A., Kreisberg, J. F., and Goldsmith, M. A., Folate receptor-alpha is a cofactor for cellular entry by Marburg and Ebola viruses, *Cell* 106(1), 117–126, 2001.
71. Simmons, G., Rennekamp, A. J., Chai, N., Vandenberghe, L. H., Riley, J. L., and Bates, P., Folate receptor alpha and caveolae are not required for Ebola virus glycoprotein-mediated viral infection, *J Virol* 77(24), 13433–13438, 2003.
72. Becker, S., Spiess, M., and Klenk, H. D., The asialoglycoprotein receptor is a potential liver-specific receptor for Marburg virus, *J Gen Virol* 76(Pt 2), 393–399, 1995.
73. Alvarez, C. P., Lasala, F., Carrillo, J., Muniz, O., Corbi, A. L., and Delgado, R., C-type lectins DC-SIGN and L-SIGN mediate cellular entry by Ebola virus in cis and in trans, *J Virol* 76(13), 6841–6844, 2002.
74. Simmons, G., Reeves, J. D., Grogan, C. C., Vandenberghe, L. H., Baribaud, F., Whitbeck, J. C., Burke, E. et al., DC-SIGN and DC-SIGNR bind Ebola glycoproteins and enhance infection of macrophages and endothelial cells, *Virology* 305(1), 115–123, 2003.
75. Lasala, F., Arce, E., Otero, J. R., Rojo, J., and Delgado, R., Mannosyl glycodendritic structure inhibits DC-SIGN-mediated Ebola virus infection in cis and in trans, *Antimicrob Agents Chemother* 47(12), 3970–3972, 2003.
76. Gupta, M., Mahanty, S., Greer, P., Towner, J. S., Shieh, W. J., Zaki, S. R., Ahmed, R., and Rollin, P. E., Persistent infection with Ebola virus under conditions of partial immunity, *J Virol* 78(2), 958–967, 2004.
77. Gibb, T. R., Norwood, D. A., Jr., Woollen, N., and Henchal, E. A., Viral replication and host gene expression in alveolar macrophages infected with Ebola virus (Zaire strain). *Clin Diagn Lab Immunol* 9(1), 19–27, 2002.
78. Wauquier, N., Becquart, P., Padilla, C., Baize, S., and Leroy, E. M., Human fatal Zaire Ebola virus infection is associated with an aberrant innate immunity and with massive lymphocyte apoptosis, *PLoS Negl Trop Dis* 4 (10), e837, 2010.
79. Villinger, F., Rollin, P. E., Brar, S. S., Chikkala, N. F., Winter, J., Sundstrom, J. B., Zaki, S. R., Swanepoel, R., Ansari, A. A., and Peters, C. J., Markedly elevated levels of interferon (IFN)-gamma, IFN-alpha, interleukin (IL)-2, IL-10, and tumor necrosis factor-alpha associated with fatal Ebola virus infection, *J Infect Dis* 179(Suppl 1), S188–S191, 1999.

80. Leroy, E. M., Baize, S., Debre, P., Lansoud-Soukate, J., and Mavoungou, E., Early immune responses accompanying human asymptomatic Ebola infections, *Clin Exp Immunol* 124(3), 453–460, 2001.
81. Leroy, E. M., Baize, S., Volchkov, V. E., Fisher-Hoch, S. P., Georges-Courbot, M. C., Lansoud-Soukate, J., Capron, M., Debre, P., McCormick, J. B., and Georges, A. J., Human asymptomatic Ebola infection and strong inflammatory response, *Lancet* 355(9222), 2210–2215, 2000.
82. Baize, S., Leroy, E. M., Georges-Courbot, M. C., Capron, M., Lansoud-Soukate, J., Debre, P., Fisher-Hoch, S. P., McCormick, J. B., and Georges, A. J., Defective humoral responses and extensive intravascular apoptosis are associated with fatal outcome in Ebola virus-infected patients, *Nat Med* 5(4), 423–426, 1999.
83. Baize, S., Leroy, E. M., Georges, A. J., Georges-Courbot, M. C., Capron, M., Bedjabaga, I., Lansoud-Soukate, J., and Mavoungou, E., Inflammatory responses in Ebola virus-infected patients, *Clin Exp Immunol* 128(1), 163–168, 2002.
84. Ksiazek, T. G., Rollin, P. E., Williams, A. J., Bressler, D. S., Martin, M. L., Swanepoel, R., Burt, F. J. et al., Clinical virology of Ebola hemorrhagic fever (EHF): Virus, virus antigen, and IgG and IgM antibody findings among EHF patients in Kikwit, Democratic Republic of the Congo, 1995, *J Infect Dis* 179(Suppl 1), S177–S187, 1999.
85. Fisher-Hoch, S. P., Platt, G. S., Neild, G. H., Southee, T., Baskerville, A., Raymond, R. T., Lloyd, G., and Simpson, D. I., Pathophysiology of shock and hemorrhage in a fulminating viral infection (Ebola). *J Infect Dis* 152(5), 887–894, 1985.
86. Connolly, B. M., Steele, K. E., Davis, K. J., Geisbert, T. W., Kell, W. M., Jaax, N. K., and Jahrling, P. B., Pathogenesis of experimental Ebola virus infection in guinea pigs, *J Infect Dis* 179(Suppl 1), S203–S217, 1999.
87. Volchkov, V. E., Chepurnov, A. A., Volchkova, V. A., Ternovoj, V. A., and Klenk, H. D., Molecular characterization of guinea pig-adapted variants of Ebola virus, *Virology* 277(1), 147–155, 2000.
88. Ryabchikova, E., Kolesnikova, L., Smolina, M., Tkachev, V., Pereboeva, L., Baranova, S., Grazhdantseva, A., and Rassadkin, Y., Ebola virus infection in guinea pigs: Presumable role of granulomatous inflammation in pathogenesis, *Arch Virol* 141(5), 909–921, 1996.
89. Ryabchikova, E., Smolina, M., Grajdantseva, A., and Rassadkin, J., Ebola virus infection in the guinea pig, in *Ebola and Marburg Viruses: Molecular and Cellular Biology*, H. Klenk and H. Feldmann, eds., Horizon Biosciences, Norfolk, 2004, pp. 239–253.
90. Hevey, M., Negley, D., Pushko, P., Smith, J., and Schmaljohn, A., Marburg virus vaccines based upon alphavirus replicons protect guinea pigs and nonhuman primates, *Virology* 251(1), 28–37, 1998.
91. Robin, Y., Bres, P., and Camain, R., Passage of Marburg virus in guinea pigs, in *Marburg Virus Disease*, G. A. Martini and R. Siegert, eds., Springer-Verlag, New York, 1971, pp. 117–122.
92. Ryabchikova, E., Strelets, L., Kolesnikova, L., P'yankov, O., and Sergeev, A., Respiratory Marburg virus infection in guinea pigs, *Arch Virol* 141(11), 2177–2190, 1996.
93. Lub, M. Y., Sergeev, A. N., P'yankov, O. V., P'yankova, O. G., Petrishchenko, V. A., and Kotlyarov, L. A., Clinical and virological characteriztion of the disease in guinea pigs aerogenically infected with marburg virus [in Russian], *Vopr Virusol* 3, 119–121, 1995.
94. Ignat'ev, G. M., Strel'tsova, M. A., Agafonov, A. P., Prozorovskii, N. S., Zhukova, N. A., Kashentseva, E. A., and Vorob'eva, M. S., The immunological indices of guinea pigs modelling Marburg hemorrhagic fever [in Russian], *Vopr Virusol* 39(4), 169–171, 1994.
95. Jaax, N. K., Davis, K. J., Geisbert, T. J., Vogel, P., Jaax, G. P., Topper, M., and Jahrling, P. B., Lethal experimental infection of rhesus monkeys with Ebola-Zaire (Mayinga) virus by the oral and conjunctival route of exposure, *Arch Pathol Lab Med* 120(2), 140–155, 1996.

96. Geisbert, T. W., Pushko, P., Anderson, K., Smith, J., Davis, K. J., and Jahrling, P. B., Evaluation in nonhuman primates of vaccines against Ebola virus, *Emerg Infect Dis* 8(5), 503–507, 2002.
97. Geisbert, T., Jahrling, P., Larsen, T., Davis, K., and Hensley, L. E., Filovirus pathogenesis in nonhuman primates, in *Ebola and Marburg Viruses: Molecular and Cellular Biology*, H. Klenk and H. Feldmann, eds., Horizon Bioscience, Norfolk, 2004, pp. 203–238.
98. Schou, S. and Hansen, A. K., Marburg and Ebola virus infections in laboratory nonhuman primates: A literature review, *Comp Med* 50(2), 108–123, 2000.
99. Bowen, E. T., Platt, G. S., Simpson, D. I., McArdell, L. B., and Raymond, R. T., Ebola haemorrhagic fever: Experimental infection of monkeys, *Trans R Soc Trop Med Hyg* 72(2), 188–191, 1978.
100. Bowen, E. T., Lloyd, G., Platt, G. S., McArdell, L. B., Webb, P. A., and Simpson, D. I. H., Virological studies on a case of Ebola virus infection in man and in monkeys, in *Ebola Virus Haemorrhagic Fever*, S. R. Pattyn, ed., Elsevier/North-Holland Biomedical Press, Amsterdam, The Netherlands, 1978, pp. 95–100.
101. Bowen, E. T. W., Lloyd, G., Harris, W. J., Platt, G. S., Baskerville, A., and Vella, E. E., Viral haemorrhagic fever in southern Sudan and northern Zaire. Preliminary studies on the aetiological agent, *Lancet* 1(8011), 571–573, 1977.
102. Davis, K. J., Anderson, A. O., Geisbert, T. W., Steele, K. E., Geisbert, J. B., Vogel, P., Connolly, B. M., Huggins, J. W., Jahrling, P. B., and Jaax, N. K., Pathology of experimental Ebola virus infection in African green monkeys. Involvement of fibroblastic reticular cells, *Arch Pathol Lab Med* 121(8), 805–819, 1997.
103. Ryabchikova, E. I., Current concepts of filovirus pathogenesis, *Symposium on Marburg and Ebola Viruses* Marburg, Germany, 2000.
104. Ryabchikova, E. I., Kolesnikova, L. V., and Netesov, S. V., Animal pathology of filoviral infections, *Curr Top Microbiol Immunol* 235, 145–173, 1999.
105. Ellis, D. S., Bowen, E. T., Simpson, D. I., and Stamford, S., Ebola virus: A comparison, at ultrastructural level, of the behaviour of the Sudan and Zaire strains in monkeys, *Br J Exp Pathol* 59(6), 584–593, 1978.
106. Geisbert, T. W., Daddario-DiCaprio, K. M., Williams, K. J., Geisbert, J. B., Leung, A., Feldmann, F., Hensley, L. E., Feldmann, H., and Jones, S. M., Recombinant vesicular stomatitis virus vector mediates postexposure protection against Sudan Ebola hemorrhagic fever in nonhuman primates, *J Virol* 82(11), 5664–5668, 2008.
107. Geisbert, T. W., Geisbert, J. B., Leung, A., Daddario-DiCaprio, K. M., Hensley, L. E., Grolla, A., and Feldmann, H., Single-injection vaccine protects nonhuman primates against infection with marburg virus and three species of Ebola virus, *J Virol* 83(14), 7296–7304, 2009.
108. Pratt, W. D., Wang, D., Nichols, D. K., Luo, M., Woraratanadharm, J., Dye, J. M., Holman, D. H., and Dong, J. Y., Protection of nonhuman primates against two species of Ebola virus infection with a single complex adenovirus vector, *Clin Vaccine Immunol* 17(4), 572–581, 2010.
109. Swenson, D. L., Wang, D., Luo, M., Warfield, K. L., Woraratanadharm, J., Holman, D. H., Dong, J. Y., and Pratt, W. D., Vaccine to confer to nonhuman primates complete protection against multistrain Ebola and Marburg virus infections, *Clin Vaccine Immunol* 15(3), 460–467, 2008.
110. Fisher-Hoch, S. P., Brammer, T. L., Trappier, S. G., Hutwagner, L. C., Farrar, B. B., Ruo, S. L., Brown, B. G. et al., Pathogenic potential of filoviruses: Role of geographic origin of primate host and virus strain, *J Infect Dis* 166(4), 753–763, 1992.
111. Fisher-Hoch, S. P., Perez-Oronoz, G. I., Jackson, E. L., Hermann, L. M., and Brown, B. G., Filovirus clearance in non-human primates, *Lancet* 340(8817), 451–453, 1992.

112. Ikegami, T., Miranda, M. E., Calaor, A. B., Manalo, D. L., Miranda, N. J., Niikura, M., Saijo, M. et al., Histopathology of natural Ebola virus subtype Reston infection in cynomolgus macaques during the Philippine outbreak in 1996, *Exp Anim* 51(5), 447–455, 2002.
113. Jahrling, P. B., Geisbert, T. W., Jaax, N. K., Hanes, M. A., Ksiazek, T. G., and Peters, C. J., Experimental infection of cynomolgus macaques with Ebola-Reston filoviruses from the 1989–1990 U.S. epizootic, *Arch Virol Suppl* 11, 115–134, 1996.
114. Formenty, P., Hatz, C., Le Guenno, B., Stoll, A., Rogenmoser, P., and Widmer, A., Human infection due to Ebola virus, subtype Cote d'Ivoire: Clinical and biologic presentation, *J Infect Dis* 179(Suppl 1), S48–S53, 1999.
115. Geisbert, T. W., Daddario-Dicaprio, K. M., Geisbert, J. B., Reed, D. S., Feldmann, F., Grolla, A., Stroher, U. et al., Vesicular stomatitis virus-based vaccines protect nonhuman primates against aerosol challenge with Ebola and Marburg viruses, *Vaccine* 26(52), 6894–6900, 2008.
116. Leffel, E. K. and Reed, D. S., Marburg and Ebola viruses as aerosol threats, *Biosecur Bioterror* 2(3), 186–191, 2004.
117. Spiridonov, V. A., Bazhutin, N. B., Belanov, E. F., Voitenko, A. V., Zolin, V. V., Krivenchuk, N. A., Omel'chenko, N. I., Polikanov, V. P., Tereshchenko, A., and Khomichev, V. V., Changes in the blood serum aminotransferase activity in the experimental infection of *Cercopithecus aethiops* monkeys with the Marburg virus [in Russian], *Vopr Virusol* 37(3), 156–157, 1992.
118. Alves, D. A., Glynn, A. R., Steele, K. E., Lackemeyer, M. G., Garza, N. L., Buck, J. G., Mech, C., and Reed, D. S., Aerosol exposure to the Angola strain of Marburg virus causes lethal viral hemorrhagic fever in cynomolgus macaques, *Vet Pathol* 47(5), 831–851, 2010.
119. Reed, D. S., Hensley, L. E., Geisbert, J. B., Jahrling, P. B., and Geisbert, T. W., Depletion of peripheral blood T lymphocytes and NK cells during the course of Ebola hemorrhagic fever in cynomolgus macaques, *Viral Immunol* 17(3), 390–400, 2004.
120. Simpson, D. I., Zlotnik, I., and Rutter, D. A., Vervet monkey disease: Experimental infection of guinea pigs and monkeys with the causative agents, *Br J Exp Pathol* 49, 458–464, 1968.
121. Simpson, D. I., Marburg agent disease, *Trans R Soc Trop Med Hyg* 63, 303–309, 1969.
122. Haas, R. and Maass, G., Experimental infection of monkeys with the Marburg virus, in *Marburg Virus Disease*, G. A. Martini and R. Siegert, eds., Springer-Verlag, New York, 1971, pp. 136–143.
123. Zlotnik, I., Marburg agent disease: Pathology, *Trans R Soc Trop Med Hyg* 63, 310–323, 1969.
124. Oehlert, W., The morphologic picture in livers, spleens, and lymph nodes of monkeys and guinea pigs after infections with the "Vervet agent," in *Marburg Virus Disease*, G. A. Martini and R. Siegert, eds., Springer-Verlag, New York, 1971, pp. 144–156.
125. Murphy, F. A., Simpson, D. I., Whitfield, S. G., Zlotnik, I., and Carter, G. B., Marburg virus infection in monkeys. Ultrastructural studies, *Lab Invest* 24(4), 279–291, 1971.
126. Geisbert, T. W. and Jaax, N. K., Marburg hemorrhagic fever: Report of a case studied by immunohistochemistry and electron microscopy, *Ultrastruct Pathol* 22(1), 3–17, 1998.
127. Geisbert, T. W., Hensley, L. E., Gibb, T. R., Steele, K. E., Jaax, N. K., and Jahrling, P. B., Apoptosis induced *in vitro* and *in vivo* during infection by Ebola and Marburg viruses, *Lab Invest* 80(2), 171–186, 2000.
128. Fisher-Hoch, S. P., Platt, G. S., Lloyd, G., Simpson, D. I., Neild, G. H., and Barrett, A. J., Haematological and biochemical monitoring of Ebola infection in rhesus monkeys: Implications for patient management, *Lancet* 2(8358), 1055–1058, 1983.
129. Hart, M. K., Vaccine research efforts for filoviruses, *Internat J Parasitol* 33, 583–595, 2003.
130. Jahrling, P. B., Geisbert, T. W., Dalgard, D. W., Johnson, E. D., Ksiazek, T. G., Hall, W. C., and Peters, C. J., Preliminary report: Isolation of Ebola virus from monkeys imported to USA, *Lancet* 335(8688), 502–505, 1990.

131. Ignatiev, G. M., Dadaeva, A. A., Luchko, S. V., and Chepurnov, A. A., Immune and pathophysiological processes in baboons experimentally infected with Ebola virus adapted to guinea pigs, *Immunol Lett* 71(2), 131–140, 2000.
132. Luchko, S. V., Dadaeva, A. A., Ustinova, E. N., Sizikova, L. P., Riabchikova, E. I., and Sandakhchiev, L. S., Experimental study of Ebola hemorrhagic fever in baboon models [in Russian], *Biull Eksp Biol Med* 120(9), 302–304, 1995.
133. Mikhailov, V. V., Borisevich, I. V., Chernikova, N. K., Potryvaeva, N. V., and Krasnianskii, V. P., The evaluation in hamadryas baboons of the possibility for the specific prevention of Ebola fever [in Russian], *Vopr Virusol* 39(2), 82–84, 1994.
134. Sullivan, N. J., Sanchez, A., Rollin, P. E., Yang, Z. Y., and Nabel, G. J., Development of a preventive vaccine for Ebola virus infection in primates, *Nature* 408(6812), 605–609, 2000.
135. Bray, M., Pathogenesis of filovirus infection in mice, in *Ebola and Marburg Viruses: Molecular and Cellular Biology*, H. Klenk and H. Feldmann, eds., Horizon Biosciences, Norfolk, 2004, pp. 255–277.
136. Bray, M., Davis, K., Geisbert, T., Schmaljohn, C., and Huggins, J., A mouse model for evaluation of prophylaxis and therapy of Ebola hemorrhagic fever, *J Infect Dis* 178(3), 651–661, 1998.
137. Bray, M., Hatfill, S., Hensley, L., and Huggins, J. W., Haematological, biochemical and coagulation changes in mice, guinea-pigs and monkeys infected with a mouse-adapted variant of Ebola Zaire virus, *J Comp Pathol* 125(4), 243–253, 2001.
138. Bray, M. and Huggins, J., Studies of the pathogenesis of filovirus infection using a mouse-adapted variant of Ebola Zaire virus, in *Symposium on Marburg and Ebola Viruses*, Marburg, Germany, 2000, p. 23.
139. Bray, M., Davis, K., Geisbert, T., Schmaljohn, C., and Huggins, J., A mouse model for evaluation of prophylaxis and therapy of Ebola hemorrhagic fever, *J Infect Dis* 179(Suppl 1), S248–S258, 1999.
140. Warfield, K. L., Alves, D. A., Bradfute, S. B., Reed, D. K., Vantongeren, S., Kalina, W. V., Olinger, G. G., and Bavari, S., Development of a model for Marburgvirus based on severe-combined immunodeficiency mice, *Virol J* 4(1), 108, 2007.
141. Warfield, K. L., Bradfute, S. B., Wells, J., Lofts, L., Cooper, M. T., Alves, D. A., Reed, D. K., VanTongeren, S. A., Mech, C. A., and Bavari, S., Development and characterization of a mouse model for Marburg hemorrhagic fever, *J Virol* 83(13), 6404–6415, 2009.
142. Lofts, L. L., Wells, J. B., Bavari, S., and Warfield K. L., Key genomic changes necessary for an in vivo lethal mouse marburgvirus variant selection process. *J Virol* 85(8), 3905–3917, 2011.
143. Bente, D., Gren, J., Strong, J. E., and Feldmann, H., Disease modeling for Ebola and Marburg viruses, *Dis Model Mech* 2(1–2), 12–17, 2009.
144. Gupta, M., Mahanty, S., Bray, M., Ahmed, R., and Rollin, P. E., Passive transfer of antibodies protects immunocompetent and imunodeficient mice against lethal Ebola virus infection without complete inhibition of viral replication, *J Virol* 75(10), 4649–4654, 2001.
145. Mahanty, S., Gupta, M., Paragas, J., Bray, M., Ahmed, R., and Rollin, P. E., Protection from lethal infection is determined by innate immune responses in a mouse model of Ebola virus infection, *Virology* 312(2), 415–424, 2003.
146. Gibb, T. R., Bray, M., Geisbert, T. W., Steele, K. E., Kell, W. M., Davis, K. J., and Jaax, N. K., Pathogenesis of experimental Ebola Zaire virus infection in BALB/c mice, *J Comp Pathol* 125(4), 233–242, 2001.
147. Warfield, K. L., Bosio, C. M., Welcher, B. C., Deal, E. M., Mohamadzadeh, M., Schmaljohn, A., Aman, M. J., and Bavari, S., Ebola virus-like particles protect from lethal Ebola virus infection, *Proc Natl Acad Sci U S A* 100(26), 15889–15894, 2003.
148. Warfield, K. L., Perkins, J. G., Swenson, D. L., Deal, E. M., Bosio, C. M., Aman, M. J., Yokoyama, W. M., Young, H. A., and Bavari, S., Role of natural killer cells in innate protection against lethal Ebola virus infection, *J Exp Med* 200(2), 169–179, 2004.

149. Huggins, J., Zhang, Z. X., and Bray, M., Antiviral drug therapy of filovirus infections: S-adenosylhomocysteine hydrolase inhibitors inhibit Ebola virus *in vitro* and in a lethal mouse model, *J Infect Dis* 179(Suppl 1), S240–S247, 1999.
150. Pushko, P., Bray, M., Ludwig, G. V., Parker, M., Schmaljohn, A., Sanchez, A., Jahrling, P. B., and Smith, J. F., Recombinant RNA replicons derived from attenuated Venezuelan equine encephalitis virus protect guinea pigs and mice from Ebola hemorrhagic fever virus, *Vaccine* 19(1), 142–153, 2000.
151. Vanderzanden, L., Bray, M., Fuller, D., Roberts, T., Custer, D., Spik, K., Jahrling, P., Huggins, J., Schmaljohn, A., and Schmaljohn, C., DNA vaccines expressing either the GP or NP genes of Ebola virus protect mice from lethal challenge, *Virology* 246(1), 134–144, 1998.
152. Rao, M., Matyas, G. R., Grieder, F., Anderson, K., Jahrling, P. B., and Alving, C. R., Cytotoxic T lymphocytes to Ebola Zaire virus are induced in mice by immunization with liposomes containing lipid A., *Vaccine* 17(23–24), 2991–2998, 1999.
153. Rao, M., Bray, M., Alving, C. R., Jahrling, P., and Matyas, G. R., Induction of immune responses in mice and monkeys to Ebola virus after immunization with liposome-encapsulated irradiated Ebola virus: Protection in mice requires CD4(+) T cells, *J Virol* 76(18), 9176–9185, 2002.
154. Warfield, K. L., Olinger, G. G., Deal, E. M., Swenson, D. L., Bailey, M., Negley, D. L., Hart, M. K., and Bavari, S., Induction of humoral and CD8+ T cell responses are required for protection against lethal Ebola virus infection, 175(2), 1184–1191, 2005.
155. Bray, M., Driscoll, J., and Huggins, J. W., Treatment of lethal Ebola virus infection in mice with a single dose of an S-adenosyl-L-homocysteine hydrolase inhibitor, *Antiviral Res* 45(2), 135–147, 2000.
156. SoRelle, R., Antibodies that protect mice against Ebola virus hold promise of vaccine and therapy for disease, *Circulation* 101(10), e9020, 2000.
157. Bray, M., Raymond, J. L., Geisbert, T., and Baker, R. O., 3-Deazaneplanocin A induces massively increased interferon-alpha production in Ebola virus-infected mice, *Antiviral Res* 55 (1), 151–9, 2002.
158. Zlotnik, I. and Simpson, D. I., The pathology of experimental Vervet monkey disease in Hamsters, *Br J Exp Pathol* 50, 393–399, 1969.
159. Ignatev, G. M., Agafonov, A. P., Strel'tsova, M. A., Kuz'min, V. A., Mainagasheva, G. I., Spirin, G. V., and Chernyi, N. B., A comparative study of the immunological indices in guinea pigs administered an inactivated Marburg virus [in Russian], *Vopr Virusol* 36, 421–423, 1991.
160. Hevey, M., Negley, D., Geisbert, J., Jahrling, P., and Schmaljohn, A., Antigenicity and vaccine potential of Marburg virus glycoprotein expressed by baculovirus recombinants, *Virology* 239(1), 206–216, 1997.
161. Riabchikova, E. I., Baranova, S. G., Tkachev, V. K., and Grazhdantseva, A. A., The morphological changes in Ebola infection in guinea pigs [in Russian], *Vopr Virusol* 38(4), 176–179, 1993.
162. Ryabchikova, E. I., Kolesnikova, L. V., and Luchko, S. V., An analysis of features of pathogenesis in two animal models of Ebola virus infection, *J Infect Dis* 179(Suppl 1), S199–S202, 1999.
163. Solcher, H., Neuropathological findings in experimentally infected guinea pigs, in *Marburg Virus Disease*, G. A. Martini and R. Siegert, eds., Springer-Verlag, New York, 1971, pp. 125–128.
164. Tikunova, N. V., Kolokol'tsov, A. A., and Chepurnov, A. A., Recombinant monoclonal human antibodies against Ebola virus, *Dokl Biochem Biophys* 378, 195–197, 2001.
165. Maruyama, T., Parren, P. W., Sanchez, A., Rensink, I., Rodriguez, L. L., Khan, A. S., Peters, C. J., and Burton, D. R., Recombinant human monoclonal antibodies to Ebola virus, *J Infect Dis* 179(Suppl 1), S235–S239, 1999.

166. Jahrling, P. B., Geisbert, J., Swearengen, J. R., Jaax, G. P., Lewis, T., Huggins, J. W., Schmidt, J. J., LeDuc, J. W., and Peters, C. J., Passive immunization of Ebola virus-infected cynomolgus monkeys with immunoglobulin from hyperimmune horses, *Arch Virol Suppl* 11, 135–140, 1996.
167. Jahrling, P. B., Geisbert, T. W., Geisbert, J. B., Swearengen, J. R., Bray, M., Jaax, N. K., Huggins, J. W., LeDuc, J. W., and Peters, C. J., Evaluation of immune globulin and recombinant interferon-alpha2b for treatment of experimental Ebola virus infections, *J Infect Dis* 179(Suppl 1), S224–S234, 1999.
168. Parren, P. W., Geisbert, T. W., Maruyama, T., Jahrling, P. B., and Burton, D. R., Pre- and postexposure prophylaxis of Ebola virus infection in an animal model by passive transfer of a neutralizing human antibody, *J Virol* 76(12), 6408–6412, 2002.
169. Wilson, J. A., Hevey, M., Bakken, R., Guest, S., Bray, M., Schmaljohn, A. L., and Hart, M. K., Epitopes involved in antibody-mediated protection from Ebola virus, *Science* 287(5458), 1664–1666, 2000.
170. Emonet, S. F., de la Torre, J. C., Domingo, E., and Sevilla, N., Arenavirus genetic diversity and its biological implications, *Infect Genet Evol* 9(4), 417–429, 2009.
171. Charrel, R. N. and de Lamballerie, X., Zoonotic aspects of arenavirus infections, *Vet Microbiol* 140(3–4), 213–220, 2010.
172. Downs, W., Anderson, C., Spence, L., Aitken, T., and Greenhall, A., Tacaribe virus, a new agent isolated from Artibeus bats and mosquitos in Trinidad, West Indies, *Am J Trop Med Hyg* 12, 640–646, 1963.
173. Buchmeier, M. J., De la Torre, J. C., and Peters, C. J., Arenaviridae: The viruses and their replication, in *Fields Virology*, 5th edition, D. M. Knipe, P. M. Howley, D. E. Griffin, and E. Andal, eds., Lippincott Williams & Wilkins, Philadelphia, 2006, pp. 1791–1827.
174. Carey, D. E., Kemp, G. E., White, H. A., Pinneo, L., Addy, R. F., Fom, A. L., Stroh, G., Casals, J., and Henderson, B. E., Lassa fever: Epidemiological aspects of the 1970 epidemic, Jos, Nigeria, *Trans R Soc Trop Med Hyg* 66(3), 402–408, 1972.
175. White, H. A., Lassa fever. A study of 23 hospital cases, *Trans R Soc Trop Med Hyg* 66(3), 390–401, 1972.
176. Monath, T. P., Mertens, P. E., Patton, R., Moser, C. R., Baum, J. J., Pinneo, L., Gary, G. W., and Kissling, R. E., A hospital epidemic of Lassa fever in Zorzor, Liberia, March–April 1972, *Am J Trop Med Hyg* 22(6), 773–779, 1973.
177. Peters, C. J., Kuehne, R. W., Mercado, R. R., Le Bow, R. H., Spertzel, R. O., and Webb, P. A., Hemorrhagic fever in Cochabamba, Bolivia, 1971, *Am J Epidemiol* 99(6), 425–433, 1974.
178. McCormick, J. B., Webb, P. A., Krebs, J. W., Johnson, K. M., and Smith, E. S., A prospective study of the epidemiology and ecology of Lassa fever, *J Infect Dis* 155(3), 437–444, 1987.
179. Johnson, K. M., McCormick, J. B., Webb, P. A., Smith, E. S., Elliott, L. H., and King, I. J., Clinical virology of Lassa fever in hospitalized patients, *J Infect Dis* 155(3), 456–464, 1987.
180. McCormick, J. B., King, I. J., Webb, P. A., Johnson, K. M., O'Sullivan, R., Smith, E. S., Trippel, S., and Tong, T. C., A case-control study of the clinical diagnosis and course of Lassa fever, *J Infect Dis* 155(3), 445–455, 1987.
181. McCormick, J. B., Epidemiology and control of Lassa fever, *Curr Top Microbiol Immunol* 134, 69–78, 1987.
182. McCormick, J. B., King, I. J., Webb, P. A., Scribner, C. L., Craven, R. B., Johnson, K. M., Elliott, L. H., and Belmont-Williams, R., Lassa fever. Effective therapy with ribavirin, *N Engl J Med* 314(1), 20–26, 1986.
183. Centers for Disease Control and Prevention (CDC), Management of patients with suspected viral hemorrhagic fever, *MMWR Morb Mortal Wkly Rep* 37(Suppl 3), 1–16, 1988.

184. McCormick, J. B. and Fisher-Hoch, S. P., Lassa fever, *Curr Top Microbiol Immunol* 262, 75–109, 2002.
185. Peters, C. J., Jahrling, P. B., Liu, C. T., Kenyon, R. H., McKee, K. T., Jr., and Barrera Oro, J. G., Experimental studies of arenaviral hemorrhagic fevers, *Curr Top Microbiol Immunol* 134, 5–68, 1987.
186. Evseev, A. A., Dvoretskaia, V. I., Bogatikov, G. V., Pshenichnov, V. A., and Mustafin, R. M., Experimental Lassa fever in hamadryas baboons [in Russian], *Vopr Virusol* 36(2), 150–152, 1991.
187. Stephenson, E. H., Larson, E. W., and Dominik, J. W., Effect of environmental factors on aerosol-induced Lassa virus infection, *J Med Virol* 14(4), 295–303, 1984.
188. Baize, S., Marianneau, P., Loth, P., Reynard, S., Journeaux, A., Chevallier, M., Tordo, N., Deubel, V., and Contamin, H., Early and strong immune responses are associated with control of viral replication and recovery in lassa virus-infected cynomolgus monkeys, *J Virol* 83(11), 5890–5903, 2009.
189. Jahrling, P. B., Hesse, R. A., Eddy, G. A., Johnson, K. M., Callis, R. T., and Stephen, E. L., Lassa virus infection of rhesus monkeys: Pathogenesis and treatment with ribavirin, *J Infect Dis* 141(5), 580–589, 1980.
190. Callis, R. T., Jahrling, P. B., and DePaoli, A., Pathology of Lassa virus infection in the rhesus monkey, *Am J Trop Med Hyg* 31(5), 1038–1045, 1982.
191. Walker, D. H., Johnson, K. M., Lange, J. V., Gardner, J. J., Kiley, M. P., and McCormick, J. B., Experimental infection of rhesus monkeys with Lassa virus and a closely related arenavirus, Mozambique virus, *J Infect Dis* 146(3), 360–368, 1982.
192. Fisher-Hoch, S. P., Mitchell, S. W., Sasso, D. R., Lange, J. V., Ramsey, R., and McCormick, J. B., Physiological and immunologic disturbances associated with shock in a primate model of Lassa fever, *J Infect Dis* 155(3), 465–474, 1987.
193. Carrion, R., Jr., Brasky, K., Mansfield, K., Johnson, C., Gonzales, M., Ticer, A., Lukashevich, I., Tardif, S., and Patterson, J., Lassa virus infection in experimentally infected marmosets: liver pathology and immunophenotypic alterations in target tissues, *J Virol* 81(12), 6482–6490, 2007.
194. Walker, D. H. and Murphy, F. A., Pathology and pathogenesis of arenavirus infections, *Curr Top Microbiol Immunol* 133, 89–113, 1987.
195. Walker, D. H., Wulff, H., and Murphy, F. A., Experimental Lassa virus infection in the squirrel monkey, *Am J Pathol* 80(2), 261–278, 1975.
196. Barkar, N. D. and Lukashevich, I. S., Lassa and Mozambique viruses: cross protection in experiments on mice and action of immunosuppressants on experimental infections [in Russian], *Vopr Virusol* 34(5), 598–603, 1989.
197. Lukashevich, I. S., Orlova, S. V., Mar'iankova, R. F., and Barkar, N. D., Pathogenicity of the Lassa virus for laboratory mice [in Russian], *Vopr Virusol* 30(5), 595–599, 1985.
198. Flatz, L., Rieger, T., Merkler, D., Bergthaler, A., Regen, T., Schedensack, M., Bestmann, L. et al., T cell-dependence of Lassa fever pathogenesis, *PLoS Pathog* 6(3), e1000836, 2010.
199. Jahrling, P. B., Smith, S., Hesse, R. A., and Rhoderick, J. B., Pathogenesis of Lassa virus infection in guinea pigs, *Infect Immun* 37(2), 771–778, 1982.
200. Walker, D. H., Wulff, H., Lange, J. V., and Murphy, F. A., Comparative pathology of Lassa virus infection in monkeys, guinea pigs, and *Mastomys natalensis*, *Bull World Health Organ* 52, 535–545, 1975.
201. Jahrling, P. B., Frame, J. D., Smith, S. B., and Monson, M. H., Endemic Lassa fever in Liberia. III. Characterization of Lassa virus isolates, *Trans R Soc Trop Med Hyg* 79(3), 374–379, 1985.
202. Jahrling, P. B., Protection of Lassa virus-infected guinea pigs with Lassa-immune plasma of guinea pig, primate, and human origin, *J Med Virol* 12(2), 93–102, 1983.
203. Huggins, J. W., Prospects for treatment of viral hemorrhagic fevers with ribavirin, a broad-spectrum antiviral drug, *Rev Infect Dis* 11(Suppl 4), S750–S761, 1989.

204. Pushko, P., Geisbert, J., Parker, M., Jahrling, P., and Smith, J., Individual and bivalent vaccines based on alphavirus replicons protect guinea pigs against infection with Lassa and Ebola viruses, *J Virol* 75(23), 11677–11685, 2001.
205. Charrel, R. N. and de Lamballerie, X., Arenaviruses other than Lassa virus, *Antiviral Res* 57(1–2), 89–100, 2003.
206. Enria, D. A., Briggiler, A. M., and Sanchez, Z., Treatment of Argentine hemorrhagic fever, *Antiviral Res* 78(1), 132–139, 2008.
207. Maiztegui, J. I., McKee, K. T. Jr., Barrera Oro, J. G., Harrison, L. H., Gibbs, P. H., Feuillade, M. R., Enria, D. A. et al., Protective efficacy of a live attenuated vaccine against Argentine hemorrhagic fever. AHF Study Group, *J Infect Dis* 177(2), 277–283, 1998.
208. Barry, M., Russi, M., Armstrong, L., Geller, D., Tesh, R., Dembry, L., Gonzalez, J. P., Khan, A. S., and Peters, C. J., Brief report: Treatment of a laboratory-acquired Sabia virus infection, *N Engl J Med* 333(5), 294–296, 1995.
209. McKee, K. T., Jr., Mahlandt, B. G., Maiztegui, J. I., Eddy, G. A., and Peters, C. J., Experimental Argentine hemorrhagic fever in rhesus macaques: Viral strain-dependent clinical response, *J Infect Dis* 152(1), 218–221, 1985.
210. Green, D. E., Mahlandt, B. G., and McKee, K. T., Jr., Experimental Argentine hemorrhagic fever in rhesus macaques: Virus-specific variations in pathology, *J Med Virol* 22(2), 113–133, 1987.
211. McKee, K. T., Jr., Mahlandt, B. G., Maiztegui, J. I., Green, D. E., and Peters, C. J., Virus-specific factors in experimental Argentine hemorrhagic fever in rhesus macaques, *J Med Virol* 22(2), 99–111, 1987.
212. Kenyon, R. H., McKee, K. T., Jr., Zack, P. M., Rippy, M. K., Vogel, A. P., York, C., Meegan, J., Crabbs, C., and Peters, C. J., Aerosol infection of rhesus macaques with Junin virus, *Intervirology* 33(1), 23–31, 1992.
213. Wagner, F. S., Eddy, G. A., and Brand, O. M., The African green monkey as an alternate primate host for studying Machupo virus infection, *Am J Trop Med Hyg* 26(1), 159–162, 1977.
214. McLeod, C. G., Jr., Stookey, J. L., White, J. D., Eddy, G. A., and Fry, G. A., Pathology of Bolivian hemorrhagic fever in the African green monkey, *Am J Trop Med Hyg* 27(4), 822–826, 1978.
215. Eddy, G. A., Scott, S. K., Wagner, F. S., and Brand, O. M., Pathogenesis of Machupo virus infection in primates, *Bull World Health Organ* 52(4–6), 517–521, 1975.
216. Terrell, T. G., Stookey, J. L., Eddy, G. A., and Kastello, M. D., Pathology of Bolivian hemorrhagic fever in the rhesus monkey, *Am J Pathol* 73(2), 477–494, 1973.
217. Webb, P. A., Justines, G., and Johnson, K. M., Infection of wild and laboratory animals with Machupo and Latino viruses, *Bull World Health Organ* 52(4–6), 493–499, 1975.
218. Kastello, M. D., Eddy, G. A., and Kuehne, R. W., A rhesus monkey model for the study of Bolivian hemorrhagic fever, *J Infect Dis* 133(1), 57–62, 1976.
219. McLeod, C. G., Stookey, J. L., Eddy, G. A., and Scott, K., Pathology of chronic Bolivian hemorrhagic fever in the rhesus monkey, *Am J Pathol* 84(2), 211–224, 1976.
220. Scott, S. K., Hickman, R. L., Lang, C. M., Eddy, G. A., Hilmas, D. E., and Spertzel, R. O., Studies of the coagulation system and blood pressure during experimental Bolivian hemorrhagic fever in rhesus monkeys, *Am J Trop Med Hyg* 27(6), 1232–1239, 1978.
221. Tesh, R. B., Jahrling, P. B., Salas, R., and Shope, R. E., Description of Guanarito virus (Arenaviridae: Arenavirus), the etiologic agent of Venezuelan hemorrhagic fever, *Am J Trop Med Hyg* 50(4), 452–459, 1994.
222. Weissenbacher, M. C., Calello, M. A., Colillas, O. J., Rondinone, S. N., and Frigerio, M. J., Argentine hemorrhagic fever: A primate model, *Intervirology* 11(6), 363–365, 1979.
223. Molinas, F. C., Giavedoni, E., Frigerio, M. J., Calello, M. A., Barcat, J. A., and Weissenbacher, M. C., Alteration of blood coagulation and complement system in neotropical primates infected with Junin virus, *J Med Virol* 12(4), 281–292, 1983.

224. Gonzalez, P. H., Laguens, R. P., Frigerio, M. J., Calello, M. A., and Weissenbacher, M. C., Junin virus infection of *Callithrix jacchus*: Pathologic features, *Am J Trop Med Hyg* 32(2), 417–423, 1983.
225. Molinas, F. C., Paz, R. A., Rimoldi, M. T., and de Bracco, M. M., Studies of blood coagulation and pathology in experimental infection of guinea pigs with Junin virus, *J Infect Dis* 137(6), 740–746, 1978.
226. Oubina, J. R., Carballal, G., Videla, C. M., and Cossio, P. M., The guinea pig model for Argentine hemorrhagic fever, *Am J Trop Med Hyg* 33(6), 1251–1257, 1984.
227. Kenyon, R. H., Green, D. E., Maiztegui, J. I., and Peters, C. J., Viral strain dependent differences in experimental Argentine hemorrhagic fever (Junin virus) infection of guinea pigs, *Intervirology* 29(3), 133–143, 1988.
228. Yun, N. E., Linde, N. S., Dziuba, N., Zacks, M. A., Smith, J. N., Smith, J. K., Aronson, J. F. et al., Pathogenesis of XJ and Romero strains of Junin virus in two strains of guinea pigs, *Am J Trop Med Hyg* 79(2), 275–282, 2008.
229. Hall, W. C., Geisbert, T. W., Huggins, J. W., and Jahrling, P. B., Experimental infection of guinea pigs with Venezuelan hemorrhagic fever virus (Guanarito): A model of human disease, *Am J Trop Med Hyg* 55(1), 81–88, 1996.
230. Hollidge, B. S., Gonzalez-Scarano, F., and Soldan, S. S., Arboviral encephalitides: Transmission, emergence, and pathogenesis, *J Neuroimmune Pharmacol* 5(3), 428–442, 2010.
231. Jonsson, C. B., Figueiredo, L. T., and Vapalahti, O., A global perspective on hantavirus ecology, epidemiology, and disease, *Clin Microbiol Rev* 23(2), 412–441, 2010.
232. Wilson, M. L., Chapman, L. E., Hall, D. B., Dykstra, E. A., Ba, K., Zeller, H. G., Traore-Lamizana, M., Hervy, J. P., Linthicum, K. J., and Peters, C. J., Rift Valley fever in rural northern Senegal: Human risk factors and potential vectors, *Am J Trop Med Hyg* 50(6), 663–675, 1994.
233. Shope, R. E., Peters, C. J., and Davies, F. G., The spread of Rift Valley fever and approaches to its control, *Bull World Health Organ* 60(3), 299–304, 1982.
234. Jouan, A., Coulibaly, I., Adam, F., Philippe, B., Riou, O., Leguenno, B., Christie, R., Ould Merzoug, N., Ksiazek, T., and Digoutte, J. P., Analytical study of a Rift Valley fever epidemic, *Res Virol* 140(2), 175–186, 1989.
235. Daubney, R., Hudson, J., and Garnham, P., Enzootic hepatitis or Rift Valley fever: An undescribed virus disease of sheep, cattle, and man from East Africa, *J Path Bact* 34, 545–579, 1931.
236. Findlay, G. M., Rift Valley fever or enzootic hepatitis, *Trans R Soc Trop Med Hyg* 25, 229, 1932.
237. Turell, M. J. and Kay, B. H., Susceptibility of selected strains of Australian mosquitoes (Diptera: Culicidae) to Rift Valley fever virus, *J Med Entomol* 35(2), 132–135, 1998.
238. Gargan, T. P., 2nd, Clark, G. G., Dohm, D. J., Turell, M. J., and Bailey, C. L., Vector potential of selected North American mosquito species for Rift Valley fever virus, *Am J Trop Med Hyg* 38(2), 440–446, 1988.
239. Gear, J. H. S., Rift Valley fever, in *CRC Handbook of Viral and Rickettsial Hemorrhagic Fevers*, J. H. S. Gear, ed., CRC Press, Inc., Boca Raton, FL, 2000.
240. El-Karamany, R., Imam, I., and Farid, A., Production of inactivated RVF vaccine, *J Egypt Publ Health Assoc* 56, 495–525, 1981.
241. Niklasson, B., Peters, C. J., Bengtsson, E., and Norrby, E., Rift Valley fever virus vaccine trial: Study of neutralizing antibody response in humans, *Vaccine* 3(2), 123–127, 1985.
242. Pittman, P. R., Liu, C. T., Cannon, T. L., Makuch, R. S., Mangiafico, J. A., Gibbs, P. H., and Peters, C. J., Immunogenicity of an inactivated Rift Valley fever vaccine in humans: A 12-year experience, *Vaccine* 18(1–2), 181–189, 1999.
243. Morrill, J. C. and Peters, C. J., Pathogenicity and neurovirulence of a mutagen-attenuated Rift Valley fever vaccine in rhesus monkeys, *Vaccine* 21(21–22), 2994–3002, 2003.

244. Harrington, D. G., Lupton, H. W., Crabbs, C. L., Peters, C. J., Reynolds, J. A., and Slone, T. W., Jr., Evaluation of a formalin-inactivated Rift Valley fever vaccine in sheep, *Am J Vet Res* 41(10), 1559–1564, 1980.
245. Bird, B. H., Ksiazek, T. G., Nichol, S. T., and Maclachlan, N. J., Rift Valley fever virus, *J Am Vet Med Assoc* 234(7), 883–893, 2009.
246. Peters, C. J. and Slone, T. W., Inbred rat strains mimic the disparate human response to Rift Valley fever virus infection, *J Med Virol* 10(1), 45–54, 1982.
247. McIntosh, B. M., Dickinson, D. B., and dos Santos, I., Rift Valley fever. 3. Viraemia in cattle and sheep. 4. The susceptibility of mice and hamsters in relation to transmission of virus by mosquitoes, *J S Afr Vet Assoc* 44(2), 167–169, 1973.
248. Morrill, J. C., Jennings, G. B., Cosgriff, T. M., Gibbs, P. H., and Peters, C. J., Prevention of Rift Valley fever in rhesus monkeys with interferon-alpha, *Rev Infect Dis* 11(Suppl 4), S815–S825, 1989.
249. Morrill, J. C., Knauert, F. K., Ksiazek, T. G., Meegan, J. M., and Peters, C. J., Rift Valley fever infection of rhesus monkeys: Implications for rapid diagnosis of human disease, *Res Virol* 140(2), 139–146, 1989.
250. Morrill, J. C., Jennings, G. B., Johnson, A. J., Cosgriff, T. M., Gibbs, P. H., and Peters, C. J., Pathogenesis of Rift Valley fever in rhesus monkeys: Role of interferon response, *Arch Virol* 110(3–4), 195–212, 1990.
251. Morrill, J. C., Czarniecki, C. W., and Peters, C. J., Recombinant human interferon-gamma modulates Rift Valley fever virus infection in the rhesus monkey, *J Interferon Res* 11(5), 297–304, 1991.
252. Peters, C. J., Jones, D., Trotter, R., Donaldson, J., White, J., Stephen, E., and Slone, T. W., Jr., Experimental Rift Valley fever in rhesus macaques, *Arch Virol* 99(1–2), 31–44, 1988.
253. Peters, C. J., Liu, C. T., Anderson, G. W., Jr., Morrill, J. C., and Jahrling, P. B., Pathogenesis of viral hemorrhagic fevers: Rift Valley fever and Lassa fever contrasted, *Rev Infect Dis* 11(Suppl 4), S743–S749, 1989.
254. Rossi, C. A. and Turell, M. J., Characterization of attenuated strains of Rift Valley fever virus, *J Gen Virol* 69(Pt 4), 817–823, 1988.
255. Easterday, B. C., Rift Valley fever, *Adv Vet Sci* 10, 65–127, 1965.
256. Mason, P. J. and Mims, C. A., Rift Valley fever virus in mice. V. The properties of a haemagglutinin present in infective serum, *Br J Exp Pathol* 37(5), 423–433, 1956.
257. Mims, C. A., The coagulation defect in Rift Valley fever and yellow fever virus infections, *Ann Trop Med Parasitol* 50(2), 147–149, 1956.
258. Mims, C. A., Rift Valley fever virus in mice. IV. Incomplete virus; its production and properties, *Br J Exp Pathol* 37(2), 129–143, 1956.
259. Mims, C. A., Rift Valley fever virus in mice. III. Further quantitative features of the infective process, *Br J Exp Pathol* 37(2), 120–8, 1956.
260. Mims, C. A., Rift Valley fever virus in mice. II. Adsorption and multiplication of virus, *Br J Exp Pathol* 37(2), 110–119, 1956.
261. Mims, C. A., Rift Valley fever virus in mice. I. General features of the infection, *Br J Exp Pathol* 37(2), 99–109, 1956.
262. Smith, D. R., Steele, K. E., Shamblin, J., Honko, A., Johnson, J., Reed, C., Kennedy, M., Chapman, J. L., and Hensley, L. E., The pathogenesis of Rift Valley fever virus in the mouse model, *Virology* 407(2), 256–267.
263. Peters, C. J., Reynolds, J. A., Slone, T. W., Jones, D. E., and Stephen, E. L., Prophylaxis of Rift Valley fever with antiviral drugs, immune serum, an interferon inducer, and a macrophage activator, *Antiviral Res* 6(5), 285–297, 1986.
264. Kende, M., Lupton, H. W., Rill, W. L., Levy, H. B., and Canonico, P. G., Enhanced therapeutic efficacy of poly(ICLC) and ribavirin combinations against Rift Valley fever virus infection in mice, *Antimicrob Agents Chemother* 31(7), 986–990, 1987.

265. Kende, M., Lupton, H. W., Rill, W. L., Gibbs, P., Levy, H. B., and Canonico, P. G., Ranking of prophylactic efficacy of poly(ICLC) against Rift Valley fever virus infection in mice by incremental relative risk of death, *Antimicrob Agents Chemother* 31(8), 1194–1198, 1987.
266. Bennett, D. G., Jr., Glock, R. D., and Gerone, P. J., Protection of mice and lambs against pantropic Rift Valley fever virus, using immune serum, *Am J Vet Res* 26, 57–61, 1965.
267. Higashihara, M., Heat- and acid-labile virus-inhibiting factor or interferon induced by Rift Valley fever virus in mice, *Jpn J Microbiol* 15(5), 482–484, 1971.
268. Kasahara, S. and Koyama, H., Long term existence of Rift Valley fever virus in immune mice, *Kitasato Arch Exp Med* 46(3–4), 105–112, 1973.
269. Tomori, O. and Kasali, O., Pathogenicity of different strains of Rift Valley fever virus in Swiss albino mice, *Br J Exp Pathol* 60(4), 417–422, 1979.
270. Canonico, P. G., Pannier, W. L., Huggins, J. W., and Rienehart, K. L., Inhibition of RNA viruses *in vitro* and in Rift Valley fever-infected mice by didemnins A and B, *Antimicrob Agents Chemother* 22(4), 696–697, 1982.
271. Anderson, A. O., Snyder, L. F., Pitt, M. L., and Wood, O. L., Mucosal priming alters pathogenesis of Rift Valley fever, *Adv Exp Med Biol* 237, 717–723, 1988.
272. Vialat, P., Billecocq, A., Kohl, A., and Bouloy, M., The S segment of Rift Valley fever phlebovirus (Bunyaviridae) carries determinants for attenuation and virulence in mice, *J Virol* 74(3), 1538–1543, 2000.
273. Anderson, G. W., Jr., Slone, T. W., Jr., and Peters, C. J., Pathogenesis of Rift Valley fever virus (RVFV) in inbred rats, *Microb Pathog* 2(4), 283–293, 1987.
274. Anderson, G. W., Jr. and Peters, C. J., Viral determinants of virulence for Rift Valley fever (RVF) in rats, *Microb Pathog* 5(4), 241–250, 1988.
275. Anderson, G. W., Jr., Rosebrock, J. A., Johnson, A. J., Jennings, G. B., and Peters, C. J., Infection of inbred rat strains with Rift Valley fever virus: Development of a congenic resistant strain and observations on age-dependence of resistance, *Am J Trop Med Hyg* 44(5), 475–480, 1991.
276. Ritter, M., Bouloy, M., Vialat, P., Janzen, C., Haller, O., and Frese, M., Resistance to Rift Valley fever virus in *Rattus norvegicus*: Genetic variability within certain "inbred" strains, *J Gen Virol* 81(Pt 11), 2683–2688, 2000.
277. Anderson, G. W., Jr., Lee, J. O., Anderson, A. O., Powell, N., Mangiafico, J. A., and Meadors, G., Efficacy of a Rift Valley fever virus vaccine against an aerosol infection in rats, *Vaccine* 9(10), 710–714, 1991.
278. Rosebrock, J. A., Schellekens, H., and Peters, C. J., The effects of ageing *in vitro* and interferon on the resistance of rat macrophages to Rift Valley Fever virus, *Anatomical Record* 205, A165–A166, 1983.
279. Rosebrock, J. A. and Peters, C. J., Cellular resistance to Rift Valley fever virus (RVFV) infection in cultured macrophages and fibroblasts from genetically resistant and susceptible rats, *In Vitro* 18, 308, 1982.
280. Niklasson, B. S., Meadors, G. F., and Peters, C. J., Active and passive immunization against Rift Valley fever virus infection in Syrian hamsters, *Acta Pathol Microbiol Immunol Scand [C]* 92(4), 197–200, 1984.
281. Anderson, G. W., Jr., Slone, T. W., Jr., and Peters, C. J., The gerbil, Meriones unguiculatus, a model for Rift Valley fever viral encephalitis, *Arch Virol* 102(3–4), 187–196, 1988.
282. Johnson, B. K., Gitau, L. G., Gichogo, A., Tukei, P. M., Else, J. G., Suleman, M. A., Kimani, R., and Sayer, P. D., Marburg, Ebola and Rift Valley fever virus antibodies in East African primates, *Trans R Soc Trop Med Hyg* 76(3), 307–310, 1982.
283. World Health Organization. Prevention and control of dengue and dengue haemorrhagic fever: Comprehensive guidelines. WHO Regional publication, SEARO. Nov. 29, 1999.

284. Pinheiro, F. and Corber, S. J., Global situation of dengue and dengue hemorrhagic fever and its emergence in the Americas, *World Health Stat Q* 50, 161–168, 1997.
285. Ho, L. J., Wang, J. J., Shaio, M. F., Kao, C. L., Chang, D. M., Han, S. W., and Lai, J. H., Infection of human dendritic cells by dengue virus causes cell maturation and cytokine production, *J Immunol* 166, 1499–1506, 2001.
286. Malavige, G. N., Fernando, S., Fernando, D. J., and Seneviratne, S. L., Dengue viral infections, *Postgrad Med* 80, 588–601, 2004.
287. Kroeger, A., Nathan, M., and Hombach, J., Dengue, *Nature Rev Microbiol* 2, 360–361, 2004.
288. Henchal, E. A., McCown, J. M., Seguin, M. C., Gentry, M. K., and Brandt, W. E., Rapid identification of dengue virus isolates by using monoclonal antibodies in an indirect immunofluorescence assay, *Am J Trop Med Hyg* 32, 164–169, 1983.
289. Guzman, M. G. and Kouri, G., Dengue: An update, *Lancet Infect Dis* 2, 33–42, 2002.
290. Rottman, A. L., Dengue: Defining protective versus pathologic immunity, *J Clin Invest* 113(7), 946–951, 2004.
291. Halstead, S. B., Pathogenesis of dengue: Challenges to molecular biology, *Science* 239, 476–481, 1988.
292. Littaua, R., Kurane, I., and Ennis, F. A., Human IgG Fc receptor II mediates antibody-dependent enhancement of dengue virus infection, *J Immunol* 144, 3183–3186, 1990.
293. Perikov, Y., Development of dengue vaccines, *Dengue Bull* 24, 71–76, 2000.
294. Brueckner, A. L., Reagan, R. L., and Yancey, F. S., Studies of dengue fever virus (Hawaii mouse adapted) in lactating hamsters, *Am J Trop Med Hyg* 5(5), 809–811, 1956.
295. Tarr, G. C. and Lubiniecki, A. S., Chemically-induced temperature-sensitive mutants of dengue virus type 2. I. Isolation and partial characterization, *Arch Virol* 48, 279–287, 1975.
296. Tarr, G. C. and Lubiniecki, A. S., Chemically induced temperature-sensitive mutants of dengue virus type 2. II. Comparison of temperature sensitivity *in vitro* with infectivity in suckling mice, hamsters, and rhesus monkeys, *Infect Immun* 13(3), 688–695, 1976.
297. Meiklejohn, G., England, B., and Lennette, E. H., Adaptation of dengue virus strains in unweaned mice, *Am J Trop Med Hyg* 1, 51–58, 1952.
298. Johnson, A. J. and Roehrig, J. T., New mouse model for dengue virus vaccine testing, *J Virol* 73, 783–786, 1999.
299. An, J., Kimura-Kuroda, J., Hirabayashi, Y., and Yasui, K., Development of a novel mouse model for dengue virus infection, *Virology* 263, 70–77, 1999.
300. Lin, H. S., Liao, C. L., Chen, L. K., Yeh, C. T., Liu, C. I., Ma, S. H., Huang, Y. Y., Kao, C. L., and King, C. C., Study of dengue virus infection in SCID mice engrafted with human K562 cells, *J Virol* 72, 9729–9737, 1998.
301. Wu, S. J., Hayes, C. G., Dubois, D. R., Windheuser, M. G., Kang, Y. H., Watts, D. M., and Sieckmann, D. G., Evaluation of the severe combined immunodeficient (SCID) mouse as an animal model for dengue viral infection, *Am J Trop Med Hyg* 52(5), 468–476, 1995.
302. Jaiswal, S., Pearson, T., Friberg, H., Shultz, L. D., Greiner, D. L., Rothman, A. L., and Mathew, A., Dengue virus infection and virus-specific HLA-A2 restricted immune responses in humanized NOD-SCID IL2rgamma null mice, *PLoS One* 4(10), e7251, 2009.
303. Huang, K.-J., Li, S. J., Chen, S.-C., Liu, H.-S., Lin, Y.-S., Yeh, T.-M., Liu, C.-C., and Lei, H.-Y., Manifestation of thrombocytopenia in dengue-2-virus-infected mice, *J Gen Virol* 81, 2177–2182, 2000.
304. Balsitis, S. J., Williams, K. L., Lachica, R., Flores, D., Kyle, J. L., Mehlhop, E., Johnson, S., Diamond, M. S., Beatty, P. R., and Harris, E., Lethal antibody enhancement of dengue disease in mice is prevented by Fc modification, *PLoS Pathog* 6(2), e1000790, 2010.

305. Tan, G. K., Ng, J. K., Trasti, S. L., Schul, W., Yip, G., and Alonso, S., A non mouse-adapted dengue virus strain as a new model of severe dengue infection in AG129 mice, *PLoS Negl Trop Dis* 4(4), e672, 2010.
306. Halstead, S. B., Shotwell, H., and Casals, J., Studies on the pathogenesis of dengue infection in monkeys. I. Clinical laboratory responses to primary infection, *J Infect Dis* 128(1), 7–14, 1973.
307. Halstead, S. B., Shotwell, H., and Casals, J., Studies on the pathogenesis of dengue infection in monkeys. II. Clinical laboratory responses to heterologous infection, *J Infect Dis* 128(1), 15–22, 1973.
308. Marchette, N. J., Halstead, S. B., Falkler Jr., W. A., Stenhouse, A., and Nash, D., Studies on the pathogenesis of dengue infection in monkeys. III. Sequential distribution of virus in primary and heterologous infections, *J Infect Dis* 128(1), 23–30, 1973.
309. Scherer, W. F., Russell, P. K., Rosen, L., and Dickerman, R. W., Experimental infection of chimpanzees with dengue viruses, *Am J Trop Med Hyg* 27(3), 590–599, 1978.
310. Angsubhakorn, S., Moe, J. B., Marchetter, N. J., Latendresse, J. R., Palumbo, N. E., Yakson, S., and Bhamarapravati, N., Neurovirulence detection of dengue virus using rhesus and cynomolgus monkeys, *J Virol Methods* 18(1), 13–24, 1987.
311. Malinoski, F. J., Hasty, S. E., Ussery, M. A., and Dalrymple, J. M., Prophylactic ribavirin treatment of dengue type 1 infection in rhesus monkeys, *Antiviral Res* 13(3), 139–149, 1990.
312. Eckels, K. H., Dubois, D. R., Summers, P. L., Schlesinger, J. J., Shelly, M., Cohen, S., Zhang, Y. M. et al., Immunization of monkeys with baculovirus-dengue type-4 recombinants containing envelope and nonstructural proteins: Evidence of priming and partial protection, *Am J Trop Med Hyg* 50(4), 472–478, 1994.
313. Onlamoon, N., Noisakran, S., Hsiao, H. M., Duncan, A., Villinger, F., Ansari, A. A., and Perng, G. C., Dengue virus-induced hemorrhage in a nonhuman primate model, *Blood* 115(9), 1823–1834.

15 Botulinum Toxins

Stephen B. Greenbaum, Jaime B. Anderson, and Frank J. Lebeda

CONTENTS

15.1 INTRODUCTION

The bacterial neurotoxins expressed by *Clostridium botulinum* are among the most toxic proteins known. The botulinum neurotoxins (BoNTs) are the causative agents of botulism, a potentially lethal disease typically associated with the ingestion of contaminated food products. Each BoNT molecule is initially synthesized as a single-chain polypeptide that undergoes a posttranslational cleavage forming light- and heavy-chain components linked by a disulfide bond [1]. Reduction of the disulfide bond is required for activation of the light chain's proteolytic function. Both components play critical roles in toxicity. The heavy chain (MW ~100 kDa) contains the neurotoxin's translocation and binding domains which mediate its uptake into peripheral cholinergic nerve termini. The light chain (MW ~50 kDa) inhibits the exocytosis of the neurotransmitter acetylcholine. Each serotype-specific light chain has a unique, zinc-dependent endoprotease activity at targeted substrates within the nerve terminals [2–8]. The seven presently known, antigenically distinct neurotoxins, types A through G, are produced by the corresponding *C. botulinum* serotypes. BoNT types E and F are also expressed by some strains of *C. butyricum* [9] and *C. baratii* [10], respectively. Type G toxin is expressed by *C. argentinense* [11]. Type C toxin is usually designated as type C1 to distinguish it from the ADP-ribosyltransferases, the C2 and C3 toxins, that are also produced by this serotype [12], which are neither homologous to the BoNTs nor specifically neurotoxic. On the other hand, tetanus neurotoxins (TeNTs) from *C. tetani* are structurally and functionally similar to BoNTs [13].

Much of the early work on the toxins focused on characterizing the molecular structure, function, and pathogenic effects of type A toxin. The crystalline form of this toxin has a MW ~900 kDa and a 19S sedimentation constant [14] and reflects the weights of additional proteins. In contrast, purified BoNT/A (MW ~150 kDa) has a corresponding sedimentation constant of 7S. The large, crystalline progenitor toxin consists of the "derivative" 7S neurotoxin component along with two or more noncovalently linked, nontoxic accessory proteins. The nontoxic-associated proteins (NAPs) [15] of the progenitor toxin complex include one to three separate hemagglutinins (e.g., HA17, HA33, HA70) and a single nontoxic nonhemagglutinin (NTNH) protein [1]. The NTNH protein is synthesized by all *C. botulinum* serotypes [16] and is generally found in all neurotoxin complexes with sedimentation constants of 12S or greater. Only certain serotypes produce HA-containing progenitor toxins and these multimeric complexes typically have sedimentation constants of 16S or higher. Type A toxins are synthesized in 900-kDa (19S), 500-kDa (16S), and 300-kDa (12S) forms. The type B, C, and D toxins are produced in 500-kDa (16S) or 300-kDa (12S) forms [17–20]. Serotypes E and F synthesize only the 300-kDa (12S) toxin, and toxin G is only synthesized in the 500-kDa (16S) form [21]. Various purification procedures can also be used to isolate the 7S neurotoxin with or without the associated nontoxic components of the larger multimeric complexes [15,20].

Because the pure neurotoxins are relatively sensitive to proteolytic degradation and the low pH environments of the gastrointestinal tract, the function of the HA and NTNH accessory proteins may be to protect the ingested neurotoxins and to facilitate absorption [22]. Interestingly, TeNT is not accompanied by any NAPs [22], which may explain why the BoNTs act as food poisons but the TeNTs do not. The auxiliary HA and NTNH proteins within the multimeric progenitor complex help maintain the stability of the progenitor toxin in the low pH region of the stomach [17,23–27]. The activity of the BoNTs at peripheral nerve terminals is not dependent on the accessory proteins. Whether active or inactive, the undissociated progenitor toxin is absorbed through the upper intestine (duodenum, pH 7) into the lymphatics (pH 8.2), the compartment in which dissociation occurs [25]. Thus, the HA and NTNH components are likely to be dispensable for disease pathogenesis after parenteral or respiratory exposure, where the toxins bypass the harsh conditions of the gastrointestinal tract.

15.2 MECHANISMS OF ACTION

The botulinum toxins function as powerful neuromuscular poisons, and numerous mammalian species are known to be at least somewhat susceptible to their activity. The initial cellular and molecular mechanisms leading to BoNT neurotoxicity involve toxin absorption, transit to specific target tissues, and nerve terminal uptake and retention. Certain steps in the pathogenic process have been characterized in detail using numerous *in vivo* and tissue culture models. After ingestion, the botulinum toxins are first absorbed into the lymphatics and circulation via receptor-mediated endocytosis and transcytosis across intestinal epithelial cells [1]. Similar events control toxin uptake at peripheral cholinergic nerve endings, as binding to high-affinity synaptic membrane receptors leads to toxin endocytosis and low pH-induced endosomal translocation. Ion channel formation by the N-terminal half of the heavy chain is also thought to be involved in this translocation process [28]. Binding at the nerve synapse is mediated by the C-terminal region of the heavy chain, whereas the N-terminal heavy-chain domain controls translocation into the cytosol [1,29,30]. The light chain then functions as a zinc-dependent endoprotease within the presynaptic nerve terminal, cleaving at least one of several synaptic proteins involved in neurotransmitter release [2–8]. Acetylcholine release from the cholinergic nerve terminal occurs through the formation of a fusion complex between acetylcholine-containing vesicles and the synaptic cell membrane [31]. This synaptic fusion complex contains several SNARE proteins (soluble N-ethylmaleimide-sensitive factor (NSF)-attachment protein receptor), including 25-kDa synaptosomal associated protein (SNAP-25), vesicle-associated membrane protein (VAMP or synaptobrevin), and syntaxin. Cleavage of any of these SNARE proteins by the neurotoxins disrupts the formation of the synaptic fusion complex, thereby blocking acetylcholine release and paralyzing the affected tissues. Botulinum toxins A and E irreversibly cleave SNAP-25, whereas toxins B, D, F, and G act on VAMP, and BoNT/C1 cleaves both SNAP-25 and syntaxin [2–8].

15.3 BOTULISM AS A CLINICAL DISEASE

BoNTs produce lethal disease in humans and numerous animal species. Botulism is characterized as an acute, descending, symmetric paralysis involving multiple cranial nerve palsies [31]. Six different clinical forms of botulism have been described in humans: food-borne botulism, infant botulism (an infectious intestinal toxicosis), wound botulism, an adult form of infant botulism, inadvertent systemic botulism, and inhalational botulism [31–33]. Infant botulism comprises the majority (72%) of reported human botulism cases in the United States, whereas most of the remaining cases involve classic food-borne botulism [34]. Infant and wound botulism are the prevalent infectious forms of the disease, whereas food-borne, inadvertent, and inhalational botulism result from exposure to preformed toxin. Food-borne botulism outbreaks in humans are typically associated with the consumption of toxin-contaminated home-prepared or home-preserved foods [35]. The vast majority of food-borne botulism cases are attributed to toxin serotypes A, B, or E. Outbreaks of type F and G botulism are rare [35,36], and only isolated anecdotal reports of human type C1 and D botulism appear in the published literature [37].

The times-to-onset of illness is dependent on toxin dose and ranges from several hours to a few days [31,38]. In contrast, the therapeutic doses of these toxins in treating dystonic and patients with other indications are much smaller and result in longer times-to-peak effect that can range up to 3 weeks [39,40]. Prominent signs and symptoms of food-borne botulism in humans include dysphagia, dry mouth, double vision, dysarthria, fatigue, ptosis, constipation, limb weakness, gaze paralysis, blurred vision, diminished gag reflex, nausea, facial palsy, dyspnea, vomiting, tongue weakness, sore throat, dizziness, pupil dilation or fixation, abdominal cramping, altered reflex response, nystagmus, diarrhea, ataxia, and paresthesia [31]. Inhalational exposure to the botulinum toxins also produces clinical disease in humans and several experimental species. Although some differences have been identified in the pathogenesis associated with respiratory versus gastrointestinal intoxication, many of the primary neurophysiological signs and symptoms of inhalational botulism parallel those observed in food-borne botulism [1,19,21,31,33].

The extremely high potency, ease of production, and stability of the botulinum toxins, along with their inclusion in various state-sponsored weaponization programs, underscore the threat associated with their potential use as agents of biological warfare and bioterrorism. This threat, along with the sparseness of toxicity data in humans, has driven extensive research on both food-borne and inhalational botulism in mice, guinea pigs, and rhesus monkeys. Many questions remain to be answered to fully characterize these toxins as potent biological agents and to develop effective medical countermeasures against accidental and intentional exposure. Future efforts to resolve these questions will rely heavily on the prominent animal models used in past studies; thus, it is critical to evaluate the ability of these models to adequately reflect toxicity and pathogenesis in humans.

15.4 ANIMAL MODELS FOR ORAL INTOXICATION

The pathogenesis of oral botulinum intoxication has been investigated extensively in a number of animal models, including the mouse, rat, guinea pig, rabbit, and nonhuman primate. Studies in mice have largely focused on the quantitative oral potency of various toxin serotypes, strains, and preparations [17,41–45]. These investigations have shown that the larger progenitor toxin complexes are typically more stable and, therefore, exhibit higher gastrointestinal toxicity than the purified derivative forms. As mentioned above, the HA and NTNH components of the multimeric toxin complexes are thought to confer increased protection against the harsh conditions of the gastrointestinal environment [17,23–27]. The effects of the HA and NTNH components on toxin stability are not restricted to the mouse model. The larger toxin complexes are less susceptible than the derivative forms to *in vitro* inactivation when incubated in intestinal juices from other experimental species [24,27,46,47].

Studies on toxin binding and uptake within peripheral nerve terminals have frequently incorporated rodent *in vivo* and tissue culture models [48–52]. Toxin binding interactions at central and peripheral nerve terminals appear to be quite similar in mice and rats but have not been extensively investigated in other animal species [53–56]. The mechanisms of action and paralytic effects of the botulinum toxins have also been extensively characterized in mice and are generally thought to parallel those in humans [2–8,31]. High interspecies sequence identity has been reported within the toxin cleavage sites of the SNARE proteins involved in neurotransmitter exocytosis, and similar nerve terminal recovery patterns following toxin-induced paralysis have been observed in mouse and human target tissues [43,57].

Studies on the quantitative potency of the botulinum toxins frequently incorporate the mouse model. The intraperitoneal mouse lethality assay remains the standard means for determining toxin concentrations in laboratory preparations and experimental samples, as mice are highly susceptible to systemic intoxication. In contrast, lethal oral doses in mice are typically several thousand times higher than lethal parenteral doses for the same toxins [17,23–25,42,43]. Certain other experimental species such as guinea pigs display much lower ratios of oral to systemic toxicity [58–60]. The early clinical presentation of food-borne botulism can be difficult to evaluate adequately in mice and other small-animal models. The later stages of lethal oral intoxication are more apparent and are associated with similar clinical signs in mice and humans, as well as several other mammalian species [31,48,61–63]. These acute signs are directly related to the paralytic effects of BoNT at peripheral nerve terminals and include impaired limb function, prostration, and labored breathing.

Whereas the mouse model has frequently been used to investigate the oral potency of the botulinum toxins, the rat model has been more widely incorporated in studies of toxin stability, persistence, and absorption from the gastrointestinal tract [23,24–26,46,64,65]. Collectively, these rodent studies have provided the foundation for our current understanding of the relationship between the stability of the toxins and their corresponding potency after gastrointestinal administration. Investigations of toxin absorption from various regions of the gastrointestinal tract have relied heavily on the rat ligated intestinal model [23–26,46]. This model allows for determinations of toxin absorption from isolated intestinal sections *in vivo* on the basis of the appearance of toxin in the lymph after gastrointestinal exposure. Regional patterns of toxin absorption from the rat gastrointestinal tract parallel those observed in several other experimental animal species [23,25,46–48,62,66–70]. The rat model has also been used to investigate toxin persistence in the lymphatics after gastrointestinal absorption. Although the kinetics of toxin appearance and removal from the lymph of rats are similar to those observed in certain other animal species, the degree to which these patterns reflect toxin behavior following systemic absorption in humans is not clear.

The rat model has also been incorporated in many of the *in vivo* and tissue culture studies on botulinum toxin binding and uptake in target tissues. Several groups have demonstrated binding of various toxin serotypes to rat nerve synapses and rat-derived adrenergic cell lines [48–53,55,71]. Toxin binding at central and peripheral nerve terminals appears to be similar in mice and rats. Thus far, these binding interactions have not been extensively studied in other prominent experimental species. Moreover, the fidelity of rodent whole-animal and tissue culture models in reflecting toxin binding at human neuromuscular junctions has yet to be determined.

The mechanisms involved in toxin-induced paralysis in rats are thought to be similar to those occurring in mice and humans. The cholinergic synapses of rats and humans share the same presynaptic protein targets and are, therefore, subject to similar patterns of toxin-induced paralysis [2–8,57,72,73]. Despite these similarities, a key difference has been identified in the susceptibility of rats versus certain other mammalian species to type B toxin. Rats are considered less suitable than other rodent species for modeling type B intoxication because of a lack of sequence identity at the BoNT/B cleavage site in an isoform of the VAMP substrate [53].

Although the rat model has been used extensively to study botulinum toxin absorption from the gastrointestinal tract, detailed oral toxicity data are not available for many serotypes. Rats are considered somewhat more resistant than other common rodent species to oral intoxication [60,62]. The literature provides no explanation for this resistance, but the relatively high susceptibility of rats to parenterally administered toxin may be indicative of poor gastrointestinal absorption efficiency rather than low systemic sensitivity [61,74]. High toxin levels can be detected in the lymph of rats after gastrointestinal toxin administration, implicating other undefined factors in the reported resistance of rats to oral intoxication [23,25,26,46]. The low susceptibility of rats to oral intoxication would argue against their use for quantitative evaluations of gastrointestinal toxicity in humans. In contrast, the systemic pathogenesis and clinical presentation associated with botulinum intoxication are often similar in rats and other experimental species, with the possible exception that rats reportedly develop bloody tears and a more defined muscular weakness than that observed in mice and rabbits [75].

The rabbit model has also been used in several studies on oral botulinum intoxication. Whereas rabbits are quite sensitive to systemic intoxication, they are significantly less susceptible to oral intoxication than are guinea pigs [58,60,76,77]. Rabbits are also less sensitive to repeated sublethal type A toxin doses than are guinea pigs [78], despite similar kinetics of circulating toxin in both species [58]. Other than reduced oral sensitivity to type A and E toxins, no other unique characteristics have been demonstrated for the rabbit model in the context of oral botulinum intoxication. Rabbits and nonhuman primates display comparable gastrointestinal absorption kinetics for type E toxin [77]. Tissue culture studies have demonstrated similar nerve terminal responses to type D toxin in rabbit isolated ileum, guinea pig vas deferens, and cat tail arrectores pilorum muscles [79]. The clinical signs of oral intoxication are analogous between rabbits and guinea pigs, but guinea pigs display a more rapid disease progression than rabbits and other small animal models [58]. Collectively, these studies provide no evidence that the rabbit model carries any distinct advantages in reflecting the human condition. Thus, the bias toward the guinea pig rather than the rabbit model in oral botulinum intoxication studies may be largely attributable to its higher susceptibility, reduced cost, and ease of breeding.

The guinea pig represents one of the most common animal models for oral botulinum intoxication and has been incorporated in numerous studies on toxin binding, absorption, and potency following gastrointestinal exposure. Toxin binding activity has been demonstrated with epithelial cells of the guinea pig's upper small intestine [27], the same region where toxin absorption is known to occur in a number of other experimental species. Following gastrointestinal absorption in guinea pigs, toxin appears transiently in the blood and in the lymphatics of the small intestine [27,58,60]. The clinical signs of oral botulinum intoxication in guinea pigs generally parallel those seen in other experimental species and include breathing difficulty, generalized weakness, and finally respiratory paralysis and death [63]. Although other rodent species also show some susceptibility to the botulinum toxins, the guinea pig appears to be particularly sensitive to oral intoxication. Importantly, the guinea pig lethal dose values determined for type A toxin in several oral toxicity studies are significantly lower than those reported for mice [17,41,43–45,58,59,80]. Guinea pigs show high oral susceptibility to type C1 and D toxins [59,76], which are not typically associated with food-borne illness in humans. The physiological basis for the high susceptibility of guinea pigs to ingested toxins has not been defined, but their sensitivity to oral intoxication is often provided as justification for their use in modeling human food-borne botulism.

Relatively few published studies have evaluated oral botulinum intoxication in nonhuman primates. Toxin behavior in the gastrointestinal tract and circulation after oral exposure does not appear to be significantly different in monkeys as compared to other animal models [77]. The limited oral toxicity data available for nonhuman primates indicate that monkeys are moderately susceptible to oral intoxication.

Several early studies reported relatively low LD_{50} values for type E toxin, intermediate toxicities for types A, F, and G, and higher LD_{50} values for types C and D [63,76]. These reports also described clinical signs of intoxication that are highly analogous to those observed in human food-borne botulism patients, including diplopia, ptosis, muscular weakness, difficulty swallowing, reduced food and water intake, and respiratory distress [63,81]. These signs appear in an ordered sequence very similar to the human clinical presentation. The only consistent deviations from the human condition reported for intoxicated monkeys are the presence of oral and nasal discharge, a lack of gastrointestinal distress, and no significant alterations in pupil size or shape [63,81,82].

Three nonhuman primate species have been examined in these experimental oral intoxication studies: rhesus, cynomolgus, and squirrel monkeys [63,76,77,81,82]. Other reports have provided valuable insight into the epidemiology of natural food-borne botulism outbreaks in other nonhuman primate species, including tamarins, marmosets, capuchins, gibbons, and baboons [82,83]. Outbreaks of type C botulism are frequently discussed in these reports, indicating one of the few potential deviations of the nonhuman primate in modeling the human disease condition. Several experimental animal models also appear to be at least somewhat susceptible to oral type C and D intoxication [59,76,84,85]. These serotypes are rarely implicated in human botulism cases. One early study briefly mentioned two food-borne outbreaks one of type C botulism and another of type D botulism [25]. A later article reported a case of infant type C botulism [86]. The low number of reports for establishing the potential for type C and D toxins to cause disease in humans emphasizes that more study is required. As supporting evidence, tissue culture experiments have demonstrated that isolated human neuromuscular junctions and neuroblastoma cells are susceptible to type C toxin [87,88]. Moreover, *in vivo* studies on the therapeutic potential of BoNT/C1 have shown that type A and C toxins exert similar paralytic effects within the extensor digitorum brevis muscles of human subjects [89].

An *in vitro* study of toxin transport across human gut epithelial cell lines provided another potential explanation for the lack of reported type C botulism cases in humans [90]. Whereas both type A and B toxins are efficiently transcytosed across two different human colon carcinoma cell lines, only minimal transcytosis of type C was observed. These findings indicate that the lack of human type C botulism cases might be a result of poor gastrointestinal absorption of the ingested toxin. Alternatively, type C toxin absorption *in vivo* might occur across select human gut epithelial regions not adequately modeled by the colon carcinoma cell monolayers. Thus, the potential roles of toxin absorption and nonphysiological factors such as species-specific eating habits in the scarcity of human type C and D botulism cases are not yet defined.

Several generalizations can be derived from oral botulinum intoxication studies in various experimental species. For instance, pathogenic parameters such as biological stability, gastrointestinal absorption, and oral toxicity are heavily dependent on the specific toxin preparations, strains, and serotypes used in the various animal models. The larger multimeric progenitor toxin complexes are considerably more stable in the gastrointestinal tract than the derivative neurotoxins and, therefore, possess higher oral toxicities in various animal models [23–27,42,46,91]. The higher oral potency of the progenitor toxins may be related to an increased ability to resist the harsh gastrointestinal conditions, along with a direct contribution of the HA component to the binding and absorption of certain toxins [27,84,92]. The different *C. botulinum* strains also appear to have unique properties that influence the oral potency of the toxins they produce. Whereas some of the more toxic strains have been identified in mouse experiments [17,44,93,94], oral toxicity comparisons between different strains have not been reported for other experimental species.

The absorption of botulinum toxins from the gastrointestinal tract has been evaluated in numerous *in vivo* and tissue culture studies. Ligated intestinal models have demonstrated that toxin absorption is generally most efficient within the upper small intestine of several experimental species [17,25–27,46,47,62,67,95]. These findings indicate that at least some fraction of ingested toxin must survive intact in the gastric environment and enter the small intestine for subsequent absorption. Although ligated intestinal models are useful in characterizing toxin behavior within isolated gastrointestinal regions, their design provides only limited resolution in evaluating the absorption of ingested toxins. These models do not address the possibility that some ingested toxin might be absorbed before reaching the gastrointestinal tract. They are also inadequate in identifying the specific gastrointestinal cell types involved in toxin absorption. Identification of these cell populations will be important in establishing the most appropriate cell culture models and in understanding the specific mechanisms governing the absorption of ingested toxins.

While numerous animal models have been used to investigate oral botulinum intoxication, interspecies comparisons of disease pathogenesis after gastrointestinal exposure remain limited. The clinical signs of intoxication following oral exposure are similar across the prominent animal models [58,61,63,77,81,82,96]. The onset and severity of the clinical presentation, as well as the time to death in lethally intoxicated animals, are typically dose-dependent and variable across individuals of the same model species [31,38,63,77,81,96], including humans [39]. Thus, very little evidence is available in the current literature to suggest that one animal model is significantly more relevant than others in reflecting oral intoxication in humans. In the absence of such data, the relative susceptibility of the different experimental species to oral intoxication may provide the most constructive foundation for comparison with humans. Unfortunately, susceptibility comparisons are further limited by the negligible amount of quantitative data on oral toxicity in humans and in some of the more common animal models such as rats, rabbits, and nonhuman primates. Surprisingly, most case studies on human food-borne botulism fail to indicate either the estimated toxin dose ingested or the toxin concentration in the contaminated foods. In the future, accurate susceptibility comparisons between humans and experimental animal species will require more thorough investigation and quantitative reporting of oral toxicity.

15.5 ANIMAL MODELS FOR INHALATIONAL INTOXICATION

Experimental respiratory exposure to the botulinum toxins produces disease and lethality in several animal species. The inadvertent exposure of three laboratory workers in Germany to aerosolized type A toxin several decades ago provided evidence for the feasibility of inhalational botulism in humans. The clinical presentations reported for these human subjects generally paralleled those seen in experimental animals [31,33]. The intoxicated laboratory workers suffered from the common signs and symptoms of botulism but recovered from respiratory exposure to undetermined amounts of type A toxin. Because the quantitative potency of inhaled botulinum toxin aerosols in humans is completely unknown, estimates for human respiratory toxicity must, therefore, rely exclusively on data obtained from animal models for inhalational intoxication. The respiratory potency of select toxin serotypes has been investigated in several experimental species including mice, guinea pigs, and monkeys. As expected, questions remain about the relationship between the inhalational intoxication seen in common animal species and in humans.

The earliest inhalational botulinum toxin exposure studies typically used larger laboratory species. More recently, the mouse model emerged as a valuable experimental system for respiratory intoxication. A study on the potency of BoNT/A in mice demonstrated the absorption, toxicity, and immunogenicity of the purified neurotoxin following intranasal installation in mice [97]. The experimental results of this study addressed qualitative rather than quantitative issues; this animal model, nevertheless, demonstrated the general capacity for the absorption of the neurotoxin components after intranasal administration. Other studies on respiratory tract morphology and model particle deposition have revealed several general limitations of the mouse model in adequately and quantitatively reflecting inhalational exposure in humans. For example, mice and other rodent species are obligate nasal breathers with complex nasal passages, and, as such, they display higher upper respiratory tract retention and lower alveolar deposition of inhaled particles than do monkeys and humans [98,99]. The quantity of particles deposited per gram of lung over a given duration is also much higher in mice than in humans, as well as most other experimental species [99].

These discrepancies underscore the relatively low fidelity of the mouse model to human respiratory anatomy and physiology but do not preclude the use of mice in certain inhalational exposure studies. For example, there is no strong evidence to indicate that the biological stability and persistence of the botulinum toxins after respiratory absorption would be dramatically different in mice as compared to that in other experimental species. In addition, there are no apparent inconsistencies between mice and humans with respect to toxin uptake, mechanisms of action, and paralytic effects at cholinergic nerve terminals. Thus, the mouse model should continue to prove useful in characterizing systemic pathogenesis and toxin activity at peripheral target tissues. Similar immunological responses in mice and humans also support the continued use of murine models in characterizing existing or novel vaccines prior to testing in higher animal species. The mouse model, therefore, remains a viable option in addressing certain questions related to respiratory intoxication. At this time there is a lack of quantitative data to support the comparative analysis of inhalational responses to BoNT in mice and humans.

Inhalational exposure to the botulinum toxins has been more extensively investigated in guinea pigs than in smaller rodent species such as mice and rats. The guinea pig is highly susceptible to botulinum intoxication, and lethal respiratory dose values are generally lower in guinea pigs than in many other experimental models such as rabbits and monkeys [58–60,76,77]. The high sensitivity of guinea pigs to the botulinum toxins is further emphasized by their short incubation periods before disease progression. Once signs of illness appear in guinea pigs, they generally parallel those seen in other animal species and include muscular weakness, unresponsiveness, breathing difficulty, and flaccid paralysis [58]. Guinea pigs also produce foamy or bloody sputum from respiratory intoxication [58]. Oral and nasal secretions have also been reported in monkeys and certain other animal models after gastrointestinal and systemic toxin exposure [63,81,100–102] but are not typically associated with human food-borne botulism cases, which are commonly characterized by dry mouth [31].

The clinical presentation following experimental respiratory exposure in monkeys is quite similar to that seen after inadvertent respiratory or gastrointestinal exposure in humans [31,102]. Some of the more common signs observed in both monkeys and humans, such as ptosis and dysphagia, are not generally reported in guinea pigs and other common experimental species. Few published studies have evaluated respiratory toxicity in nonhuman primates, and no quantitative data on human susceptibility are available. Such information would facilitate more detailed comparisons between monkeys and guinea pigs as the prominent animal models for respiratory botulinum toxin exposure in humans.

The relevance of these animal models in reflecting inhalational intoxication in humans should also be evaluated in more general terms with respect to their comparative respiratory anatomy and physiology. As with most other small experimental species, guinea pigs are nasal breathers with complex nasal passages that enhance the upper respiratory retention of inhaled particles [98,99,103,104]. Alveolar deposition patterns in the distal lung are much more sensitive to changes in particle size in guinea pigs than in monkeys and humans [98]. Alveolar deposition decreases in guinea pigs with increasing particle size from around 0.5 to 2.5 μm, whereas the opposite trend has been observed in monkeys and humans [92,104]. These deposition patterns generally translate into reduced alveolar deposition and higher total retention of inhaled particles in guinea pigs compared to monkeys and humans [98,104]. Guinea pigs also have higher respiratory rates and minute volumes, as well as significantly lower tidal volumes, than both monkeys and humans [99,104]. These disparities collectively result in the deposition of much higher quantities of inhaled particles per gram of lung per unit time in guinea pigs than in either monkeys or humans [104]. Such differences indicate that the behavior of inhaled substances in the human respiratory tract is generally most closely reflected in the nonhuman primate model. Because most of these studies incorporate benign model aerosols, caution must be exercised when using such data to predict respiratory deposition patterns specifically for the botulinum toxins.

Both guinea pigs and monkeys appear to be quite susceptible to inhalational intoxication, yet respiratory lethal dose values have not been thoroughly investigated in nonhuman primates. The guinea pig model should continue to prove adequate for certain studies, depending on the specific questions to be addressed, the experimental design, and the availability of resources. Guinea pigs may be sufficient for investigating certain cellular and molecular aspects of pathogenesis after respiratory absorption, as no clear differences have been defined regarding toxin activity in the circulation and target tissues of guinea pigs and primates. In contrast, quantitative respiratory deposition, absorption, and toxicity data from nonhuman primate studies should be more representative of human intoxication than those derived from the guinea pig model.

Although *in vivo* animal models remain critical tools for investigating inhalational botulinum intoxication, certain aspects of disease pathogenesis can be more specifically addressed using tissue culture systems. Pulmonary adenocarcinoma cell lines and primary alveolar epithelial cells were used to investigate toxin transcytosis within the respiratory tract [97]. Transcytosis of both BoNT/A and purified type A heavy chain was observed across alveolar epithelial monolayers. The relevance of this culture model depends on the adequacy of the assumption that toxin deposition and absorption after inhalational exposure primarily occurs within the distal respiratory tract. Previous studies on the respiratory deposition of model particle aerosols provide some support for this speculation, but additional work is needed to more specifically investigate the behavior of inhaled botulinum toxins. Significant toxin deposition and absorption may also occur within the nasal passages, the airway branches, and the bronchioles. Future studies could incorporate labeled toxin preparations along with real-time imaging systems to evaluate deposition and absorption after respiratory exposure in the relevant animal models. Such experiments would facilitate more detailed animal model comparisons and support the development of the most appropriate tissue culture models for respiratory toxin absorption.

As indicated in the above summary of results, different methods and models have been applied to study the effects of inhalational exposure to botulinum toxins. Previously, the interconnections among *in vivo*, *ex vivo*, *in vitro*, and *in silico* studies have been considered [105]. Evaluation of the correlative relationships between animal models and clinical studies, as represented in Figure 4.3 in the study by Anderson and Tucker [105], is essential when attempting to extrapolate results, using different methods, from one species to another. One published study used a theoretical approach to compare the results from different animal and clinical studies of botulinum toxin-induced paralysis [106]. Experimental kinetic data were examined for the development of paralysis from *in vivo* studies in mice [107,108] and from the isolated phrenic nerve-diaphragm preparation in rats [109]. These data were compared to clinical results in which the toxin was injected into a dystonic patient and the progressive relief of symptoms was periodically monitored [110]. Results from these studies indicated that the times-to-peak effect varied significantly depending on the model, ranging from several hours *ex vivo* to 3 days in mice and 2 weeks in humans. To establish commonality, a single *in silico* model was constructed to simulate all three time courses and to show the relationships among the *ex vivo* and *in vivo* results from animal models and the clinic. This example also raises the point that other roles for botulinum toxin are being examined in the laboratory and in the clinic. Some of the more recent studies are listed in Table 15.1 [111–137].

15.6 DISCUSSION

The botulinum toxins are extremely potent neuromuscular poisons capable of causing disease and lethality in humans and numerous other mammalian species.

TABLE 15.1
Recent Publications of Animal Models Serving Different Functional Roles for Botulinum Neurotoxin

Animal Model	Functional Role of Model	Preparation	Selected References
Guinea pig	5HT release	*In vitro*	[111]
Mouse	Antiallodynic efficacy (neuropathic pain)	*In vivo*	[112]
Mouse	Antibody protection	*In vivo*	[113]
Mink, guinea pig	Antitoxin titers	*In vivo*	[114]
Mouse	Bone loss	*In vivo*	[115]
Mouse	Ca^{2+} entry, pancreatic acinar cells	*In vitro*	[116]
Mouse	candidate drug screen, NMJ; toe spread	*Ex vivo, in vivo*	[117]
Mouse	Candidate drug screen—spinal cord cells	*In vitro*	[118]
Chick embryo	Candidate drug screen—cultured neurons	*In vitro*	[119]
Guinea pig	Changes in AChE levels—nasal glands	*In vivo, in vitro*	[120]
Mouse	Detect antibodies	*In vitro*	[121]
Cow	Detect toxin	*Ex vivo*	[122]
Cow	Detect toxin in milk	*In vivo*	[123]
Fresh water fish	Detection and toxicity	*In vivo*	[124]
Mouse	Diffusion of toxin	*In vivo*	[125]
Mouse	Dry eye model (keratoconjunctivitis sicca)	*In vivo*	[126]
Horse	Immune response	*In vivo*	[127]
Mouse	Joint pain; arthritis	*In vivo*	[128]
Mouse	Neuritogenesis (sprouting) motor neurons	*In vitro*	[129]
Mouse	Neurotoxicity	*Ex vivo*	[130]
Mouse	Paralysis and recovery	*In vivo*	[131]
Mouse	Recovery from paralysis	*In vivo*	[132]
Mouse	Spontaneous recurrent seizures	*In vivo*	[133]
Chicken	Toxicoinfectious botulism	*In vivo*	[134]
Channel catfish	Toxicosis	*In vivo*	[135]
Mouse	Toxin effects on muscle ultrastructure	*In vivo*	[136]
Mouse	Wound botulism bioassay	*In vivo*	[137]

The toxins traditionally affect humans and animals through food-borne exposure to contaminated food products. Their oral toxicity is unique among protein toxins, as their association with various nontoxic polypeptide components in multimeric complexes provides significant protection from the harsh conditions of the gastrointestinal tract. The progenitor toxin complexes possess sufficient biological stability to resist proteolysis and allow for intestinal absorption of the intact neurotoxins into the lymphatics and circulation. Once absorbed, the neurotoxins rapidly target peripheral neuromuscular junctions and inhibit neurotransmitter release, leading to paralysis of the affected tissues.

This pathogenic process has been characterized to some extent in several experimental animal models with a wide range of susceptibilities to oral intoxication. Rodent species such as mice, rats, and guinea pigs have been used extensively to investigate the gastrointestinal absorption, mechanisms of action, and oral potencies of the botulinum toxins. Nonhuman primates have been incorporated to a lesser extent because of factors such as high cost, low availability, and regulatory constraints. Collectively, these studies have revealed a number of advantages and shortcomings associated with the different animal models. Such findings will prove useful in identifying focus areas for additional research and in determining the most appropriate animal models to use in future investigations. Additional information on food-borne botulism has been derived from case studies of human outbreaks, which also provide some basis for comparison of the prominent animal models for the human condition. Yet many questions remain to be addressed regarding the susceptibility of humans to ingested botulinum toxins and the pathogenesis of disease reflected in the various animal models. Although numerous studies have evaluated lethal dose values in experimental animals, minimal quantitative data are available on oral toxicity in humans. In addition, relatively little is known about the mechanisms involved in the intestinal binding, absorption, and systemic transit of the botulinum toxins to target tissues after oral exposure in experimental animals and humans. Such information would facilitate more detailed comparisons of the prominent animal models in the context of future studies.

Although the botulinum toxins have historically been associated primarily with food-borne illness, they also represent potential inhalational exposure threats. The respiratory potency of the toxins has been demonstrated in several experimental animal models. Recent *in vivo* and tissue culture studies have also established the capacity of the purified neurotoxins to be absorbed intact from the respiratory tract. The resulting pathogenesis after respiratory absorption is generally considered to be similar to that observed after gastrointestinal toxin absorption, as circulating toxin rapidly targets peripheral nerve terminals and inhibits presynaptic neurotransmitter release. Unique requirements for toxin binding and absorption may be associated with the respiratory exposure route. Moreover, the correlates for protection against and treatment of inhalational intoxication may be different from those associated with food-borne botulism. Whereas medical countermeasures against food-borne botulism have been fairly well characterized in both humans and animal models, the efficacy of such approaches in treating inhalational intoxication in humans remains unknown. Future efforts to define the pathogenesis, prevention, and therapy of inhalational intoxication will require the most appropriate animal models for the human condition. Respiratory toxicity and pathogenesis data have been generated in several existing animal models, including mice, guinea pigs, and nonhuman primates. Our understanding of inhalational intoxication in humans will benefit from further characterization of these whole-animal models and the development of new tissue culture systems for respiratory exposure to the BoNTs.

15.7 CONCLUSIONS

This chapter and previous reviews indicate that the mouse model is still prominent and of significant value for many types of botulinum toxin research topics. From a biodefense perspective, the most recent data continue to support the mouse model for studies on the toxin as a biothreat agent and for countermeasure evaluation. Coordination of *in vivo* and *in vitro* animal studies, integrated with *in silico* studies, is needed to understand further the effects of different toxin types, preparations, exposure routes, environmental conditions, and individual variability, in order to make accurate extrapolations to humans.

DISCLAIMER

The opinions, interpretations, conclusions, and recommendations are those of the author and are not necessarily endorsed by the U.S. Army.

REFERENCES

1. Simpson, L. L. Identification of the major steps in botulinum toxin action. *Annu. Rev. Pharmacol. Toxicol.*, 44, 167–193, 2004.
2. Schiavo, G., Rossetto, O., Santucci, A., DasGupta, B. R., and Montecucco, C. Botulinum neurotoxins are zinc proteins. *J. Biol. Chem.*, 267, 23479–23483, 1992.
3. Schiavo, G., Rossetto, O., Catsicas, S., Polverino de Laureto, P., DasGupta, B. R., Benfenati, F., and Montecucco, C. Identification of the nerve terminal targets of botulinum neurotoxin serotypes A, D, and E. *J. Biol. Chem.*, 268, 23784–23787, 1993.
4. Schiavo, G., Shone, C. C., Rossetto, O., Alexander, F. C., and Montecucco, C. Botulinum neurotoxin serotype F is a zinc endopeptidase specific for VAMP/synaptobrevin. *J. Biol. Chem.*, 268, 11516–11519, 1993.
5. Schiavo, G., Malizio, C., Trimble, W. S., Polverino de Laureto, P., Milan, G., Sugiyama, H., Johnson, E. A., and Montecucco, C. Botulinum G neurotoxin cleaves VAMP/synaptobrevin at a single Ala-Ala peptide bond. *J. Biol. Chem.*, 269, 20213–20216, 1994.
6. Blasi, J., Chapman, E. R., Link, E., Binz, T., Yamasaki, S., De Camilli, P., Sudhof, T. C., Niemann, H., and Jahn, R. Botulinum neurotoxin A selectively cleaves the synaptic protein SNAP-25. *Nature*, 365, 160–163, 1993.
7. Yamasaki, S., Baumeister, A., Binz, T., Blasi, J., Link, E., Cornille, F., Roques, B. et al., Cleavage of members of the synaptobrevin/VAMP family by types D and F botulinal neurotoxins and tetanus toxin. *J. Biol. Chem.*, 269, 12764–12772, 1994.
8. Schiavo, G., Shone, C. C., Bennett, M. K., Scheller, R. H., and Montecucco, C. Botulinum neurotoxin type C cleaves a single Lys-Ala bond within the carboxyl-terminal region of syntaxins. *J. Biol. Chem.*, 270, 10566–10570, 1995.
9. Aureli, P., Fenicia, L., Pasolini, B., Gianfranceschi, M., McCroskey, L. M., and Hatheway, C. L. Two cases of type E infant botulism caused by neurotoxigenic *Clostridium butyricum* in Italy. *J. Infect. Dis.*, 154, 207–211, 1986.
10. Harvey, S. M., Sturgeon, J., and Dassey, D. E. Botulism due to Clostridium baratii type F toxin. *J. Clin. Microbiol.*, 40, 2260–2262, 2002.
11. Allen, S. D., Emery, C. L., and Siders, J. A. Clostridium, in *Manual of Clinical Microbiology*, P. R. Murray, E. J. Barron, M. Q. Pfaller, F. C. Tenover, and R. H. Yolken, eds., pp. 654–671, American Society for Microbiology, Washington, D.C., 1999.

12. Barth, H., Aktories, K., Popoff, M. R., and Stiles, B. G. Binary bacterial toxins: biochemistry, biology, and applications of common Clostridium and Bacillus proteins. *Microbiol. Mol. Biol. Rev.*, 68, 373–402, 2004.
13. Fairweather, N. F., and Lyness, V. A. The complete nucleotide sequence of tetanus toxin. *Nucleic Acids Res.*, 14, 7809–7812, 1986.
14. Simpson, L. L. The origin, structure, and pharmacological activity of botulinum toxin. *Pharmacol. Rev.*, 33, 155–188, 1981.
15. Sharma, S. K., Ramzan, M. A., and Singh, B. R. Separation of the components of type A botulinum neurotoxin complex by electrophoresis. *Toxicon*, 41, 321–331, 2003.
16. Krebs, K. M., and Lebeda, F. J. Comparison of the structural features of botulinum neurotoxin and NTNH, a non-toxic accessory protein of the progenitor complex. *The Botulinum Journal*, 1, 116–134, 2008.
17. Ohishi, I., Sugii, S., and Sakaguchi, G. Oral toxicities of *Clostridium botulinum* toxins in response to molecular size. *Infect. Immun.*, 16, 107–109, 1977.
18. Inoue, K., Fujinaga, Y., Watanabe, T., Ohyama, T., Takeshi, K., Moriishi, K., Nakajima, H., and Oguma, K. Molecular composition of Clostridium botulinum type A progenitor toxins. *Infect. Immun.*, 64, 1589–1594, 1996.
19. Hines, H. B., Lebeda, F. J., Hale, M., and Brueggemann, E. E. Characterization of botulinum progenitor toxins by mass spectrometry. *Appl. Environ. Microbiol.*, 71, 4478–4486, 2005.
20. Cheng, L. W., Onisko, B., Johnson, E. A., Reader, J. R., Griffey, S. M., Larson, A. E., Tepp, W. H., Stanker, L. H., Brandon, D. L., and Carter, J. M. Effects of purification on the bioavailability of botulinum neurotoxin type A. *Toxicology*, 249, 123–129, 2008.
21. Schiavo, G., Matteoli, M., and Montecucco, C. Neurotoxins affecting neuroexocytosis. *Physiol. Rev.*, 80, 717–766, 2000.
22. Singh, B. R., Li, B., and Read, D. Botulinum versus tetanus neurotoxins: Why is botulinum neurotoxin but not tetanus neurotoxin a food poison? *Toxicon*, 33, 1541–1547, 1995.
23. Sugii, S., Ohishi, I., and Sakaguchi, G. Oral toxicities of *Clostridium botulinum* toxins. *Jpn. J. Med. Sci. Biol.*, 30, 70–73, 1977.
24. Sugii, S., Ohishi, I., and Sakaguchi, G. Correlation between oral toxicity and *in vitro* stability of *Clostridium botulinum* type A and B toxins of different molecular sizes. *Infect. Immun.*, 16, 910–914, 1977.
25. Sugii, S., Ohishi, I., and Sakaguchi, G. Intestinal absorption of botulinum toxins of different molecular sizes in rats. *Infect. Immun.*, 17, 491–496, 1977.
26. Ohishi, I. Absorption of *Clostridium botulinum* type B toxins of different molecular sizes from different regions of rat intestine. *FEMS Microbiol. Lett.*, 16, 257–260, 1983.
27. Fujinaga, Y., Inoue, K., Watanabe, S., Yokota, K., Hirai, Y., Nagamachi, E., and Oguma, K. The haemagglutinin of *Clostridium botulinum* type C progenitor toxin plays an essential role in binding of toxin to the epithelial cells of guinea pig small intestine, leading to the efficient absorption of the toxin. *Microbiology*, 143 (Pt 12), 3841–3847, 1997.
28. Lebeda, F. J., and Singh, B. R. Membrane channel activity and translocation of tetanus and botulinum neurotoxins. *J. Toxicol.—Toxin Rev.*, 18, 45–76, 1999.
29. Daniels-Holgate, P. U., and Dolly, J. O. Productive and non-productive binding of botulinum neurotoxin A to motor nerve endings are distinguished by its heavy chain. *J. Neurosci. Res.*, 44, 263–271, 1996.
30. Lalli, G., Herreros, J., Osborne, S. L., Montecucco, C., Rossetto, O., and Schiavo, G. Functional characterisation of tetanus and botulinum neurotoxins binding domains. *J. Cell Sci.*, 112, 2715–2724, 1999.
31. Arnon, S. S., Schechter, R., Inglesby, T. V., Henderson, D. A., Bartlett, J. G., Ascher, M. S., Eitzen, E. et al., Botulinum toxin as a biological weapon: medical and public health management. *J. Amer. Med. Assoc.*, 285, 1059–1070, 2001.
32. Cherington, M. Clinical spectrum of botulism. *Muscle Nerve*, 21, 701–710, 1998.

33. Middlebrook, J., and Franz, D. Botulinum toxins, in *Textbook of Military Medicine: Medical Aspects of Chemical and Biological Defense*, R. Zajtchuk, ed., Office of the Surgeon General, Department of the Army, Washington, DC, 1997.
34. Mackle, I. J., Halcomb, E., and Parr, M. J. Severe adult botulism. *Anaesth. Intensive Care*, 29, 297–300, 2001.
35. Maselli, R. A. Pathogenesis of human botulism. *Ann. N. Y. Acad. Sci.*, 841, 122–139, 1998.
36. Sonnabend, O., Sonnabend, W., Heinzle, R., Sigrist, T., Dirnhofer, R., and Krech, U. Isolation of *Clostridium botulinum* type G and identification of type G botulinal toxin in humans: Report of five sudden unexpected deaths. *J. Infect. Dis.*, 143, 22–27, 1981.
37. Lamanna, C. The most poisonous poison. *Science*, 130, 763–772, 1959.
38. Lecour, H., Ramos, H., Almeida, B., and Barbosa, R. Food-borne botulism. A review of 13 outbreaks. *Arch. Intern. Med.*, 148, 578–580, 1988.
39. Lebeda, F. J., Cer, R. Z., Stephens, R., and Mudunuri, U. Temporal characteristics of botulinum neurotoxin therapy. *Expert Rev. Neurother.*, 10, 93–103, 2010.
40. Lebeda, F. J., Cer, R. Z., Stephens, R., and Mudunuri, U. Time course of the effects of botulinum neurotoxin, in *Dystonia: Causes, Symptoms and Treatment*, J. Kurstot and M. Forsström, eds., pp. 179–198, Nova Science Publishers, Inc., Hauppauge, NY, 2010.
41. Bulatova, T. I., and Kaulen, D. R. Natural resistance of irradiated animals to botulin toxins. *J. Hyg. Epidemiol. Microbiol. Immunol.*, 10, 67–73, 1966.
42. Sakaguchi, G., and Sakaguchi, S. Oral toxicities of *Clostridium botulinum* type E toxins of different forms. *Jpn. J. Med. Sci. Biol.*, 27, 241–244, 1974.
43. Sugiyama, H., DasGupta, B. R., and Yang, K. H. Toxicity of purified botulinal toxin fed to mice. *Proc. Soc. Exp. Biol. Med.*, 147, 589–591, 1974.
44. Ohishi, I. Oral toxicities of *Clostridium botulinum* type A and B toxins from different strains. *Infect. Immun.*, 43, 487–490, 1984.
45. Nukina, M., Miyata, T., Sakaguchi, S., and Sakaguchi, G. Detection of neutral sugars in purified type G botulinum progenitor toxin and the effects of some glycolytic enzymes on its molecular dissociation and oral toxicity. *FEMS Microbiol. Lett.*, 63, 159–164, 1991.
46. Ohishi, I., Sugii, S., and Sakaguchi, G. Absorption of botulinum type B progenitor and derivative toxins through the intestine [proceedings]. *Jpn. J. Med. Sci. Biol.*, 31, 161–163, 1978.
47. Miyazaki, S., and Sakaguchi, G. Experimental botulism in chickens: The cecum as the site of production and absorption of botulinum toxin. *Jpn. J. Med. Sci. Biol.*, 31, 1–15, 1978.
48. Zacks, S., and Sheff, M., Biochemistry and mechanism of action of toxic proteins. U.S. Army Edgewood Arsenal, Final Progress Report, DTIC AD816140, pp. 1–8, 1969.
49. Williams, R. S., Tse, C. K., Dolly, J. O., Hambleton, P., and Melling, J. Radioiodination of botulinum neurotoxin type A with retention of biological activity and its binding to brain synaptosomes. *Eur. J. Biochem.*, 131, 437–445, 1983.
50. Agui, T., Syuto, B., Oguma, K., Iida, H., and Kubo, S. Binding of *Clostridium botulinum* type C neurotoxin to rat brain synaptosomes. *J. Biochem.*, 94, 521–527, 1983.
51. Evans, D. M., Williams, R. S., Shone, C. C., Hambleton, P., Melling, J., and Dolly, J. O. Botulinum neurotoxin type B. Its purification, radioiodination and interaction with rat-brain synaptosomal membranes. *Eur. J. Biochem.*, 154, 409–416, 1986.
52. Yokosawa, N., Kurokawa, Y., Tsuzuki, K., Syuto, B., Fujii, N., Kimura, K., and Oguma, K. Binding of *Clostridium botulinum* type C neurotoxin to different neuroblastoma cell lines. *Infect. Immun.*, 57, 272–277, 1989.
53. Bakry, N. M., Kamata, Y., and Simpson, L. L. Expression of botulinum toxin binding sites in Xenopus oocytes. *Infect. Immun.*, 65, 2225–2232, 1997.
54. Hirokawa, N., and Kitamura, M. Binding of *Clostridium botulinum* neurotoxin to the presynaptic membrane in the central nervous system. *J. Cell Biol.*, 81, 43–49, 1979.

55. Black, J. D., and Dolly, J. O. Selective location of acceptors for botulinum neurotoxin A in the central and peripheral nervous systems. *Neuroscience*, 23, 767–779, 1987.
56. Herreros, J., Marti, E., Ruiz-Montasell, B., Casanova, A., Niemann, H., and Blasi, J. Localization of putative receptors for tetanus toxin and botulinum neurotoxin type A in rat central nervous system. *Eur. J. Neurosci.*, 9, 2677–2686, 1997.
57. Foran, P. G., Mohammed, N., Lisk, G. O., Nagwaney, S., Lawrence, G. W., Johnson, E., Smith, L., Aoki, K. R., and Dolly, J. O. Evaluation of the therapeutic usefulness of botulinum neurotoxin B, C1, E, and F compared with the long lasting type A. Basis for distinct durations of inhibition of exocytosis in central neurons. *J. Biol. Chem.*, 278, 1363–1371, 2003.
58. Sergeyeva, T. Detection of botulinal toxin and type A microbe in the organism of sick animals and in the organs of cadavers. *Zhurnal Mikrobiologii i Imniunobiologii (English translation: DTIC AD639403)*, 97–101, 1962.
59. Cardella, M., Resistance of guinea pigs immunized with botulinum toxoids to aerogenic challenge with toxin. U.S. Army Biological Laboratories Report Test No. 60-TE-1323, DTIC AD0404870, pp. 1–8, 1963.
60. Sergeyeva, T. Detection of type E botulin toxin in an organism. *Zhurnal Mikrobiologii Epidemiologii i Imniunobiologii (English translation: DTIC AD651368)* 4, 54–59, 1966.
61. Moll, T., and Brandly, C. A. Botulism in the mouse, mink, and ferret with special reference to susceptibility and pathological alterations. *Am. J. Vet. Res.*, 12, 355–363, 1951.
62. May, A. J., and Whaler, B. C. The absorption of *Clostridium botulinum* type A toxin from the alimentary canal. *Br. J. Exp. Pathol.*, 39, 307–316, 1958.
63. Ciccarelli, A. S., Whaley, D. N., McCroskey, L. M., Gimenez, D. F., Dowell, V. R., Jr., and Hatheway, C. L. Cultural and physiological characteristics of *Clostridium botulinum* type G and the susceptibility of certain animals to its toxin. *Appl. Environ. Microbiol.*, 34, 843–848, 1977.
64. Heckly, R. J., Hildebrand, G. J., and Lamanna, C. On the size of the toxic particle passing the intestinal barrier in botulism. *J. Exp. Med.*, 111, 745–759, 1960.
65. Coulston, F., Annual Report, DTIC AD0634374, pp. 1–18, 1966.
66. Dack, G., and Gibbard, J. Studies on botulinum toxin in the alimentary track of dogs, rabbits, guinea pigs and mice. *J. Infect. Dis.*, 39, 173, 1926.
67. Dack, G., and Gibbard, J. Permeability of the small intestine of rabbits and hogs to botulinum toxin. *J. Infect. Dis.*, 39, 181, 1926.
68. Haerem, S., Dack, G., and Dragstedt, L. Acute intestinal obstruction. II. The permeability of obstructed bowel segments of dogs to *Clostridium botulinum* toxin. *Surgery*, 3, 339–350, 1938.
69. Dack, G., and Hoskins, D. Absorption of botulinum toxin from the colon of *Macaca mulatta*. *J. Infect. Dis.*, 71, 260–263, 1942.
70. Coleman, I. W. Studies on the oral toxicity of *Clostridium botulinum* toxin, type A. *Can. J. Biochem. Physiol.*, 32, 27–34, 1954.
71. Dong, M., Richards, D. A., Goodnough, M. C., Tepp, W. H., Johnson, E. A., and Chapman, E. R. Synaptotagmins I and II mediate entry of botulinum neurotoxin B into cells. *J. Cell Biol.*, 162, 1293–1303, 2003.
72. Sellin, L. C., Kauffman, J. A., and Dasgupta, B. R. Comparison of the effects of botulinum neurotoxin types A and E at the rat neuromuscular junction. *Med. Biol.*, 61, 120–125, 1983.
73. Billante, C. R., Zealear, D. L., Billante, M., Reyes, J. H., Sant'Anna, G., Rodriguez, R., and Stone, R. E., Jr. Comparison of neuromuscular blockade and recovery with botulinum toxins A and F. *Muscle Nerve*, 26, 395–403, 2002.
74. Kauffman, J. A., Way, J. F., Jr., Siegel, L. S., and Sellin, L. C. Comparison of the action of types A and F botulinum toxin at the rat neuromuscular junction. *Toxicol. Appl. Pharmacol.*, 79, 211–217, 1985.

75. Biskup, R., Snodgrass, H., and Vocci, F., The constancy of the mouse unit in the bioassay of type A botulinum toxin. Edgewood Arsenal Technical Report EATR No. 4260, pp. 1–34, 1969.
76. Dolman, C., and Murakami, L. *Clostridium botulinum* type F with recent observations on other types. *J. Infect. Dis.*, 109, 107, 1961.
77. Iida, H., Ono, T., Karashimada, T., and Ando, Y. Studies on the serum therapy of type E botulism: Absorption of toxin from the gastrointestinal tract. *Jpn. J. Med. Sci. Biol.*, 23, 282–285, 1970.
78. Matveev, K. Effect of sublethal doses of botulinal toxin on the organism following multiple administrations. *Zhurnal Mikrobiologii*, 30, 71–79, 1959.
79. Rand, M. J., and Whaler, B. C. Impairment of sympathetic transmission by botulinum toxin. *Nature*, 206, 588–591, 1965.
80. Lamanna, C., and Meyers, C. E. Influence of ingested foods on the oral toxicity in mice of crystalline botulinal type A toxin. *J. Bacteriol.*, 79, 406–410, 1960.
81. Herrero, B. A., Ecklung, A. E., Streett, C. S., Ford, D. F., and King, J. K. Experimental botulism in monkeys—A clinical pathological study. *Exp. Mol. Pathol.*, 6, 84–95, 1967.
82. Lewis, J. C., Smith, G. R., and White, V. J. An outbreak of botulism in captive hamadryas baboons (*Papio hamadryas*). *Vet. Rec.*, 126, 216–217, 1990.
83. Petit, T. Seasonal outbreaks of botulism in captive South American monkeys. *Vet. Rec.*, 128, 311–312, 1991.
84. Mahmut, N., Inoue, K., Fujinaga, Y., Hughes, L., Arimitsu, H., Sakaguchi, Y., Ohtsuka, A., Murakami, T., Yokota, K., and Oguma, K. Characterisation of monoclonal antibodies against haemagglutinin associated with *Clostridium botulinum* type C neurotoxin. *J. Med. Microbiol.*, 51, 286–294, 2002.
85. Mahmut, N., Inoue, K., Fujinaga, Y., Arimitsu, H., Sakaguchi, Y., Hughes, L., Hirst, R. et al., Mucosal immunisation with *Clostridium botulinum* type C 16 S toxoid and its non-toxic component. *J. Med. Microbiol.*, 51, 813–820, 2002.
86. Oguma, K., Yokota, K., Hayashi, S., Takeshi, K., Kumagai, M., Itoh, N., Tachi, N., and Chiba, S. Infant botulism due to *Clostridium botulinum* type C toxin. *Lancet*, 336, 1449–1450, 1990.
87. Coffield, J. A., Bakry, N., Zhang, R. D., Carlson, J., Gomella, L. G., and Simpson, L. L. *In vitro* characterization of botulinum toxin types A, C and D action on human tissues: combined electrophysiologic, pharmacologic and molecular biologic approaches. *J. Pharmacol. Exp. Ther.*, 280, 1489–1498, 1997.
88. Purkiss, J. R., Friis, L. M., Doward, S., and Quinn, C. P. *Clostridium botulinum* neurotoxins act with a wide range of potencies on SH-SY5Y human neuroblastoma cells. *Neurotoxicology*, 22, 447–453, 2001.
89. Eleopra, R., Tugnoli, V., Rossetto, O., Montecucco, C., and De Grandis, D. Botulinum neurotoxin serotype C: A novel effective botulinum toxin therapy in human. *Neurosci. Lett.*, 224, 91–94, 1997.
90. Maksymowych, A. B., Reinhard, M., Malizio, C. J., Goodnough, M. C., Johnson, E. A., and Simpson, L. L. Pure botulinum neurotoxin is absorbed from the stomach and small intestine and produces peripheral neuromuscular blockade. *Infect. Immun.*, 67, 4708–4712, 1999.
91. Chen, F., Kuziemko, G. M., and Stevens, R. C. Biophysical characterization of the stability of the 150-kilodalton botulinum toxin, the nontoxic component, and the 900-kilodalton botulinum toxin complex species. *Infect Immun*, 66, 2420–2425, 1998.
92. Fujinaga, Y., Inoue, K., Nomura, T., Sasaki, J., Marvaud, J. C., Popoff, M. R., Kozaki, S., and Oguma, K. Identification and characterization of functional subunits of *Clostridium botulinum* type A progenitor toxin involved in binding to intestinal microvilli and erythrocytes. *FEBS Lett.*, 467, 179–183, 2000.

93. Sakaguchi, G., Oishi, I., Kozaki, S., Sakaguchi, S., and Kitamura, M. Molecular structures and biological activities of *Clostridium botulinum* toxins. *Jpn. J. Med. Sci. Biol.*, 27, 95–99, 1974.
94. Mills, D. C., and Sugiyama, H. Comparative potencies of botulinum toxin for infant and adult mice. *Curr. Microbiol.*, 6, 239–242, 1981.
95. Miyazaki, S., and Sakaguchi, G. Pathogenisis of chicken botulism. *Jpn. J. Med. Sci. Biol.*, 32, 129, 1979.
96. Ono, T., Karashimada, T., and Iida, H. Studies of the serum therapy of type E botulism. 3. *Jpn. J. Med. Sci. Biol.*, 23, 177–191, 1970.
97. Park, J. B., and Simpson, L. L. Inhalational poisoning by botulinum toxin and inhalation vaccination with its heavy-chain component. *Infect. Immun.*, 71, 1147–1154, 2003.
98. Palm, P. E., McNerney, J. M., and Hatch, T. Respiratory dust retention in small animals; a comparison with man. *AMA Arch. Ind. Health*, 13, 355–365, 1956.
99. Schlesinger, R. Deposition and clearance of inhaled particles, in *Concepts in Inhalation Toxicology*, 2nd edition, Chapter 8, R. McClellan and R. Henderson, eds., Taylor and Francis Group, Washington, D.C., 1995.
100. Oberst, F., Cresthull, P., Crook, J. W., and House, M. J., Botulinum antitoxin as a therapeutic agent in monkeys with experimental botulism. Edgewood Report CRDLR 3331, DTIC AD0627996, pp. 1–15, 1965.
101. Oberst, F., Cresthull, P., Crook, J.W., and House, M.J. Artificial-respiration studies in monkeys incapacitated by experimental botulism. Edgewood Report CRDLR 3346, DTIC AD0482889, pp. 1–37, 1965.
102. Franz, D., Pitt, L. M., Hanes, M. A., and Rose, K. J. Efficacy of prophylactic and therapeutic administration of antitoxin for inhalation botulism, in *Botulinum and Tetanus Neurotoxins: Neurotransmission and Biological Aspects*, B. Das Gupta, ed., pp. 473–476, Plenum Press, New York, 1993.
103. Warheit, D. B. Interspecies comparisons of lung responses to inhaled particles and gases. *Crit. Rev. Toxicol.*, 20, 1–29, 1989.
104. Snipes, M. B. Long-term retention and clearance of particles inhaled by mammalian species. *Crit. Rev. Toxicol.*, 20, 175–211, 1989.
105. Anderson, J. B., and Tucker, K. Development and validation of animal models, in *Biodefense: Research Methodology and Animal Models*, J. R. Swearengen, ed., pp. 41–60, CRC Press, Boca Raton, 2006.
106. Lebeda, F. J., Adler, M., Erickson, K., and Chushak, Y. Onset dynamics of type A botulinum neurotoxin-induced paralysis. *J. Pharmacokinet. Pharmacodyn.*, 35, 251–267, 2008.
107. Aoki, K. R. Botulinum toxin: A successful therapeutic protein. *Curr. Med. Chem.*, 11, 3085–3092, 2004.
108. Keller, J. E. Recovery from botulinum neurotoxin poisoning *in vivo*. *Neuroscience*, 139, 629–637, 2006.
109. Simpson, L. L. Kinetic studies on the interaction between botulinum toxin type A and the cholinergic neuromuscular junction. *J. Pharmacol. Exp. Ther.*, 212, 16–21, 1980.
110. Dressler, D., and Adib Saberi, F. Botulinum toxin: mechanisms of action. *Eur. Neurol.*, 53, 3–9, 2005.
111. Kojima, S., Ikeda, M., and Kamikawa, Y. Further investigation into the mechanism of tachykinin NK(2) receptor-triggered serotonin release from guinea-pig proximal colon. *J. Pharmacol. Sci.*, 110, 122–126, 2009.
112. Luvisetto, S., Marinelli, S., Lucchetti, F., Marchi, F., Cobianchi, S., Rossetto, O., Montecucco, C., and Pavone, F. Botulinum neurotoxins and formalin-induced pain: Central vs. peripheral effects in mice. *Brain Res.*, 1082, 124–131, 2006.
113. Cheng, L. W., Stanker, L. H., Henderson, T. D., Lou, J., and Marks, J. D. Antibody protection against botulinum neurotoxin intoxication in mice. *Infect. Immun.*, 77, 4305–4313, 2009.

114. Kolbe, D. R., and Coe Clough, N. E. Correlation of *Clostridium botulinum* type C antitoxin titers in mink and guinea pigs to protection against type C intoxication in mink. *Anaerobe*, 14, 128–130, 2008.
115. Poliachik, S. L., Bain, S. D., Threet, D., Huber, P., and Gross, T. S. Transient muscle paralysis disrupts bone homeostasis by rapid degradation of bone morphology. *Bone*, 46, 18–23, 2010.
116. Rosado, J. A., Redondo, P. C., Salido, G. M., Sage, S. O., and Pariente, J. A. Cleavage of SNAP-25 and VAMP-2 impairs store-operated Ca^{2+} entry in mouse pancreatic acinar cells. *Am. J. Physiol. Cell Physiol.*, 288, C214–221, 2005.
117. Thyagarajan, B., Krivitskaya, N., Potian, J. G., Hognason, K., Garcia, C. C., and McArdle, J. J. Capsaicin protects mouse neuromuscular junctions from the neuroparalytic effects of botulinum neurotoxin a. *J. Pharmacol. Exp. Ther.*, 331, 361–371, 2009.
118. Sheridan, R. E., Smith, T. J., and Adler, M. Primary cell culture for evaluation of botulinum neurotoxin antagonists. *Toxicon*, 45, 377–382, 2005.
119. Stahl, C., Unger, L., Mazuet, C., Popoff, M., Straub, R., and Frey, J. Immune response of horses to vaccination with the recombinant Hc domain of botulinum neurotoxin types C and D. *Vaccine*, 27, 5661–5666, 2009.
120. Chen, F., Wang, J., Liu, S., Mi, W., and Wang, Y. Botulinum toxin type A decreases the concentration of acetylcholinesterase in nasal glands of guinea pig. *Lin Chuang Er Bi Yan Hou Ke Za Zhi*, 19, 370–372, 2005.
121. Dressler, D., Lange, M., and Bigalke, H. Mouse diaphragm assay for detection of antibodies against botulinum toxin type B. *Mov. Disord.*, 20, 1617–1619, 2005.
122. Prevot, V., Tweepenninckx, F., Van Nerom, E., Linden, A., Content, J., and Kimpe, A. Optimization of polymerase chain reaction for detection of *Clostridium botulinum* type C and D in bovine samples. *Zoonoses Public Health*, 54, 320–327, 2007.
123. Moeller, R. B., Puschner, B., Walker, R. L., Rocke, T. E., Smith, S. R., Cullor, J. S., and Ardans, A. A. Attempts to identify *Clostridium botulinum* toxin in milk from three experimentally intoxicated Holstein cows. *J. Dairy Sci.*, 92, 2529–2533, 2009.
124. Yule, A. M., LePage, V., Austin, J. W., Barker, I. K., and Moccia, R. D. Repeated low-level exposure of the round goby (Neogobius melanostomas) to *Clostridium botulinum* type E neurotoxin. *J. Wildl. Dis.*, 42, 494–500, 2006.
125. Carli, L., Montecucco, C., and Rossetto, O. Assay of diffusion of different botulinum neurotoxin type a formulations injected in the mouse leg. *Muscle Nerve*, 40, 374–380, 2009.
126. Hongyok, T., Chae, J. J., Shin, Y. J., Na, D., Li, L., and Chuck, R. S. Effect of chitosan-N-acetylcysteine conjugate in a mouse model of botulinum toxin B-induced dry eye. *Arch. Ophthalmol.*, 127, 525–532, 2009.
127. Stahl, A. M., Ruthel, G., Torres-Melendez, E., Kenny, T. A., Panchal, R. G., and Bavari, S. Primary cultures of embryonic chicken neurons for sensitive cell-based assay of botulinum neurotoxin: implications for therapeutic discovery. *J. Biomol. Screen.*, 12, 370–377, 2007.
128. Krug, H. E., Frizelle, S., McGarraugh, P., and Mahowald, M. L. Pain behavior measures to quantitate joint pain and response to neurotoxin treatment in murine models of arthritis. *Pain Med.*, 10, 1218–1228, 2009.
129. Coffield, J. A., and Yan, X. Neuritogenic actions of botulinum neurotoxin A on cultured motor neurons. *J. Pharmacol. Exp. Ther.*, 330, 352–358, 2009.
130. Rasetti-Escargueil, C., Jones, R. G., Liu, Y., and Sesardic, D. Measurement of botulinum types A, B and E neurotoxicity using the phrenic nerve-hemidiaphragm: Improved precision with in-bred mice. *Toxicon*, 53, 503–511, 2009.
131. Morbiato, L., Carli, L., Johnson, E. A., Montecucco, C., Molgo, J., and Rossetto, O. Neuromuscular paralysis and recovery in mice injected with botulinum neurotoxins A and C. *Eur. J. Neurosci.*, 25, 2697–2704, 2007.

132. Kim, H. M., Galatz, L. M., Patel, N., Das, R., and Thomopoulos, S. Recovery potential after postnatal shoulder paralysis. An animal model of neonatal brachial plexus palsy. *J. Bone Joint Surg. Am.*, 91, 879–891, 2009.
133. Antonucci, F., Bozzi, Y., and Caleo, M. Intrahippocampal infusion of botulinum neurotoxin E (BoNT/E) reduces spontaneous recurrent seizures in a mouse model of mesial temporal lobe epilepsy. *Epilepsia*, 50, 963–966, 2009.
134. Trampel, D. W., Smith, S. R., and Rocke, T. E. Toxicoinfectious botulism in commercial caponized chickens. *Avian Dis.*, 49, 301–303, 2005.
135. Gaunt, P. S., Kalb, S. R., and Barr, J. R. Detection of botulinum type E toxin in channel catfish with visceral toxicosis syndrome using catfish bioassay and endopep mass spectrometry. *J. Vet. Diagn. Invest.*, 19, 349–354, 2007.
136. Velasco, E., Gledhill, T., Linares, C., and Roschman-Gonzalez, A. Ultrastructural analysis of mouse levator auris longus muscle intoxicated *in vivo* by botulinum neurotoxin type A. *Invest. Clin.*, 49, 469–486, 2008.
137. Wheeler, C., Inami, G., Mohle-Boetani, J., and Vugia, D. Sensitivity of mouse bioassay in clinical wound botulism. *Clin. Infect. Dis.*, 48, 1669–1673, 2009.

16 Ricin

Stephen B. Greenbaum and Jaime B. Anderson

CONTENTS

16.1 BACKGROUND

Ricin is a potent protein toxin synthesized in the seeds of the castor bean plant, *Ricinus communis*. Castor beans are cultivated worldwide and provide an abundant natural source for the toxin, which is readily isolated from the castor meal or cake after the oil is extracted [1,2]. Purified ricin is water-soluble and moderately heat-resistant, and retains its activity across a wide pH range [3–5]. It has long been considered a significant threat for use in biological warfare and bioterrorism, in large part due to its stability and ease of production [3–8]. Studies conducted during World War II yielded a crystalline form of ricin with much higher toxicity than any previous crude extract [4,9,10]. Although the crystalline preparation was still relatively heterogeneous, subsequent refinements in extraction methodologies have since enabled high yield production of pure ricin containing no contaminating hemagglutinating or proteolytic activity. Recombinant ricin has also been produced in transgenic plants [11–14].

The *R. communis* genome has recently been shown to encode seven full-length ricin family proteins, all of which are capable of inhibiting protein synthesis in eukaryotic cells [15]. Two naturally occurring isoforms of the toxin have been characterized; ricin D is synthesized by both small- and large-grain castor beans, and ricin E is typically produced by small-grain beans [16,17]. The two isoforms display similar effects in inhibiting protein synthesis, although their unique cellular binding affinities are thought to result in moderate differences in toxicity to certain cell lines. Ricin isoforms have also been subfractionated by chromatography into ricin I, II, and III [18]. All three subisoforms are cytotoxic when evaluated in a cell culture model, with ricin III displaying four- to eightfold higher toxicity than ricins I and II in this model. Ricin III also displays high hemagglutinating activity, whereas ricin I and II are hemagglutination-negative. Additionally, ricin III has high *in vivo* toxicity when administered to mice via the intraperitoneal route, resulting in 50% mortality after 5 days at a dose of 5 ng/kg. In contrast, no mortality or change in body weight is observed after 7 days in mice receiving ricin I and II at doses up to 20 ng/kg.

Ricin is relatively stable in food and water and represents a significant threat as an agent for intentional contamination of consumable products [6–8,19–23]. Ricin is also a powerful inhalational poison; most published studies indicate that its toxicity by the respiratory route is at least a thousand times greater than its oral potency. In fact, the toxicity of ricin by the respiratory exposure route is comparable to that associated with intraperitoneal or intravenous injection [6,7,24–28]. This property distinguishes ricin from many other biological toxins, including the botulinum toxins, which are typically more potent by parenteral administration than by inhalation exposure. The respiratory toxicity of ricin increases with decreasing particle size, and small-particle aerosols may persist for several hours after environmental dispersal [4,5,26,29]. The toxin is also relatively heat-resistant and has been reported to persist for up to 3 days under dry conditions, heightening potential concerns about its use as an aerosolized agent [7,20,21,30].

16.2 MECHANISMS OF ACTION

Ricin is a heterodimeric type II ribosome-inactivating protein (RIP) that exerts its toxic effects by inhibiting protein synthesis [31]. Ricin was the first protein in the RIP family to be cloned and structurally characterized by x-ray crystallography [32–34]. The toxin is a member of a multigene family and is transcribed as pre-proricin mRNA in the endosperm cells of maturing castor seeds [32,35]. The N-terminal signal peptide encoded by this mRNA is cleaved during translation, and the elongating polypeptide is folded and glycosylated as proricin [32,35,36]. Proricin is then cleaved into the functional toxin composed of disulfide bond-linked ricin A-chain (RCA or A-chain) and ricin B-chain (RCB or B-chain) subunits. The A-chain possesses cytotoxic activity and has, therefore, been termed the effector component [31]. The lectinic B-chain, also referred to as the haptomer unit, mediates cellular binding and endocytosis through interactions with terminal galactose residues. The 34-kDa B-chain has arisen through multiple gene duplications and is subject to more posttranslational modifications than the 32-kDa A-chain [30,37–40]. Modifications to both chains include extensive mannosylation, which has been implicated in an alternative ricin uptake pathway in certain mannose receptor-expressing cell types [41–45].

Ricin uptake into target cells is typically initiated by RCB binding to terminal galactose residues on membrane-bound glycoproteins and glycolipids [31,46–49]. The abundance of galactose residues on most mammalian cells is thought to translate into efficient ricin binding and uptake at exposed tissues. Toxin entry into cultured cells appears to be dependent on the presence of calcium, although the molecular basis for this requirement is not clear [50–52]. Cellular uptake of bound ricin occurs through vesicle-mediated endocytosis [50,53–56]. RCB subsequently dissociates from RCA and facilitates its transport from the vesicle to the cytosol [40,47]. Once in the cytosol, RCA inhibits protein synthesis by inactivating the 60S subunit of eukaryotic ribosomes [31,46,47,57–59]. The 28S ribosomal RNA (rRNA) of the eukaryotic 60S ribosomal subunit contains a highly conserved loop with a centralized GAGA sequence. This structure has been designated the sarcin/ricin loop, as it is targeted by both toxins in eukaryotic cells. RCA hydrolyzes the N-glycosidic bond at the second adenosine (residue A4324) of the GAGA sequence within the sarcin/ricin loop [57–61]. Depurination of A4324 by RCA disrupts the binding site for elongation factor 2, thereby preventing protein translation and causing cell death [46,62,63]. A single RCA molecule is reportedly capable of inactivating thousands of ribosomes within the eukaryotic cell [31,64,65].

While the cytotoxicity of ricin is attributed primarily to the inhibition of protein synthesis, additional mechanisms may contribute to the pathology associated with ricin intoxication in humans and animals. For example, the toxin has been shown to disrupt the lipid bilayer of cell membranes and to carry direct lipolytic activity [66–68]. Ricin has also been found to induce apoptotic changes in cultured cells and intoxicated animals [69,70]. Other local and systemic effects have been observed after ricin exposure, including oxidative stress and inflammatory cytokine cascades [23,71–80]. Recent tissue culture and animal studies have further investigated these effects and their potential roles as mediators of ricin-induced tissue damage and lethality.

16.3 RICIN POISONING IN HUMANS AND ANIMALS

Ricin is thought to be toxic to all vertebrates and has been shown to cause disease in humans and numerous other mammalian species. The pathology of ricin poisoning is generally dependent on the route of administration because the toxin typically exerts significant local effects within exposed tissues [6,7,9,10,23,72,76–78,81–85]. It is relatively stable and resistant to proteolysis by various physiological digestive enzymes [8,30,85–89]. The toxin, therefore, maintains significant activity within the harsh conditions of the gastrointestinal tract, and ingestion of sufficient toxin quantities can induce acute intestinal inflammation, epithelial damage, and hemorrhaging of the intestinal mucosa [7,9,10,23,81,85,90]. These pathological changes lead to a variety of related signs and symptoms after ingestion in humans, including acute abdominal pain and cramping, diarrhea, vomiting, weakness, prostration, rectal hemorrhaging, urinary retention, increased respiration rate, fever, and vascular collapse [3,7,91–93]. Although convulsions, coma, and death have resulted from severe intoxications, most clinical cases of ricin poisoning are not fatal [94]. However, relatively few fully chewed castor beans have been reported to be required for lethality in humans [94,95]. Oral toxicity is thought to be dependent on the extent of toxin release from the seeds during mastication and digestion [3,92–96].

Ricin also demonstrates significant toxicity and lethality when delivered via the respiratory route. Experimental exposure to aerosolized ricin results in extensive respiratory tract lesions, acute pulmonary edema, alveolar flooding, and mortality in various animal models [26,73,83,97–100]. Acute systemic inflammation and hemorrhaging in multiple organs have also been observed after respiratory exposure [73]. The toxic effects of inhaled ricin in humans are expected to be similar to those seen in experimental animals, based on the common mechanism of toxin action and the similar pathologies associated with oral and systemic exposure in humans and animals. However, no information on aerosol toxicity in humans is available, and data on potential respiratory exposures are very limited. Workers exposed to castor bean dust in castor oil processing plants have been reported to experience an allergic reaction with symptoms including chest tightness, nasal and throat congestion, itching of the eyes, and urticaria, as well as wheezing and asthma in more severe cases [101]. However, it has been proposed that these responses are likely to have been caused by exposure to castor bean constituents other than ricin [102]. Minor illnesses are reported to have resulted from presumed low-dose aerosol intoxications in laboratory workers during World War II [7]. Within several hours after exposure, the workers developed clinical signs and symptoms including fever, nausea, joint pain, chest tightness, tracheal inflammation and coughing, respiratory distress, and sweating. Some patients also suffered from drowsiness, loss of mental acuity, myosis, cyanosis, dehydration, urinary retention, and circulatory collapse. Affected individuals began to sweat profusely several hours later as the primary symptoms subsided. All patients subsequently recovered with no evident long-term clinical problems, although less obvious effects such as chronic lung scarring and reduced pulmonary function could have gone undetected. Additional data on respiratory toxicity and pathogenesis are not available for humans but have been generated in several experimental species.

16.4 ANIMAL MODELS FOR ORAL INTOXICATION

Small-rodent models have typically been incorporated in investigations of the gastrointestinal absorption and toxicity of ricin. Although animal studies and data from human exposures have generally established that the toxin is far less potent by the oral route than by inhalation or parenteral injection, two recent studies indicate that, at least under certain experimental conditions, ricin may be significantly more toxic by ingestion than previously reported [100,103]. Regardless, toxicity data from mice and rats, as well as toxin absorption studies in various rat models, indicate that systemic uptake of ricin from the gastrointestinal tract is relatively inefficient, and that much of the pathology associated with oral exposure is localized to the gut tissues [9,10,23,104]. These findings and related observations of toxicokinetics and disease presentation are generally consistent across a number of animal species. This section provides summary data on the following aspects of oral ricin intoxication, based on studies in the prominent animal models and cases of ricin poisoning in humans: *Binding, Absorption, and Distribution*; *Cytotoxicity and Pathology*; *Disease Presentation*; and *Oral Toxicity*.

16.4.1 Binding, Absorption, and Distribution

Ricin binding in the gastrointestinal tract has been evaluated in various tissue culture and *in vivo* models. The toxin displays direct binding, internalization, and inhibitory activity on epithelial cells from the rat's small intestine [105]. Binding is mediated by the B-chain and is sensitive to the presence of D-galactosidase, supporting a role for galactosyl residues on exposed cell membranes in toxin binding and uptake in the gastrointestinal tract. These findings and related data have also been proposed to indicate that cytotoxicity in exposed gut tissues may be attributable both to direct effects of ricin binding on cell viability and to the inhibition of protein synthesis by internalized toxin [10,23,104–106].

The rat model was used extensively in early studies on the absorption, distribution, and fate of ricin after oral exposure. Low gastrointestinal absorption efficiencies (0.015%–0.017%) were determined in rats receiving lethal oral doses of ricin [9]. The absorption efficiency of ricin in the rat everted jejunal sac model was reported to be even lower (0.006%), although the physiological relevance of this data relies on the assumption that the jejunum represents the prominent site of gastrointestinal ricin absorption. The higher absorption efficiency observed *in vivo* may be indicative of toxin absorption not only within the jejunum but also from other intestinal compartments. In fact, data obtained from early *in vivo* studies may also have underestimated ricin absorption due to limitations in the tissues evaluated, the extraction techniques used, and the capabilities available at the time for detecting and quantifying toxin content.

Additional studies using the rat model further characterized gastrointestinal ricin absorption and the systemic fate of the absorbed toxin. Toxin was primarily detected in the stomach and small intestine within 2 h after oral administration of a sublethal oral dose (10 mg/kg) of ricin D [104]. Almost half of the ricin dose was transferred from the stomach to the large intestine within 12 h, where toxin was detected for up to 3 days after exposure. Around 20% of the toxin was excreted in the feces during the 72 h after oral administration, and total ricin recovery at the end of that time period was less than half of the administered dose [104]. This data and other studies on ricin internalization and pathology suggest that, although systemic absorption of the toxin after oral exposure may be limited, significant uptake into local gut tissues occurs.

Ricin content in the blood, lymph, and organs of orally intoxicated rats was also evaluated in these experiments. One of the first published studies on ricin toxicokinetics found that ricin appeared in the lymph within 1 h after oral administration and persisted at moderate levels for at least 6 h after exposure [104]. The plasma contained notably higher concentrations than the lymph, and circulating toxin levels increased significantly throughout the 6-h period after ricin ingestion. Total toxin absorption into the lymph and blood of rats in those experiments was reported to be 0.02% and 0.27%, respectively, and ricin was apparently absorbed intact, based on the toxicity of lymph and blood samples to mice [104]. The liver was the prominent target tissue for circulating ricin after gastrointestinal exposure; lower toxin levels were found in the spleen, and ricin was not detected in the pancreas, kidneys, brain, heart, or lungs. Of note, oral ricin poisoning has also been reported to cause reversible hepatotoxicity in humans [94,107].

More recent studies on ricin distribution after oral exposure in the small rodent species have benefited from advances in sampling techniques and detection capabilities. In one such study, ricin was extracted from homogenized tissue samples from orally intoxicated rats and subsequently quantified by enzyme-linked immunosorbent assay [100]. Ricin was detected in the liver, spleen, lungs, and blood extracts, with the highest yields obtained from liver tissue. Another recent study yielded similar data on toxin distribution in rats after oral intoxication, with the liver having the highest detectable quantities, followed by the spleen, gastrointestinal tissues, kidneys, and blood [100]. This data is generally consistent with previous findings on ricin distribution after oral exposure in rats, and with clinical case observations of hepatotoxicity after ricin ingestion in humans.

Another recent study evaluated ricin absorption and distribution after intragastric instillation in mice [108]. For these experiments, an immunopolymerase chain reaction assay was used to monitor toxin content in serum and fecal samples. Ricin was detected in feces within 2 h after administration of a lethal dose and continued to persist through the 24-h test period. Serum toxicokinetics varied widely among individual animals, and the time required for systemic absorption of significant toxin quantities was reported to be 6–7 h after gastric administration. One proposed explanation for the delayed systemic absorption is that toxin-induced inflammation and tissue damage may be required for efficient permeation of ricin through the intestinal barriers and into the bloodstream [100].

The Brush Border Membrane Preparation (BBMP) isolated from the mucosal tissues of the rat small intestine has emerged as a new model for studying intestinal binding dynamics and effects. This model was recently used to evaluate binding interactions and damage resulting from intestinal exposure to ricin [81]. Binding activity and kinetics were monitored by surface plasmon resonance on BBMP that were immobilized on biosensor chips and then exposed to varying concentrations of ricin solutions. Further development of the BBMP platform may prove useful in generating quantitative data on the intestinal binding and transport of ricin, evaluating the specific molecular mechanisms of intestinal toxicity, and investigating potential inhibitors.

16.4.2 Cytotoxicity and Pathology

A number of rat *in vivo* and tissue culture studies have focused on elucidating the relationships between ricin interactions with gut tissues and the physiological and pathological changes that occur after gastrointestinal toxin exposure. The lectinic B-chain of ricin was shown to bind directly to the intestinal mucosa of rats *in vivo* and exert toxic effects on absorptive gut epithelial cells [82]. Ricin and RCB also rapidly bound to cultured rat intestinal epithelial cells, and at least some of the bound toxin was subsequently internalized intact [105]. Treatment with either ricin or RCB reduced viable rat intestinal epithelial cell numbers within 30 min of exposure. The ricin B-chain appeared to disrupt the cell membrane structure after binding to the intestinal epithelium; the A-chain then mediated cytotoxicity after internalization [105]. The primary rat intestinal epithelial cell cultures used in these experiments and the various transformed cell lines used in other studies have been reported to display similar susceptibility to ricin.

Another series of early studies in rats investigated the role of ricin-induced damage to the intestinal epithelium on the systemic absorption of the toxin from the gut [104]. Ricin absorption appears to be facilitated by toxin-mediated damage to the intestinal epithelium, as a similar but less cytotoxic control protein (castor bean hemagglutinin) was not absorbed to any significant extent after oral administration [104]. Additional rat studies examined the contributions of both cytotoxicity and lectinic activity on ricin absorption from the gut. Ricin and a lectinic derivative with significantly reduced cytotoxicity (BMH [bismaleimidohexane]-ricin) were found to bind and translocate across the absorptive villi of the rat intestinal mucosa [109]. Both the native toxin and the lectinic derivative were detected in the rat liver 48 h after ingestion. However, absorption efficiency was reported to be lower for the less cytotoxic derivative than for the native toxin, as hepatic BMH-ricin concentrations were less than half those of the native toxin [109]. NBS [N-bromosuccinimide]-ricin, a ricin derivative with minimal lectinic activity and cytotoxicity, displayed no intestinal binding or absorption.

Rabbit *ex vivo* models have also been used to evaluate the pathological changes induced by ricin in gastrointestinal tissues. The toxin has been shown to bind to the epithelial microvilli of rabbit jejunal explants and to subsequently induce substantial degeneration of the villi and intestinal epithelial cells [110]. Interestingly, significantly greater binding is detected in jejunal tissues from suckling compared to adult rabbits, and lower ricin concentrations are required for 50% protein synthesis inhibition in the suckling intestinal tissues. The increased susceptibility of suckling jejunal explants to ricin treatment may be caused by the high endocytic capacity of immature intestinal membranes.

These findings indicate that both the lectinic and cytotoxic activities of the native toxin contribute to the uptake of ricin into gastrointestinal tissues. Another study has shown that only the native toxin induces epithelial damage within the small intestine, indicating that this pathology requires fully functional RCA and RCB subunits [109]. Interestingly, the dramatic pathological changes induced within the gastrointestinal tract are not considered sufficient to cause the relatively rapid death resulting from oral intoxication in experimental animals. This hypothesis is supported by a recent study in which mice died after receiving low oral doses but displayed no detectable changes in the gut tissues [103]. However, the effects of ricin on intestinal glucose absorption and the severe hypoglycemia that has been reported after oral intoxication suggest that ricin-induced damage to intestinal tissues still contributes to lethality [9,82,111,112]. The severe liver and kidney damage occasionally seen after ricin ingestion also emphasizes the potential for at least some systemic toxicity and provides a potential alternative explanation for the observed lethality.

The mouse model has been used in a number of recent *in vivo* studies of gastrointestinal intoxication. Administration of the toxin to mice via intragastric instillation has been shown to result in time- and dose-dependent epithelial damage and inflammation in the small intestine [23]. Immunization of mice by intragastric administration of ricin toxoid prior to ricin exposure results in mucosal and systemic immune responses that protect against ricin-induced intestinal damage. This model provides a means for quantitative analysis of ricin-induced intestinal damage, further elucidating the associated inflammatory mechanisms, and evaluating the induction of protective immunity against intestinal ricin exposure. Another recent study in mice has shown that oral intoxication results in gastric necrosis and acute enteritis, as well as lymph node reactivity and necrosis similar to that seen after parenteral ricin injection [108]. Lymphoid necrosis has also been observed in lethally intoxicated dogs [113].

Pathological changes after oral ricin intoxication in the small rodent models generally include gastrointestinal congestion, distention, inflammation, and edema [9,82,108,112]. Detailed microscopic evaluations of the gastrointestinal tract reveal villus atrophy and crypt elongation, epithelial cell degeneration, reduced goblet cell numbers, neutrophil and eosinophil infiltrates, delayed absorptive epithelial cell regeneration, and jejunal dissociation of the epithelium from the lamina propria [9,10,23,103,104,109]. These morphological changes coincide with a marked impairment in glucose absorption from the small intestine [9,82].

One recent study suggests that lethality from oral ricin intoxication in mice may actually be attributable to severe hypoglycemia rather than to specific tissue damage in the intestines or other affected organs [103]. In this study, the gastrointestinal pathology observed in mice after exposure to lethal ricin doses via oral gavage is reported to be insufficient to explain the resulting lethality. A previous study correlating hypoglycemia with ricin toxicosis also supports the hypothesis that hypoglycemia may be a primary contributor to lethality from ricin intoxication [112].

Serological and pathological effects after lethal oral intoxication have also been reported for several livestock species. Elevated red blood cell counts, increased hemoglobin levels, and decreased peripheral leukocyte numbers have been detected before death in pigs, sheep, and cattle [90]. Postmortem findings in these animals include intestinal and pulmonary edema, lymphadenopathy, enlarged and hemorrhagic gastrointestinal mucous membranes, dense hemorrhagic livers, enlarged and moderately hemorrhaged kidneys, and swollen gall bladders.

16.4.3 Disease Presentation

The clinical presentation of ricin poisoning by the oral route has been well established through cases of castor bean ingestion in humans and animals. Initial symptoms usually appear within 6 h and include abdominal pain, cramping, diarrhea, vomiting, heartburn, and weakness [3,92–95,107] Extensive fluid loss may occur in more severe cases, leading to dehydration, hypotension, and circulatory collapse. Liver and kidney failure have also been reported. Death may occur within 1–7 days after ingestion of sufficient ricin doses. Lethality has been attributed to multiorgan failure, cardiovascular collapse, and hypovolemic shock, although the specific contributors remain to be fully characterized. Collapse, severe convulsions, and coma have been reported for some patients before death. Postmortem pathologies in humans typically parallel those seen in laboratory animals and include severe gastrointestinal hemorrhaging and edema, lymphoid tissue necrosis, renal degeneration, and hepatic necrosis. Similar clinical signs and dose-dependent times to death have been reported for mice, rats, dogs, and humans [94,100,104,113–116]. Livestock species such as pigs, sheep, and cattle develop weakness, drowsiness, anorexia, loss or coordination, and poor responsiveness to external stimuli [90]. Increased heart rate and respiration, along with temperature spikes, frequent urination, diarrhea, nausea, salivation, and spasms of the extremities have also been observed in these animals.

16.4.4 Oral Toxicity

The oral potency of ricin has only been quantitatively determined in a limited number of experimental models, although inadvertent ricin poisoning from castor bean ingestion has been reported in numerous animal species. The majority of the detailed oral toxicity data in the published literature has been generated in small rodent species such as mice and rats. Oral lethal dose values reported for mice and rats are typically in the range of 10–30 mg/kg, and are generally at least 1000 times higher than the parenteral lethal doses [6,7,9,19,77,82,83,85,98,104,108]. However, one recent study suggests that, at least under certain experimental conditions, the oral toxicity of ricin may be significantly higher than previously reported [103]. In this study, mice were fasted for 20 h prior to and 4 h after the administration by intragastric gavage of a bodyweight-adjusted volume of solution containing a highly purified in-house ricin preparation. This experimental approach yielded an oral median lethal dose (LD_{50}) of just 10 μg/kg, drastically lower than all previously reported values. The authors of this study suggested that historical lethal dose values may have been higher, in part, due to the use of crude ricin preparations rather than highly purified toxin. They also proposed a relationship between the oral toxicity of ricin and the amount of food present in the stomach during intoxication. Interestingly, another recent study also reported that the oral toxicity of ricin in rats was significantly higher than that anticipated based on previous data for both mice and rats [100]. In this study, an 8 mg/kg oral dose of a crude ricin preparation delivered by esophageal instillation was found to cause lethality much faster than expected and in all tested animals. Here, the authors proposed that the higher toxicity observed in these experiments may have been due to the presence of uncharacterized components in the crude ricin preparation.

A series of recent studies have evaluated the stability and oral toxicity of various ricin-containing solutions and food matrices. In one study, several liquid solutions were spiked with ricin and administered orally to mice after storage for varying periods of time [19]. This study determined oral lethal dose values of 5–20 mg/kg for ricin in phosphate-buffered saline, consistent with earlier published data and significantly higher than that recently determined in mice using the fasting approach. The oral LD_{50} values for ricin in apple juice, water, and half-and-half were somewhat higher than for the phosphate-buffered saline ricin solution. Storage of the ricin solutions for up to 11 days at 4°C had no significant effect on their oral toxicity in mice.

Since most cases of human intoxications have resulted from the ingestion of unknown quantities of ricin contained in castor beans, oral toxicity estimates for humans are generally based on the lethal dose values that have been experimentally determined in the small rodent species. Although speculation on the validity of such estimates would benefit from additional interspecies toxicity comparisons, specific oral lethal dose values are not available for many other animal species. A Soviet review of ricin cites toxic ingested doses for a variety of different domestic and wild animal species, but these values are expressed as grams of castor beans per kilogram body weight [90]. This report indicates that horses have the highest susceptibility of the reviewed species, and, therefore, the lowest toxic dose (0.1 g castor beans/kg), followed by geese (0.4), calves (0.5), rabbits (1.0), sheep (1.25), pigs (1.4), cattle (2.0), piglets (2.4), goats (5.5), and chickens (14.0). These dose values, even on a relative scale, carry little value for extrapolations to other animal species and to humans, as they may also reflect interspecies and experimental variances such as the extent of seed mastication and release of toxin. Additionally, as previously described, toxicity data from at least two recent studies call into question the relatively high lethal dose values commonly reported for mice and rats, as well as their use as the basis for estimates in humans [100,103]. Further studies are needed to determine what factors may have contributed to the significantly higher oral potency of ricin in those studies and to establish experimental parameters that are most suitable for modeling oral intoxication in humans.

16.5 ANIMAL MODELS FOR INHALATIONAL INTOXICATION

The toxicity and pathogenesis associated with inhalational ricin exposure have been studied extensively in small-rodent models, and to a lesser degree in nonhuman primates [26–29,73,76–78,83,98–100,103,117–124]. All experimental species evaluated thus far are highly susceptible to respiratory intoxication. The high respiratory potency of ricin has been attributed to the fact that it exerts significant cytotoxic activity directly within exposed tissues. Because of this local toxicity, the effects of respiratory exposure are dependent to some extent on aerosol particle size and the resulting distribution of the toxin within the respiratory tract [3,26,29,73,83,97,117,118,120,125]. Although a number of different exposure methodologies have been used to characterize respiratory intoxication, the majority of published data has been generated using three primary animal models: the mouse, rat, and nonhuman primate. This section provides summary information, based on these models, on the following aspects of respiratory ricin intoxication: *Delivery and Distribution*; *Cytotoxicity and Pathology*; *Disease Presentation*; and *Respiratory Toxicity*.

16.5.1 Delivery and Distribution

In the laboratory setting, respiratory distribution patterns are dictated to a significant extent by the procedures used to deliver the toxin to the lungs of experimental animals. Studies on the respiratory toxicity and pathology of ricin have, therefore, incorporated several different ricin exposure methodologies, including nose-only (NO), head-only (HO), and whole-body (WB) aerosol inhalation systems, as well as intratracheal toxin instillation. Although some similarities have been observed across the various exposure systems, substantial differences in toxin distribution and lung pathology patterns have also been reported.

A study published in 2003 directly evaluated the effects of different aerosol generation and exposure systems on the distribution of ricin in the respiratory tracts of mice [26]. Lung localization of ricin in mice was found to be almost identical after either WB or NO inhalation exposure. However, nasopharyngeal and tracheal toxin localization was significantly higher in WB- than in NO-exposed mice, and thus the total dose delivered to the respiratory tract of WB-exposed mice was greater than that of NO-exposed animals. The higher upper respiratory doses seen in WB-exposed mice may have been caused by the increased toxin inhalation and ingestion resulting from grooming and preening behavior among unrestrained animals. Ricin levels in the lungs of both exposure groups gradually declined with time after aerosol exposure as increasing amounts of toxin appeared in the trachea. Delayed accumulation of ricin in the trachea may result from mucociliary clearance of inhaled toxin from the lungs.

The same study evaluated the role of particle size on respiratory ricin deposition in mice using 1- and 5-μm ricin aerosols generated by Collison nebulizers or spinning-top aerosol generators, respectively [26]. Exposure to 1-μm aerosols resulted in the delivery of a significantly higher fraction of the total ricin dose to the lungs than that observed after exposure to 5-μm aerosols. Ricin deposition after inhalation of 1-μm aerosols was primarily detected within the bronchiolar epithelium, with some particles also localizing at alveolar pneumocytes.

The distribution and uptake of radiolabeled ricin preparations have also been evaluated after inhalational exposure in mice [29,120]. As expected, most of the administered radioactivity can initially be detected in the lungs and trachea after NO aerosol inhalation [29]. Interestingly, the majority of the radioactivity in the lungs localizes in the larger airways and bronchioles within 24 h after exposure [120]. The loss of detectable radioactivity in the alveoli at later time points after aerosol inhalation may indicate toxin absorption at these sites. The gastrointestinal tissues are the only other prominent sites for localization of inhaled ricin [29]. The stomach and duodenum contain substantial radioactivity soon after aerosol exposure, indicating that a significant fraction of the inhaled material may be swallowed. Radioactivity in both the stomach and the duodenum subsequently decline, whereas toxin levels in the ileum, caecum, and colon increase several hours after ricin inhalation. The kidneys, spleen, liver, testes, thymus, and blood also contain some radioactivity within a few hours after exposure.

16.5.2 Cytotoxicity and Pathology

Pathological changes after inhalational ricin intoxication are predominantly observed in the respiratory tract, and death is usually attributed to overwhelming pulmonary edema [28,29,83,98,99,117,118,120,121]. These findings have led some investigators to suggest that the lungs represent the only pathologically relevant target tissues for toxin activity after aerosol exposure. Inhaled ricin causes high permeability pulmonary edema and necrosis of the pulmonary epithelium. The cytotoxicity of ricin itself may directly induce this epithelial damage; alternatively, ricin-mediated activation of regulatory cell populations and inflammatory mediators may initiate pulmonary epithelial necrosis.

Leukocyte infiltrates that are activated and recruited after respiratory intoxication have been proposed to play a role in ricin-induced lung damage [121]. Similar mechanisms may be involved in the development of lymphatic lesions. Type I pneumocyte apoptosis has also been observed after inhalational ricin exposure in rodents and may be associated with the lethal pulmonary edema occurring after respiratory intoxication [29]. Ricin binds to human bronchial epithelial cell cultures, bovine pulmonary endothelial cell lines, and type I and II pneumocytes *in vitro*, and may have similar binding potential within the lungs after respiratory exposure [121,126,127]. Ricin has also been implicated in the pathogenesis of vascular leak syndrome, although the processes involved in the translocation of ricin across the respiratory epithelium to the vascular endothelium are not known [83].

The mouse model has been used extensively to investigate the pathological changes associated with lethal respiratory exposure to ricin [26,117,118,120]. Lung inflammation, airway epithelial necrosis, and edema of the mediastinal, perivascular, peribronchiolar, and interstitial spaces are seen within 24 h after WB aerosol exposure. Neutrophil infiltration of the pulmonary blood vessels has also been observed, along with lymphocytolysis in the thymus and tracheobronchial lymph nodes. Pulmonary lesions in mice become severe and diffuse within 48 h after exposure, subsequently developing into acute necrotizing pneumonia. Epithelial necrosis is most severe in the trachea and respiratory bronchioles, although all airways and alveoli can be affected. Type 2 epithelial cell and alveolar macrophage populations become fully depleted within 48 h after exposure, whereas neutrophils accumulate in the thickened alveolar septa, and macrophages collect in the tracheobronchial lymph nodes.

Recent studies have further characterized the inflammatory response resulting from respiratory exposure to ricin in mice [72,77,78]. An investigation of ricin-induced pulmonary inflammation in macrophage-depleted transgenic mice demonstrated roles for pulmonary macrophages and interleukin-1 signaling in the local inflammatory response to ricin in the lungs [77] Another study evaluating the expression of inflammatory genes in mice after aerosol exposure to sublethal ricin doses also identified potential roles for interleukin-1 and other cytokine pathways in the pulmonary effects of ricin [78]. Respiratory intoxication has been shown to produce a systemic inflammatory response as well. Intratracheal instillation of a lethal ricin dose into mice causes acute inflammation and hemorrhaging in multiple organs; this response is dose-dependent, as a sublethal dose induces localized inflammatory damage in the lungs but not in other organs [72].

The pathogenesis of inhalational ricin intoxication has been evaluated across seven different inbred mouse strains (BALB/C, BXSB, C57BL/6J, CBA/J, C3H/HeJ, C3H/HeN, A/J) [98]. The majority of animals in all strain groups develop chronic pulmonary inflammation and spindled cell fibroproliferation, alveolar emphysema, and renal perivascular edema. The latter pathology, which may be associated with ricin-induced vascular leak syndrome, has been observed in most mice but is most dramatic in C3H/HeJ and BALB/c strains. The most severe type II pneumocyte hyperplasia and pulmonary inflammation is seen in BXSB and C57BL/6J mice, whereas pulmonary spindled cell expansion is most prominent in C57BL/6J mice. The lungs of BALB/C and C3H/HeN mice show the most significant alveolar edema, and alveolar emphysema is most pronounced in A/J mice.

Adrenal gland damage has been observed in mice after intravenous ricin administration, and lesions of the mediastinal lymphatics have been reported after both inhalational exposure and toxin injection [24,83,128]. These common pathological findings after respiratory and parenteral ricin exposure indicate that ricin inhalation likely leads to some systemic absorption and toxicity. A role for systemic toxicity is supported by the finding that respiratory exposure to lethal toxin doses triggers a systemic inflammatory response and hemorrhaging in multiple organs in addition to the lungs [72]. Systemic effects may depend to some extent on toxin dose, and the extent to which systemic pathogenesis contributes to disease and lethality after pulmonary ricin exposure remains to be fully elucidated.

The pathological effects of inhalational ricin exposure have also been characterized in the rat model [121,129]. Major ultrastructural changes are generally isolated to the terminal bronchioles and lung alveoli. Significant necrosis of bronchiolar Clara and epithelial cells has been reported, along with necrosis and swelling of type II pneumocytes. Lung congestion, acute alveolar edema, and severe alveolar inflammation have been observed in lethally intoxicated animals. The alveolar edema seen in rats and other experimental species is thought to be associated with damage to the respiratory epithelium, and possibly to the alveolar capillary endothelium. Alveolar macrophage populations become severely depleted and apoptotic soon after aerosol exposure. Partial recovery of alveolar macrophage numbers has been observed beyond the 24-h time point; this recovery often coincides with the onset of overwhelming lethal pulmonary edema, alveolar fibrin deposition, and type II pneumocyte proliferation. Interestingly, pulmonary injury resulting from ricin inhalation can be significantly attenuated by alveolar macrophage depletion, implicating these cell populations as mediators of ricin-induced lung damage. Additional pathological changes include enlarged endothelial cells, widened endothelial tight junctions, increased pinocytic vesicles, and interstitial inflammatory cell infiltrates, including macrophages, lymphocytes, neutrophils, and eosinophils. The liver, spleen, and kidneys are also typically congested in intoxicated rats.

The pathology associated with respiratory intoxication has been evaluated to a limited extent in nonhuman primates [28,83]. Pathological findings in rhesus monkeys after exposure to lethal HO aerosolized ricin doses (21–42 μg/kg) are generally restricted to the respiratory tract and include acute airway inflammation, purulent tracheitis, fibrinopurulent pneumonia and pleuritis, peribronchovascular edema, mediastinal lymphadenitis, diffuse necrosis, and extensive alveolar flooding. Postmortem findings generally parallel those seen in other experimental models. Upper airway inflammation is generally mild, although the tracheolaryngeal regions often contain some neutrophil, lymphocyte, and macrophage infiltrates along with occasional necrotic epithelial cells. In contrast, the pulmonary airways and alveoli show significant and widespread necrosis, edema, leukocyte infiltrates, and fibrin deposits. Inflammatory cells pervade the peribronchovascular lymphatics, and reduced lymphocyte numbers are detected in the bronchus-associated lymphoid tissue. Complete degeneration of the bronchiolar and alveolar epithelium has been observed in some animals. Although isolated adrenal gland necrosis has been seen in a few intoxicated animals, severe pathological findings in monkeys are typically confined to the respiratory tract and parallel the lesions detected in rodents after aerosol exposure.

16.5.3 Disease Presentation

The clinical presentation of inhalational ricin intoxication has been evaluated across a number of different inbred mouse strains [98]. Mice from each strain exhibit lethargy, reduced responsiveness, piloerection, kyphosis, photophobia, chemosis, and loss of appetite around 30 h after WB ricin aerosol exposure. Rapid and labored breathing is often evident within 48 h, and severe piloerection is also seen in BXSB mice. Breathing difficulty and piloerection are not observed in C57BL/6J mice; these animals display only mild signs including reduced responsiveness. Relatively mild and inconsistent signs of reduced appetite and activity are seen in most strains, although BXSB mice also develop significant anorexia and poor body condition. Mice from all tested strains die within 3–5 days after receiving high aerosol doses (18.2 μg/kg), and lethality is generally attributed to cardiovascular collapse. Certain strains (C57BL/6J, A/J, and C3H/HeN) display significantly longer survival times than others (BXSB, CBA/J, and C3H/HeJ) after aerosol exposure to a lower lethal dose (8.6 μg/kg) [98]. Similar dose-dependent survival times have been reported in CD-1 mice after lethal respiratory intoxication [117,118].

Disease presentation after respiratory exposure has also been characterized in the rat model. Signs of illness are typically not observed for the first 24 h after aerosol exposure to crude in-house or commercial ricin preparations [121]. Intoxicated animals later exhibit reduced activity, piloerection, labored breathing, general malaise, minimal food and water intake, and occasional colored nasal discharge. As seen in mice, severe weight loss has been observed in rats during the course of illness, and body weight has been reported to serve as a strong predictor of survival. Intoxicated animals either die from pulmonary edema-induced hypoxia or recover gradually over the next several days.

The clinical presentation of respiratory intoxication has been described for non-human primates as well [28,83]. Rhesus monkeys develop signs of illness on the first day after HO exposure to lethal doses of aerosolized ricin. Initial clinical signs include significant fever, lethargy, depression, reduced food and fluid intake, antisocial behavior, and rapid labored breathing. As with other animal models for respiratory ricin intoxication, time to death in monkeys is dose-dependent and ranges from 1 to 3 days for inhaled doses of 3–96 μg/kg. Loss of skin elasticity indicates that the animals become dehydrated before death. A clear, frothy nasal discharge has been observed in some animals at the time of death; this clear fluid is also found in the trachea and primary bronchi.

As previously noted, data on clinical signs of respiratory intoxication in humans are very limited. The reported cases of potential inhalational exposures in humans were relatively mild and did not result in acute illness or death. It is, therefore, difficult to draw parallels with disease manifestations in lethally intoxicated animals. The signs and symptoms cited for laboratory workers in World War II generally suggested that, as with experimental animals, many of the prominent effects were localized to the respiratory tract [7]. These patients displayed chest tightness, tracheal inflammation, coughing, and respiratory distress within several hours of exposure. They did show some potential signs of inflammatory and systemic responses, including fever, nausea, sweating, dehydration, myosis, cyanosis, and circulatory collapse. All exposed individuals recovered with no identified long-term effects, however, so no data are available on pathological changes within affected tissues.

16.5.4 Respiratory Toxicity

Ricin is more potent via the respiratory exposure route than most conventional chemical agents [7]. A wide range of human inhalation LD_{50} estimates has been reported; these estimates are in some cases notably lower than the lethal respiratory dose values determined in several animal models [5–8,26–28,83,98–100,117,118,130]. In contrast, human lethal ricin concentration (LCt_{50}) estimates of 30–70 mg/min per cubic meter fall within the high end of the range of lethal concentration values determined for several animal models [4,5,7,99,121,130,131]. Importantly, significant differences in respiratory toxicity have been demonstrated between experimental species and across multiple strains of the same species. In some cases, these differences have been attributed to the variable toxin preparations and exposure protocols used in inhalation exposure studies. As mentioned previously, inhalation toxicity is also highly dependent on aerosol particle size and exposure methodology, as these parameters dictate respiratory tract deposition patterns for inhaled ricin.

The respiratory toxicity of ricin has been reported for several mouse strains [28,76,98,103,122]. BALB/C mice have been found to tolerate higher inhaled ricin doses than any other tested strains, yielding WB aerosol LD_{50} values of 11–15 μg/kg [28,76,98]. Somewhat lower inhalation LD_{50} values have been generated for various other mouse strains, including BXSB (2.8 μg/kg), Swiss Webster (4 ug/kg), NIH Swiss (4.9 μg/kg), CBA/J (5.3 μg/kg), C57BL/6J (5.3 μg/kg), C2H/HeJ (5.3 μg/kg), A/J (8.2 μg/kg), and C3H/HeN (9.0 μg/kg) [28,98,103]. An aerosol lethal concentration of 9 mg/min per cubic meter has been cited for mice, although the strain used to determine this value is not specified [131–133].

The rat model has also been used in studies on the respiratory toxicity of ricin. An LCt_{50} range of 4.5–5.9 mg/min per cubic meter has been generated in Porton strain rats exposed to HO aerosols of a commercial ricin preparation (from the *Hale Queen* castor seed variety) [129]. Comparable NO aerosol LCt_{50} values of 5.79 and 4 mg/min per cubic meter have been reported for a commercial ricin preparation in outbred Crl:CD(SD)BR rats. Significantly higher lethal concentration values of 11.21 mg/min per cubic meter (LCt_{30}), 11.9 mg/min per cubic meter (LCt_{30}), and 12.7 mg/min per cubic meter (LCt_{50}) have been determined for various in-house ricin preparations in Porton rats [99,121,134]. Collectively, these studies suggest that mice and rats show similar susceptibilities to inhalational ricin intoxication, but that quantitative toxicity determinations vary across different animal strains, ricin preparations, and aerosol exposure parameters.

Respiratory toxicity data have been reported for several other animal models, including guinea pigs, rabbits, and nonhuman primates. A guinea pig LCt_{50} of 7 mg/min per cubic meter has been cited for a commercial ricin preparation [131]. Somewhat lower lethal dose values have been indicated for the same commercial ricin in rabbits, including an inhalation LCt_{50} of 4 mg/min per cubic meter and an intratracheal LD_{50} of 0.5 μg/kg. As with many other experimental species, quantitative susceptibility in rabbits is expected to be dependent on aerosol particle size and toxin source. Although the pathogenesis of inhalational ricin exposure has been characterized in monkeys, minimal quantitative toxicity information is available for any nonhuman primate model. LD_{50} values of 5.8 and 15 μg/kg have been reported in African green monkeys and rhesus monkeys, respectively [28]. Several ricin reviews and toxicity summaries cite an LCt_{50} of 100 mg/min per cubic meter for monkeys based on early work by the U.S. military [7,132,133].

16.5.5 Vaccines and Therapeutics

Investigations of vaccine-induced protection against respiratory ricin intoxication have predominantly incorporated mouse models [99,103,122–124]. Several recent studies have evaluated the efficacy of recombinant ricin A-chain subunit vaccines against aerosol exposure in mice. Intranasal vaccination of mice with a deglycosylated recombinant ricin A-chain was found to confer limited protection against subsequent aerosol exposure to lethal ricin doses [123,124]. The protective effect was significantly enhanced by coadministration of the recombinant vaccine with select mucosal adjuvants derived from *Escherichia coli* heat-labile enterotoxin. These adjuvants dramatically increased the ricin-specific antibody response but did not mitigate the lung damage resulting from exposure to aerosolized toxin [123,124].

Another recent study demonstrated that intramuscular injection of a recombinant A-chain vaccine (RiVax) could confer protection in mice against ricin exposure by either the oral or aerosol route [103]. Interestingly, in contrast to the intranasal vaccine studies, intramuscular vaccination with RiVax also protected against ricin-induced pulmonary damage. In previous studies, RiVax was also found to protect mice against intraperitoneal challenge, and to safely induce ricin-neutralizing antibodies in human participants under a pilot clinical trial [135,136]. A subsequent study demonstrated that intramuscular and intradermal administration of RiVax both conferred protection to mice against intraperitoneal, aerosol, and oral challenge [135]. Both vaccination routes also minimized ricin-induced pulmonary damage after aerosol exposure to the toxin.

These studies are significant in demonstrating vaccine efficacy, establishing a basis for further studies on responses to intoxication and vaccination, and advancing the use of the mouse model in generating meaningful data with applicability to humans. Of note, the pilot clinical trial on RiVax showed that passive transfer of the antibodies produced in human volunteers conferred protection to mice against lethal ricin challenge [135]. A number of other recent studies have also used mouse models to evaluate the efficacy of antibody-based, chemical, and anti-inflammatory interventions for ricin intoxication [79,97,111,137–139]. These studies have investigated protective effects against ricin challenge by the respiratory route as well as oral and systemic exposure. Future vaccine and therapeutic investigations will benefit from the growing body of knowledge on the use of the mouse model, as well as other experimental species, in establishing correlates for protection in humans.

16.6 DISCUSSION

Ricin is a powerful plant-derived toxin known to cause illness and lethality in humans and numerous animal species. As a cytotoxic lectin, it readily gains access to exposed cells and subsequently inhibits protein synthesis, leading to cell death. This process generally occurs within gastrointestinal tissues because ricin poisoning cases are usually associated with castor seed ingestion. However, ricin is also capable of exerting its cytotoxic effects when introduced into the respiratory tract by aerosol inhalation or intratracheal instillation. The pathogenesis associated with oral and respiratory exposure to ricin has been investigated in several animal models but has been most extensively characterized in the small rodent species. Mice and rats show similar susceptibilities to the toxin and display many of the same clinical signs and pathological changes after experimental exposure. The limited studies available on natural and experimental ricin poisoning in other species indicate that the pathogenesis and clinical presentation are fairly consistent in most of the common animal models.

Many questions remain to be answered regarding the cellular and molecular basis of ricin poisoning, as well as the quantitative susceptibility of humans to the toxin. Although lethal dose values for several exposure routes have been determined in experimental animals, at least one recent study has demonstrated the potential for ricin to have significantly higher oral toxicity than previously established. Additionally, little is known about oral and respiratory toxicity in humans, either in quantitative terms or relative to tested animal species. Our current understanding of the potential roles for systemic absorption and toxicity after oral or respiratory ricin exposure is limited. Some investigators have proposed that toxin uptake into the circulation and lymphatics may play a major role in disease pathogenesis and lethality, despite the fact that many of the observed toxic effects occur directly within the primary exposed tissues. Clarification of this issue will be important in characterizing both the process of ricin poisoning and the potential efficacy of various medical countermeasures. The mouse model has proven valuable in recent studies on recombinant subunit vaccines, and will likely continue to play an important role in future preclinical investigations. Further development of tissue culture models for oral, systemic, and respiratory ricin poisoning should also facilitate more detailed characterizations of disease mechanisms, pathogenesis, diagnosis, and prophylaxis and treatment in humans.

16.7 CONCLUSIONS

Recent publications and reviews indicate that the mouse is currently the prominent animal model for *in vivo* studies on the pathogenesis of and potential therapeutics for ricin intoxication. The continued characterization and use of this model will be of value in future studies on the toxin as a biothreat agent. Meanwhile, traditional and technology-driven *in vitro* and *ex vivo* models provide platforms for quantitative evaluations of toxin binding, kinetics, and molecular interactions within exposed tissues. Further development of these models will be beneficial, in conjunction with continued *in vivo* studies to evaluate questions such as the influence of experimental conditions on oral toxicity data, the cause of lethality after acute oral intoxication, and the potential role of systemic inflammation in the pathogenesis of oral and respiratory intoxication. These studies will be important in establishing a more comprehensive understanding of threat-related aspects of ricin and in continuing to develop and evaluate promising medical countermeasures such as recombinant subunit vaccines for use in humans.

DISCLAIMER

The opinions, interpretations, conclusions, and recommendations are those of the authors based on analysis of the reviewed literature, and do not necessarily reflect the views of their employers or the U.S. Army.

REFERENCES

1. Ishiguro, M. et al., Biochemical studies on ricin. I. Purification of Ricin. *J Biochem*, 1964, **55**: 587–592.
2. Simmons, B.M. and J.H. Russell, A single affinity column step method for the purification of ricin toxin from castor beans (*Ricinus communis*). *Anal Biochem*, 1985, **146**(1): 206–210.
3. Audi, J. et al., Ricin poisoning: A comprehensive review. *JAMA*, 2005, **294**(18): 2342–2351.
4. Cope, A.C et al., *Chemical Warfare Agents and Related Chemical Problems—Part I: Summary Technical Report of Division 9*. Washington, DC: National Defense Research Committee, 1945, pp. 179–203.
5. Parker, D.T., A.C. Parker, and C.K. Ramachandran, *Joint Technical Data Source Book*, Vol. 6, Part 3, DPGJCP-961007. U.S. Dugway Proving Ground, Utah: Joint Contact Point Directorate, 1996, pp. 1–38.
6. U.S. Army, *Potential Military Chemical/Biological Agents and Compounds, Army Field Manual*, 1990, pp. 3–9.
7. Augerson, W., *A Review of the Scientific Literature as It Pertains to Gulf War Illnesses. Volume 5: Chemical and Biological Warfare Agents*. Santa Monica, CA: RAND National Defense Research Institute Publication, 2000, p. 59.
8. Hunt, R., Ricin, *Report for the War Department, Chemical Warfare Service, Chemical Warfare Monograph No. 37*, Washington, DC: American University Experiment Station, 1918, pp. 107–117.
9. Ishiguro, M. et al., Biochemical studies on oral toxicity of ricin. I. Ricin administered orally can impair sugar absorption by rat small intestine. *Chem Pharm Bull (Tokyo)*, 1983, **31**(9): 3222–3227.
10. Sekine, I. et al., Pathological study on mucosal changes in small intestine of rat by oral administration of ricin. I. Microscopical observation. *Acta Pathol Jpn*, 1986, **36**(8): 1205–1212.
11. Frigerio, L. et al., Free ricin A chain, proricin, and native toxin have different cellular fates when expressed in tobacco protoplasts. *J Biol Chem*, 1998, **273**(23): 14194–14199.
12. Sehnke, P.C. et al., Expression of active, processed ricin in transgenic tobacco. *J Biol Chem*, 1994, **269**(36): 22473–22476.
13. Sehnke, P.C. and R.J. Ferl, Processing of preproricin in transgenic tobacco. *Protein Expr Purif*, 1999, **15**(2): 188–195.
14. Tagge, E.P. et al., Preproricin expressed in *Nicotiana tabacum* cells in vitro is fully processed and biologically active. *Protein Expr Purif*, 1996, **8**(1): 109–118.
15. Leshin, J. et al., Characterization of ricin toxin family members from *Ricinus communis*. *Toxicon*, 2010, **55**(2–3): 658–661.
16. Oda, T., N. Komatsu, and T. Muramatsu, Cell lysis induced by ricin D and ricin E in various cell lines. *Biosci Biotechnol Biochem*, 1997, **61**(2): 291–297.
17. Woo, B.H., J.T. Lee, and K.C. Lee, Purification of Sepharose-unbinding ricin from castor beans (*Ricinus communis*) by hydroxyapatite chromatography. *Protein Expr Purif*, 1998, **13**(2): 150–154.
18. Sehgal et al., Purification, characterization and toxicity profile of ricin isoforms from castor beans. *Food Chem Toxicol*, 2010, **48**(11): 3171–3176.
19. Garber, E.A., Toxicity and detection of ricin and abrin in beverages. *J Food Prot*, 2008, **71**(9): 1875–1883.
20. Jackson, L.S., Z. Zhang, and W.H. Tolleson, Thermal stability of ricin in orange and apple juices. *J Food Sci*, 2010, **75**(4): T65–T71.

21. Jackson, L.S., W.H. Tolleson, and S.J. Chirtel, Thermal inactivation of ricin using infant formula as a food matrix. *J Agric Food Chem*, 2006, **54**(19): 7300–7304.
22. He, X. et al., Effect of food matrices on the biological activity of ricin. *J Food Prot*, 2008, **71**(10): 2053–2058.
23. Yoder, J.M., R.U. Aslam, and N.J. Mantis, Evidence for widespread epithelial damage and coincident production of monocyte chemotactic protein 1 in a murine model of intestinal ricin intoxication. *Infect Immun*, 2007, **75**(4): 1745–1750.
24. Godal, A. et al., Pharmacological studies of ricin in mice and humans. *Cancer Chemother Pharmacol*, 1984, **13**(3): 157–163.
25. Fodstad, O., S. Olsnes, and A. Pihl, Toxicity, distribution and elimination of the cancerostatic lectins abrin and ricin after parenteral injection into mice. *Br J Cancer*, 1976, **34**(4): 418–425.
26. Roy, C.J. et al., Impact of inhalation exposure modality and particle size on the respiratory deposition of ricin in BALB/c mice. *Inhal Toxicol*, 2003, **15**(6): 619–638.
27. Griffiths, G.D., G.J. Phillips, and J. Holley, Inhalation toxicology of ricin preparations: Animal models, prophylactic and therapeutic approaches to protection. *Inhal Toxicol*, 2007, **19**(10): 873–887.
28. Wannemacher, R.W. and J.B. Anderson, Inhalation ricin: Aerosol procedures, toxicology, and therapy. In: H. Salem and S. Katz (eds.), *Inhalation Toxicology*, 2nd ed., Chapter 39. Boca Raton/London/New York: CRC Press, 2006, pp. 973–980.
29. Doebler, J.A. et al., The distribution of [125I]ricin in mice following aerosol inhalation exposure. *Toxicology*, 1995, **98**(1–3): 137–149.
30. Olsnes, S. et al., Studies on the structure and properties of the lectins from *Abrus precatorius* and *Ricinus communis. Biochim Biophys Acta*, 1975, **405**(1): 1–10.
31. Olsnes, S., K. Refsnes, and A. Pihl, Mechanism of action of the toxic lectins abrin and ricin. *Nature*, 1974, **249**(458): 627–631.
32. Halling, K.C. et al., Genomic cloning and characterization of a ricin gene from *Ricinus communis. Nucleic Acids Res*, 1985, **13**(22): 8019–8033.
33. Montfort, W. et al., The three-dimensional structure of ricin at 2.8 A. *J Biol Chem*, 1987, **262**(11): 5398–5403.
34. Rutenber, E. et al., Crystallographic refinement of ricin to 2.5 A. *Proteins*, 1991, **10**(3): 240–250.
35. Lamb, F.I., L.M. Roberts, and J.M. Lord, Nucleotide sequence of cloned cDNA coding for preproricin. *Eur J Biochem*, 1985, **148**(2): 265–270.
36. Lord, J.M., Precursors of ricin and *Ricinus communis* agglutinin. Glycosylation and processing during synthesis and intracellular transport. *Eur J Biochem*, 1985, **146**(2): 411–416.
37. Lin, T.T. and S.L. Li, Purification and physicochemical properties of ricins and agglutinins from *Ricinus communis. Eur J Biochem*, 1980, **105**(3): 453–459.
38. Villafranca, J.E. and J.D. Robertus, Ricin B chain is a product of gene duplication. *J Biol Chem*, 1981, **256**(2): 554–556.
39. Rutenber, E., M. Ready, and J.D. Robertus, Structure and evolution of ricin B chain. *Nature*, 1987, **326**(6113): 624–626.
40. Refsnes, K., S. Olsnes, and A. Pihl, On the toxic proteins abrin and ricin. Studies of their binding to and entry into Ehrlich ascites cells. *J Biol Chem*, 1974, **249**(11): 3557–3562.
41. Simmons, B.M., P.D. Stahl, and J.H. Russell, Mannose receptor-mediated uptake of ricin toxin and ricin A chain by macrophages. Multiple intracellular pathways for a chain translocation. *J Biol Chem*, 1986, **261**(17): 7912–7920.
42. Magnusson, S., R. Kjeken, and T. Berg, Characterization of two distinct pathways of endocytosis of ricin by rat liver endothelial cells. *Exp Cell Res*, 1993, **205**(1): 118–125.

43. Magnusson, S. et al., Interactions of ricin with sinusoidal endothelial rat liver cells. Different involvement of two distinct carbohydrate-specific mechanisms in surface binding and internalization. *Biochem J*, 1991, **277(Pt 3)**: 855–861.
44. Magnusson, S. and T. Berg, Endocytosis of ricin by rat liver cells *in vivo* and *in vitro* is mainly mediated by mannose receptors on sinusoidal endothelial cells. *Biochem J*, 1993, **291(Pt 3)**: 749–755.
45. Riccobono, F. and M.L. Fiani, Mannose receptor dependent uptake of ricin A1 and A2 chains by macrophages. *Carbohydr Res*, 1996, **282**(2): 285–292.
46. Olsnes, S. et al., Rates of different steps involved in the inhibition of protein synthesis by the toxic lectins abrin and ricin. *J Biol Chem*, 1976, **251**(13): 3985–3992.
47. Olsnes, S. and A. Pihl, Different biological properties of the two constituent peptide chains of ricin, a toxic protein inhibiting protein synthesis. *Biochemistry*, 1973, **12**(16): 3121–3126.
48. Sandvig, K., S. Olsnes, and A. Pihl, Kinetics of binding of the toxic lectins abrin and ricin to surface receptors of human cells. *J Biol Chem*, 1976, **251**(13): 3977–3984.
49. Baenziger, J.U. and D. Fiete, Structural determinants of *Ricinus communis* agglutinin and toxin specificity for oligosaccharides. *J Biol Chem*, 1979, **254**(19): 9795–9799.
50. Sandvig, K. and S. Olsnes, Entry of the toxic proteins abrin, modeccin, ricin, and diphtheria toxin into cells. II. Effect of pH, metabolic inhibitors, and ionophores and evidence for toxin penetration from endocytotic vesicles. *J Biol Chem*, 1982, **257**(13): 7504–7513.
51. Sandvig, K. and S. Olsnes, Entry of the toxic proteins abrin, modeccin, ricin, and diphtheria toxin into cells. I. Requirement for calcium. *J Biol Chem*, 1982, **257**(13): 7495–7503.
52. Naseem, S.M., R.B. Wellner, and J.G. Pace, The role of calcium ions for the expression of ricin toxicity in cultured macrophages. *J Biochem Toxicol*, 1992, **7**(2): 133–138.
53. van Deurs, B. et al., Routing of internalized ricin and ricin conjugates to the Golgi complex. *J Cell Biol*, 1986, **102**(1): 37–47.
54. van Deurs, B. et al., Estimation of the amount of internalized ricin that reaches the trans-Golgi network. *J Cell Biol*, 1988, **106**(2): 253–267.
55. van Deurs, B. et al., Delivery of internalized ricin from endosomes to cisternal Golgi elements is a discontinuous, temperature-sensitive process. *Exp Cell Res*, 1987, **171**(1): 137–152.
56. van Deurs, B. et al., Receptor-mediated endocytosis of a ricin-colloidal gold conjugate in vero cells. Intracellular routing to vacuolar and tubulo-vesicular portions of the endosomal system. *Exp Cell Res*, 1985. **159**(2): 287–304.
57. Endo, Y. et al., The mechanism of action of ricin and related toxic lectins on eukaryotic ribosomes. The site and the characteristics of the modification in 28 S ribosomal RNA caused by the toxins. *J Biol Chem*, 1987, **262**(12): 5908–5912.
58. Endo, Y. and K. Tsurugi, RNA N-glycosidase activity of ricin A-chain. Mechanism of action of the toxic lectin ricin on eukaryotic ribosomes. *J Biol Chem*, 1987, **262**(17): 8128–8130.
59. Endo, Y. and K. Tsurugi, The RNA N-glycosidase activity of ricin A-chain. The characteristics of the enzymatic activity of ricin A-chain with ribosomes and with rRNA. *J Biol Chem*, 1988, **263**(18): 8735–8739.
60. Szewczak, A.A. et al., The conformation of the sarcin/ricin loop from 28S ribosomal RNA. *Proc Natl Acad Sci USA*, 1993, **90**(20): 9581–9585.
61. Szewczak, A.A. and P.B. Moore, The sarcin/ricin loop, a modular RNA. *J Mol Biol*, 1995, **247**(1): 81–98.
62. Endo, Y., Mechanism of action of ricin and related toxins on the inactivation of eukaryotic ribosomes. *Cancer Treat Res*, 1988, **37**: 75–89.
63. Lord, M.J. et al., Ricin. Mechanisms of cytotoxicity. *Toxicol Rev*, 2003, **22**(1): 53–64.

64. Olsnes, S. and A. Pihl, Treatment of abrin and ricin with -mercaptoethanol opposite effects on their toxicity in mice and their ability to inhibit protein synthesis in a cell-free system. *FEBS Lett*, 1972, **28**(1): 48–50.
65. Olsnes, S. et al., Ribosome inactivation by the toxic lectins abrin and ricin. Kinetics of the enzymic activity of the toxin A-chains. *Eur J Biochem*, 1975, **60**(1): 281–288.
66. Morlon-Guyot, J. et al., Identification of the ricin lipase site and implication in cytotoxicity. *J Biol Chem*, 2003, **278**(19): 17006–17011.
67. Lombard, S., M.E. Helmy, and G. Pieroni, Lipolytic activity of ricin from *Ricinus sanguineus* and *Ricinus communis* on neutral lipids. *Biochem J*, 2001, **358(Pt 3)**: 773–781.
68. Day, P.J. et al., Binding of ricin A-chain to negatively charged phospholipid vesicles leads to protein structural changes and destabilizes the lipid bilayer. *Biochemistry*, 2002, **41**(8): 2836–2843.
69. Griffiths, G.D., M.D. Leek, and D.J. Gee, The toxic plant proteins ricin and abrin induce apoptotic changes in mammalian lymphoid tissues and intestine. *J Pathol*, 1987, **151**(3): 221–229.
70. Hughes, J.N., C.D. Lindsay, and G.D. Griffiths, Morphology of ricin and abrin exposed endothelial cells is consistent with apoptotic cell death. *Hum Exp Toxicol*, 1996, **15**(5): 443–451.
71. Kumar, O., K. Sugendran, and R. Vijayaraghavan, Oxidative stress associated hepatic and renal toxicity induced by ricin in mice. *Toxicon*, 2003, **41**(3): 333–338.
72. Wong, J. et al., Proinflammatory responses of human airway cells to ricin involve stress-activated protein kinases and NF-kappaB. *Am J Physiol Lung Cell Mol Physiol*, 2007, **293**(6): L1385–L1394.
73. Wong, J. et al., Intrapulmonary delivery of ricin at high dosage triggers a systemic inflammatory response and glomerular damage. *Am J Pathol*, 2007, **170**(5): 1497–1510.
74. Korcheva, V. et al., Role of apoptotic signaling pathways in regulation of inflammatory responses to ricin in primary murine macrophages. *Mol Immunol*, 2007, **44**(10): 2761–2771.
75. Korcheva, V. et al., Administration of ricin induces a severe inflammatory response via nonredundant stimulation of ERK, JNK, and P38 MAPK and provides a mouse model of hemolytic uremic syndrome. *Am J Pathol*, 2005, **166**(1): 323–339.
76. DaSilva, L. et al., Pulmonary gene expression profiling of inhaled ricin. *Toxicon*, 2003, **41**(7): 813–822.
77. Lindauer, M.L. et al., Pulmonary inflammation triggered by ricin toxin requires macrophages and IL-1 signaling. *J Immunol*, 2009, **183**(2): 1419–1426.
78. David, J., L.J. Wilkinson, and G.D. Griffiths, Inflammatory gene expression in response to sub-lethal ricin exposure in Balb/c mice. *Toxicology*, 2009, **264**(1–2): 119–130.
79. Mabley, J.G., P. Pacher, and C. Szabo, Activation of the cholinergic antiinflammatory pathway reduces ricin-induced mortality and organ failure in mice. *Mol Med*, 2009, **15**(5–6): 166–172.
80. Naseem, S.M. and J.G. Pace, Effect of anti-inflammatory agents on ricin-induced macrophage toxicity. *J Biochem Toxicol*, 1993, **8**(3): 145–153.
81. Liu, L. et al., Analysis of intestinal injuries induced by ricin *in vitro* using SPR technology and MS identification. *Int J Mol Sci*, 2009, **10**(5): 2431–2439.
82. Ishiguro, M. et al., Effects of ricin, a protein toxin, on glucose absorption by rat small intestine. (Biochemical studies on oral toxicity of ricin. II). *Chem Pharm Bull (Tokyo)*, 1984, **32**(8): 3141–3147.
83. Wilhelmsen, C.L. and M.L. Pitt, Lesions of acute inhaled lethal ricin intoxication in rhesus monkeys. *Vet Pathol*, 1996, **33**(3): 296–302.
84. Strocchi, P. et al., Lesions caused by ricin applied to rabbit eyes. *Invest Ophthalmol Vis Sci*, 2005, **46**(4): 1113–1116.

85. Jackson, J.H., Tissue changes in alimentary canal of mouse induced by ricin poisoning. *J Physiol*, 1957, **135**(2): 30-1.
86. Moriyama, H., Studies on ricin. The first report. *Jpn. J. Exp. Med.*, 1934. **12**(395): 44.
87. Warner, J., Review of reactions to biotoxins in water, *Battelle Memorial Institute Final Report*, CBIAC Task 152. Columbus, OH: Battelle Memorial Institute, 1990.
88. Funatsu, G. and M. Funatsu, Limited hydrolysis of ricin D with alkaline protease from *Bacillus subtilis. Agric Biol Chem*, 1977, **41**: 1309–1310.
89. Yoshitake, S., K. Watanabe, and G. Funatsu, Limited hydrolysis of ricin D with trypsin in the presence of sodium dodecyl sulfate. *Agric Biol Chem*, 1979, **43**: 2193–2195.
90. Golosnitskiy, A., *Ricin poisoning* (in Russian). *Profilaktika Otravleniy Zhivotnykh Rastitel'nymi Yademi*, 1979, **128**.
91. Lim, H., H.J. Kim, and Y.S. Cho, A case of ricin poisoning following ingestion of Korean castor bean. *Emerg Med J*, 2009, **26**(4): 301–302.
92. Bradberry, S.M. et al., Ricin poisoning. *Toxicol Rev*, 2003, **22**(1): 65–70.
93. Klaim, G.J. and J.J. Jaeger, Castor seed poisoning in humans: A review. *Institute Technical Report No. 453* for USAMRDC, DTIC ADA229133. San Francisco, CA: Letterman Army Institute of Research, Division of Cutaneous Hazards, 1990.
94. Rauber, A. and J. Heard, Castor bean toxicity re-examined: A new perspective. *Vet Hum Toxicol*, 1985, **27**(6): 498–502.
95. Challoner, K.R. and M.M. McCarron, Castor bean intoxication. *Ann Emerg Med*, 1990. **19**(10): 1177–1183.
96. Balint, G.A., Ricin: The toxic protein of castor oil seeds. *Toxicology*, 1974, **2**(1): 77–102.
97. Pratt, T.S. et al., Oropharyngeal aspiration of ricin as a lung challenge model for evaluation of the therapeutic index of antibodies against ricin A-chain for post-exposure treatment. *Exp Lung Res*, 2007, **33**(8–9): 459–481.
98. Wilhelmsen, C.L., Inhaled ricin dose ranging and pathology in inbred strains of mice. *USAMRIID Technical Report*. Fort Detrick, MD: USAMRIID, U.S. Department of Defense Laboratory, 2000.
99. Griffiths, G.D. et al., Protection against inhalation toxicity of ricin and abrin by immunisation. *Hum Exp Toxicol*, 1995, **14**(2): 155–164.
100. Cook, D.L., J. David, and G.D. Griffiths, Retrospective identification of ricin in animal tissues following administration by pulmonary and oral routes. *Toxicology*, 2006, **223**(1–2): 61–70.
101. Brugsch, H.G., Toxic hazards: The castor bean. *Mass Med Soc*, 1960, **262**: 1039–1040.
102. Poli, M.A. et al., Chapter 15: Ricin, in *Textbook of Military Medicine: Medical Aspects of Biological Warfare*, Z.F. Dembek, editor. Washington, DC: Office of the Surgeon General, Department of the Army, 2007.
103. Smallshaw, J.E., J.A. Richardson, and E.S. Vitetta, RiVax, a recombinant ricin subunit vaccine, protects mice against ricin delivered by gavage or aerosol. *Vaccine*, 2007, **25**(42): 7459–7469.
104. Ishiguro, M. et al., Biochemical studies on oral toxicity of ricin. IV. A fate of orally administered ricin in rats. *J Pharmacobiodyn*, 1992, **15**(4): 147–156.
105. Ishiguro, M. et al., Interaction of toxic lectin ricin with epithelial cells of rat small intestine *in vitro*. *Chem Pharm Bull (Tokyo)*, 1992, **40**(2): 441–445.
106. Morino, H., R. Sakakibara, and M. Ishiguro, The binding of ricin to its receptor is not required for the expression of its toxicity. *Biol Pharm Bull*, 1995, **18**(12): 1770–1772.
107. Palatnick, W. and M. Tenenbein, Hepatotoxicity from castor bean ingestion in a child. *J Toxicol Clin Toxicol*, 2000, **38**(1): 67–69.
108. He, X. et al., Ricin toxicokinetics and its sensitive detection in mouse sera or feces using immuno-PCR. *PLoS One*, 2010, **5**(9): e12858.

109. Ishiguro, M. et al., Biochemical studies on oral toxicity of ricin. V. The role of lectin activity in the intestinal absorption of ricin. *Chem Pharm Bull (Tokyo)*, 1992, **40**(5): 1216–1220.
110. Olson, A.D. et al., Differential toxicity of RCAII (ricin) on rabbit intestinal epithelium in relation to postnatal maturation. *Pediatr Res*, 1985, **19**(8): 868–872.
111. Roche, J.K. et al., Post-exposure targeting of specific epitopes on ricin toxin abrogates toxin-induced hypoglycemia, hepatic injury, and lethality in a mouse model. *Lab Invest*, 2008, **88**(11): 1178–1191.
112. Pincus, S.H. et al., Identification of hypoglycemia in mice as a surrogate marker of ricin toxicosis. *Comp Med*, 2002, **52**(6): 530–533.
113. Roels, S. et al., Lethal ricin intoxication in two adult dogs: Toxicologic and histopathologic findings. *J Vet Diagn Invest*, 2010, **22**(3): 466–468.
114. Albretsen, J.C., S.M. Gwaltney-Brant, and S.A. Khan, Evaluation of castor bean toxicosis in dogs: 98 cases. *J Am Anim Hosp Assoc*, 2000, **36**(3): 229–233.
115. Fodstad, O. et al., Toxicity of abrin and ricin in mice and dogs. *J Toxicol Environ Health*, 1979, **5**(6): 1073–1084.
116. Mouser, P. et al., Fatal ricin toxicosis in a puppy confirmed by liquid chromatography/mass spectrometry when using ricinine as a marker. *J Vet Diagn Invest*, 2007, **19**(2): 216–220.
117. Vogel, P. et al., Comparison of the pulmonary distribution and efficacy of antibodies given to mice by intratracheal instillation or aerosol inhalation. *Lab Anim Sci*, 1996, **46**(5): 516–523.
118. Poli, M.A. et al., Aerosolized specific antibody protects mice from lung injury associated with aerosolized ricin exposure. *Toxicon*, 1996, **34**(9): 1037–1044.
119. Griffiths, G.D. et al., Liposomally-encapsulated ricin toxoid vaccine delivered intratracheally elicits a good immune response and protects against a lethal pulmonary dose of ricin toxin. *Vaccine*, 1997, **15**(17–18): 1933–1939.
120. Doebler, J. et al., Autoradiographic localization of [125-I]-ricin in lungs and trachea of mice following an aerosol inhalation exposure. *USAMRICD Technical Report No. TR-96-03*. Fort Detrick, MD: USAMRICD, U.S. Department of Defense Laboratory, 1996.
121. Brown, R.F. and D.E. White, Ultrastructure of rat lung following inhalation of ricin aerosol. *Int J Exp Pathol*, 1997, **78**(4): 267–276.
122. Kende, M. et al., Oral immunization of mice with ricin toxoid vaccine encapsulated in polymeric microspheres against aerosol challenge. *Vaccine*, 2002, **20**(11–12): 1681–1691.
123. Kende, M. et al., Enhancement of intranasal vaccination with recombinant chain A ricin vaccine (rRV) in mice by the mucosal adjuvants LTK63 and LTR72. *Vaccine*, 2007, **25**(16): 3219–3227.
124. Kende, M. et al., Enhancement of intranasal vaccination in mice with deglycosylated chain A ricin by LTR72, a novel mucosal adjuvant. *Vaccine*, 2006, **24**(12): 2213–2221.
125. Griffiths, G.D. et al., Local and systemic responses against ricin toxin promoted by toxoid or peptide vaccines alone or in liposomal formulations. *Vaccine*, 1998, **16**(5): 530–535.
126. Rushing, S.R., M.L. Saylor, and M.L. Hale, Translocation of ricin across polarized human bronchial epithelial cells. *Toxicon*, 2009, **54**(2): 184–191.
127. Griffiths, G.D., C.D. Lindsay, and D.G. Upshall, Examination of the toxicity of several protein toxins of plant origin using bovine pulmonary endothelial cells. *Toxicology*, 1994, **90**(1–2): 11–27.
128. Richer, G., F.K. Jansen, and P. Gros, [Immunotoxins (author's translation)]. *Nouv Presse Med*, 1982, **11**(17): 1321–1324.
129. Griffiths, G.D. et al., *The Toxicology of Ricin and Abrin Toxins—Studies on Immunisation Against Abrin Toxicity*. Porton Down, UK: Chemical and Biological Defence Establishment, 1993.

130. Bide, R. et al., Inhalation toxicologic procedures for exposure of small laboratory animals to highly toxic materials. Part B: Exposure to aerosols with notes on toxicity of ricin. *Technical Memorandum DRES TM-2000-066*. Suffield, Alberta: Defense Research Establishment Suffield, 2001.
131. Sigma Chemical Company, *Lectin from* Ricinus *Commiunis toxin RCA-60*. Material Safety Data Sheet CB-016827, 1990.
132. Lewis, R., Ricin. In: *Sax's Dangerous Properties of Industrial Materials*. New York: Wiley, 2000.
133. Gangolli, S., Ricin. In: *The Dictionary of Substances and Their Effects*. Cambridge: Royal Society of Chemistry, 2004.
134. Griffiths, G.D., P. Rice, A.C. Allenby, S.C. Bailey, and D.G. Upshall, Inhalation toxicology and histopathology of ricin and abrin toxins. *Inhalation Toxicol*, 1995, **7**(1): 269–288.
135. Vitetta, E.S. et al., A pilot clinical trial of a recombinant ricin vaccine in normal humans. *Proc Natl Acad Sci USA*, 2006, **103**(7): 2268–2273.
136. Smallshaw, J.E. et al., Preclinical toxicity and efficacy testing of RiVax, a recombinant protein vaccine against ricin. *Vaccine*, 2005, **23**(39): 4775–4784.
137. Beyer, N.H. et al., A mouse model for ricin poisoning and for evaluating protective effects of antiricin antibodies. *Clin Toxicol (Phila)*, 2009, **47**(3): 219–225.
138. Neal, L.M. et al., A monoclonal immunoglobulin G antibody directed against an immunodominant linear epitope on the ricin A chain confers systemic and mucosal immunity to ricin. *Infect Immun*, 2010, **78**(1): 552–561.
139. Stechmann, B. et al., Inhibition of retrograde transport protects mice from lethal ricin challenge. *Cell*, 2010, **141**(2): 231–242.

17 Staphylococcal and Streptococcal Superantigens

In Vitro *and* In Vivo *Assays*

Teresa Krakauer and Bradley G. Stiles

CONTENTS

17.1 INTRODUCTION

Staphylococcus aureus and *Streptococcus pyogenes* represent ubiquitous, formidable pathogens linked to many human and animal diseases [1–3]. These facultative, β-hemolytic bacteria readily colonize skin and various mucosal surfaces through numerous virulence factors that facilitate their survival and dissemination. In addition to the staphylococcal enterotoxins (SEs) and toxic shock syndrome toxin-1 (TSST-1) that interact with specific subsets of T cells [4–6], *S. aureus* also produces protein A, coagulases, hemolysins, and leukocidins [7,8]. A sobering societal reality involves the ever-increasing resistance of *S. aureus* toward antibiotics like methicillin [9], and now vancomycin which still represents our major last line of antibiotic defense [10,11]. It is estimated that ~50 million dollars are spent annually in Canada for managing antibiotic-resistant *S. aureus* in hospitals, and costs for the dairy industry are even higher [12]. Indeed, *S. aureus* represents an important health and economic concern throughout the world [13–16].

The SEs (A–U, and counting) are associated with one of the most prevalent forms of food poisoning found throughout the world [2,17–21]. It is evident that various populations are naturally exposed to these toxins, as demonstrated by SEB seroconversion rates [22]. However, whether humans develop toxin-specific antibodies following ingestion of contaminated food and/or colonization by a toxin-producing strain of *S. aureus* is unknown. The first definitive report of human staphylococcal food poisoning was in 1914 after consumption of milk from a cow with *S. aureus*-induced mastitis. SE poisoning typically occurs after ingesting processed meats or dairy products previously contaminated by improper handling and storage at temperatures conducive to *S. aureus* growth and production of one or more SEs. Only microgram quantities of consumed toxin are needed to cause emesis and diarrhea within ~4 h, and one may still experience a general malaise 24–72 h later [19]. Food poisoning by the SEs, with SEA representing the most commonly implicated serotype [23], is rarely fatal for healthy individuals; however, children and the elderly do represent the highest-risk groups. Host-derived inflammatory compounds such as prostaglandins and leukotrienes may mediate the enteric effects [24,25]. In addition to causing food poisoning, the SEs (particularly SEB) are potential nefarious agents for biological warfare and bioterrorism [26].

In contrast with the SEs and food-borne illness, toxic shock syndrome (TSS) caused by *S. aureus* TSST-1 was first described in 1978 among children [27], and later linked to menstruation and use of highly absorbent tampons by women [28–30]. Increased levels of protein, carbon dioxide, oxygen, a neutral pH, as well as removal of magnesium near vaginally adherent *S. aureus* are all factors implicated in increased growth of the bacterium and subsequent production of TSST-1 *in vivo* [17,31–33]. In the early literature, TSST-1 was originally described as an enterotoxin called SEF [34]. However, this later proved incorrect as homogeneous SEF (TSST-1) demonstrably lacks enterotoxicity in nonhuman primates [35]. The symptoms of TSS are intimately linked to an altered immune response involving elevated serum levels of proinflammatory cytokines [36–38], rash, hypotension, fever, and multiorgan failure [39,40]. Although less common, a nonmenstrual form of TSS is also attributed to SEB and SEC1 from *S. aureus* growing on other body sites [30,41,42]. Unlike decreasing menstrual TSS cases reported since the early 1980s, a result of increased public awareness and reformulated tampons, cases of nonmenstrual TSS remain relatively constant. All TSS patients may suffer recurring bouts unless the offending strain of *S. aureus* is eliminated or kept at a minimal growth rate. However, bacterial control becomes more difficult with increasing resistance to existing antibiotics and few alternatives that are now clinically available.

Antibodies play an important role in susceptibility to TSST-1-induced TSS [43–45]. Therefore, individuals not seroconverting toward the offending toxin due to toxin-induced hyporesponsive T cells [46] and/or T-cell-dependent B-cell apoptosis [47] are more likely to experience TSS relapses. Perhaps these findings emphasize a need for vaccines that may break tolerance toward TSST-1, and other bacterial superantigens, especially among high-risk populations [48–56]. The therapeutic use of immunoglobulins can be useful for preventing staphylococcal- or streptococcal-related shock. Intravenous immunoglobulin, pooled from human donors, is particularly beneficial in the clinic but there are problems of quality control (i.e., batch-to-batch variation) and characterization (i.e., how such a reagent affords protection) [57,58]. Use of multiple humanized monoclonal antibodies targeting unique epitopes on the SEs, TSST-1, and SPEs represents a logical step forward [59], akin to that described for *Clostridium botulinum* neurotoxin A [60].

A microbial relative of *S. aureus* is *S. pyogenes*, a group A streptococcus as defined by the carbohydrate-based, serotyping system developed by Rebecca Lancefield during the 1930s [61]. Normal niches for *S. pyogenes*, like *S. aureus*, include the skin and mucosal surfaces of a host. Group A steptococci can cause various human diseases such as pharyngitis, impetigo, necrotizing fasciitis, scarlet fever, and rheumatic fever [62]. In similar fashion as *S. aureus*, *S. pyogenes* possesses potent virulence factors that include protein toxins, antiphagocytic properties involving a capsule and M protein, as well as a protease that cleaves the C5a component of complement [63].

The term "superantigen," commonly used in conjunction with SEs, TSST-1, and SPEs, originated in the late 1980s [64,65] to define microbial proteins that activate a large population (5–30%) of specific T cells at picogram levels. These molecules are in contrast with "conventional" antigens that normally stimulate <0.01% of T cells at much higher concentrations [64–69]. Interactions of superantigens with host cells differ from conventional antigens, in that the former: (1) directly bind outside the peptide-binding groove of major histocompatibility complex (MHC) class II; (2) exert biological effects as an intact molecule without internalization and antigen processing; (3) are not MHC class II restricted, but differences do indeed exist between alleles (i.e., human HLA-DR, -DQ, -DP or murine IA and IE) and superantigen presentation to T cells [64,65]. Additionally, recognition of a superantigen–MHC class II complex by the T-cell receptor (TCR) depends on the variable region within a TCR β chain (Vβ), and not a Vα–Vβ chain combination commonly used by conventional peptide antigens [1,64,70]. Microbial superantigens produced by various Gram-positive and Gram-negative bacteria [71–78], as well as viruses [79–84], are listed in Table 17.1 and reviewed elsewhere in more depth [74,85].

17.2 PHYSICAL CHARACTERISTICS OF SEs, TSST-1, AND SPEs

The SEs, TSST-1, and SPEs are 22–30-kDa single-chain proteins secreted by staphylococci or streptococci that form distinct homology groups based on amino acid sequence [64,76,86]. Historically, it is possible that *S. aureus* and *S. pyogenes* obtained common DNA that ultimately yielded the divergently evolving, yet closely related, superantigens recognized today. Clearly, such proteins are evolutionary successes as per their ubiquitous presence in different microbial species (Table 17.1). The staphylococcal and streptococcal superantigens are encoded on plasmids, bacteriophage DNA, or mobile genetic elements and are synthesized during the late logarithm to stationary phases [17,87]. Among the different SE "serotypes," SEA, SED, and SEE share the highest amino acid sequence homology that ranges from 53% to 81%. The SEB molecule is 50–66% homologous with SECs (1, 2, and 3 subtypes), whereas among the SPEs the A serotype is most similar to SEB with 51% homology [88].

TABLE 17.1
Bacterial and Viral Superantigens

Bacterial superantigens
Mycoplasma arthritidis mitogen (MAM) [71]
Mycobacterium tuberculosis superantigen (MTS) [72]
Pseudomonas aeruginosa exotoxin A [73]
Staphylococcus aureus SEA-SEV, TSST-1 [64,68,74,75]
Streptococcus pyogenes SPEA, C, G, H, I, J, L, and M [75]
Streptococcal mitogenic exotoxin Z (SMEZ); streptococcal superantigen (SSA) [76]
Yersinia enterocolitica superantigen [77]
Yersinia pseudotuberculosis-derived mitogen (YPM) [78]
Viral superantigens
Cytomegalovirus (specific identity unknown) [79]
Epstein-Barr virus (specific identity unknown) [80]
Herpes virus Saimari (HVS) 14 protein [81]
Human immunodeficiency virus Nef protein [82]
Mouse mammary tumor virus (MMTV) superantigens [83]
Rabies virus nucleocapsid protein [84]

Despite varying sequences, structural studies and x-ray crystallography of SEA, SEB, SEC2, TSST-1, SPEA, and SPEC reveal quite conserved conformations with two tightly packed domains containing β-sheet plus α-helix structures separated by a shallow groove representing the TCR-binding site [89–94]. Structure-function studies with site-directed mutagenesis and overlapping peptides of these toxins, along with crystallographic analysis of toxin/HLA-DR complexes, provide further clues regarding specific residues critical for binding to MHC class II and TCR [95–97]. The SEs, TSST-1, and SPEs additionally share similar structures as evidenced by cross-reactivity and neutralization with antibodies [51,55, 98–102]. Historically, the SEs were considered serologically distinct by immunodiffusion assays. However, subsequent studies using more sensitive enzyme-linked immunosorbent assay technology with polyclonal and monoclonal antibodies clearly show that common epitopes exist among these toxins [99–102].

17.3 BINDING OF SEs, TSST-1, AND SPEs TO MHC CLASS II AND TCR

The staphylococcal and streptococcal superantigens bind to conserved elements of MHC class II with high affinity (K_d = 1–100 nM) [2,103–105]. However, each toxin preferentially binds to distinct alleles that suggest different contact sites with MHC class II [106–109]. The HLA-DR molecule generally interacts better with SEs and TSST-1 than with HLA-DP or -DQ, whereas the preferential binding of SPEA to HLA-transfected L cells is HLA-DQ > -DR > -DP [103,110]. Competitive binding studies reveal at least two different binding sites on MHC class II for the SEs and TSST-1 [111].

Upon comparing the binding attributes of staphylococcal superantigens, SEA has the highest affinity for HLA-DR mediated by two binding sites [109,112–114]. The higher affinity site is located within the C-terminus and binds to HLA-DR β chain in a zinc-dependent manner [112,113]. The second binding site for HLA-DR on SEA is similar to that for SEB and located within the N-terminus, which interacts with the α chain of HLA-DR [114]. Studies indicate that one SEA molecule does not interact with the α and β chains from the same MHC class II molecule [115]. The cross-linking of two MHC class II molecules by SEA is necessary for cytokine expression in monocytes [116], and cross-linking of MHC class II by either SEB or TSST-1 may also play an important role in toxin-induced activity [117,118]. Cross-linking allows SEA to persist on the surface of antigen presenting cells (APCs), thus prolonging toxin exposure and effects on T cells [119].

Akin to the N-termini of the SEA toxin family, analogous regions of SEB, TSST-1, and SPEA also bind to MHC class II as per studies using recombinantly modified toxins and monoclonal antibodies [51,55,96,97,120]. The co-crystal structures of SEB or TSST-1 complexed with HLA-DR1 and associated antigen peptide clearly reveal distinct binding differences [117,118]. For example, SEB interacts exclusively with the α chain of HLA-DR1 and is unaffected by the HLA-associated peptide. The SPEC molecule forms dimers in solution and exclusively interacts with the β chains of two MHC class II molecules via zinc [121]. This mode of cross-linking MHC class II is also evident with SPEJ [105]. Perhaps, as further evidence of evolutionary relatedness between the streptococcal and staphylococcal superantigens, the zinc-dependent interactions of SPEH with MHC class II are mediated by a "hybrid" SPEH consisting of an N-terminal domain related to the SEB homology group and a C-terminus resembling SPEC [95]. Overall, it is clear that diverse methods exist for binding of SEs, TSST-1, and SPEs to both MHC class II and TCR.

The groove formed between conserved domains of staphylococcal and streptococcal superantigens represents an important interaction site for the TCR Vβ chain [90,96,97,117]. Each toxin binds to a distinct repertoire of Vβ-bearing T cells, thus possessing a unique biological "fingerprint" [105,122,123]. Mutations within the MHC class II binding domains of SEA differentially affect binding to TCR Vβ [124], as evidenced by a small increase in superantigen affinity for MHC class II, thus overcoming a large decrease in affinity for TCR Vβ [125]. A superantigen–MHC class II complex binds directly to main-chain, but not side-chain, residues of soluble Vβ chain from TCR [70,126–128]. Each toxin possesses unique MHC class II/TCR contact sites, and binding affinity of TCR with toxin is strengthened by an MHC class II–toxin complex [129,130].

17.4 CELLULAR RESPONSES TO SEs, TSST-1, AND SPEs

Recognition of the superantigen–MHC class II complex by TCR molecules ultimately results in cell signaling and proliferation [131]. Incubation of TSST-1 or SEB with nonproliferating T cells can also increase phosphatidylinositol levels and intracellular calcium movement, activating the protein kinase C (PKC) pathway important for interleukin (IL)-2 expression [132]. These superantigens also activate the protein tyrosine kinase (PTK) pathway and transcriptional factors like NF-κβ and

AP-1, resulting in elevated expression of proinflammatory cytokines [131,133–135]. The biological effects of superantigens are induced at low, nonsaturating occupancy rates that indicate "low affinity" binding to MHC class II most relevant for T-cell activation.

Human whole blood and purified peripheral blood mononuclear cells (PBMCs) are commonly used *in vitro* to study cell activation by staphylococcal superantigens, as well as potential therapeutic agents against these toxins [136–143]. PBMCs secrete a number of proteins following SE, TSST-1, or SPE exposure and include: IL-1, IL-2, IL-6, tumor necrosis factor (TNF)-α, interferon (IFN)-γ, macrophage inflammatory protein (MIP)-1α, MIP-1β, and monocyte chemoattractant protein (MCP)-1. Although monocytes alone can produce many chemokines as well as proinflammatory cytokines like IL-1, IL-6, and TNF-α, T cells enhance these mediator levels [136,144]. In the literature, there are contradictory reports regarding APC and T-cell responses to these bacterial toxins without the other cell type, as evidenced by cytokine/chemokine production by human monocytic lines or fresh isolates devoid of T cells [145]. However, others have found that monocyte-derived IL-1 and TNF-α following SEA exposure is strictly a T-cell dependent event [146]. Purified human T cells increase mRNA expression of several cytokines after superantigen exposure without APCs, but secretion of these cytokines and T-cell proliferation is dependent on MHC class II-bearing cells [147]. MHC class II-linked stimulation of T cells by the SEs is a general requirement, but those cells possessing certain TCR Vβs can independently respond with less efficiency [148]. However, presentation of superantigen to T cells without MHC class II also induces anergy [149].

Additional cell types that respond to superantigens include B, nasal, intestinal, and vaginal epithelial cells, as well as intestinal fibroblasts and synovial myofibroblasts [150,151]. The cross-linking of TCR with MHC class II by superantigen triggers B-cell proliferation and differentiation into immunoglobulin-producing cells in a dose-dependent manner, but high concentrations of superantigen inhibit immunoglobulin production [152]. The CD28 costimulatory pathway plays a prominent role in superantigen-induced differentiation of B cells. Suppression of immunoglobulin secretion by TSST-1 reportedly occurs via apoptosis [47], which can clearly hamper development of protective immunity against this toxin [6]. Such an effect on B cells is likely linked to recurring susceptibility of TSS among patients [43–45]. In addition to B cells, the direct stimulation of synovial fibroblasts by superantigens also induces expression of chemokine genes. This raises an important issue of autoreactivity and chemotactic responses that initiate, or augment, a chronic inflammatory process like arthritis [153,154]. Upon activation by SEB, nasal epithelial cells produce granulocyte colony-stimulating factor and various chemokines including MCP-1 [151]. Transcytosis of SEB across intestinal epithelial cells has been observed *in vitro* [155] and *in vivo* the toxin penetrates the gut lining, evoking local and systemic immune responses [156]. The binding of a dodecapeptide (YNKKKATVQELD), conserved in superantigens, to vaginal epithelial cells effectively competes with TSST-1 for cell-surface receptor [157]. In response to SEA, but not SEB, human intestinal myofibroblasts elicit IL-6, IL-8, and MCP-1 [158].

17.5 *IN VIVO* EFFECTS OF SEs, TSST-1, AND SPEs

17.5.1 Nonhuman Primates: The Historical "Gold Standard" Animal Model

The SEs readily induce an emetic response in primates when ingested in low microgram quantities, and pending the dose, there may be more severe consequences that progress into toxic shock [19,159,160]. Additionally, the consumption of *S. aureus* with only cell-associated SEA can also cause emesis in nonhuman primates [161]. For many years, the classic primate studies for SEs have been done by various groups and are considered a "gold standard" for *in vivo* work. However, these experiments have become increasingly more expensive, politically sensitive (i.e., animal rights issues), and thus difficult to implement which altogether fuels the need for alternative animal models. In contrast to the SEs, TSST-1 does not elicit emesis in nonhuman primates after ingestion but it naturally causes TSS in humans and animals via *S. aureus* growth on mucosal surfaces [2,35,40].

Unlike many other bacterial enterotoxins, specific cells and receptors in the intestinal tract have not been clearly associated with SE intoxication. The latter seemingly requires a complex interplay between immunological and nonimmunological mechanisms involving multiple cell types. SEB stimulation of mast cells and subsequent release of cysteinyl leukotrienes elicits emesis and skin reactions in primates [25,150,162]. Oral administration of SEB induces activation and expansion of murine Vβ8 T-cells in Peyer's patches accompanied by increased IFN-γ and IL-2 mRNA expression [163], which might contribute to the intestinal effects of SEs. When given intrarectally to mice, SEA or SEB elicits an inflammatory intestinal response and exacerbates a pre-existing, microbial-based syndrome called inflammatory bowel disease that further suggests an immune-based response by animals [164]. An enteric immune link may also explain earlier results by Sugiyama et al. [165] showing that nonhuman primates, when orally administered a specific SE, become transiently resistant to a subsequent higher dose of homologous, but not heterologous, SE. The immunologically based results from studies within the intestine are likely connected to toxin-specific stimulation of unique Vβ-bearing T cells.

In addition to toxin-specific resistance elicited by a single oral dose of SE, chronic intravenous exposure to SEA can virtually delete all Vβ-reactive T cells in mice, therefore, representing a potentially dangerous scenario for the host [166]. Increased frequency of FoxP3 CD4 regulatory T cells has been observed in TCR transgenic mice repeatedly exposed to SEB [167]. Another study suggests that regulatory T cells become highly suppressive and nonproliferating after repeated SEB exposure *in vivo* [168]. Preactivation of endogenous regulatory T cells with a superantigen *in vivo* enhance the suppressive potency of regulatory T cells [169]. Footpad injections of SEB in mice also elicit a dose-related tolerance toward SEB among Vβ8 T-cells, with a low toxin concentration imparting a transient effect versus that more lasting following a high dose [170]. Another study shows that mice intranasally administered SEA (1 μg every week for 3 weeks), but not a recombinant SEA lacking superantigenicity, become resistant to a subsequent lethal challenge of SEA but not TSST-1 [171]. This "tolerant" state is evidently not due to toxin-specific antibody or deletion/anergy of SEA-reactive T cells. However, a significant increase in serum IL-10 levels among these animals correlates with previous *in vitro* and *in vivo* findings revealing protection against SE-induced effects [137,172,173]. Table 17.2 lists animal models used to study the enteric effects of SEs.

Studies with human Caco-2 monolayers reveal transcytosis of SEA, SEB, as well as TSST-1, and ingested SEB enters the bloodstream of mice more readily than SEA [155]. These data suggest that SEs cross the gastric mucosa and circulate throughout the body. *In vitro*, the SEs are not cytotoxins that directly disrupt human intestinal cells [174]. However, SEB affects the gut mucosa as evidenced by increased ion permeability in human T84 colonic cell monolayers incubated with SEB-stimulated PBMCs [175]. The interactions of most superantigens with epithelial and endothelial cells are indirect via release of IL-1, TNF-α, and IFN-γ from superantigen-activated APCs and T cells [176,177].

Although a very debatable topic, it appears that binding to MHC class II may not play a role in SE enteric effects as recombinant variants of SEA (L48G) and SEB (F44S) devoid of MHC binding and T-cell mitogenic properties remain emetic [178]. The disulfide loop of various SEs, not present in TSST-1, is implicated in emesis, but that too remains equivocal [179,180]. Carboxymethylation or tyrosine replacement of histidines on SEA [181] and SEB [182] generates molecules that remain superantigenic, but devoid of: (1) enterotoxicity; (2) lethal effects; and (3) skin reactivity

TABLE 17.2
Enteric and Other *In Vivo* Models for Bacterial Superantigen-Induced Diseases

Animal	Inducing Agents	Route	Mediators, Symptoms, Pathology
		Mouse	
BALB/c	SEB	i.g.	Activation and expansion of Vβ8 T cells at 4 h, deletion 7–10 days [166]; IFN-γ and IL-2 increase in mucosal lymphoid tissue at 4 h [163]
C57BL/6	SEB	i.p.	Acute lung inflammation, leukocyte infiltration, capillary leakage, and endothelial cell injury by 6 h [192]
BALB/c-SCID	SEA or SEB	i.r.	Exacerbation of IBD [164]
		Monkey	
Cynomolgus	SEB	i.g.	Immediate-type skin reaction, emesis, and biphasic cysteinyl leukotriene generation at 1 and 3 h [162]
	SEA, SEB, or SEC1	i.g. or i.v.	Emesis at 3 h, followed by diarrhea [19]
	SEB	i.d.	Immediate-type skin hypersensitivity, cutaneous mast cell degranulation, and emesis [150]

Note: IBD, inflammatory bowel disease; i.d., intradermal; IFN, interferon; i.g., intragastric; IL, interleukin; i.p., intraperitoneal; i.r., intrarectal; i.v., intravenous.

[162,183,184]. Chemically modified SEB also inhibits the emetic/diarrheic effects of wild-type SEB in nonhuman primates when given concomitantly, suggesting competition for a common receptor(s) [183]. Lack of enterotoxicity with carboxymethyl-modified SEA is not due to an altered conformation and/or increased susceptibility to degradation by gastric proteases [184]. Analysis of each histidine regarding SEA-induced emesis and superantigenicity reveals that H61 is important for the former, but not latter, property and thus further demonstrates that emesis and superantigenicity are distinct molecular properties [184]. Modification of H44, H50, H114, or H187 on SEA generates variants that retain both emetic and superantigenic properties. Antibodies against a peptide region of SEA (121–180), lacking the disulfide loop (C91–C105) and histidines, prevent SEA-induced emesis by steric hindrance of toxin with an ill-defined receptor(s) in the intestinal tract [185].

As stated before, affinity for MHC class II molecules and specific TCR Vβ enables superantigenic toxins to perturb the immune system and induce high levels of proinflammatory cytokines [1,139–146,186,187]. The SEs, TSST-1, and SPEs are pyrogenic in primates as well as rabbits [31,37,188–190], a likely result of elevated proinflammatory cytokine levels that include the synergistic acting IL-1 and TNF-α from PBMCs [190,191]. Both of these cytokines are endogenous pyrogens that induce fever via the hypothalamus [191]. In addition, the circulating levels of other cytokines such as IFN-γ, IL-2, and IL-6 also increase after toxin exposure. IFN-γ augments immunological responses by increasing: (1) the expression of MHC class II by APCs and epithelial/endothelial cells; (2) the proinflammatory actions of IL-1 and TNF-α. Systemic administration of SEB causes acute lung injury characterized by increased: (1) expression of adhesion molecules like intercellular adhesion molecule-1 and vascular cell adhesion molecule; (2) neutrophil and mononuclear cell infiltration; (3) endothelial cell injury; and (4) vascular permeability [192]. Intranasal delivery of SEB induces a prolonged lung injury still evident even after 3–4 days of steroid treatment [193]. Exudates from superantigen-injected air pouches predominantly contain neutrophils with some macrophages [194]. Endothelial cells surrounding air pouches express intercellular adhesion molecule-1, TNF-α, MIP-2 (an IL-8-related protein in mice), and other chemokines. Superantigenic shock results from various biological effects elicited by proinflammatory cytokines that, when present in high levels, adversely affect different organs like the lungs [192].

17.5.2 Mice: Not Perfect but Often the Preferred Model

In addition to nonhuman primates, mice have historically been used by various groups as an alternative model for studying superantigen-mediated shock [186,195–201]. From a cost perspective, it is very feasible to use mice for basic toxin studies involving the discovery of potential vaccines and therapeutics. However, these animals lack an emetic response and are thus less appropriate for studying food poisoning aspects of the SEs. Additionally, mice are naturally less susceptible than primates to SEs, TSST-1, and SPEs because of decreased toxin affinity for MHC class II [4,200]. To overcome this last caveat, potentiating agents such as D-galactosamine, actinomycin D, lipopolysaccharide (LPS), viruses, or *Trypanosoma cruzi* have been used by various groups [186,195–198,200–204]. These agents amplify superantigen toxicity in mice so that practical, lower amounts elicit biological effects that include toxic shock.

Many of our *in vivo* endeavors with SEs and TSST-1 have been accomplished by an LPS-potentiated mouse model with a lethal endpoint, as it has been well established by many laboratories through various *in vitro* and *in vivo* studies that a natural synergy exists between these bacterial exotoxins and LPS [195,196,200–203,205–209]. As little as 2 μg of LPS in humans causes endotoxic shock [210]. Because bacterial superantigens like the SEs, TSST-1, and SPEA synergistically augment the effects of LPS many log-fold, only picogram quantities of LPS in conjunction with a superantigen can cause severe effects [188,211]. Upon considering basic microbiology and the number of Gram-negative bacteria in normal intestinal flora, along with a recognized increase in these microbes among TSS patients, the odds of this superantigen-LPS synergy naturally occurring are rather high [188,212]. All of these studies reveal a correlation between elevated serum levels of various proinflammatory cytokines (IL-1, IL-2, TNF-α, and/or IFN-γ) with SEA-, SEB-, or TSST-1-induced shock [1,48,66,186,196]. A recent study analyzed the interdependent effects of SEB used alone, and together with LPS (*Escherichia coli* O55:B5), on serum levels of cytokines/chemokines in BALB/c mice commonly used in LPS-potentiated shock models [211]. *In vivo*, SEB alone induces only moderate levels of IL-2 and MCP-1 with all mice surviving a high dose (100 μg/animal). In these same studies, LPS (80 μg/mouse) alone causes 48% lethality and induces high levels of IL-6 and MCP-1. SEB induces low levels of TNF-α, IL-1, IFN-γ, and MIP-2 but LPS addition increases expression of these cytokines, as well as IL-6 and MCP-1. Importantly, the synergistic action of SEB and LPS results in lethal shock and hypothermia not evident with SEB only or low LPS doses (1 or 10 μg). Investigation of cytokine serum levels with survival status in the SEB plus LPS group reveals significantly higher TNF-α, IL-6, MIP-2, and MCP-1 concentrations in nonsurvivors early after SEB administration. In addition to these cytokines and chemokines, significantly higher serum concentrations of IFN-γ and IL-2 are observed at later times in nonsurvivors of toxic shock as compared with survivors. Thus, the synergistic action of SEB and LPS promotes early TNF-α release and prolongs IL-6, IFN-γ, IL-2, MIP-2, and MCP-1 release in nonsurvivors. Overall, the elevated prolonged levels of these key cytokines lead to lethal toxic shock within 48 h when LPS is used with SEB. Although there is a general agreement that Th1 cytokines, typified by IFN-γ, are important in these potentiated models of SEB-induced shock, the role of Th2 cytokines cannot be overlooked. For instance IL-10, a prototypic Th2 cytokine, is detected *in vivo* after repeated superantigen stimulation [173,213,214]. IL-10-deficient mice have increased serum levels of IL-2, IFN-γ, plus TNF-α after SEB stimulation, and become more susceptible to lethal shock [213]. Additionally, these efforts correlate nicely with others using SEA and genetic knockouts lacking IFN-γ, IL – 2 or a TNF receptor, p55 [172,215]. Table 17.3 summarizes results for SE or TSST-1 intoxication in knockout mice deficient in IL-10, TNF receptor type I (TNF-RI or p55) or type II (TNF-RII or p75), CD28, CD54, CD43, CD95, or perforin [172,216–225].

Transgenic mice expressing human HLA-DQ6 and CD4 succumb to normally sublethal amounts of SEB (with D-galactosamine potentiation), and the serum levels of TNF-α correlate with onset of lethal shock [226]. Transgenic mice expressing

TABLE 17.3
Effects of Specific Genes on Susceptibility of Knockout Mice to Staphylococcal Superantigens

Targeted Gene	Effect on Susceptibility to Superantigen-Induced Shock
IL-10	Increased susceptibility to SEA or SEB-induced shock, and higher serum levels of TNF-α, IL-1, IL-2, IL-6, IL-12, MIP-1α, MIP-2, and IFN-γ [172,217]
TNF-RI	Protection against SEA or SEB-induced shock [172,218]
TNF-RII	Slightly decreased susceptibility to SEA-mediated shock [172]
CD28	Protection against TSST-1-induced TSS [219]
	Protection against lethal toxic shock induced by second injection of SEB, and decreased serum levels of TNF-α [220]
CD54	Protection against SEB-induced shock in D-galactosamine-sensitized mice [221]
CD43	Increased T-cell proliferation *in vitro*, and enhanced homotypic adhesion by SEB [222]
CD95	In MRL-lpr/lpr mice, increased susceptibility to SEB-induced shock [223]
Perforin	Decreased lysis of MHC class II-positive APC by SEA-activated CD8 T cells [224]

Source: With kind permission from Springer Science + Business Media: *Immunol. Res.*, Immune response to staphylococcal superantigens, 20, 1999, 163, Krakauer, T.

Note: APC, antigen presenting cell; IFN, interferon; IL, interleukin; MHC, major histocompatibility complex; MIP, macrophage inflammatory protein; SEA/SEB, staphylococcal enterotoxin A/B; TNF, tumor necrosis factor; TSST, toxic shock syndrome toxin.

human HLA-DR3 and CD4 lethally respond to SEs without a potentiating agent, thus providing a "simpler" model for future *in vivo* toxin studies [227]. PBMCs isolated from these animals and then incubated with SEB markedly produce IL-6 and IFN-γ, as compared with those from BALB/c mice, thus suggesting that proinflammatory cytokines also play a key role in this murine shock model. Similar studies have also been done with SPEA and mice expressing human HLA-DQ8 and CD4 [228]. In HLA-DQ8 CD4 transgenic mice, aerosolized SEB elevates serum IFN-γ, IL-2, and IL-6 levels but not TNF-α [229]. Following toxin exposure, the lung lesions in these transgenics, temperature fluctuations, and lethality starting at ≥96 h are similar to those in nonhuman primates exposed to a lethal aerosol dose of SEB. Other investigations [230] suggest that two high doses of SEB (30 and 100 μg/mouse) are necessary to induce toxic shock in these transgenic mice that still includes a sensitizing agent (D-galactosamine) [226]. Other transgenics that over-express murine TCR Vβ3 also have increased mortality linked to elevated TNF and IFN-γ levels following infection by SEA-producing *S. aureus* [231]. Clearly, genetically manipulated mice that express human HLA and CD4, or those possessing increased levels of specific murine TCR, will no doubt benefit future endeavors in this field and provide a clearer understanding of superantigen-mediated toxicity. It is evident that various mouse models exist for the staphylococcal and streptococcal superantigens, as shown in Table 17.4.

In addition to lethality as an endpoint, temperature has been used for studying SE- and TSST-1-induced shock in LPS-potentiated mice. These studies were

TABLE 17.4
Toxic Shock Models for Bacterial Superantigen-Induced Effects

Animal	Inducing Agents	Route	Mediators, Symptoms, Pathology
			Mouse
BALB/c	TSST-1 + LPS	i.v.	TNF-α peaks at 1–2 h, lethal shock [197, 200]
	SEB + LPS	i.p.	TNF-α peaks at 1 h; IFN-γ, IL-1, and IL-6 increase at 2 h; lethal shock [196], hypothermia [172]
	SEB + LPS	SEB oral/LPS i.p.	TNF-α, IFN-γ, IL-1, and IL-6 increase at 6 h, lung injury, lethal shock [208]
	D-galactosamine + SEB	i.p.	High levels of TNF-α, IFN-γ, and IL-2 by 2 h, lethal shock [186,207], hepatic necrosis and diffuse hyperemia [207], gut epithelial cell apoptosis, reduction in goblet cells [218]
	Actinomycin D + SEB	i.p.	Blood congestion in lungs and intestine by 4 h, PPMCs in lungs, spleen, and liver, alveolar septa thickening at 8 h, lethal shock at 2–4 days [197]
C3H/HEJ	SEB + SEB	i.n.; i.n. + i.p.	Bronchiolar epithelial degeneration, lung neutrophilic infiltration, IL-2, IL-6 and MCP-1 in serum and lung, lethal shock at 96 h [235]
Transgenic DR3	SEB	i.n.	Neutrophilic infiltration, TNF-α, IFN-γ, IL-6, IL-12, and MCP-1 increase at 3 h [239]
Transgenic DQ8	SPEA	i.p.	TNF-α, IFN-γ, and IL-6 increase at 4 h, lethal shock at 60 h [228]
			Rat
Sprague-Dawley	Catheterized; SEB + LPS	i.v.	TNF-α increases at 1.5 h, IFN-γ at 4 h, hepatic injury and dysfunction [205,206]
			Rabbit
Dutch belted	TSST-1 + LPS	i.v.	TNF-α peaks at 4 h, lethal shock [200]
	SEC + LPS		Fever at 4 h, hypothermia, labored breathing, diarrhea, vascular collapse, lethality by 24 h [209]
New Zealand white	SEA	i.v.	TNF-α, IFN-γ and IL-2 increase at 1–2 h, peak at 3–5 h, febrile reaction evident at 1 h [248]
			Monkey
Rhesus	SEB	Aerosol	Leukocyte infiltration, intra-alveolar edema, parenchymal cell degeneration, lymphocyte necrosis, and lethal shock [26]

Note: IFN, interferon; IL, interleukin; i.n., intranasal; i.p., intraperitoneal; i.v., intravenous; LPS, lipopolysaccharide; MCP, monocyte chemoattractant protein; PPMC, pulmonary parenchymal mononuclear cell; SEA/SEB/SEC, staphylococcal enterotoxin A/B/C; SPEA, streptococcal pyrogenic exotoxin A; TNF, tumor necrosis factor; TSST, toxic shock syndrome toxin.

accomplished by implanting a subcutaneous transponder [172] or intraperitoneal telemetry device [232], in which the latter also measures movement. Results from these investigations reveal a rapid (within 10 h) temperature decrease readily evident among intoxicated mice, thus providing a rapid nonlethal model. Interestingly, none of these studies detected a temperature increase like that evident with nonhuman primates [233], thus suggesting a very rapid onset of shock in this murine model.

Recently, intranasal administration of SEB has been used in lethal murine models [234,235]. A high (72 μg) intranasal dose of SEB is apparently lethal in C3H/HeJ mice, a TLR4-defective strain, but the mechanism of intoxication remains unclear [234]. However, in contrast, a different group reveals that this SEB dose does not mediate shock in these animals, although two low doses (5 and 2 μg) can be lethal [235]. This two-hit (or dual-dosing) model requires that SEB is given 2 h apart with the first dose delivered intranasally and the subsequent dose administered either intranasally or intraperitoneally. Increased serum levels of IL-2, IL-6, and MCP-1, accompanied by an early presence of elevated lung MCP-1 concentrations, are evident in this dual-dosing model [235]. MCP-1, a potent activator and chemotactic factor for T cells as well as monocytes, probably contributes to early leukocyte recruitment into the lung. Pathological lesions, temperature fluctuations, and time course of lethality also resemble those in transgenic mice, nonhuman primates, and humans [229,236]. Intranasal administration of SEA to C57BL/6 mice causes accumulation of CD8 Vβ3 T-cells and CD11c MHC class II cells in the lungs mediated by IFN-γ [237]. Other studies indicate that airway inflammation is dependent on superantigen dose [238,239]. A low dose (20 ng) of intranasal SEB induces airway inflammation and eosinophil degranulation, whereas higher concentrations (2000 ng/mouse) result in neutrophilic airway inflammation, permanent airway destruction, toxic shock, and mortality in HLA-DR3 transgenics [238,239].

Injection of SEB into mice induces apoptosis and T-cell anergy, which is likely linked to a rapid (within 1 h) loss of L-selectin on specific Vβ-bearing T cells and signal transduction [240,241]. Others have discovered that, through endocytosis, surface levels of TCR-CD3 decrease ~50% among Vβ-reactive T cells within just 30 min after SEB exposure [242]. The rapid hyperactivation and proliferation of T cells in mice after an SEB injection is transient, as within 48 h most proliferating T cells are eliminated by activation-induced cell death [67,243]. These effects can render an animal incapable of primary immune responses against other antigens (perhaps a microbial pathogen?), even if Freund's adjuvant is given 3 days after SEB exposure [199]. The CD95 receptor for adhesion plays an important role in eliminating activated T cells whereas those remaining are functionally unresponsive and essentially "dead." After injection of SEB into mice, splenic Vβ8 T-cells are physically deleted or nonresponsive (a.k.a. anergic) to homologous toxin and produce less IL-2 and IFN-γ [243]. In contrast, others report that these anergic cells can secrete more IFN-γ which then mediates toxic shock following a subsequent dose of SEB [244]. An evident paradox is that an anti-inflammatory cytokine like IL-10, which protects against SE-induced shock [172], is also produced by SEB-primed T cells [244]. This effect perhaps reflects an attempt, a feeble one in TSS cases, by the host to counter proinflammatory IFN-γ. It is likely that SEB-induced anergy differentially affects CD4 and CD8 T-cells, with the former becoming more susceptible [244]. This may also explain why cytotoxic

CD8, not CD4, T-cells are activated by superantigens that represent potential antitumor reagents perhaps clinically useful in the future [245].

17.5.3 Rabbit Models

In addition to mice, rabbits have also afforded a reliable *in vivo* model for SE-, SPE- or TSST-1-induced shock as determined by temperature and lethal endpoints [200,209,246–253]. Some of these models for TSST-1 and SPEA use an implanted infusion pump that delivers toxin over time, thus mimicking more naturally an infection, toxin release, and subsequent TSS [246,249]. As evidenced in mice with the various staphylococcal and streptococcal superantigens, different rabbit strains also possess varying susceptibility toward TSST-1, as seen in New Zealand white rabbits being more susceptible than Dutch belted rabbits [251]. As witnessed in humans with TSS, rabbits given TSST-1 or SEB experience elevated levels of circulating LPS eliminated by, along with the clinical signs of TSS, polymyxin B [203,251,252]. Increased levels of circulating LPS may be due to impaired liver clearance induced by these protein toxins [209,253], and further liver damage/dysfunction is evident with both LPS and superantigen in a rat model [205,206].

17.5.4 Ferret and Shrew Models

In addition to nonhuman primates, mice, rats, and rabbits, other less-well-defined models for SE intoxication have been described in the literature. For example, goats have been used for studying *in vivo* effects (fever) of TSST-1 and SEB after intravenous administration [254]. There is also a ferret model for oral SEB intoxication, which elicits emesis and rapid fever [255]; however, this latter model uses milligram, and not microgram, quantities of toxin used in either murine or nonhuman primate models.

Finally, another emetic model more recently described for the SEs is that using a rather unusual laboratory animal: the house musk shrew [256]. Much lower amounts of SE are required in the shrew model, as compared with the ferret model, via intraperitoneal or oral routes of intoxication. However, an obvious and less-than-pleasant caveat with any emetic (or diarrheic) model is quantitation, as per volume (which requires collection and measurement) and/or number of events.

Basic aspects of intoxication have been investigated in each animal model listed above, but it is clear that additional work must be done in the future with regard to their use for vaccine and therapeutic discovery. Clearly, studies involving the SEs, TSST-1, and SPEs are not lacking for available animal models that can answer some (but not all!) questions linked to biological effects of these superantigens.

17.6 THERAPEUTICS AND VACCINES AGAINST SEs, TSST-1, AND SPEs

To date, there are neither small-molecular weight therapeutics nor vaccines against SEs, TSST-1, or SPEs approved for human use by the United States Food and Drug Administration (FDA). Upon understanding the intoxication process, potential therapies/vaccines toward these toxins should target at least one of three important steps: (1) TCR–toxin–MHC class II interactions; (2) accessory, costimulatory, or adhesion molecules involved in activation of T cells; and (3) cytokine release by activated T cells and APCs [216]. *In vitro* and *in vivo* inhibition of the above targets has been reported by various groups. For example, steroids and IL-10 were investigated as possible agents for inhibiting proinflammatory cytokines and T-cell proliferation after TSST-1 stimulation of human PBMCs *in vitro* [137]. Arad et al. [257] discovered that a conserved region (residues 150–161: YNKKKATVQELD) from SEB prevents SEB-, as well as SEA-, TSST-1-, or SPEA-, induced lethal shock in mice when given 30 min following toxin. This peptide, not located within the classically defined MHC class II or TCR-binding domains, prevents transcytosis of various SEs and TSST-1 across a human colonic cell (T84) monolayer and may block costimulatory signaling necessary for T-cell activation [155]. However, a subsequent study indicates that these peptides were ineffective inhibitors of SEB-induced effects both *in vitro* and *in vivo* [258]. Another recent study with a different peptide (designated as P72: DLADKYKDKYVDVFG), which does not bind MHC class II, reveals inhibition of SEA-, SEB-, and SEC-mediated responses [259]. A different approach for blocking receptor interactions of SEB uses a bispecific chimeric inhibitor composed of the DRα1 domain of MHC class II and TCR Vβ connected by a flexible $(GSTAPPA)_2$ linker [260]. This chimera prevents cellular activation and IL-2 release in SEB-stimulated PBMCs. The drawback of this approach is that individual chimeras must be constructed for each SE, as TCR Vβ preferences differ.

Blockade of SEB signal pathways in a host cell represents a different mode of intervention, as these transduction events occur after exposure and will likely work for other SEs. Nuclear factor kappa B (NFkB) is an attractive therapeutic target as its activation leads to the inducible expression of many mediators in inflammatory diseases. *In vitro* and *in vivo* studies have shown that many of the inflammation-associated genes implicated in superantigen-induced lethal shock contain NFκB-binding sites in the promoter/enhancer region [261,262]. A cell-penetrating cyclic peptide (cSN50: AAVALLPAVLLALLAPCYVQRKRQKLMPC) targeting NFκB nuclear transport attenuates SEB-induced T-cell responses and diminishes serum inflammatory cytokine levels [263]. Liver apoptosis, hemorrhagic necrosis, and mortality are also reduced in mice given cSN50 before D-galactosamine [264]. Intraperitoneal administration of cSN50 30 min prior to an intranasal SEB dose in BALB/c mice reduces proinflammatory cytokines and chemokines in the bronchoalveolar space, as well as attenuates neutrophil and monocyte infiltration into the lungs and vascular injury [264]. Bortezomib, a dipeptidyl boronic acid (384 Da) that inhibits NFκB, decreases SEB-induced serum cytokine/chemokine levels but has no effect on mortality and liver necrosis *in vivo* [265]. Another potent NFκB inhibitor is dexamethasone, a well-known immunosuppressive drug used clinically to treat various inflammatory diseases. *In vitro*, dexamethasone potently inhibits staphylococcal exotoxin-induced T-cell proliferation, cytokine release, and activation markers in human PBMCs [266]. *In vivo*, dexamethasone also significantly reduces serum levels of cytokines and protects mice from SEB-induced shock in the two-hit SEB-only, as well as SEB plus LPS, models [196,267]. Furthermore, dexamethasone (1.25–5 mg/kg) attenuates the hypothermic response to SEB in both toxic shock models and improves survival of mice by 100% even when administered 2–3 h after SEB.

17.6.1 Efficacy Studies: Characterized Animal Models are Essential

Several *in vivo* models have been used to study potential therapies that prevent superantigen-induced shock. Therapeutic agents such as nitric oxide inhibitors decrease SEA and SEB effects by inhibiting IL-1, IL-2, IL-6, TNF, and IFN-γ production [268,269]. Blockade of the CD28 costimulatory receptor by its synthetic ligand, CTLA4 immunoglobulin, prevents TSST-1-induced proliferation of T cells *in vitro* and lethal TSS *in vivo* [270]. Neutralizing antibodies against TNF-α also prevent SEB-induced lethality [186], and IL-10 blocks production of various cytokines like IL-1, TNF-α, as well as IFN-γ with a resultant reduction in lethality from superantigen-induced toxic shock [173]. A novel nasal application of SEA in mice induces subsequent tolerance toward SEA, but not TSST-1 [171]. This phenomenon is evidently linked to increased serum levels of IL-10, but not depletion of SEA reactive T-cells or development of toxin-specific antibodies. Anti-inflammatory agents such as indomethacin, dexamethasone, and the antipyretic acetaminophen also effectively lower the febrile response in rabbits given SEA by diminishing serum concentrations of IL-1, IL-6, TNF-α, and IFN-γ [250]. Studies with human PBMCs *in vitro* and an LPS-potentiated mouse model show that either pentoxifylline or pirfenidone lower proinflammatory cytokine expression, thus abrogating the ill effects of SEB or TSST-1 [140,141]. Another group has shown that IFN-γ production by SEB-stimulated lymphocytes from Peyer's patches significantly decreases after oral administration of tryptanthrin, an anti-inflammatory compound derived from a medicinal plant found throughout Asia [271]. SEA also causes lung inflammation through IL-8 produced by alveolar macrophages, but a hexapeptide inhibitor (RRWWCR) of this cytokine decreases neutrophil influx into this organ in a rabbit model by preventing IL-8 binding to neutrophils [272]. The release of proinflammatory cytokines due to SEB or TSST-1 stimulation is also diminished in mice by soluble β-glucans [273]. However, this mechanism is not well characterized to date. More recently, another FDA-approved immunosuppressive drug (rapamycin) was shown to protect SEB-induced shock even when administered 24 h after SEB [274]. Rapamycin inhibits proinflammatory cytokines and T-cell proliferation *in vitro* that likely impacts other T-cell signaling pathways induced by SEB [274].

In addition to therapeutics, various groups have also developed different experimental vaccines for the staphylococcal and streptococcal superantigens. This approach for protection is logical, as pre-existing antibodies against the SEs, TSST-1, and SPEs clearly play an important role in disease outcome [43,44,275], and the use of intravenous immunoglobulin has also proven useful in humans following the onset of TSS [276–278]. Experimentally, passive transfer of SEB-specific antibodies to naïve rhesus monkeys upto 4 h after a SEB aerosol also prevents lethal shock [279]. Recombinantly attenuated mutants of SEA, SEB, TSST-1, SPEA, and SPEC that do not bind MHC class II and/or specific Vβ TCR molecules represent successful experimental vaccines for preventing toxic shock in different animal models [48–55,280–283]. When given either parenterally or mucosally, these vaccines do not cause ill effects and are efficacious against a toxin challenge or *S. aureus* infection (Table 17.5). Other murine and nonhuman primate studies have used formaldehyde toxoids of SEA, SEB, or SEC1 as effective immunogens that protect against a

TABLE 17.5
Vaccine Studies for Bacterial Superantigens

Animal	Immunogen/Adjuvant	Route	Results
			Mouse
BALB/c and CD1	SEB formaldehyde toxoid in proteosomes/aluminum hydroxide used for i.m. route only	i.n., i.m.	The i.n. or i.m. vaccinations yielded 53%–60% protection (BALB/c) toward a lethal SEB challenge (i.m.). An i.m. prime with i.n. boost yielded 80% protection. CD1 mice (i.m. vaccinated) were 100% protected toward an i.m. challenge [287]
BALB/c	SEB (N23K or F44S mutants)/aluminum hydroxide	i.p.	80% protection against 30 LD_{50} SEB challenge (i.p.) among vaccinated animals versus 7% protection for adjuvant-only controls. Sera from vaccinated mice protected naïve animals against lethal SEB challenge [280]
	SEB (L45R,Y89A,Y94A triple mutant)/aluminum hydroxide (i.p. route) or cholera toxin (i.n. and oral routes)	i.p., i.n., oral	Among i.p./i.n. vaccinated mice, there was 100% protection against either an 8 LD_{50} (aerosol) or 30 LD_{50} (i.p.) SEB challenge. Oral vaccination yielded 38% and 75% protection rates toward an i.p. or aerosol challenge, respectively. Only 0%–10% of adjuvant-only controls were protected against either SEB challenge [49]
	TSST-1 (H135A mutant)/aluminum hydroxide	s.c.	Lethal *S. aureus* (i.v.) challenge resulted in 0% survival among adjuvant-only controls versus 60% protection for H135A-vaccinated animals [56]
	TSST-1 (H135A mutant)/RIBI	i.p.	Among the H135A-vaccinated animals, 67% were protected against a 15 LD_{50} challenge (i.p.) of TSST-1 versus 8% for adjuvant-only controls [48]
C3H/HeJ	SEB (H12Y, H32Y, H105Y, H121Y mutant)/aluminum hydroxide	i.p., i.n.	100% of vaccinated mice were protected against an i.p. or i.n. lethal SEB challenge versus 20% survival of BSA-vaccinated controls [281]
NMRI	SEA (L48R,Y92A, D70R triple mutant)/Freund's	s.c.	Vaccinated mice challenged with *S. aureus* (i.v.) had a delayed time to death and decreased weight loss versus BSA-vaccinated controls. Hyperimmune serum protected naïve animals [50]
Transgenic for human HLA-DR3 and CD4	SEB (L45R, Y89A, Y94A triple mutant)/RIBI	i.p.	100% protection against a 10 μg SEB challenge (i.p.) and markedly decreased IFN/IL-6 levels in immunized versus adjuvant-control, animals [227]

continued

TABLE 17.5 (continued)
Vaccine Studies for Bacterial Superantigens

Animal	Immunogen/Adjuvant	Route	Results
			Rabbit
New Zealand white	TSST-1 (H135A mutant)/Freund's	s.c.	100% protection against lethal *S. aureus* (s.c.) challenge [282]
	TSST-1 (G31R, H135A mutant) or formaldehyde toxoid of wild type/ aluminum hydroxide	s.c.	Mutant or toxoid afforded 100% protection toward a lethal TSST-1 challenge (i.v.) versus 0% amongst controls [54]
Dutch belted	SPEC (Y15A, N38D and Y15A, H35A, N38D mutants)/Freund's	s.c.?	100% protection from lethal SPEC challenge (miniosmotic pump, s.c.) versus 0% among controls [52]
	SPEA (N20D, C98S, N20D, D45N, C98S, Q19H, N20D, L41A, L42A, D45N, C98S mutants)/Freund's	s.c.	All mutants were 100% protective against lethal SPEA challenge (miniosmotic pump, s.c.) and fever, unlike controls respectively experiencing 0% and 10% protection [53]
			Monkey
Rhesus	SEB (L45R, Y89A, Y94A triple mutant)/aluminum hydroxide	i.m.	Depending on the dose (5 vs. 20 μg) and injections (2 vs. 3), there was 60%–100% protection against lethality toward a 75 LD_{50} (aerosol) dose of SEB [287]. 0% survival for adjuvant-only controls. A 20 μg dose given 3 times protected against SEB-induced hyperthermia, unlike adjuvant-only controls [233]
	SEB formaldehyde toxoid in proteosomes/aluminum hydroxide used for i.m. route only	i.m., i.m. + i.t.	100% protection against 15 LD_{50} SEB challenge (aerosol) in both groups versus 0% survival among controls [286]
			Piglets
Crossbred	SEB (L45R, Y89A)/cholera toxin	Oral	No ill effects with vaccine. Toxin-specific serum IgG and fecal IgA detected but cholera toxin did not enhance antibody response. No efficacy challenge results [284]

Note: BSA, bovine serum albumin; IFN, interferon; IgA/IgG, immunoglobulin A/G; IL, interleukin; i.m., intramuscular; i.n., intranasal; i.p., intraperitoneal; i.t., intratracheal; i.v., intravenous; LD_{50}, mean lethal dose; *S. aureus*, *Staphylococcus aureus*; SEA/SEB, staphylococcal enterotoxin A/B/; s.c., subcutaneous; SPEA/SPEC, streptococcal pyrogenic exotoxin A/C; TSST, toxic shock syndrome toxin.

homologous toxin challenge after parenteral or mucosal vaccination [285–288]. Formaldehyde treatment of proteins has been used to generate successful toxoids of the SEs, and many other antigens, throughout time. However, such treatment can adversely affect antigen processing and subsequent presentation to the immune system [289], especially when antigen is administered mucosally [290].

17.7 CONCLUSIONS

Staphylococcus aureus and *Streptococcus pyogenes* produce various superantigenic toxins representing important virulence factors that interact with MHC class II and TCR molecules on host cells. Through an insidious twist of fate, the host's abnormally elevated immune response toward SEs, TSST-1, or SPEs via various proinflammatory cytokines can trigger severe lethal shock. Similar sequence homologies, structural conformations, and biological activities among this family of protein exotoxins suggest a common pathway through divergent and/or convergent evolution. With time, and through natural evolutionary processes, more of these fascinating microbial toxins will undoubtedly be discovered and novel biological properties elucidated by future investigators. As superantigens can evidently afford an advantage to a pathogen, such as delayed clearance from the host [291], there is biological justification for the energy expended during transcription and translation of these particular genes. After an early cytokine "burst" from SE-, TSST-1-, or SPE-activated T cells, subsequent immunosuppression likely aids microbial survival. To discover more effective means of controlling staphylococci, streptococci, and their associated toxins, the animal models described in this chapter represent a necessary step toward developing better therapeutic and vaccine strategies. Clearly, there is an inherent urgency for more work in this field that extends well beyond biodefense. A formidable, constantly evolving nemesis for us is seen on a daily basis in clinics throughout the world. This involves bacterial pathogens such as *S. aureus* and *S. pyogenes* with increasing antibiotic resistance. How humans respond now to these and other dynamic microbial threats will have lasting consequences for subsequent generations. Animal models, when used by various laboratory groups around the world, will no doubt continue to play a pivotal role in fighting staphylococcal, streptococcal, and other bacterial diseases.

REFERENCES

1. Kotzin, B.L. et al. Superantigens and their potential role in human disease, *Adv. Immunol.*, 54, 99, 1993.
2. Monday, S.R. and Bohach, G.A. Properties of *Staphylococcus aureus* enterotoxins and toxic shock syndrome toxin-1, in J.E. Alouf and J.H. Freer, eds., *The Comprehensive Sourcebook of Bacterial Protein Toxins*, Academic Press, London, 1999, Chapter 33, pp. 589–610.
3. Schuberth, H.-J. et al. Characterization of leukocytotoxic and superantigen-like factors produced by *Staphylococcus aureus* isolates from milk of cows with mastitits, *Vet. Microbiol.*, 82, 187, 2000.
4. Fleischer, B. et al. An evolutionary conserved mechanism of T cell activation by microbial toxins. Evidence for different affinities of T cell receptor-toxin interaction, *J. Immunol.*, 146, 11, 1991.

5. Smith, B.G. and Johnson, H. The effect of staphylococcal enterotoxins on the primary *in vitro* immune response, *J. Immunol.*, 115, 575, 1975.
6. Poindexter, N.J. and Schlievert, P.M. Toxic-shock-syndrome toxin 1-induced proliferation of lymphocytes: Comparison of the mitogenic response of human, murine, and rabbit lymphocytes, *J. Infect. Dis.*, 153, 772, 1986.
7. Bhakdi, S., Muhly, M., and Fussle, R. Correlation between toxin binding and hemolytic activity in membrane damage by staphylococcal alpha-toxin, *Infect. Immun.*, 46, 318, 1984.
8. Bhakdi, S. and Muhly, M. Decomplementation antigen, a possible determinant of staphylococcal pathogenicity, *Infect. Immun.*, 47, 41, 1985.
9. Kreiswirth, B. et al. Evidence for a clonal origin of methicillin resistance in *Staphylococcus aureus*, *Science*, 259, 227, 1993.
10. Centers for Disease Control and Prevention (CDC). Reduced susceptibility of *Staphylococcus aureus* to vancomycin—Japan, 1996, *Morb. Mortal. Weekly Rep.*, 46, 624, 1997.
11. Centers for Disease Control and Prevention (CDC). *Staphylococcus aureus* resistant to vancomycin—United States, 2002, *Morb. Mortal. Weekly Rep.*, 51, 565, 2002.
12. Lang, A., Ward, S., and Michie, C.A. In brief, *Trends Immunol.*, 23, 389, 2002.
13. Vriens, M. et al. Costs associated with a strict policy to eradicate methicillin-resistant *Staphylococcus aureus* in a Dutch university medical center: A 10-year survey, *Eur. J. Clin. Microbiol. Infect. Dis.*, 21, 782, 2002.
14. Carmeli, Y. et al. Health and economic outcomes of vancomycin-resistant enterococci, *Arch. Intern. Med.*, 162, 2223, 2002.
15. Capitano, B. et al. Cost effect of managing methicillin-resistant *Staphylococcus aureus* in a long-term care facility, *J. Am. Geriatr. Soc.*, 51, 10, 2003.
16. Kern, W.V. Management of *Staphylococcus aureus* bacteremia and endocarditis: Progresses and challenges, *Curr. Opin. Infect. Dis.*, 23, 346, 2010.
17. Dinges, M.M., Orwin, P.M., and Schlievert, P.M. Exotoxins of *Staphylococcus aureus*, *Clin. Microbiol. Rev.*, 13, 16, 2000.
18. Omoe, K. et al. Identification and characterization of a new staphylococcal enterotoxin-related putative toxin encoded by two kinds of plasmids, *Infect. Immun.*, 71, 6088, 2003.
19. Bergdoll, M.S. Monkey feeding test for staphylococcal enterotoxin, *Meth. Enzymol.*, 165, 324, 1988.
20. Loir, L., Baron, F., and Gautier, M. *Staphylococcus aureus* and food poisoning, *Genet. Mol. Res.*, 2, 63, 2003.
21. Bennett, R.W. Staphylococcal enterotoxin and its rapid identification in foods by enzyme-linked immunosorbent assay-based methodology, *J. Food Prot.*, 68, 1264, 2005.
22. McGann, V.G., Rollins, J.B., and Mason, D.W. Evaluation of resistance to staphylococcal enterotoxin B: Naturally acquired antibodies of man and monkey, *J. Infect. Dis.*, 124, 206, 1971.
23. Holmberg, S.D. and Blake, P.A. Staphylococcal food poisoning in the United States. New facts and old misconceptions, *JAMA*, 251, 487, 1984.
24. Jett, M. et al. *Staphylococcus aureus* enterotoxin B challenge of monkeys: Correlation of plasma levels of arachidonic acid cascade products with occurrence of illness, *Infect. Immun.*, 58, 3494, 1990.
25. Scheuber, P.H. et al. Cysteinyl leukotrienes as mediators of staphylococcal enterotoxin B in the monkey, *Eur. J. Clin. Invest.*, 17, 455, 1987.
26. Ulrich, R.G. et al. Staphylococcal enterotoxin B and related pyrogenic toxins, in F.R. Sidell, E.T. Takafuji, and D.R. Franz, eds., *Textbook of Military Medicine: Medical Aspects of Chemical and Biological Warfare*, Office of the Surgeon General, Department of the Army, United States of America, 1997, Chapter 31, pp. 621–630.
27. Todd, J., Fishaut, M., Kapral, F., and Welch, T. Toxic-shock syndrome associated with phage-group-I staphylococci, *Lancet*, 2, 1116, 1978.

28. Shands, K.N. et al. Toxic-shock syndrome in menstruating women: Association with tampon use and *Staphylococcus aureus* and clinical features in 52 cases, *N. Eng. J. Med.*, 303, 1436, 1980.
29. Schlievert, P.M. et al. Identification and characterization of an exotoxin from *Staphylococcus aureus* associated with toxic-shock syndrome, *J. Infect. Dis.*, 143, 509, 1981.
30. Crass, B.A. and Bergdoll, M.S. Involvement of staphylococcal enterotoxins in nonmenstrual toxic shock syndrome, *J. Clin. Microbiol.*, 23, 1138, 1986.
31. McCormick, J.K., Yarwood, J.M., and Schlievert, P.M. Toxic shock syndrome and bacterial superantigens: An update, *Ann. Rev. Microbiol.*, 55, 77, 2001.
32. Mills, J. et al. Control of production of toxic-shock-syndrome toxin-1 (TSST-1) by magnesium ion, *J. Infect. Dis.*, 151, 1158, 1985.
33. Schlievert, P.M., Blomster, D.A., and Kelly, J.A. Toxic shock syndrome *Staphylococcus aureus*: Effect of tampons on toxic shock syndrome toxin 1 production, *Obstet. Gynecol.*, 64, 666, 1984.
34. Bergdoll, M.S. et al. A new staphylococcal enterotoxin, enterotoxin F, associated with toxic-shock-syndrome *Staphylococcus aureus* isolates, *Lancet*, 1, 1017, 1981.
35. Reiser, R.F. et al. Purification and some physicochemical properties of toxic-shock toxin, *Biochemistry*, 22, 3907, 1983.
36. Schlievert, P.M. Alteration of immune function by staphylococcal pyrogenic exotoxin type C: Possible role in toxic-shock syndrome, *J. Infect. Dis.*, 147, 391, 1983.
37. Ikejima, T. et al. Induction by toxic-shock-syndrome toxin-1 of a circulating tumor necrosis factor-like substance in rabbits and of immunoreactive tumor necrosis factor and interleukin-1 from human mononuclear cells, *J. Infect. Dis.*, 158, 1017, 1988.
38. Parsonnet, J. Mediators in the pathogenesis of toxic shock syndrome: Overview, *Rev. Infect. Dis.*, 11, S263, 1989.
39. Freedman, J.D. and Beer, D.J. Expanding perspectives on the toxic shock syndrome, *Adv. Intern. Med.*, 36, 363, 1991.
40. Bohach, G.A. et al. Staphylococcal and streptococcal pyrogenic toxins involved in toxic shock syndrome and related illnesses, *Crit. Rev. Microbiol.*, 17, 251, 1990.
41. Garbe, P.L. et al. *Staphylococcus aureus* isolates from patients with nonmenstrual toxic shock syndrome. Evidence for additional toxins, *JAMA*, 252, 2538, 1985.
42. Andrews, M.-M. et al. Recurrent nonmenstrual toxic shock syndrome: Clinical manifestations, diagnosis, and treatment, *Clin. Infect. Dis.*, 32, 1470, 2001.
43. Bonventre, P.F. et al. Antibody responses to toxic-shock-syndrome (TSS) toxin by patients with TSS and by healthy staphylococcal carriers, *J. Infect. Dis.*, 150, 662, 1984.
44. Vergeront, J.M. et al. Prevalence of serum antibody to staphylococcal enterotoxin F among Wisconsin residents: Implications for toxic-shock syndrome, *J. Infect. Dis.*, 148, 692, 1983.
45. Notermans, S. et al. Serum antibodies to enterotoxins produced by *Staphylococcus aureus* with special reference to enterotoxin F and toxic shock syndrome, *J. Clin. Microbiol.*, 18, 1055, 1983.
46. Mahlknecht, U. et al. The toxic shock syndrome toxin-1 induces anergy in human T cells in vivo, *Human Immunol.*, 45, 42, 1996.
47. Hofer, M.F. et al. Differential effects of staphylococcal toxic shock syndrome toxin-1 on B cell apoptosis, *Proc. Natl. Acad. Sci. USA*, 93, 5425, 1996.
48. Stiles, B.G., Krakauer, T., and Bonventre, P.F. Biological activity of toxic shock syndrome toxin 1 and a site-directed mutant, H135A, in a lipopolysaccharide-potentiated mouse lethality model, *Infect. Immun.*, 63, 1229, 1995.
49. Stiles, B.G. et al. Mucosal vaccination with recombinantly attenuated staphylococcal enterotoxin B and protection in a murine model, *Infect. Immun.*, 69, 2031, 2001.
50. Nilsson, I.M. et al. Protection against *Staphylococcus aureus* sepsis by vaccination with recombinant staphylococcal enterotoxin A devoid of superantigenicity, *J. Infect. Dis.*, 180, 1370, 1999.

51. Bavari, S., Dyas, B., and Ulrich, R.G. Superantigen vaccines: A comparative study of genetically attenuated receptor-binding mutants of staphylococcal enterotoxin A, *J. Infect. Dis.*, 174, 338, 1996.
52. McCormick, J.K. et al. Development of streptococcal pyrogenic exotoxin C vaccine toxoids that are protective in the rabbit model of toxic shock syndrome, *J. Immunol.*, 165, 2306, 2000.
53. Roggiani, M. et al. Toxoids of streptococcal pyrogenic exotoxin A are protective in rabbit models of streptococcal toxic shock syndrome, *Infect. Immun.*, 68, 5011, 2000.
54. Gampfer, J. et al. Double mutant and formaldehyde inactivated TSST-1 as vaccine candidates for TSST-1-induced toxic shock syndrome, *Vaccine*, 20, 1354, 2002.
55. Ulrich, R.G., Olson, M.A., and Bavari, S. Development of engineered vaccines effective against structurally related bacterial superantigens, *Vaccine*, 16, 1857, 1998.
56. Hu, D.L. Vaccination with nontoxic mutant toxic shock syndrome toxin-1 protects against *Staphylococcus aureus* infection, *J. Infect. Dis.*, 188, 743, 2003.
57. Schlievert, P. M. Use of intravenous immunoglobulin in the treatment of staphylococcal and streptococcal toxic shock syndromes and related illnesses, *J. Allergy Clin. Immunol.*, 108, S107, 2001.
58. Barry, W., Hudgins, L., Donta, S.T., and Pesanti, E.L. Intravenous immunoglobulin therapy for toxic shock syndrome, *JAMA*, 267, 3315, 1992.
59. Tilahun, M. E. et al. Potent neutralization of SEB by synergistic action of chimeric antibodies, *Infect. Immun.*, 78, 2801, 2010.
60. Nowakowski, A. et al. Potent neutralization of botulinum neurotoxin by recombinant oligoclonal antibody, *Proc. Nat. Acad. Sci. USA*, 99, 11346, 2002.
61. Lancefield, R.C. A serological differentiation of human and other groups of hemolytic streptococci, *J. Exp. Med.*, 57, 571, 1933.
62. Stevens, D.L. Invasive group A streptococcus infections, *Clin. Infect. Dis.*, 14, 2, 1992.
63. Stevens, D.L. The toxins of group A streptococcus, the flesh eating bacteria, *Immunol. Invest.*, 26, 129, 1997.
64. Marrack, P. and Kappler, J. The staphylococcal enterotoxins and their relatives, *Science*, 248, 705, 1990.
65. Choi, Y. et al. Interaction of *Staphylococcus aureus* toxin "superantigens" with human T cells, *Proc. Natl. Acad. Sci. USA*, 86, 8941, 1989.
66. Webb, S.R. and Gascoigne, N.R. T-cell activation by superantigens, *Curr. Opin. Immunol.*, 6, 467, 1994.
67. Blackman, M.A. and Woodland, D.L. *In vivo* effects of superantigens, *Life Sci.*, 57, 1717, 1995.
68. Johnson, H.M., Torres, B.A., and Soos, J.M. Superantigens: Structure and relevance to human disease, *Proc. Soc. Exp. Biol. Med.*, 212, 99, 1996.
69. Florquin, S. and Aaldering, L. Superantigens: A tool to gain new insight into cellular immunity, *Res. Immunol.*, 148, 373, 1997.
70. Li, H. et al. The structural basis of T cell activation by superantigens, *Ann. Rev. Immunol.*, 17, 435, 1999.
71. Ribeiro-Dias, F. et al. *Mycoplasma arthritidis* superantigen (MAM)-induced macrophage nitric oxide release is MHC class II restricted, interferon gamma dependent, and toll-like receptor 4 independent, *Exp. Cell Res.*, 286, 345, 2003.
72. Ohmen, J. D. et al. Evidence for a superantigen in human tuberculosis, *Immunity*, 1, 35, 1994.
73. Legaard, P.K., LeGrand, R.D., and Misfeldt, M.L. The superantigen *Pseudomonas* exotoxin A requires additional functions from accessory cells for T lymphocyte proliferation, *Cell Immunol.*, 135, 372, 1991.
74. Proft, T. and Fraser, J.D. Bacterial superantigens, *Clin. Exp. Immunol.*, 133, 299, 2003.

75. Proft, T. et al. Superantigens and streptococcal toxic shock syndrome, *Emerg. Infect. Dis.*, 9, 1211, 2003.
76. Proft, T. and Fraser, J.D. Streptococcal superantigenic toxins, in J.E. Alouf and M.R. Popoff, eds., *The Comprehensive Sourcebook of Bacterial Protein Toxins*, 3rd ed., Academic Press, Paris, 2006, Chapter 51, pp. 844–861.
77. Stuart, P.M. et al. Characterization of human T-cell responses to *Yersinia enterocolitica* superantigen, *Hum. Immunol.*, 43, 269, 1995.
78. Abe, J. et al. Pathogenic role of a superantigen in *Yersinia pseudotuberculosis* infection, *Adv. Exp. Med. Biol.*, 529, 459, 2003.
79. Dobrescu, D. et al. Enhanced HIV-1 replication in V beta 12 T cells due to human cytomegalovirus in monocytes: Evidence for a putative herpesvirus superantigen, *Cell*, 82, 753, 1995.
80. Sutkowski, N. et al. An Epstein-Barr virus-associated superantigen, *J. Exp. Med.*, 184, 971, 1996.
81. Yao, Z. et al. Herpesvirus saimiri open reading frame 14, a protein encoded by T lymphotropic herpesvirus, binds to MHC class II molecules and stimulates T cell proliferation, *J. Immunol.*, 156, 3260, 1996.
82. Torres, B.A. et al. Characterization of Nef-induced CD4 T cell proliferation, *Biochem. Biophys. Res. Commun.*, 225, 54, 1996.
83. Acha-Orbea, H. and MacDonald, H.R. Superantigens of mouse mammary tumor virus, *Ann. Rev. Immun.*, 13, 459, 1995.
84. Lafon, M. Rabies virus superantigen, *Res. Immunol.*, 144, 209, 1993.
85. Alouf, J.E. and Muller-Alouf, H. What are Superantigens?, In J.E. Alout and M.R. Popoff, eds., *The Comprehensive Sourcebook of Bacterial Protein Toxins*, Academic Press, Paris, 2006, Chapter 49, pp. 821–829.
86. Uchiyama, T., Imanishi, K., Miyoshi-Akiyama, T., and Kata, H. Staphylococcal superantigens and the diseases they cause, in J.E. Alouf and M.R. Popoff, eds., *The Comprehensive Sourcebook of Bacterial Protein Toxins*, 3rd ed., Academic Press, Paris, 2006, Chapter 50, pp. 830–843.
87. Betley, M.J. et al. Staphylococcal enterotoxin A gene is associated with a variable genetic element, *Proc. Natl. Acad. Sci. USA*, 81, 5179, 1984.
88. Betley, M.J., Borst, D.W., and Regassa, L.B. Staphylococcal enterotoxins, toxic shock syndrome toxin and streptococcal pyrogenic exotoxins: A comparative study of their molecular biology, *Chem. Immunol.*, 55, 1, 1992.
89. Singh, B.R., Fen-Ni, F., and Ledoux, D.N. Crystal and solution structures of superantigenic staphylococcal enterotoxins compared, *Struct. Biol.*, 1, 358, 1994.
90. Swaminathan, S. et al. Crystal structure of staphylococcal enterotoxin B, a superantigen, *Nature*, 359, 801, 1992.
91. Chen, J. et al. MMDB: Entrez's 3D-structure database, *Nucl. Acid Res.*, 31, 474, 2003.
92. Papageorgiou, A.C., Tranter, H.S., and Acharya, K.R. Crystal structure of microbial superantigen staphylococcal enterotoxin B at 1.5 angstrom resolution: Implications for superantigen recognition by MHC class II molecules and T cell receptors, *J. Mol. Biol.*, 277, 61, 1998.
93. Papageorgiou, A. C. et al. The refined crystal structure of toxic shock syndrome toxin-1 at 2.07 angstrom resolution, *J. Mol. Biol.*, 260, 553, 1996.
94. Papageorgiou, A.C. et al. Structural basis for the recognition of superantigen streptococcal pyrogenic exotoxin A (SpeA1) by MHC class II molecules and T-cell receptors, *EMBO J.*, 18, 9, 1999.
95. Arcus, V.L. et al. Conservation and variation in superantigen structure and activity highlighted by the three-dimensional structures of two new superantigens from *Streptococcus pyogenes*, *J. Mol. Biol.*, 299, 157, 2000.

96. Hurley, J.M. et al. Identification of class II major histocompatibility complex and T cell receptor binding sites in the superantigen toxic shock syndrome toxin 1, *J. Exp. Med.*, 181, 2229, 1995.
97. Kappler, J.W. et al. Mutations defining functional regions of the superantigen staphylococcal enterotoxin B, *J. Exp. Med.*, 175, 387, 1992.
98. Kum, W.W. and Chow, A.W. Inhibition of staphylococcal enterotoxin A-induced superantigenic and lethal activities by a monoclonal antibody to toxic shock syndrome toxin-1, *J. Infect. Dis.*, 183, 1739, 2001.
99. Bavari, S., Ulrich, R.G., and LeClaire, R.D. Cross-reactive antibodies prevent the lethal effects of *Staphylococcus aureus* superantigens, *J. Infect. Dis.*, 180, 1365, 1999.
100. Thompson, N.E., Ketterhagen, M.J., and Bergdoll, M.S. Monoclonal antibodies to staphylococcal enterotoxin B and C: Cross-reactivity and localization of epitopes on tryptic fragments, *Infect. Immun.*, 45, 281, 1984.
101. Spero, L., Morlock, B.A., and Metzger, J.F. On the cross-reactivity of staphylococcal enterotoxins A, B, and C, *J. Immunol.*, 120, 86, 1978.
102. Bohach, G.A. et al. Cross-neutralization of staphylococcal and streptococcal pyrogenic toxins by monoclonal and polyclonal antibodies, *Infect. Immun.*, 56, 400, 1988.
103. Kline, J.B. and Collins, C.M. Analysis of the superantigenic activity of mutant and allelic forms of streptococcal pyrogenic exotoxin A, *Infect. Immun.*, 64, 861, 1996.
104. Mollick, J.A. et al. Staphylococcal exotoxin activation of T cells. Role of exotoxin-MHC class II binding affinity and class II isotype, *J. Immunol.*, 146, 463, 1991.
105. Proft, T. et al. Immunological and biochemical characterization of streptococcal pyrogenic exotoxins I and J (SPE-I and SPE-J) from *Streptococcus pyogenes*, *J. Immunol.*, 166, 6711, 2001.
106. Yagi, J., Rath, J.S., and Janeway, C.A. Control of T cell responses to staphylococcal enterotoxins by stimulator cell MHC class II polymorphism, *J. Immunol.*, 147, 1398, 1991.
107. Herrmann, T., Acolla, R.S., and MacDonald, H.R. Different staphylococcal enterotoxins bind preferentially to distinct major histocompatibility complex class II isotypes, *Eur. J. Immunol.*, 19, 2171, 1989.
108. Herman, A. et al. HLA-DR alleles differ in their ability to present staphylococcal enterotoxins to T cells, *J. Exp. Med.*, 172, 709, 1990.
109. Chintagumpala, M.M., Mollick, J.A., and Rich, R.R. Staphylococcal toxins bind to different sites on HLA-DR, *J. Immunol.*, 147, 3876, 1991.
110. Imanishi, K., Igarashi, H., and Uchiyama, T. Relative abilities of distinct isotypes of human major histocompatibility complex class II molecules to bind streptococcal pyrogenic exotoxin types A and B, *Infect. Immun.*, 60, 5025, 1992.
111. See, R.H., Krystal, G., and Chow, A.W. Receptors for toxic shock syndrome toxin-1 and staphylococcal enterotoxin A on human blood monocytes, *Can. J. Microbiol.*, 38, 937, 1992.
112. Hudson, K.R. et al. Staphylococcal enterotoxin A has two cooperative binding sites on major histocompatibility complex class II, *J. Exp. Med.*, 182, 711, 1995.
113. Tiedemann, R.E. and Fraser, J.D. Cross-linking of MHC class II molecules by staphylococcal enterotoxin A is essential for antigen-presenting cell and T cell activation, *J. Immunol.*, 157, 3958, 1996.
114. Thibodeau, J. et al. Molecular characterization and role in T cell activation of staphylococcal enterotoxin A binding to the HLA-DR alpha-chain, *J. Immunol.*, 158, 3698, 1997.
115. Ulrich, R.G., Bavari, S., and Olson, M.A. Staphylococcal enterotoxins A and B share a common structural motif for binding class II major histocompatibility complex molecules, *Nature Struct. Biol.*, 2, 554, 1995.
116. Mehindate, K. et al. Cross-linking of major histocompatibility complex class II molecules by staphylococcal enterotoxin A superantigen is a requirement for inflammatory cytokine gene expression, *J. Exp. Med.*, 182, 1573, 1995.

117. Jardetzky, T.S. et al. Three-dimensional structure of a human class II histocompatibility molecule complexed with superantigen, *Nature*, 368, 711, 1994.
118. Kim, J. et al. Toxic shock syndrome toxin-1 complexed with a class II major histocompatibility molecule HLA-DR1, *Science*, 266, 1870, 1994.
119. Pless, D.D. et al. Persistence of zinc-binding bacterial superantigens at the surface of antigen-presenting cells contributes to the extreme potency of these superantigens as T-cell activators, *Infect. Immun.*, 73, 5358, 2005.
120. Kum, W.W., Wood, J.A., and Chow, A.W. A mutation at glycine residue 31 of toxic shock syndrome toxin-1 defines a functional site critical for major histocompatiblity complex class II binding and superantigenic activity, *J. Infect. Dis.*, 174, 1261, 1996.
121. Li, P.L. et al. The superantigen streptococcal pyrogenic exotoxin C (SPE-C) exhibits a novel mode of action, *J. Exp. Med.*, 186, 375, 1997.
122. Kappler, J. et al. V beta-specific stimulation of human T cells by staphylococcal toxins, *Science*, 244, 811, 1989.
123. Choi, Y. et al. Selective expansion of T cells expressing V beta 2 in toxic shock syndrome, *J. Exp. Med.*, 172, 981, 1990.
124. Newton, D.W. et al. Mutations in the MHC class II binding domains of staphylococcal enterotoxin A differentially affect T cell receptor Vbeta specificity, *J. Immunol.*, 157, 3988, 1996.
125. Leder, L. et al. A mutational analysis of the binding of staphylococcal enterotoxins B and C3 to the T cell receptor beta chain and major histocompatiblity complex class II, *J. Exp. Med.*, 187, 823, 1998.
126. Gascoigne, N.R. and Ames, K.T. Direct binding of secreted T-cell receptor beta chain to superantigen associated with class II major histocompatiblity complex protein, *Proc. Natl. Acad. Sci. USA*, 88, 613, 1991.
127. Fields, B.A. et al. Crystal structure of a T-cell receptor beta-chain complexed with a superantigen, *Nature*, 384, 188, 1996.
128. Li, H. et al. Three-dimensional structure of the complex between a T cell receptor beta chain and the superantigen staphylococcal enterotoxin B, *Immunity*, 9, 807, 1998.
129. Seth, A. et al. Binary and ternary complexes between T-cell receptor, class II MHC and superantigen in vitro, *Nature*, 369, 324, 1994.
130. Redpath, S. et al. Cutting edge: Trimolecular interaction of TCR with MHC class II and bacterial superantigen shows a similar affinity to MHC:peptide ligands, *J. Immunol.*, 163, 6, 1999.
131. Chatila, T. and Geha, R.S. Signal transduction by microbial superantigens via MHC class II molecules, *Immunol. Rev.*, 131, 43, 1993.
132. Chatila, T. et al. Toxic shock syndrome toxin-1 induces inositol phospholipid turnover, protein kinase C translocation, and calcium mobilization in human T cells, *J. Immunol.*, 140, 1250, 1988.
133. Scholl, P.R. et al. Role of protein tyrosine phosphorylation in monokine induction by the staphylococcal superantigen toxic shock syndrome toxin-1, *J. Immunol.*, 148, 2237, 1992.
134. Trede, N.S. et al. Transcriptional activation of the human TNF-alpha promoter by superantigen in human monocytic cells: Role of NF-kappa B, *J. Immunol.*, 155, 902, 1995.
135. Sundstedt, A. et al. *In vivo* anergized CD4+ T cells express perturbed AP-1 and NF-kappa B transcription factors, *Proc. Natl. Acad. Sci. USA*, 93, 979, 1996.
136. Krakauer, T. Induction of CC chemokines in human peripheral blood mononuclear cells by staphylococcal exotoxins and its prevention by pentoxifylline, *J. Leuk. Biol.*, 66, 158, 1999.
137. Krakauer, T. Inhibition of toxic shock syndrome toxin-1 induced cytokine production and T cell activation by interleukin-10, interleukin-4, and dexamethasone, *J. Infect. Dis.*, 172, 988, 1995.

138. Jupin, C. et al. Toxic shock syndrome toxin 1 as an inducer of human tumor necrosis factors and gamma interferon, *J. Exp. Med.*, 167, 752, 1988.
139. Grossman, D. et al. Dissociation of the stimulatory activities of staphylococcal enterotoxins for T cells and monocytes, *J. Exp. Med.*, 172, 1831, 1990.
140. Krakauer, T. and Stiles, B.G. Pentoxifylline inhibits superantigen-induced toxic shock and cytokine release, *Clin. Diagn. Lab. Immunol.*, 6, 594, 1999.
141. Hale, M.L. et al. Pirfenidone blocks the *in vitro* and *in vivo* effects of staphylococcal enterotoxin B, *Infect. Immun.*, 70, 2989, 2002.
142. Langezaal, I. et al. Evaluation and prevalidation of an immunotoxicity test based on human whole-blood cytokine release, *Altern. Lab. Anim.*, 30, 581, 2002.
143. Hermann, C. et al. A model of human whole blood lymphokine release for *in vitro* and *ex vivo* use, *J. Immunol. Meth.*, 275, 69, 2003.
144. Carlsson, R., Fischer, H., and Sjogren, H.O. Binding of staphylococcal enterotoxin A to accessory cells is a requirement for its ability to activate human T cells, *J. Immunol.*, 140, 2484, 1988.
145. Trede, N.S., Geha, R.S., and Chatila, T. Transcriptional activation of IL-1 beta and tumor necrosis factor-alpha genes by MHC class II ligands, *J. Immunol.*, 146, 2310, 1991.
146. Fischer, H. et al. Production of TNF-alpha and TNF-beta by staphylococcal enterotoxin A activated human T cells, *J. Immunol.*, 144, 4663, 1990.
147. Lagoo, A. et al. IL-2, IL-4, and IFN-gamma gene expression versus secretion in superantigen-activated T cells. Distinct requirement for costimulatory signals through adhesion molecules, *J. Immunol.*, 152, 1641, 1994.
148. Lando, P. et al. Regulation of superantigen-induced T cell activation in the absence and the presence of MHC class II, *J. Immunol.*, 157, 2857, 1996.
149. Hewitt, C. et al. Major histocompatibility complex independent clonal T cell anergy by direct interaction of *Staphylococcus aureus* enterotoxin B with the T cell antigen receptor, *J. Exp. Med.*, 175, 1493, 1992.
150. Scheuber, P.H. et al. Staphylococcal enterotoxin B as a nonimmunological mast cell stimulus in primates: The role of endogenous cysteinyl leukotrienes, *Int. Arch. Allergy Appl. Immunol.*, 82, 289, 1987.
151. Huvenne, W. et al. *Staphylococcus aureus* enterotoxin B augments granulocyte migration and survival via airway epithelial cell activation, *Allergy*, 65, 1013, 2010.
152. Stohl, W., Elliott, J.E., and Linsley, P.S. Human T cell-dependent B cell differentiation induced by staphylococcal superantigens, *J. Immunol.*, 153, 117, 1994.
153. Mourad, W. et al. Engagement of major histocompatibility complex class II molecules by superantigen induces inflammatory cytokine gene expression in human rheumatoid fibroblast-like synoviocytes, *J. Exp. Med.*, 175, 613, 1992.
154. Wooley, P.H. and Cingel, B. Staphylococcal enterotoxin B increases the severity of type II collagen induced arthritis in mice, *Ann. Rheum. Dis.*, 54, 298, 1995.
155. Hamad, A.R., Marrack, P., and Kappler, J.W. Transcytosis of staphylococcal superantigen toxins, *J. Exp. Med.*, 185, 1447, 1997.
156. Shupp, J.W., Jett, M., and Pontzer, C.H. Identification of a transcytosis epitope on staphylococcal enterotoxins, *Infect. Immun.*, 70, 2178, 2002.
157. Peterson, M.L. et al. The innate immune system is activated by stimulation of vaginal epithelial cells with *Staphylococcus aureus* and toxic shock syndrome toxin 1, *Infect. Immun.*, 73, 2164, 2005.
158. Pinchuk, I.V. et al. Monocyte chemoattractant protein-1 production by intestinal myofibroblasts in response to staphylococcal enterotoxin A: Relevance to staphylococcal enterotoxigenic disease, *J. Immunol.*, 178, 8097, 2007.
159. Hodoval, L.F. et al. Pathogenesis of lethal shock after intravenous staphylococcal enterotoxin B in monkeys, *Appl. Microbiol.*, 16, 187, 1968.

160. Raj, H.D. and Bergdoll, M.S. Effect of enterotoxin B on human volunteers, *J. Bacteriol.*, 98, 833, 1969.
161. Adesiyun, A.A. and Tatini, S.R. Biological activity of cell-associated staphylococcal enterotoxin, *J. Med. Primatol.*, 11, 163, 1982.
162. Scheuber, P.H. et al. Skin reactivity of unsensitized monkey upon challenge with staphylococcal enterotoxin B: A new approach for investigating the site of toxin action, *Infect. Immun.*, 50, 869, 1985.
163. Spiekermann, G.M. and Nagler-Anderson, C. Oral administration of the bacterial superantigen staphylococcal enterotoxin B induces activation and cytokine production by T cells in murine gut-associated lymphoid tissue, *J. Immunol.*, 161, 5825, 1998.
164. Lu, J. et al. Colonic bacterial superantigens can evoke an inflammatory response and exaggerate disease in mice recovering from colitis, *Gastroenterology*, 125, 1785, 2003.
165. Sugiyama, H., Bergdoll, M.S., and Dack, G.M. Early development of a temporary resistance to the emetic action of staphylococcal enterotoxin, *J. Infect. Dis.*, 111, 233, 1962.
166. McCormack, J.E. et al. Profound deletion of mature T cells *in vivo* by chronic exposure to exogenous superantigen, *J. Immunol.*, 150, 3785, 1993.
167. Eroukhmanoff, L. et al. T-cell tolerance induced by repeated antigen stimulation: Selective loss of Foxp3-conventional CD4 T cells and induction of CD4 T-cell anergy, *Eur. J. Immunol.*, 39, 1078, 2009.
168. Schartner, J.M. et al. Recurrent superantigen exposure *in vivo* leads to highly suppressive $CD4^+CD25^+$ and $CD4^+CD25^-$ T cells with anergic and suppressive genetic signatures, *Clin. Exp. Immunol.*, 155, 348, 2009.
169. Tanriver, Y. et al. Superantigen-activated regulatory T cells inhibit the migration of innate immune cells and the differentiation of naive T cells, *J. Immunol.*, 183, 2946, 2009.
170. Miethke, T. et al. Exogenous superantigens acutely trigger distinct levels of peripheral T cell tolerance/immunosuppression: Dose-response relationship, *Eur. J. Immunol.*, 24, 1893, 1994.
171. Collins, L.V. et al. Mucosal tolerance to a bacterial superantigen indicates a novel pathway to prevent toxic shock, *Infect. Immun.*, 70, 2282, 2002.
172. Stiles, B.G. et al. Correlation of temperature and toxicity in murine studies of staphylococcal enterotoxins and toxic shock syndrome toxin 1, *Infect. Immun.*, 67, 1521, 1999.
173. Bean, A.G. et al. Interleukin 10 protects mice against staphylococcal enterotoxin B-induced lethal shock, *Infect. Immun.*, 61, 4937, 1993.
174. Buxser, S. and Bonventre, P.F. Staphylococcal enterotoxins fail to disrupt membrane integrity or synthetic functions of Henle 407 intestinal cells, *Infect. Immun.*, 31, 929, 1981.
175. Lu, J. et al. Epithelial ion transport and barrier abnormalities evoked by superantigen-activated immune cells are inhibited by interleukin-10 but not interleukin-4, *J. Pharm. Exper. Ther.*, 287, 128, 1998.
176. McKay, D.M. Bacterial superantigens: Provocateurs of gut dysfunction and inflammation? *Trends Immunol.*, 22, 497, 1997.
177. Krakauer, T. Stimulant-dependent modulation of cytokines and chemokines by airway epithelial cells: Cross-talk between pulmonary epithelial and peripheral blood mononuclear cells, *Clin. Diagn. Lab. Immunol.*, 9, 126, 2002.
178. Harris, T.O. Lack of complete correlation between emetic and T-cell-stimulatory activities of staphylococcal enterotoxins, *Infect. Immun.*, 61, 3175, 1993.
179. Spero, L. and Morlock, B.A. Biological activities of the peptides of staphylococcal enterotoxin C formed by limited tryptic hydrolysis, *J. Biol. Chem.*, 253, 8787, 1978.
180. Hovde, C.J. Investigation of the role of the disulphide bond in the activity and structure of staphylococcal enterotoxin C1, *Mol. Microbiol.*, 13, 897, 1994.
181. Stelma, G.N. and Bergdoll, M.S. Inactivation of staphylococcal enterotoxin A by chemical modification, *Biochem. Biophys. Res. Commun.*, 105, 121, 1982.

182. Alber, G., Hammer, D.K., and Fleischer, B. Relationship between enterotoxic- and T lymphocyte-stimulating activity of staphylococcal enterotoxin B, *J. Immunol.*, 144, 4501, 1990.
183. Reck, B. et al. Protection against the staphylococcal enterotoxin-induced intestinal disorder in the monkey by anti-idiotypic antibodies, *Proc. Natl. Acad. Sci. USA*, 85, 3170, 1988.
184. Hoffman, M. et al. Biochemical and mutational analysis of the histidine residues of staphylococcal enterotoxin A, *Infect. Immun.*, 64, 885, 1996.
185. Hu, D-L. et al. Analysis of the epitopes on staphylococcal enterotoxin A responsible for emetic activity, *J. Vet. Med. Sci.*, 63, 237, 2001.
186. Miethke, T. et al. T cell-mediated lethal shock triggered in mice by the superantigen staphylococcal enterotoxin B: Critical role of tumor necrosis factor, *J. Exp. Med.*, 175, 91, 1992.
187. Muller-Alouf, H. et al. Human pro- and anti-inflammatory cytokine patterns induced by *Streptococcus pyogenes* erythrogenic (pyrogenic) exotoxin A and C superantigens, *Infect. Immun.*, 64, 1450, 1996.
188. Schlievert, P.M. Role of superantigens in human disease, *J. Infect. Dis.*, 167, 997, 1993.
189. McCormick, J.K. et al. Functional characterization of streptococcal pyrogenic exotoxin J, a novel superantigen, *Infect. Immun.*, 69, 1381, 2001.
190. Okusawa, S. et al. Interleukin 1 induces a shock-like state in rabbits. Synergism with tumor necrosis factor and the effect of cyclooxygenase inhibition, *J. Clin. Invest.*, 81, 1162, 1988.
191. Krakauer, T., Vilcek, J., and Oppenheim, J.J. Proinflammatory cytokines: TNF and IL-1 families, chemokines, TGFβ and others, in W.E. Paul, ed., *Fundamental Immunology*, Raven Press, New York, 1998, Chapter 21, pp. 775–811.
192. Neumann, B. et al. Induction of acute inflammatory lung injury by staphylococcal enterotoxin B, *J. Immunol.*, 158, 1862, 1997.
193. Krakauer, T., Buckley, M., Huzella, L.M., and Alves, D. Critical timing, location and duration of glucocorticoid administration rescues mice from superantigen-induced shock and attenuates lung injury, *Int. Immunopharmacol.*, 9, 1168, 2009.
194. Tessier, P.A. et al. Induction of acute inflammation *in vivo* by staphylococcal superantigens. II. Critical role for chemokines, ICAM-1, and TNF-alpha, *J. Immunol.*, 161, 1204, 1998.
195. Sugiyama, H. et al. Enhancement of bacterial endotoxin lethality by staphylococcal enterotoxin, *J. Infect. Dis.*, 114, 111, 1964.
196. Stiles, B.G. et al. Toxicity of staphylococcal enterotoxins potentiated by lipopolysaccharide: Major histocompatibility complex class II molecule dependency and cytokine release, *Infect. Immun.*, 61, 5333, 1993.
197. Chen, J.Y. et al. Increased susceptibility to staphylococcal enterotoxin B intoxication in mice primed with actinomycin D, *Infect. Immun.*, 62, 4626, 1994.
198. Sarawar, S.R., Blackman, M.A., and Doherty, P.C. Superantigen shock in mice with an inapparent viral infection, *J. Infect. Dis.*, 170, 1189, 1994.
199. Marrack, P. et al. The toxicity of staphylococcal enterotoxin B in mice is mediated by T cells, *J. Exp. Med.*, 171, 455, 1990.
200. Dinges, M.M. and Schlievert, P.M. Comparative analysis of lipopolysaccharide-induced tumor necrosis factor alpha activity in serum and lethality in mice and rabbits pretreated with the staphylococcal superantigen toxic shock syndrome toxin 1, *Infect. Immun.*, 69, 7169, 2001.
201. Dalpke, A.H. and Heeg, K. Synergistic and antagonistic interactions between LPS and superantigens, *J. Endotoxin Res.*, 9, 51, 2003.
202. Dinges, M.M. and Schlievert, P.M. Role of T cells and gamma interferon during induction of hypersensitivity to lipopolysaccharide by toxic shock syndrome toxin 1 in mice, *Infect. Immun.*, 69, 1256, 2001.

203. Stone, R.L. and Schlievert, P.M. Evidence for the involvement of endotoxin in toxic shock syndrome, *J. Infect. Dis.*, 155, 682, 1987.
204. Paiva C.N. et al. *Trypanosoma cruzi* sensitizes mice to fulminant SEB-induced shock: Overrelease of inflammatory cytokines and independence of Chagas' disease or TCR Vβ-usage, *Shock*, 19, 163, 2003.
205. Beno, D.W. et al. Differential induction of hepatic dysfunction after intraportal and intravenous challenge with endotoxin and staphylococcal enterotoxins B, *Shock*, 19, 352, 2003.
206. Beno, D.W. et al. Chronic staphylococcal enterotoxins B and lipopolysaccharide induce a bimodal pattern of hepatic dysfunction and injury, *Crit. Care Med.*, 31, 1154, 2003.
207. Nagaki, M. et al. Hepatic injury and lethal shock in galactosamine-sensitized mice induced by the superantigen staphylococcal enterotoxin B, *Gastroenterology*, 106, 450, 1994.
208. LeClaire, R.D. et al. Potentiation of inhaled staphylococcal enterotoxin B-induced toxicity by lipopolysaccharide in mice, *Toxicol. Path.*, 24, 619, 1996.
209. Schlievert, P.M. Enhancement of host susceptibility to lethal endotoxin shock by staphylococcal pyrogenic exotoxin type C, *Infect. Immun.*, 36, 123, 1982.
210. Sauter, C. and Wolfensberger, C. Interferon in human serum after injection of endotoxin, *Lancet*, 2, 852, 1980.
211. Krakauer, T., Buckley, M., and Fisher, D. Proinflammatory mediators of toxic shock and their correlation to lethality, *Mediators Inflamm.* 2010, 517594, 2010.
212. Chow, A. W. et al. Vaginal colonization with *Staphylococcus aureus*, positive for toxic-shock marker protein, and *Escherichia coli* in healthy women, *J. Infect. Dis.*, 150, 80, 1984.
213. Florquin, S., Amraoui, Z., Abramowicz, D., and Goldman, M. Systemic release and protective role of IL-10 in staphylococcal enterotoxin B-induced shock in mice, *J. Immunol.*, 153, 2618, 1994.
214. Sundstedt, A. et al. Immunoregulatory role of IL-10 during superantigen-induced-hyporesponsiveness in vivo, *J. Immunol.*, 158, 180, 1997.
215. Khan, A.A., Priya, S., and Saha, B. IL-2 regulates SEB induced toxic shock syndrome in BALB/c mice, *PLoS One*, 4, e8473, 2009.
216. Krakauer, T. Immune response to staphylococcal superantigens, *Immunol. Res.*, 20, 163, 1999.
217. Hasko, G. et al. The crucial role of IL-10 in the suppression of the immunological response in mice exposed to staphylococcal enterotoxin B, *Eur. J. Immunol.*, 28, 1417, 1998.
218. Blank, C. et al. Superantigen and endotoxin synergize in the induction of lethal shock, *Eur. J. Immunol.*, 27, 825, 1997.
219. Saha, B. et al. Protection against lethal toxic shock by targeted disruption of the CD28 gene, *J. Exp. Med.*, 183, 2675, 1996.
220. Mittrucker, H.W. et al. Induction of unresponsiveness and impaired T cell expansion by staphylococcal enterotoxin B in CD28-deficient mice, *J. Exp. Med.*, 183, 2481, 1996.
221. Xu, H. et al. Leukocytosis and resistance to septic shock in intercellular adhesion molecule 1-deficient mice, *J. Exp. Med.*, 180, 95, 1994.
222. Manjunath, N. et al. Negative regulation of T-cell adhesion and activation by CD43, *Nature*, 377, 535, 1995.
223. Mountz, J.D. et al. Increased susceptibility of fas mutant MRL-lpr/lpr mice to staphylococcal enterotoxin B-induced septic shock, *J. Immunol.*, 155, 4829, 1995.
224. Sunstedt, A., Grundstrom, S., and Dohlsten, M. T cell- and perforin-dependent depletion of B cells *in vivo* by staphylococcal enterotoxin A, *Immunol.*, 95, 76, 1998.
225. Miethke, T. et al. Bacterial superantigens induce rapid and T cell receptor V beta-selective down-regulation of L-selectin (gp90Mel-14) in vivo, *J. Immunol.*, 151, 6777, 1993.

226. Yeung, R.S. et al. Human CD4 and human major histocompatibility complex class II (DQ6) transgenic mice: Supersensitivity to superantigen-induced septic shock, *Eur. J. Immun.*, 26, 1074, 1996.
227. DaSilva, L. et al. Human-like immune responses of human leukocyte antigen-DR3 transgenic mice to staphylococcal enterotoxins: A novel model for superantigen vaccines, *J. Infect. Dis.*, 185, 1754, 2002.
228. Welcher, B.C. et al. Lethal shock induced by streptococcal pyrogenic exotoxin A in mice transgenic for human leukocyte antigen-DQ8 and human CD4 receptors: Implications for development of vaccines and therapeutics, *J. Infect. Dis.*, 186, 501, 2002.
229. Roy, C.J. et al. Human leukocyte antigen-DQ8 transgenic mice: A model to examine the toxicity of aerosolized staphylococcal enterotoxin B, *Infect. Immun.*, 73, 2452, 2005.
230. Rajagopalan, G., Sen, M.M., and David, C.S. *In vitro* and *in vivo* evaluation of staphylococcal superantigen peptide antagonists, *Infect. Immun.*, 72, 6733, 2004.
231. Zhao, Y.-X. et al. Overexpression of the T-cell receptor V beta 3 in transgenic mice increases mortality during infection by enterotoxin A-producing *Staphylococcus aureus*, *Infect. Immun.*, 63, 4463, 1995.
232. Vlach, K.D., Boles, J.W., and Stiles, B.G. Telemetric evaluation of body temperature and physical activity as predictors of mortality in a murine model of staphylococcal enterotoxic shock, *Comp. Med.*, 50, 160, 2000.
233. Boles, J.W. et al. Correlation of body temperature with protection against staphylococcal enterotoxin B exposure and use in determining vaccine dose-schedule, *Vaccine*, 21, 2791, 2003.
234. Savransky, V. et al. Murine lethal toxic shock caused by intranasal administration of staphylococcal enterotoxin B, *Toxicol. Pathol.*, 31, 373, 2003.
235. Huzella, L.M. et al. Central roles for IL-2 and MCP-1 following intranasal exposure to SEB: A new mouse model, *Vet. Res. Sci.*, 86, 241, 2009.
236. Mattix, M.E. et al. Aerosolized staphylococcal enterotoxin B-induced pulmonary lesions in rhesus monkeys (*Macaca mulatta*), *Toxicol. Pathol.*, 23, 262, 1995.
237. Muralimohan, G. et al. Inhalation of *Staphylococcus aureus* enterotoxin A induces IFN-gamma and CD8 T cell-dependent airway and interstitial lung pathology in mice, *J. Immunol.*, 181, 3698, 2008.
238. Rajagopalan, G. et al. Intranasal exposure to bacterial superantigens induces airway inflammation in HLA class II transgenic mice, *Infect Immun.*, 74, 1284, 2006.
239. Rajagopalan, G. et al. Intranasal exposure to staphylococcal enterotoxin B elicits an acute systemic inflammatory response, *Shock*, 25, 647, 2006.
240. Hamel, M. et al. Activation and re-activation potential of T cells responding to staphylococcal enterotoxin B, *Int. Immunol.*, 7, 1065, 1995.
241. Niedergang, F. et al. The *Staphylococcus aureus* enterotoxin B superantigen induces specific T cell receptor down-regulation by increasing its internalization, *J. Biol. Chem.*, 270, 12839, 1995.
242. MacDonald, H.R. et al. Peripheral T-cell reactivity to bacterial superantigens in vivo: The response/anergy paradox, *Immunol. Rev.*, 133, 105, 1993.
243. Sundstedt, A. and Dohlsten, M. *In vivo* anergized $CD4^+$ T cells have defective expression and function of the activating protein-1 transcription factor, *J. Immunol.*, 161, 5930, 1998.
244. Florquin, S., Amraoui, Z., and Goldman, M. T cells made deficient in interleukin-2 production by exposure to staphylococcal enterotoxin B *in vivo* are primed for interferon-gamma and interleukin-10 secretion, *Eur. J. Immunol.*, 25, 1148, 1995.
245. Hedlund, G. et al. Superantigen-based tumor therapy: *In vivo* activation of cytotoxic T cells, *Cancer Immunol. Immunother.*, 36, 89, 1993.
246. Parsonnet, J. et al. A rabbit model of toxic shock syndrome that uses a constant, subcutaneous infusion of toxic shock syndrome toxin 1, *Infect. Immun.*, 55, 1070, 1987.

247. Kim, Y.B. and Watson, D.W. A purified group A streptococcal pyrogenic exotoxin. Physiochemical and biological properties including the enhancement of susceptibility to endotoxin lethal shock, *J. Exp. Med.*, 131, 611, 1970.
248. Huang, W.T., Lin, M.T., and Won, S.J. Staphylococcal enterotoxin A-induced fever is associated with increased circulating levels of cytokines in rabbits, *Infect. Immun.*, 65, 2656, 1997.
249. Lee, P.K. and Schlievert, P.M. Quantification and toxicity of group A streptococcal pyrogenic exotoxins in an animal model of toxic shock syndrome-like illness, *J. Clin. Microbiol.*, 27, 1890, 1989.
250. Huang, W.T., Wang, J.J., and Lin, M.T. Antipyretic effect of acetaminophen by inhibition of glutamate release after staphylococcal enterotoxin A fever in rabbits, *Neurosci. Lett.*, 355, 33, 2004.
251. De Azavedo, J.C.S. and Arbuthnott, J.P. Toxicity of staphylococcal toxic shock syndrome toxin 1 in rabbits, *Infect. Immun.*, 46, 314, 1984.
252. Pettit, G.W., Elwell, M.R., and Jahrling, P.B. Possible endotoxemia in rabbits after intravenous injection of *Staphylococcus aureus* enterotoxin B, *J. Infect. Dis.*, 135, 646, 1977.
253. Fujikawa, H. et al. Clearance of endotoxin from blood of rabbits injected with staphylococcal toxic shock syndrome toxin-1, *Infect. Immun.*, 52, 134, 1986.
254. Van Miert, A., Van Duin, C., and Schotman, A. Comparative observations of fever and associated clinical hematological and blood biochemical changes after intravenous administration of staphylococcal enterotoxins B and F (toxic shock syndrome toxin-1) in goats, *Infect. Immun.*, 46, 354, 1984.
255. Wright, A., Andrews, P., and Titball, R.W. Induction of emetic, pyrexic, and behavioral effects of *Staphylococcus aureus* enterotoxin B in the ferret, *Infect. Immun.*, 68, 2386, 2000.
256. Hu, D.-L. et al. Induction of emetic response to staphylococcal enterotoxins in the house musk shrew (*Suncus murinus*), *Infect. Immun.*, 71, 567, 2003.
257. Arad, G. et al. Superantigen antagonist protects against lethal shock and defines a new domain for T-cell activation, *Nat. Med.*, 6, 414, 2000.
258. Visvanathan, K. et al. Inhibition of bacterial superantigens by peptides and antibodies, *Infect. Immun.*, 69, 875, 2001.
259. Wang, S., Li, Y., Xiong, H., and Cao, J. A broad-spectrum inhibitory peptide against staphylococcal enterotoxin superantigen SEA, SEB and SEC, *Immunol. Lett.*, 121, 167, 2008.
260. Geller-Hong, E., Möllhoff, M., Shiflett, P.R., and Gupta, G. Design of chimeric receptor mimics with different TcRVβ isoforms: Type-specific inhibition of superantigen pathogenesis, *J. Biol. Chem.*, 279, 5676, 2004.
261. Vallabhapurapu, S. and Karin, M. Regulation and function of NFκB transcription factors in the immune system, *Ann. Rev. Immunol.*, 27, 693, 2009.
262. Krakauer, T. Molecular therapeutic targets in inflammation: Cyclooxygenase and NF-κB. *Curr. Drug Targets—Inflam. Allergy*, 3, 317, 2004.
263. Liu, D. et al. Suppression of staphylococcal enterotoxin B-induced toxicity by a nuclear import inhibitor, *J. Biol. Chem.*, 279, 19239, 2004.
264. Liu, D., Zienkiewicz, J., DiGiandomenico, A., and Hawiger, J. Suppression of acute lung inflammation by intracellular peptide delivery of a nuclear import inhibitor, *Molec. Therapy*, 17, 796, 2009.
265. Tilahun, A.Y. et al. Detrimental effect of the proteasome inhibitor, bortezomib in bacterial superantigen- and lipopolysaccharide-induced systemic inflammation, *Mol. Ther.*, 18, 1143, 2010.
266. Krakauer, T. A sensitive ELISA for measuring the adhesion of leukocytic cells to human endothelial cells, *J. Immunol. Meth.*, 177, 207, 1994.
267. Krakauer, T. and Buckley, M. Dexamethasone attenuates staphylococcal enterotoxin B-induced hypothermic response and protects mice from superantigen-induced toxic shock, *Antimicrob. Agents Chemother.*, 50, 391, 2006.

268. LeClaire, R.D. et al. Protective effects of niacinamide in staphylococcal enterotoxin-B-induced toxicity, *Toxicology*, 107, 69, 1996.
269. Won, S.-J. et al. Staphylococcal enterotoxin A acts through nitric oxide synthase mechanisms in human peripheral blood mononuclear cells to stimulate synthesis of pyrogenic cytokines, *Infect. Immun.*, 68, 2003, 2000.
270. Saha, B. et al. Toxic shock syndrome toxin-1 induced death is prevented by CTLA4Ig, *J. Immunol.*, 157, 3869, 1996.
271. Takei, Y. et al. Tryptanthrin inhibits interferon-γ production by Peyer's patch lymphocytes derived from mice that had been orally administered staphylococcal enterotoxin, *Biol. Pharm. Bull.*, 26, 365, 2003.
272. Miller, E.J., Cohen, A.B., and Peterson, B.T. Peptide inhibitor of interleukin-8 (IL-8) reduces staphylococcal enterotoxin-A (SEA) induced neutrophil trafficking to the lung, *Inflamm. Res.*, 45, 393, 1996.
273. Soltys, J. and Quinn, M.T. Modulation of endotoxin- and enterotoxin-induced cytokine release by *in vivo* treatment with beta-(1,6)-branched beta-(1,3)-glucan, *Infect. Immun.*, 67, 244, 1999.
274. Krakauer, T., Buckley, M., Issaq, H.J., and Fox, S.D. Rapamycin protects mice from staphylococcal enterotoxin B-induced toxic shock and blocks cytokine release *in vitro* and *in vivo*, *Antimicrob. Agents Chemother.*, 54, 1125, 2010.
275. Eriksson, B.K. et al. Invasive group A streptococcal infections: T1M1 isolates expressing pyrogenic exotoxins A and B in combination with selective lack of toxin-neutralizing antibodies are associated with increased risk of streptococcal toxic shock syndrome, *J. Infect. Dis.*, 180, 410, 1999.
276. Norrby-Teglund, A. et al. Plasma from patients with severe invasive group A streptococcal infections treated with normal polyspecific IgG inhibits streptococcal superantigen-induced T cell proliferation and cytokine production, *J. Immunol.*, 156, 3057, 1996.
277. Barry, W. et al. Intravenous immunoglobulin therapy for toxic shock syndrome, *JAMA*, 267, 3315, 1992.
278. Stegmayer, B. et al. Septic shock induced by group A streptococcal infection: Clinical and therapeutic aspects, *Scand. J. Infect. Dis.*, 24, 589, 1992.
279. LeClaire, R.D., Hunt, R.E., and Bavari, S. Protection against bacterial superantigen staphylococcal enterotoxin B by passive vaccination, *Infect. Immun.*, 70, 2278, 2002.
280. Woody, M.A. et al. Differential immune responses to staphylococcal enterotoxin B mutations in a hydrophobic loop dominating the interface with major histocompatibility complex class II receptors, *J. Infect. Dis.*, 177, 1013, 1998.
281. Savransky, V. et al. Immunogenicity of the histidine-to-tyrosine staphylococcal enterotoxin B mutant protein in C3H/HeJ mice, *Toxicon*, 43, 433, 2004.
282. Bonventre, P.F. et al. A mutation at histidine residue 135 of toxic shock syndrome toxin yields an immunogenic protein with minimal toxicity, *Infect. Immun.*, 63, 509, 1995.
283. Boles, J.W. et al. Generation of protective immunity by inactivated recombinant staphylococcal enterotoxin B vaccine in nonhuman primates and identification of correlates of immunity, *Clin. Immunol.*, 108, 51, 2003.
284. Inskeep, T.K. et al. Oral vaccine formulations stimulate mucosal and systemic antibody responses against staphylococcal enterotoxin B in a piglet model, *Clin. Vac. Immunol.*, 17, 1163, 2010.
285. Bergdoll, M.S. Immunization of rhesus monkeys with enterotoxoid B, *J. Infect. Dis.*, 116, 191, 1966.
286. Lowell, G.H. et al. Immunogenicity and efficacy against lethal aerosol staphylococcal enterotoxin B challenge in monkeys by intramuscular and respiratory delivery of proteosome-toxoid vaccines, *Infect. Immun.*, 64, 4686, 1996.

287. Lowell, G.H. et al. Intranasal and intramuscular proteosome-staphylococcal enterotoxin B (SEB) toxoid vaccines: Immunogenicity and efficacy against lethal SEB intoxication in mice, *Infect. Immun.*, 64, 1706, 1996.
288. Tseng, J. et al. Humoral immunity to aerosolized staphylococcal enterotoxin B (SEB), a superantigen, in monkeys vaccinated with SEB toxoid-containing microspheres, *Infect. Immun.*, 63, 2880, 1995.
289. di Tommaso, A. et al. Formaldehyde treatment of proteins can constrain presentation to T cells by limiting antigen processing, *Infect. Immun.*, 62, 1830, 1994.
290. Cropley, I. et al. Mucosal and systemic immunogenicity of a recombinant, non-ADP-ribosylating pertussis toxin: Effects of formaldehyde treatment, *Vaccine*, 13, 1643, 1995.
291. Rott, O. and Fleischer, B. A superantigen as virulence factor in an acute bacterial infection, *J. Infect. Dis.*, 169, 1142, 1994.

Index

C

D

E

F

G

H

I

J

L

M

N

O

P

Q

R

S

T

U

V

W

Y

Z